D1162984

Coal Energy Systems

Academic Press *Sustainable World* Series

Series Editor
Richard C. Dorf

University of California, Davis

The *Sustainable World* series concentrates on books that deal with the physical and biological basis of the world economy and our dependence on the tools, devices, and systems used to control, develop and exploit nature. Engineering is the key element in developing and implementing the technologies necessary to plan for a sustainable world economy. If the industrialization of the world is to continue as a positive force, the creation and application of environmentally friendly technologies should be one of the highest priorities for technological innovation in the present and future.

This series includes titles on all aspects of the technology, planning, economics, and social impact of sustainable technologies. Please contact the editor or the publisher if you are interested in more information on the titles in this new series, or if you are interested in contributing to the series.

Current published titles:

Technology, Humans and Society: Towards a Sustainable World, edited by Richard C. Dorf, 500 pages, published in 2001.

Wind Power in View: Energy Landscapes in a Crowded World, edited by Martin J. Pasqualetti, Paul Gipe, Robert W. Righter, 234 pages, published in 2002.

Coal Energy Systems

Bruce G. Miller

ELSEVIER
ACADEMIC PRESS

Amsterdam Boston Heidelberg London New York Oxford
Paris San Diego San Francisco Singapore Sydney Tokyo

Elsevier Academic Press
30 Corporate Drive, Suite 400, Burlington, MA 01803, USA
525 B Street, Suite 1900, San Diego, California 92101-4495, USA
84 Theobald's Road, London WC1X 8RR, UK

This book is printed on acid-free paper. ∞

Library of Congress Cataloging-in-Publication Data
Application Sumitted

British Library Cataloguing in Publication Data
A catalogue record for this book is available from the British Library

ISBN: 0-12-497451-1

For all information on all Elsevier Academic Press publications visit our Web site at www.elsevier.books.com

Printed in the United States of America
04 05 06 07 08 09 9 8 7 6 5 4 3 2 1

Dedication

For my family, Sharon, Konrad, and Anna, for their patience and support during the writing of this book.

Contents

Preface **xiii**

1. Introduction to Coal **1**
What is Coal? 1
Origin of Coal 1
Coalification 2
Classification of Coal 4
Basic Coal Analysis 5
Rank of Coal 6
Coal Type 7
Grade of Coal 8
Classification Systems 8
Coal Distribution and Resources 12
Coal Reserves in the World 13
Major Coal-Producing Regions in the World 17
References 26

2. Past, Present, and Future Role of Coal **29**
The Use of Coal in the Pre-Industrial Revolution Era 29
Early History of United States Coal Mining and Use 31
The Use of Coal during the Industrial Revolution 31
Post-Industrial Revolution Use of Coal 33
Overview of Energy in the United States 33
Coal Production in the United States 39
Synthetic Coal 48
Coal Consumption in the United States 48
U.S. Coal Exports and Imports 50
World Primary Energy Production and Consumption 51
World Primary Energy Production 54
World Primary Energy Consumption 58
Future Projections of Energy Use and Coal's Contribution to the Energy Mix 61

World Energy Consumption of Oil 61
World Energy Consumption of Natural Gas 63
World Energy Consumption of Coal 63
World Energy Consumption of Nuclear Energy 69
World Energy Consumption of Renewable Energy 69
Energy Outlook for the United States 70
Role of Coal in the United States' 2001 Energy Policy 72
References 75

3. The Effect of Coal Usage on Human Health and the Environment 77
Coal Mining 78
Underground Mining 79
Surface Mining 86
Legislation/Reclamation 89
Coal Preparation 90
Water Contamination from Preparation Plants 92
Air Contamination from Preparation Plants 92
Refuse Contaminants from Preparation Plants 93
Health and Safety Issues 93
Coal Transportation 95
Coal Combustion By-Products (CCB) 95
Emissions from Coal Combustion 97
Sulfur Oxides 97
Nitrogen Oxides (NO_x) 100
Particulate Matter (PM) 103
Organic Compounds 105
Carbon Monoxide 106
Trace Elements 107
Greenhouse Gases: Carbon Dioxide 114
References 118

4. Coal-Fired Emissions and Legislative Action in the United States 123
Major Coal-Related Health Episodes 123
Pre-Industrial Revolution 124
Post-Industrial Revolution 124
History of Legislative Action for Coal-Fired Power Plants 125
Pre-1970 Legislation 126
Clean Air Act Amendments of 1970 128
Clean Air Act Amendments of 1977 and Prevention of Significant Deterioration 139
Clean Air Act Amendments of 1990 141
Additional NO_x Regulations and Trading Programs 149
New Source Review 152

Impending Legislation and Pollutants under Consideration for Regulation 155
Emissions Legislation in other Countries 162
Sulfur Dioxide 162
Nitrogen Oxides 164
Particulate Matter 169
Trace Elements/Mercury 170
Carbon Dioxide 171
Air Quality and Coal-Fired Emissions 173
Six Principal Pollutants 175
Acid Rain 184
Hazardous Air Pollutants 187
Carbon Dioxide (CO_2) 189
References 191

5. Technologies for Coal Utilization 195
Coal Combustion 195
Brief History of Boilers and Coal Combustion Systems 196
Basic Steam Fundamentals and Their Application to Boiler Development 204
Chemistry of Coal Combustion 207
Coal Combustion Systems 212
Influence of Coal Properties on Utility Boiler Design 229
Carbonization 237
Brief History of Carbonization (High-Temperature) 238
Coking Processes 239
Coal Properties for Coke Production 241
Coking Conditions 242
Low-Temperature Carbonization 243
Gasification 246
Brief History of Coal Gasification 247
Principles of Coal Gasification 248
Gasifier Types 249
Influence of Coal Properties on Gasification 253
Regional Distribution of Gasification Systems 255
Commercial Gasification Systems 256
Liquefaction 267
The Beginning of the Synthetic Fuel Industry 269
Indirect Liquefaction: Fischer–Tropsch Synthesis 271
Direct Liquefaction 273
References 278

6. Emissions Control Strategies for Power Plants 283
Currently Regulated Emissions 284
Sulfur Dioxide (SO_2) 284

Nitrogen Oxides (NO_x) 322
Particulate Matter 347
Pollutants with Pending Compliance Regulation 369
Mercury 369
Potential Future Regulated Emissions 375
Carbon Dioxide 376
Multipollutant Control 382
ECO Process 383
Airborne Process 384
LoTOx Process 384
Mobotec Systems 385
Others 385
References 385

7. Future Power Generation (Near-Zero Emissions During Electricity Generation) 393
Clean Coal Technology Demonstration Program 395
Clean Coal Technology Program Evolution 395
CCT Program Funding and Costs 396
CCT Program Projects 397
CCT Program Accomplishments 418
Power Plant Improvement Initiative (PPII) 422
PPII Projects 423
Benefits of the PPII 424
Clean Coal Power Initiative (CCPI) 424
Program Importance 425
Round 1 CCPI Projects 427
CCPI Benefits 431
Vision 21 432
Vision 21 Technologies 434
Vision 21 Benefits 435
FutureGen 435
Benefits of the DOE's Clean Coal Power Program/Demonstrations 437
References 439

8. Coal's Role in Providing United States Energy Security 445
Overview of U.S. Energy Security Issues 446
National Energy Plan and Coal Utilization 448
Energy Conservation/Efficiency 449
Diversity of Fuel Sources 449
Environmental Protection 450
Technological Innovations 450
Global Alliances and Markets 450
The Role of Coal in the National Energy Plan 450

Energy and the Economy 451
Natural Gas Use in Power Generation 454
The Potential of Coal to Reduce U.S. Dependency on Imported Crude Oil 456
The Resurgence of Coal in Electric Power Generation 458
Production of Hydrogen from Coal 459
The Role of Coal in Providing Security to the U.S. Food Supply 462
Coal's Role in International Energy Security and Sustainable Development 464
Concluding Statements 467
References 468

Appendix A. Coal-Fired Emission Factors **473**
Appendix B. Original List of Hazardous Air Pollutants **499**
Appendix C. Initial 263 Units Identified in Phase I (SO_2) of the Acid Rain Program **505**
Appendix D. Commercial Gasification Facilities Worldwide **509**

Index **513**

Preface

Coal is currently a major energy source in the United States as well as throughout the world, especially among many developing countries, and will continue to be so for many years. Fossil fuels will continue to be the dominant energy source for fueling the U.S. economy, with coal playing a major role for decades. Coal provides stability in price and availability, will continue to be a major source of electricity generation, will be the major source of hydrogen for the coming hydrogen economy, and has the potential to become an important source of liquid fuels. Conservation and renewable/sustainable energy are important in the overall energy picture but will play a lesser role in helping us satisfy our energy demands.

It is recognized in the energy industry that the manner in which coal is used must, and will, change. Concerns over the environmental effects of coal utilization are resulting in better methods for controlling emissions during combustion, as well as more research and development into technologies to utilize coal more efficiently especially in non-traditional (*i.e.*, direct combustion) methods. While major advances have been made in reducing the environmental impact when using coal, we have other technologies in hand, either near commercialization or under development, that will allow coal to be used in an even more environmentally friendly manner. The roadblocks to implementing these technologies are the financial risks associated with new technologies and the resulting higher costs of energy to the consumers. Consumers in the United States, for example, have become accustomed to low energy prices and are reluctant to pay more for their energy, whether it be transportation fuels for their vehicles, natural gas or propane for domestic heating, or electricity for their homes. The implementation of these technologies that increase energy efficiency or reduce pollution will be driven by legislative mandate and, to a lesser extent, the willingness of the consumer to pay more for energy.

The importance of coal to the economy is very evident—on a local level as well as a global scale. Growing up in rural Beulah, North Dakota, which is located in the heart of the northern lignite fields, I saw how important the energy industry—mining, power generation, and methane production from coal gasification—is to a local economy. Located within a radius of about 100 miles from Beulah in the 1970s were six mines, eight power plants, and a coal gasification plant. Coal was in my veins, as several relatives worked at the mines or plants. One mine, in particular—the Knife River Coal Mine—helped me through my college years by providing me employment during the summer and allowed me to work with my father, who spent most of his career at the mine until his retirement. Later in my education and over the next nearly 25 years of my career, I came to realize the economic importance of coal to larger regions such as the state of North Dakota, the other coal states of the nation, and the world. Coal is widely dispersed throughout the world, unlike petroleum and natural gas; most countries (except for parts of the Middle East) contain coal reserves, thereby allowing them the opportunity to be energy self-sufficient or, at the very least, providing options for relying on domestic rather than imported energy.

It is my intention with this book to illustrate the importance of coal as an energy source both in the United States and in the world. The book begins with an introduction to coal and its distribution and reserves in the world to provide the reader with basic coal information as a prelude to the subsequent chapters. The second chapter presents a brief history of coal use, its current status as an energy source, and the future role of coal. Coal is compared to other energy sources, including oil, natural gas, nuclear, and renewables. While coal has been instrumental in the advancement of civilization and will continue to be a major fuel source for several decades, the value of coal is partially offset by the environmental issues it raises. These issues are discussed in the third chapter, where some of these issues also have impacts on human health. The fourth chapter presents a history of legislative action in the United States as it pertains to coal-fired power plants and discusses impending legislation. A brief discussion of emissions and legislation from other countries and how they compare to the United States is also provided. Technologies used for generating power, heat, coke, and chemicals from coal are discussed in the fifth chapter and include combustion, carbonization, gasification, and liquefaction. The emphasis in this chapter is on coal combustion, as this is currently the single largest use of coal. The sixth chapter provides an in-depth discussion of emissions control strategies for power plants, as electricity generation is the single largest use for coal today. The progress that has been made over the last approximately 30 years in reducing emissions from power plants is discussed, as are commercial control strategies currently used and under development. Future power generation, with the goal of near-zero emissions, is discussed in the seventh chapter. Major research and development programs, sponsored primarily by the U.S. Department of Energy in partnership with industry, are also discussed in this chapter, as well as developing technologies to achieve near-zero emissions power and clean fuel plants with carbon dioxide management capability. The book concludes by discussing the role of coal in providing energy security to the United States, as well as its role in providing international energy security and sustainable development.

I will conclude by first stating that all errors or omissions are entirely my own. I also want to express my thanks for all those who helped make this book a reality. First and foremost, I want to thank my wife, Sharon, and children, Konrad and Anna, for supporting me these last 12 to 14 months while I spent long hours writing and too few hours with them (missing family events and forgoing vacations). Thanks go to Harold Schobert for encouraging me to undertake this project; David Tillman, for constantly hounding me to keep at it; and Donna Baney, for typing some of the tables in the appendices. A very special thank you goes to Ruth Krebs for her work on the figures. I would like to also thank my parents, Pearl (a schoolteacher her entire career) and Fred (a coal miner for most of his career), for recognizing the value of an education and encouraging me to pursue various interests. And, finally, I thank God for providing me with the talent, ambition, and drive to achieve all that I have accomplished.

Bruce G. Miller

CHAPTER 1

Introduction to Coal

This chapter presents an introductory overview of coal that includes a description of coal along with discussions of how it is formed, coal resources, and recoverable reserves in the world, with an emphasis on the United States' coals and coalfields, the types and characteristics of coal, and coal classification systems relevant to commercial coal use. The purpose of this chapter is to provide the reader with basic coal information as a prelude to the subsequent chapters.

What Is Coal?

An encompassing description of coal has been given by van Krevelen [1], in which he states: "Coal is a rock, a sediment, a conglomerate, a biological fossil, a complex colloidal system, an enigma in solid-state physics and an intriguing object for chemical and physical analyses." In short, coal is a chemically and physically heterogeneous, "combustible," sedimentary rock consisting of both organic and inorganic material. Organically, coal consists primarily of carbon, hydrogen, and oxygen, with lesser amounts of sulfur and nitrogen. Inorganically, coal consists of a diverse range of ash-forming compounds distributed throughout the coal. The inorganic constituents can vary in concentration from several percentage points down to parts per billion of the coal. Coal is the most abundant fossil fuel in the United States, as well as the world. At the end of 2000, recoverable coal reserves in the United States, which contains the world's largest coal reserves, totaled 274 billion short tons compared to a total world reserve of 1083 billion short tons [2]. On an oil-equivalent basis, there is approximately twice as much recoverable coal in the world as oil and natural gas combined [3]; consequently, coal has been and will continue to be a major economic/energy resource, a topic that will be discussed in detail in subsequent chapters.

Origin of Coal

Coal is found in deposits called seams that originated through the accumulation of vegetation that has undergone physical and chemical changes.

These changes include decaying of the vegetation, deposition and burying by sedimentation, compaction, and transformation of the plant remains into the organic rock found today. Coals differ throughout the world in the kinds of plant materials deposited (type of coal), in the degree of metamorphism or coalification (rank of coal), and in the range of impurities included (grade of coal).

There are two main theories for the accumulation of the vegetal matter that gives rise to coal seams [4]. The first theory, and the one most accepted as it explains the origin of most coals, is that the coal formed *in situ* (that is, where the vegetation grew and fell), and such a deposit is said to be autochthonous in origin. The beginning of most coal deposits started with thick peat bogs where the water was nearly stagnant and plant debris accumulated. Vegetation tended to grow for many generations, with plant material settling on the swamp bottom and converted into peat by microbiological action. After some time, the swamps became submerged and were covered by sedimentary deposits, and a new future coal seam was formed. When this cycle was repeated, over hundreds of thousands of years, additional coal seams were formed. These cycles of accumulation and deposition were followed by diagenetic (*i.e.*, biological) and tectonic (*i.e.*, geological) actions and, depending upon the extent of temperature, time, and forces exerted, formed the different ranks of coal observed today.

While the formation of most coals can be explained by the autochthonous process, some deposits are not easily explained by this model. Some coals appear to have been formed through the accumulation of vegetal matter that has been transported by water. According to this theory (*i.e.*, allochthonous origin), the fragments of plants have been carried by streams and deposited on the bottom of the sea or in lakes where they build up strata, which later become compressed into solid rock.

Major coal deposits formed in every geological period since the Upper Carboniferous Period, 350 to 270 million years ago; the main coal-forming periods are shown in Figure 1-1 [5], which shows the relative ages of the world's major coal deposits. The considerable diversity of various coals is due to the differing climatic and botanical conditions that existed during the main coal-forming periods along with subsequent geophysical actions.

Coalification

The geochemical process that transforms plant material into coal is called coalification and is often expressed as:

$$\text{peat} \rightarrow \text{lignite} \rightarrow \text{subbituminous coal} \rightarrow \text{bituminous coal} \rightarrow \text{anthracite}$$

This is a simplistic classification; more elaborate systems have evolved and are discussed in the next section. Coalification can be described geochemically as consisting of three processes: the microbiological degradation

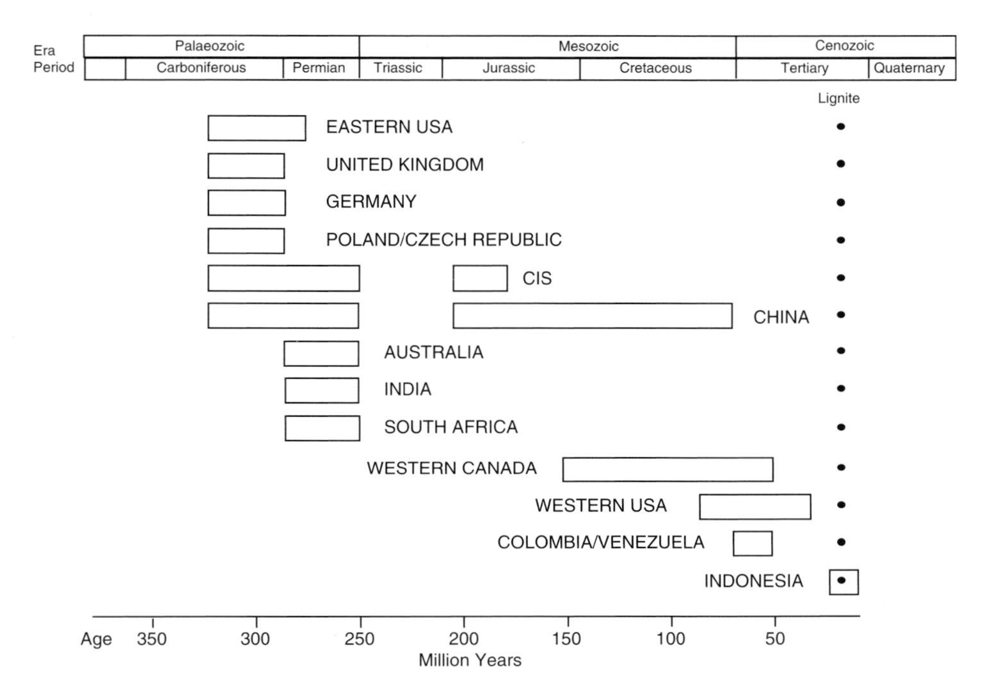

FIGURE 1-1. Comparison of the geological ages of the world's hard coal and lignite deposits. (From Walker, S., *Major Coalfields of the World*, IEA Coal Research, London, 2000. With permission.)

of the cellulose of the initial plant material, the conversion of the lignin of the plants into humic substances, and the condensation of these humic substances into larger coal molecules [6]. The kind of decaying vegetation, conditions of decay, depositional environment, and movements of the Earth's crust are important factors in determining the nature, quality, and relative position of the coal seams [1]. Of these, the physical forces exerted upon the deposits play the largest role in the coalification process. Variations in the chemical composition of the original plant material contributed to the variability in coal composition [1,7]. The vegetation of various geologic periods differed biologically and chemically. The conditions under which the vegetation decayed are also important. The depth, temperature, degree of acidity, and natural movement of water in the original swamp are important factors in the formation of the coal [1,8].

The geochemical phase of the coalification process is the application of temperature and pressure over millions of years and is the most important factor of the coalification process. While there is some disagreement as to which has been more important in promoting the chemical and physical changes—high pressures exerted by massive overburdening strata or time-temperature factors—the changes are characterized physically by decreasing

Materials	Partial Processes	Main Chemical Reactions
Decaying Vegetation		
↓	Peatification	Bacterial and fungal life cycles
Peat		
↓	Lignitification	Air oxidation, followed by decarboxylation and dehydration
Lignite		
↓	Bituminization	Decarboxylation and hydrogen disproportioning
Bituminous coal		
↓	Preanthracitization	Condensation to small aromatic ring systems
Semianthracite		
↓	Anthracitization	Condensation of small aromatic ring systems to larger ones; dehydrogenation
Anthracite	Graphitization	Complete carbonification

FIGURE 1-2. The coalification process. (From Van Krevelen, D. W., *Coal: Typology–Physics–Chemistry–Constitution*, Third ed., Elsevier Science, Amsterdam, 1993. With permission).

porosity and increasing gelification and vitrification [9]. Chemically, there is a decrease in moisture and volatile matter (*i.e.*, methane, carbon dioxide) content, as well as an increase in the percentage of carbon, a gradual decrease in the percentage of oxygen, and, ultimately, as the anthracitic stage is approached, a marked decrease in the content of hydrogen [7,9]. For example, carbon content (on a dry, mineral-matter-free basis) increases from approximately 50% in herbaceous plants and wood to 60% in peat, 70% in lignite, 75% in subbituminous coal, 80 to 90% in bituminous coal, and >90% in anthracite [7,10–12]. This change in carbon content is known as carbonification. The coalification/carbonization process is shown in Figure 1-2, where some of the main chemical reactions that occur during coalification are listed [1].

Classification of Coal

Efforts to classify coals began over 175 years ago and were prompt[illegible] the need to establish some order to the confusion of different coal[illegible] ypes of classification systems arose. Some schemes are intended [illegible] ien- tific studies, and other systems are designed to assist coal p[illegible] and users. The scientific systems of classification are concerned [illegible] in, composition, and fundamental properties of coals, while the [illegible] al syste ess trade and market issues, utilization, technological proper-ties, ability for certain end uses. It is the latter classification systems that iscussed here. Excellent discussions on scientific classifications are g ewhere [1,10].

Bas Analysis

Pri cussing the rank, type, grade, and classification systems of coal, a b ription of basic coal analyses, upon which classification schemes are s provided. These analyses do not yield any information on coal str ut do provide important information on coal behavior and are use marketing of coals. Three analyses are used in classifying coal, two ch are chemical analyses and one is a calorific determination. The chemical analyses include proximate and ultimate analysis. The proximate analysis gives the relative amounts of moisture, volatile matter, ash (*i.e.*, inorganic material left after all the combustible matter has been burned off), and, indirectly, the fixed carbon content of the coal. The ultimate analysis gives the amounts of carbon, hydrogen, nitrogen, sulfur, and oxygen comprising the coal. Oxygen is typically determined by difference—that is, by subtracting the total percentages of carbon, hydrogen, nitrogen, and sulfur from 100—because of the complexity in determining oxygen directly; however, this technique accumulates all the errors that occur when determining the other elements into the calculated value for oxygen. The third important analysis, the calorific value, also known as heating value, is a measure of the amount of energy that a given quantity of coal will produce when burned.

Because moisture and mineral matter (or ash) are extraneous to the coal substance, analytical data can be expressed on several different bases to reflect the composition of as-received, air-dried, or fully water-saturated coal or the composition of dry, ash-free (daf), or dry, mineral-matter-free (dmmf) coal. The most commonly used bases in the various classification schemes are shown in Figure 1-3 [13]. The most commonly used bases can be described as follows [1]:

- *As-received*—Data are expressed as percentages of the coal with the moisture. This category is also sometimes referred to as *as-fired* and is commonly used by the combustion engineer to monitor operations and for performing calculations as it is the whole coal that is being utilized;
- *Dry basis* (db)—Data are expressed as percentages of the coal after the moisture has been removed;
- *Dry, ash-free* (daf)—Data are expressed as percentages of the coal with the moisture and ash removed;
- *Dry, mineral-matter-free* (dmmf)—The coal is assumed to be free of both moisture and mineral matter, and the data are a measure of only the organic portion of the coal;

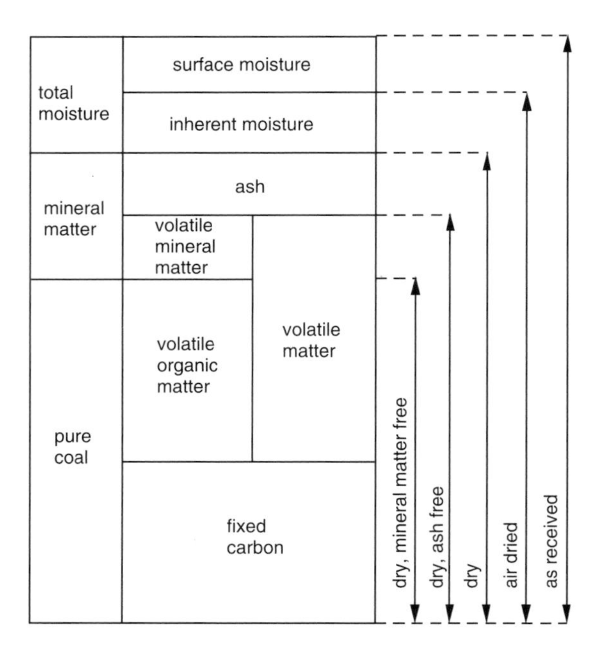

FIGURE 1-3. Relationship of different analytical bases to coal components. (From Ward, C. R., Ed., *Coal Geology and Coal Technology*, Blackwell Scientific, Melbourne, 1984, p. 66. With permission.)

- *Moist, ash-free* (maf)—The coal is assumed to be free of ash but still contains moisture;
- *Moist, mineral-matter-free* (mmmf)—The coal is assumed to be free of mineral matter but still contains moisture.

Rank of Coal

The degree of coal maturation is known as the *rank* of coal and is an indication of the extent of metamorphism the coal has undergone. Rank is also a measure of carbon content as the percentage of fixed carbon increases with extent of metamorphism. In the United States, lignites and subbituminous coals are referred to as being low in rank, while bituminous coals and anthracites are classified as high-rank coals. Figure 1-4 illustrates the relationship between rank and fixed-carbon content [14]. The fixed-carbon content shown in Figure 1-4 is calculated on a dry, mineral-matter-free basis. Figure 1-4 also shows the comparison between heating value and rank; the heating value is calculated on a moist, mineral matter-free basis. Note that the heating value increases with increasing rank but begins to decrease with semi-anthracitic and higher rank coals. This decrease in heating value is due to the significant decrease in volatile matter, which is shown in Figure 1-4 [14].

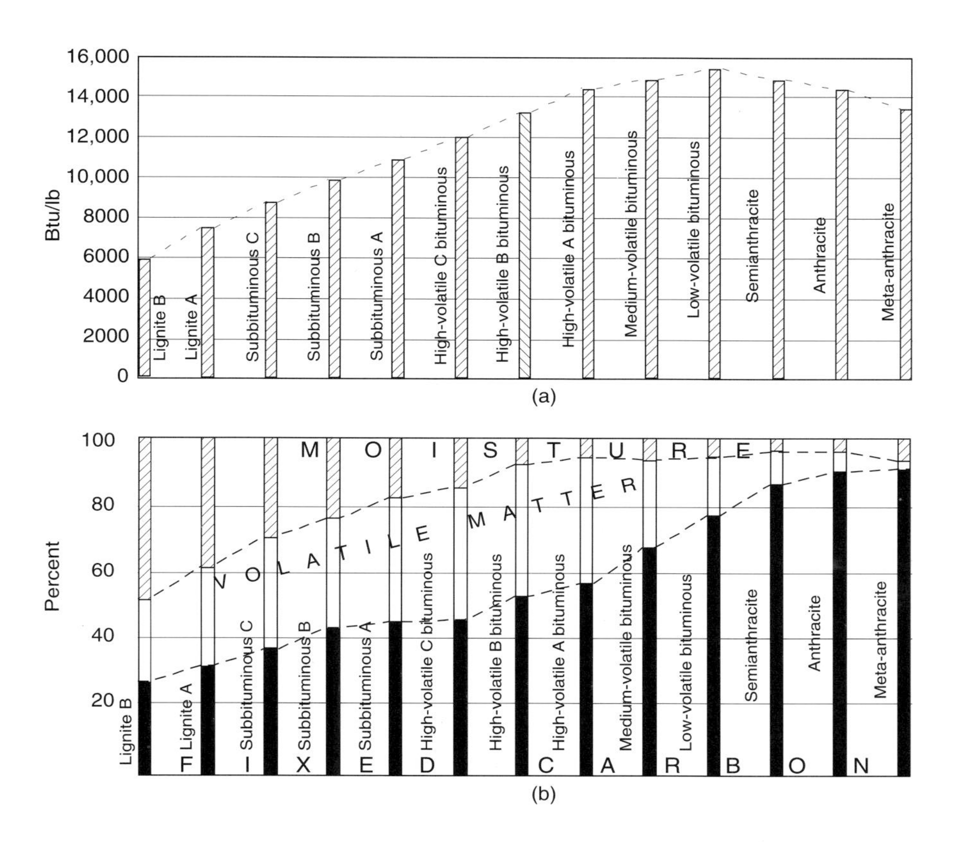

FIGURE 1-4. Comparison of heating values (on a moist, mineral-matter-free basis) and proximate analyses of coals of different ranks. (From Averitt, P., Coal resources of the U.S., January 1, 1974, *U.S. Geological Survey Bulletin*, No. 1412, 1975 [reprinted 1976].)

Coal Type

The ultimate microscopic constituents of coal are called *macerals*. The three main groups are characterized by their appearance, chemical composition, and optical properties. In most cases, the constituents can be traced back to specific components of the plant debris from which the coal formed [10]. The three maceral groups are vitrinite, exinite (sometimes also referred to as liptinite), and inertinite, which in turn can be subdivided into finer classifications. Only the three maceral groups will be introduced here, as extensive discussions of petrography and its relevance to industrial processes can be found elsewhere [1,8,10].

Vitrinite group macerals are derived from the humification of woody tissues and can either possess remnant cell structures or be structureless [8]. Vitrinite contains more oxygen than the other macerals at any given rank level. Exinite group macerals were derived from plant resins, spores, cuticles, and algal remains, which are fairly resistant to bacterial and fungal decay. Exinite group macerals exhibit higher hydrogen content than the other macerals, especially at lower rank. The inertinite group macerals were derived mostly from woody tissues, plant degradation products, or fungal remains. These are characterized by a high inherent carbon content that resulted from thermal or biological oxidation [8]. Petrographic analysis has many uses. Initially it was primarily used to characterize and correlate seams and resolve questions about coal diagenesis and metamorphism, but later it influenced developments in coal preparation (*i.e.*, crushing, grinding, and removal of mineral constituents) and conversion technologies [10]. Industrially, petrographic analysis can provide insight into the hardness of a coal (*i.e.*, its mechanical strength) as well as the thermoplastic properties of a particular coal, which are of significant importance in the coking industry.

Grade of Coal

The grade of coal refers to the amount of mineral matter that is present in the coal and is a measure of coal quality. Sulfur content, ash fusion temperatures (*i.e.*, measurement of the behavior of ash at high temperatures), and the quantity of trace elements in coal are also used to grade coal. Formal classification systems have not been developed with regard to the grading of coal; however, grade is important to the coal user. Mineral matter may occur as finely dispersed or in discrete partings in the coal. Some of the inorganic matter and trace elements are derived from the original vegetation, but the majority is introduced during coalification by wind or water to the peat swamp or through movement of solutions in cracks, fissures, and cavities [15]. Coal mineralogy can affect the ability to remove minerals during coal preparation/cleaning, the coal combustion and conversion (*i.e.*, production of liquid fuels or chemicals) characteristics, and metallurgical coke properties.

Classification Systems

An excellent discussion of the many classification systems, scientific as well as commercial, is provided by van Krevelen [1]. The commercial systems, which will be discussed here, consist of two primary systems—the American Society for Testing Materials (ASTM) system used in the United States/North America and an international Economic Commission for Europe (ECE) codification system developed in Europe. It is interesting that, in all countries, the classification systems used commercially are primarily based on the content of volatile matter [1]. In some countries, a second parameter is also used; in the United States, for example, this parameter is the heating value (see Figure 1-4). For many European countries, this parameter is either the caking or the coking properties. Caking coals are coals that pass through a plastic state upon heating during which they soften, swell, and resolidify into a coherent carbonaceous matrix, while noncaking coals do not become plastic when heated and produce a weakly coherent char residue. Coking coals are strongly caking coals that exhibit characteristics that make them suitable for conversion into metallurgical and other industrial cokes [10].

ASTM Classification System

The ASTM classification system (ASTM D388) distinguishes among four coal classes, each of which is subdivided into several groups (see Table 1-1). As previously mentioned, high-rank coals (*i.e.*, medium volatile bituminous coals or those of higher rank) are classified based on their fixed-carbon and volatile-matter contents (expressed on a dmmf basis), while low-rank coals are classified in terms of their heating value (expressed on a mmmf basis). This classification system was developed for commercial applications but has proved to be satisfactory for certain scientific uses as well [9]. If a given

TABLE 1-1
ASTM Coal Classification by Rank

Class/Group	*Fixed Carbon*[a] (%)	*Volatile Matter*[b] (%)	*Heating Value*[b] (Btu/lb)
Anthracitic			
Meta-anthracite	>98	<2	—
Anthracite	92–98	2–8	—
Semi-anthracite	86–92	8–14	—
Bituminous			
Low-volatile	78–86	14–22	—
Medium-volatile	69–78	22–31	—
High-volatile A	<69	>31	>14,000
High-volatile B	—	—	13,000–14,000
High-volatile C	—	—	10,500–13,000[c]
Subbituminous			
Subbituminous A	—	—	10,500–11,500[c]
Subbituminous B	—	—	9500–10,500
Subbituminous C	—	—	8300–9500
Lignitic			
Lignite A	—	—	6300–8300
Lignite B	—	—	<6300

[a]Calculated on dry, mineral-matter-free coal. Correction from ash to mineral matter is made by means of the Parr formula: mineral matter = 1.08[percent ash + 0.55(percent sulfur)]. Ash and sulfur are on a dry basis.

[b]Calculated on mineral-matter-free coal with bed moisture content.

[c]Coals with heating values between 10,500 and 11,500 Btu/lb are classified as high volatile C bituminous if they possess caking properties or as subbituminous A if they do not.

Source: Berkowitz, N., *An Introduction to Coal Technology*, Academic Press, New York, 1979. With permission.

coal is described as being a certain rank, then an estimate of some properties can be made; for example, if the coal is classified as subbituminous/lignitic or anthracitic, then it would not be considered for certain applications, such as for coke production.

International Classification/Codification Systems

Because of the increasing amount of coal trade in the world, the ECE Coal Committee developed a new classification system in 1988 for higher rank coals [1]. A shortcoming of the original international system was that it was primarily developed for trading Northern Hemisphere coals, which have distinctly different characteristics than those from the Southern Hemisphere (*e.g.*, Australia and South Africa). As trade of Southern Hemisphere coals increased, it became apparent that a new classification system was needed. This new system, which in reality is a system of codes, is better known as the Codification System. The Codification System for hard coals, combined with the International Organization for Standardization (ISO) Codification of Brown Coals and Lignites (which was established in 1974), provides a complete codification for coals in the international trade. The ISO Codification of Brown Coals and Lignites is given in Table 1-2 [1]. Total moisture content of run-of-mine coal and tar yield (*i.e.*, determination of the yields of tar, water, gas, and coke residue by low-temperature distillation) are the two parameters coded. The ECE International Codification of Higher Rank Coals is much more complicated and is provided in Table 1-3. Eight basic parameters define the main properties of the coal, represented by a 14-digit code number. The codification is commercial; includes petrographic, rank, grade, and environmental information; is for medium- and high-rank coals only; is for blends and single coals; is for raw and washed coals; and is for all end-use applications [1]. The major drawback of this system is that it is complicated.

TABLE 1-2
Codification of Brown Coals and Lignites

Parameter	*Total Moisture Content (Run-of-Mine Coal)*		*Tar Yield (Dry, Ash Free)*	
Digit	1		2	
Coding	*Code*	*Weight %*	*Code*	*Weight %*
	1	≤20	0	≤10
	2	>20–30	1	>10–15
	3	>30–40	—	—
	4	>40–50	2	>15–20
	5	>50–60	3	>20–25
	6	>60	4	>25

Source: Van Krevelen, D. W., *Coal: Typology–Physics–Chemistry–Constitution,* Third ed., Elsevier Science, Amsterdam, 1993. With permission.

TABLE 1-3
International Codification of Higher Rank Coals[a]

Parameter	Vitrinite Reflectance (Mean Random)		Characteristics of Reflectogram[b]				Maceral Group Composition (mmf) Inertinite[c]		Liptinite		
Digit	1, 2		3				4		5		
Coding	Code	R_{random} (%)	Code	Standard Deviation		Type	Code	Vol.%	Code	Vol.%	Petrographic Tests
	02	0.2–0.29	0	≤1	No gap	Seam coal	0	0 to <10	1	0 to <5	
	03	0.3–0.39	1	>0.1≤0.2	No gap	Simple blend	1	10 to <20	2	5 to <10	
	04	0.4–0.49	2	>0.2	No gap	Complex blend	2	20 to <30	3	10 to <15	
	—	—	3	—	1 gap	Blend with 1 gap	—	—	—	—	
	48	4.8–4.89	4	—	2 gaps	Blend with 2 gaps	7	70 to <80	7	30 to <35	
	49	4.9–4.99	5	—	>2 gaps	Blend with >2 gaps	8	80 to <90	8	35 to <40	
	50	≥5.0	—	—	—	—	9	≥90	9	≥40	

Parameter	Crucible Swelling No.		Volatile Matter[d] (daf)		Ash, Dry		Total Sulfur, Dry		Gross Calorific Value (daf)		
Digit	6		7, 8		9, 10		11, 12		13, 14		
Coding	Code	Number	Code	Wt.%	Code	Wt.%	Code	Wt.%	Code	MJ/kg	Technological Tests
	0	0–0.5	48	≥48	00	0 to <1	00	0 to <0.1	21	<22	
	1	1–1.5	46	46 to <48	01	1 to <2	01	0.1 to <0.2	22	22 to <23	
	2	2–2.5	44	44 to <46	02	2 to <3	02	0.2 to <0.3	23	23 to <24	
	—	—	—	—	—	—	—	—	—	—	
	7	7–7.5	12	12 to <14	20	20 to <21	29	2.9 to <3.0	37	37 to <38	
	8	8–8.5	10	10 to <12	—	—	30	3.0 to <3.1	38	38 to <39	
	9	9–9.5	09	9 to <10	—	—	—	—	39	≥39	
	—	—	—	—	—	—	—	—	—	—	
	—	—	02	2 to <3	—	—	—	—	—	—	
	—	—	01	1 to <2	—	—	—	—	—	—	

[a]Higher rank coals are coals with gross calorific value (maf) of ≥24 MJ/kg, and those with gross calorific value (maf) of <24 MJ/kg provided mean random vitrinitic reflectance ≥0.6%. To convert from MJ/kg to Btu/lb, multiply by 429.23.

[b]A reflectogram as characterized by code number 2 can also result from a high rank seam coal.

[c]It should be noted that some of the inertinite may be reactive.

[d]Where the ash content of the coal is more than 10%, it must be reduced before analysis to below 10% by dense medium separation. In these cases, the cutting density and resulting ash content should be noted.

Source: Van Krevelen, D. W., *Coal: Typology–Physics–Chemistry–Constitution*, Third ed., Elsevier Science, Amsterdam, 1993. With permission.

Coal Distribution and Resources

Coal deposits are broadly categorized into *resources* and *reserves*. Resources refer to the quantity of coal that may be present in a deposit or coalfield but may not take into account the feasibility of mining the coal economically. Reserves generally tend to be classified as proven or measured and probable or indicated, depending on the level of exploration of the coalfield. The basis for computing resources and reserves varies between countries, which makes it difficult for direct comparisons. Walker [5] discusses the various measurement criteria used by the major coal-producing countries in the world in detail. Figure 1-5 illustrates the relationship between resources and reserves in the United States as of January 1, 1997 [16]. The United States has a total of nearly 4000 billion short tons of coal resources, with approximately 19 billion short tons classified as recoverable reserves at active mines out of 275 billion short tons that are economically recoverable.

The various categories shown in Figure 1-5 are defined as:

- *Total resources*—Coal that can currently, or potentially may, be extracted economically;
- *Measured resources*—Quantity of coal that has been determined to a high degree of geologic assurance;
- *Indicated resources*—Quantity of coal that has been determined to a moderate degree of geological assurance;

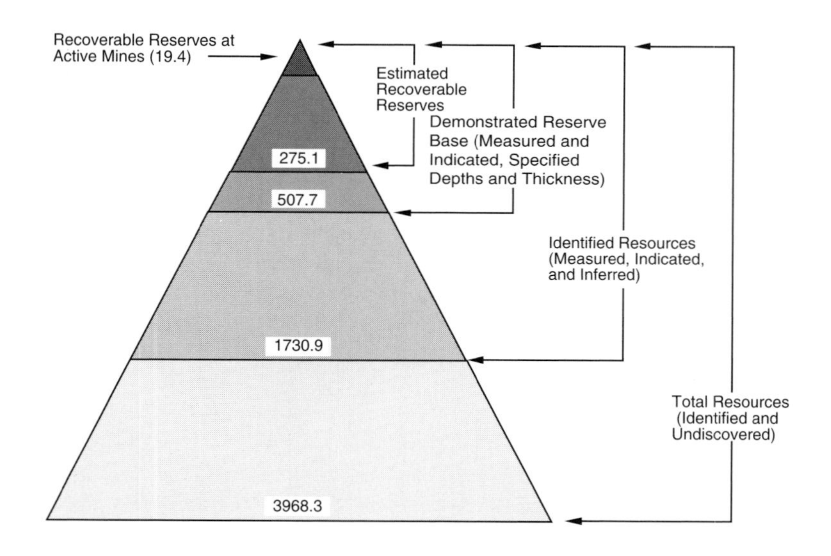

FIGURE 1-5. United States coal resources and reserves in billion short tons. (From EIA, *U.S. Coal Reserves: 1997 Update*, U.S. Department of Energy, Energy Information Administration, Washington, D.C., February 1999, p. 5, Appendix A.)

- *Inferred resources*—Quantity of coal that has been determined with a low degree of geologic assurance;
- *Recoverable reserves*—Coal that can be recovered economically with technology currently available or in the foreseeable future.

Terminology also varies among countries and can contribute to confusion when comparing coal resources and reserves. For purposes of discussion in this chapter and to lessen confusion, recoverable coal reserves will primarily be used when comparing world coal deposits.

Coal Reserves in the World

Coal is the most abundant fossil fuel in the world. Grimston [3] reported that at the end of 1998, oil reserves were 143 gigatons (Gt) representing a reserve-to-production (R/P) ratio of 41 years while natural gas reserves were 132 gigatons of oil equivalent (Gtoe) with a R/P ratio of 63 years. Coal was reported to have reserves of 486 Gtoe and a R/P ratio of 218 years, roughly double that of oil and natural gas combined. Coal reserves are also more widely distributed throughout the world, as shown in Figure 1-6. All major regions of the world contain coal, except for the Middle East, which contains almost two-thirds of the world oil reserves and, along with the states of the Former Soviet Union (FSU), contains more than two-thirds of the natural gas reserves [3]. The Energy Information Administration (EIA) estimated the reserves of recoverable coal at over 1083 billion short tons as of January 1, 2001 [2]. According to the EIA [17], this is enough coal to last approximately

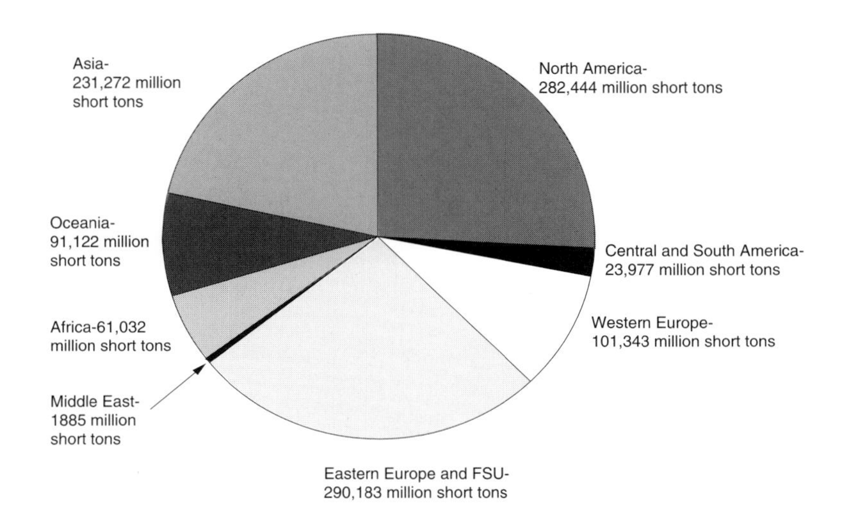

FIGURE 1-6. Distribution of recoverable coal reserves in the world.

230 years at current consumption levels, which is a projection similar to that reported by Grimston [3,17].

A detailed breakdown of the EIA's estimate of recoverable world coal reserves of 1083 billion short tons is provided in Table 1-4, which classifies the recoverable coal reserves for the major regions and countries of

TABLE 1-4
World Estimated Recoverable Coal Reserves (million short tons)

Region/Country	*Anthracite and Bituminous*	*Lignite and Subbituminous*	*Total*
North America			
Canada	3826	3425	7251
Greenland	0	202	202
Mexico	948	387	1335
United States	126,804	146,852	273,656
Total	131,579	150,866	282,444
Central and South America			
Argentina	0	474	474
Bolivia	1	0	1
Brazil	0	13,149	13,149
Chile	34	1268	1302
Colombia	6908	420	7328
Ecuador	0	26	26
Peru	1058	110	1168
Venezuela	528	0	528
Total	8530	15,448	23,977
Western Europe			
Austria	0	28	28
Croatia	7	36	43
France	24	15	40
Germany	25,353	47,399	72,753
Greece	0	3168	3168
Ireland	15	0	15
Italy	0	37	37
Netherlands	548	0	548
Norway	0	1	1
Portugal	3	36	40
Slovenia	0	303	303
Spain	220	507	728
Sweden	0	1	1
Turkey	306	3760	4066
United Kingdom	1102	551	1653
Yugoslavia	71	17,849	17,919
Total	27,650	73,693	101,343

(*continued*)

TABLE 1-4
(continued)

Region/Country	*Anthracite and Bituminous*	*Lignite and Subbituminous*	*Total*
Eastern Europe and Former USSR			
Bulgaria	14	2974	2988
Czech Republic	2330	3929	6259
Hungary	0	1209	1209
Kazakhstan	34,172	3307	34,479
Kyrgyzstan	0	895	895
Poland	22,377	2050	24,427
Romania	1	1605	1606
Russia	54,110	118,964	173,074
Slovakia	0	190	190
Ukraine	17,939	19,708	37,647
Uzbekistan	1102	3307	4409
Total	132,046	158,138	290,183
Middle East			
Iran	1885	0	1885
Total	1885	0	1885
Africa			
Algeria	44	0	44
Botswana	4740	0	4740
Central African Republic	0	3	3
Congo (Kinshasa)	97	0	97
Egypt	0	24	24
Malawi	0	2	2
Mozambique	234	0	234
Niger	77	0	77
Nigeria	23	186	209
South Africa	54,586	0	54,586
Swaziland	229	0	229
Tanzania	220	0	220
Zambia	11	0	11
Zimbabwe	553	0	553
Total	60,816	216	61,032
Asia and Oceania			
Afghanistan	73	0	73
Australia	46,903	43,585	90,489
Burma	2	0	2
China	68,564	57,651	126,215
India	90,826	2205	93,031
Indonesia	871	5049	5919
Japan	852	0	852
Korea, North	331	331	661

(continued)

TABLE 1-4
(continued)

Region/Country	*Anthracite and Bituminous*	*Lignite and Subbituminous*	*Total*
Korea, South	86	0	86
Malaysia	4	0	4
Nepal	2	0	2
New Caledonia	2	0	2
New Zealand	36	594	631
Pakistan	0	2497	2497
Philippines	0	366	366
Taiwan	1	0	1
Thailand	0	1398	1398
Vietnam	165	0	165
Total	208,719	113,675	322,394
World Total	571,224	512,035	1,083,259

Source: EIA, *International Energy Annual 2001*, U.S. Department of Energy, Energy Information Administration, Washington, D.C., 2003, pp. 114–115.

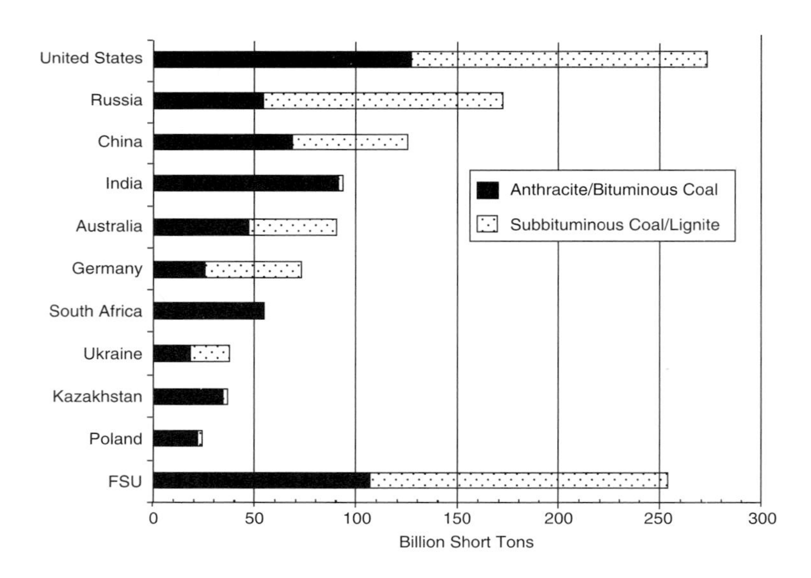

FIGURE 1-7. Countries with the largest recoverable coal reserves. (Note that the bottom bar shows the total for the states of the Former Soviet Union for comparison.)

the world into two major categories: (1) recoverable anthracite and bituminous coal (*i.e.*, hard coal) and (2) recoverable lignite and subbituminous coal. Figure 1-7 shows the ten countries with the largest recoverable coal reserves. The United States contains the largest quantity of recoverable coal reserves,

~274 billion short tons, or ~25% of the world's total. The country with the second largest quantity of recoverable coal reserves is Russia, which contains approximately 16% of the world's total reserves. If the Russian reserves are combined with those of the other FSU countries, as is commonly done, the states of the FSU contain nearly 23% of the world's total reserves, or 254 billion short tons (Figure 1-7). What is especially worth noting is that more than 70 countries contain recoverable coal but the ten shown in Figure 1-7 contain more than 983 billion short tons, or more than 90% of the world's total.

Major Coal-Producing Regions in the World

Coal is found on all inhabited continents of the world. It is very likely that coal is also on Antarctica, particularly when one looks at the coal-forming periods in history and the corresponding locations of the present-day continents. A review of the major coal-producing countries in the world is provided here, summarized by coal-producing region; specifically, the ten countries listed in Figure 1-7 are highlighted.

North America

The recoverable coal reserves of North America are the second largest in the world, with more than 282,000 million short tons identified (see Figure 1-7). Coal is found in the United States, Canada, Mexico, and, to a much lesser extent, Greenland (see Table 1-4).

United States The coal reserves of the United States are the largest of any country in the world: about 274,000 million short tons as of January 1, 2001 [2]. Recoverable coal reserves are found in 32 of the states, and the major coalfields are shown in Figure 1-8. The ten states with the largest recoverable coal reserves (listed in Table 1-5) contain approximately 89% of the total coal in the United States [16]. The top five states contain more than 70% of the total recoverable coal reserves in the United States.

Estimated low-sulfur recoverable coal reserves make up the largest portion of the total, at 36% [16]. Low-sulfur coal is defined as less than 0.8 and 0.5% by weight (as-received) sulfur for high-grade bituminous coal and high-grade lignite, respectively. These sulfur contents are a quantitative rating and have been correlated with U.S. sulfur emissions regulations from coal-fired power plants and the various stages of control that are required [16]. Estimated recoverable medium-sulfur reserves (0.8–2.2% for bituminous coal and 0.5–1.3% for lignite) and high-sulfur reserves (>2.2% for bituminous coal and >1.3% for lignite) account for 31 and 33% of the total, respectively.

The U.S. Geological Survey has divided the reserves into seven provinces: (1) Eastern Province; (2) Interior Province; (3) Gulf Province;

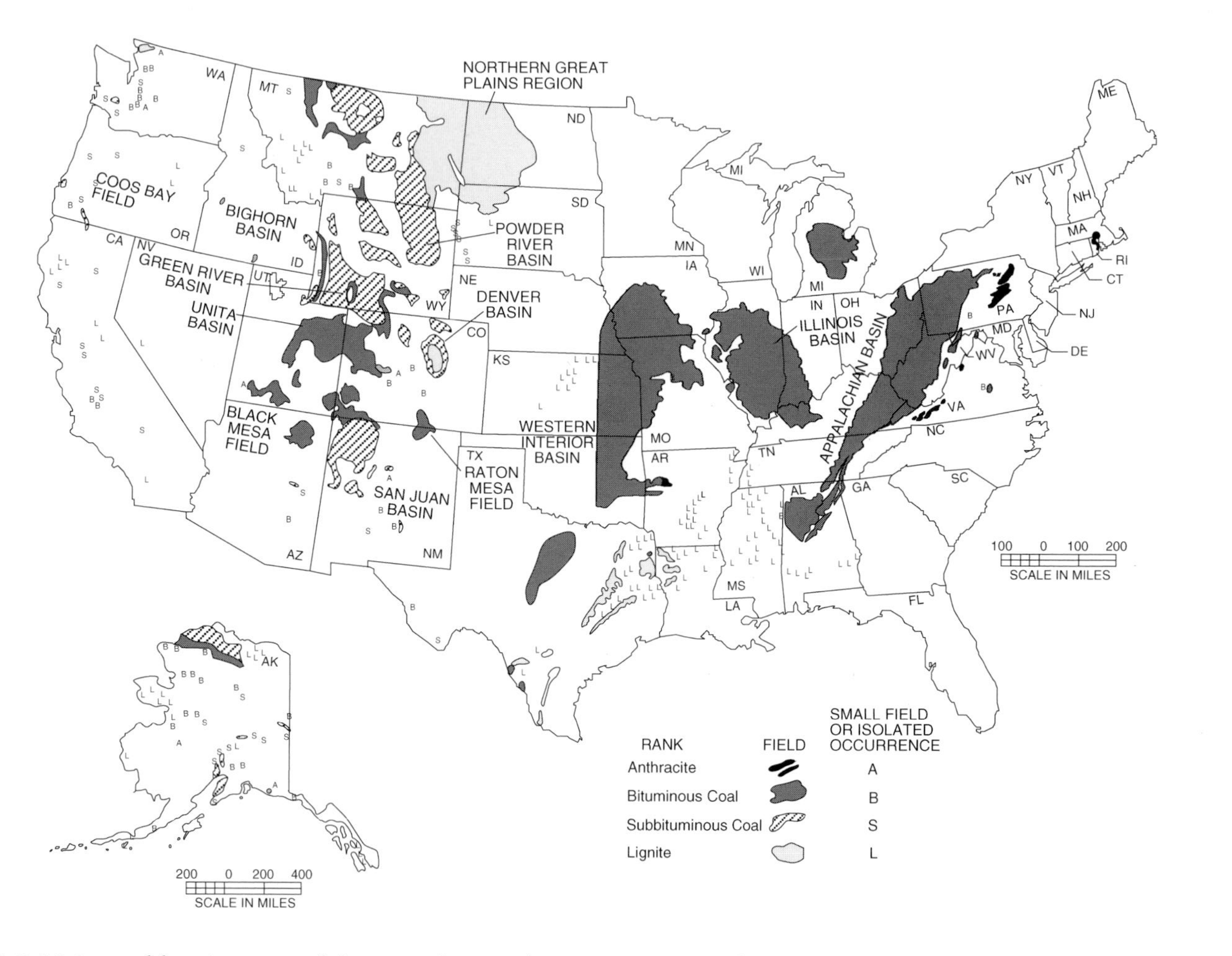

FIGURE 1-8. Major coal-bearing areas of the United States. (From EIA, U.S. Coal Reserves: 1997 Update, U.S. Department of Energy, Energy Information Administration, Washington, D.C., February 1999, p. 5, Appendix A.)

TABLE 1-5
Top Ten States with the Largest Coal Reserves as of January 1, 1997

State	*Reserves (Million Short Tons)*
Montana	75,309.7
Wyoming	44,813.4
Illinois	38,205.6
West Virginia	19,322.0
Kentucky	15,976.6
Pennsylvania	12,397.3
Ohio	11,671.9
Colorado	10,044.9
Texas	9953.9
North Dakota	7167.2
Total	244,862.5
Percentage of U.S. Total	89.5%

(4) Northern Great Plains Province; (5) Rocky Mountain Province; (6) Pacific Coast Province; and (7) Alaskan Province. The provinces are further subdivided into regions, fields, and districts. Carboniferous coal deposits in the eastern United States occur in a band of coal-bearing sediments that include the Appalachian and Illinois basins. Coal deposits in the western United States range from Upper Jurassic to Tertiary in age.

The Eastern Province includes the anthracite regions of Pennsylvania and Rhode Island, the Atlantic Coast region of middle Virginia and North Carolina, and the vast Appalachian basin, which extends from Pennsylvania through eastern Ohio, eastern Kentucky, West Virginia, western Virginia, Tennessee, and into Alabama. The Eastern Province is about 900 miles long and 200 miles wide at its broadest point [7]. This province also contains the greatest reserves of anthracite in the United States, with more than 760 million short tons in eastern Pennsylvania.

The Appalachian basin contains the largest deposits of bituminous coal in the United States. In the northern region of the Appalachian basin, the coal rank ranges from high-volatile bituminous coal in the west to low-volatile bituminous coal in the east. In the central region of the basin, the coal includes low- to high-volatile bituminous rank. In the southern region, the coals are mainly of high-volatile bituminous rank with some medium- and low-volatile bituminous coals [5]. Coals are used for steam production, electricity generation, and metallurgical coke production. These coals have high heating values, low- to medium-ash contents (up to 20%), and variable sulfur contents, with much of the coal containing 2 to 4% sulfur.

The Interior Province is subdivided into three regions: the Northern region, consisting of Michigan; the Eastern region or Illinois basin, consisting of Illinois, southern Indiana, and western Kentucky; and the Western region, consisting of Iowa, Missouri, Nebraska, Kansas, Oklahoma,

Arkansas, and western Texas. The Eastern region is the most important region of this province as it has vast reserves contained in Illinois (*i.e.*, more than 38,000 million short tons) and western Kentucky (nearly 7000 million short tons of the approximately 16,000 million short tons listed in Table 1-5). The coal in the Interior Province is mainly bituminous in rank and tends to be lower in rank and higher in sulfur than the Eastern Province bituminous coals. Coals are used for steam production, electricity generation, and metallurgical coke production. Coal composition in this province is quite variable, with coals from the Illinois basin being noted for having a high sulfur content (3–7%). The ash content is variable.

The Gulf Province consists of the Mississippi region in the east and the Texas region in the west. The coals in this province, which extends from Alabama through Mississippi, Louisiana, and into Texas, are lignitic in rank and are the lowest rank coals in the United States, having moisture contents up to 40%.

The Northern Great Plains Province contains the large lignite deposits of North Dakota, South Dakota, and eastern Montana, along with the subbituminous fields of northern and eastern Montana and northern Wyoming. These lignite deposits are contained in the Fort Union Region and are the largest lignite deposits in the world [7]. The coals are used primarily as power station fuels. The lignite has a high-moisture (38%), low-ash (6%), and medium-sulfur (<1%) content and a heating value of approximately 6800 Btu/lb. The Northern Great Plains Province also contains extensive subbituminous coal reserves from the Powder River basin [18]. Wyoming and Montana are the states with the largest recoverable coal reserves in the United States. Wyoming's coal reserves are split between the Northern Great Plains Province and the Rocky Mountain Province. The Powder River basin coals are used primarily as power station fuels and average about 1% sulfur with generally low ash content (3–10%).

The Rocky Mountain Province includes the coalfields of the mountainous districts of Montana, Wyoming, Utah, Colorado, and New Mexico. The coals range in rank from lignite through anthracite in this province. The most important Rocky Mountain Province coals are the coals from Wyoming, primarily those from the Green River, Hanna, and Hanna Fork coalfields. These coals are subbituminous in rank, typically contain low sulfur, and are used in power generation stations.

The Pacific Coast Province is limited to small deposits in Washington, Oregon, and California. The coals range in rank from lignite to anthracite. The fields are small and scattered and are not being utilized to any great extent.

The Alaskan Province contains coal in several regions [11]. These coals vary in rank from lignite to bituminous with a small amount of anthracite. The total reserves are estimated to be 15% bituminous coal and 85% subbituminous coal and lignite; however, extensive mining is not performed due to the low population density and pristine wilderness environment.

Only fields close to main lines of transportation have been developed. The coals are used primarily as power station fuels.

Canada Canada has about 7300 million short tons of recoverable coal ranging in rank from anthracite to lignite. The coal deposits formed in late Jurassic, Cretaceous, and early Tertiary times. Most of the recoverable reserves are in British Columbia, Alberta, and Saskatchewan, which is an extension of the Great Plains province coals from the United States. Coals from western Canada tend to be low in sulfur; those from Alberta and Saskatchewan are used as power station fuels, while British Columbia metallurgical coal is exported to the Far East. Coals from eastern Canada, primarily the Cape Breton Island coalfield in Nova Scotia, are the most important in the Atlantic region. The coals are of high-volatile bituminous rank and vary from medium to high sulfur. Coal production in Nova Scotia is a small percentage of the national output and is expected to decline further [5].

Eastern Europe and the States of the Former Soviet Union

Eastern Europe and the FSU contain extensive recoverable coal reserves totaling some 290,000 million short tons, or 27% of the world's total. Four of the countries listed in Table 1-4 contain over 90% of the recoverable reserves for this region: Russia, Ukraine, Kazakhstan, and Poland.

Russia Russia has extensive coal reserves, more than 173,000 million short tons (~16% of the world total), of which 119,000 million short tons are subbituminous and lignitic in rank. The coal resources in eastern Siberia and the Russian Far East remain largely unused because of their remoteness and lack of infrastructure [5]. Russia's main coal basins contain coals ranging from Carboniferous to Jurassic in age. Most hard coal reserves are in numerous coalfields in European and central Asian Russia, particularly in the Kuznetsk and Pechora basins and the Russian sector of the Dontesk basin. The Kansk–Achinsk basin in eastern Siberia is the country's main source of subbituminous coal. The Moscow basin contains significant lignite reserves but production has virtually stopped [5].

The Kuznetsk basin, which is located to the east of Novosibirsk, contains coals exhibiting a wide range in quality and rank from brown coal to semi-anthracite. The ash content of the coal is variable, and the sulfur content is generally low. High-quality coals with low moisture, ash, and sulfur contents are used for coking and steam coal production. This basin is now the largest single producer in Russia providing coking and steam coal.

The Pechora basin is located in the extreme northeast of European Russia. The coal rank in the basin increases from brown coal in the west to bituminous coal and anthracite in the east. Ash content varies considerably from 9 to 43%, while sulfur content, for the most part, does not exceed 1.5%. This basin is the principal supplier of coking coal.

The Dontesk basin is located in eastern Russia and western Ukraine and contains the entire range of coal rank from brown coal to anthracite, which increases in quantity toward the basin's central and eastern sections. These coals tend to have ash contents of 15 to 20% and sulfur contents of 2 to 4% and are used as coking and steam coals.

The Kansk-Achinsk basin, located adjacent to the east side of the Kuznetsk basin, contains brown coals that are described as lignites or subbituminous coals; however, their heating value is higher than that of most lignites. These coals have low to medium ash contents (6–20%) and low sulfur contents (<1%), which make them attractive for power station fuels.

Ukraine The Ukraine has significant coal reserves totaling approximately 37,600 million short tons, which is nearly evenly split between hard coal (bituminous and anthracite) and brown coal as shown in Table 1-4 [2]. Most of the coal resources are found in two coal basins: the Donetsk and Dneiper basins. The Donetsk basin, which is Carboniferous in age, is located in the east (and crosses over into Russia) and contains most of the country's hard coal resources. These coals contain medium ash (15–20%) and medium to high sulfur (2–4%) contents. These coals are used for steam production, power station fuels, and metallurgical applications. The Dneiper basin is adjacent to the eastern edge of the Donetsk basin and stretches across much of central Ukraine. This basin contains Ukraine's brown coal reserves and currently is of relatively minor importance [5].

Kazakhstan Kazakhstan contains similar total recoverable coal reserves as the Ukraine: approximately 37,500 million short tons; however, unlike the Ukraine, most of Kazakhstan's reserves are hard coals that total more than 34,000 million short tons. The coal deposits are late Carboniferous and Jurassic in age and are located mainly in the Karaganda and Ekibastuz basins, which produce hard coal. The coal deposits of these basins lie along the southern edge of the Siberian platform [5]. In the Karaganda basin, coking and steam coals are produced that have sulfur contents ranging from 1.5 to 2.5% and high ash content (20 to 35%). Coals from the Ekibastuz basin typically have high ash (39% on average) and low sulfur (<1%) contents and are predominantly used for thermal power generation.

Poland Poland contains recoverable coal reserves of more than 24,000 million short tons, of which more than 22,000 million short tons are hard coal. The hard-coal deposits are found in three main basins located in the southern half of the country: the Upper Silesian, Lower Silesian, and Lublin basins. These basins are of Carboniferous age. Poland uses its hard coal in world export markets. Poland's lignite deposits are found in a number of Tertiary basins across the central and southwestern parts of the country.

Poland ranks fourth in world lignite production and is the second largest European producer after Germany. The lignite is used as a fuel for electricity generation. Polish lignite has variable ash contents (4–25%) and low to medium sulfur contents (0.2–1.7%).

Asia

Asia contains significant recoverable coal reserves totaling over 231,000 million short tons, or approximately 21% of the world total. Two countries comprise most of this total: China (126,000 million short tons) and India (93,000 million short tons).

China China contains more than 126,000 million short tons of recoverable coal reserves in the world, third behind only the United States and Russia [2]. These recoverable reserves are nearly equally divided between hard coal and lignite deposits (*i.e.*, 68,500 and 57,700 million short tons, respectively), with the hard coals being of Carboniferous, Permian, and Jurassic age and the lignite of Tertiary age. Coalfields are scattered throughout China, with the largest deposits being found in western China, stretching from north to south with most of the reserves in the northern part, specifically in the Inner Mongolia, Shanxi, and Shaanxi provinces. Significant anthracite deposits are found in the Shanxi and Guizhou provinces. Bituminous coal deposits occur in the Heilongjiang, Shanxi, Jiangxi, Shandong, Henan, Anhui, and Guizhou provinces [5]. China is the world's largest coal producer; most of the coal is used internally for industry and electricity generation. The hard-coal rank appears to increase slightly northward from the Yangtze River, while locally seam quality is very variable [5].

India India's recoverable coal reserves rank fourth in the world with more than 93,000 million short tons. These reserves vary in rank from lignites to bituminous coal, with most of it being hard coal (*i.e.*, nearly 91,000 million short tons), although coal quality is generally poor. India's coalfields are located mainly in the east in the states of Assam, Bihar, Uttar Pradesh, Madhya Pradesh, Andhra Pradesh, Orissa, and West Bengal [5]. India's coals are principally of Permian age with some of Tertiary age.

The most significant deposits are in the Raniganj and Jharia basins of northeast India. In the Raniganj basin, the rank increases from noncaking bituminous coal in the east to medium coking coal in the west. Ash content varies, though, from 15 to 35%; sulfur content is low (<1%). The Jharia coalfield is India's major source of prime coking coal, although it also contains significant non-coking coal as well. As with Raniganj basin coals, ash content varies from 15 to 35% and the coal has low sulfur contents in the Jharia basin. Most of India's lignite mining occurs in southern India in the Neyveli coalfield, although other areas contain larger resources. The lignite is low ash (2–12%) and low sulfur (<1%); however, the moisture content is high, varying between 45 and 55%. India's coal is used primarily for power production. Although India has substantial recoverable resources, coal imports are steadily rising to meet demands for coking coal as well as for steam coal as new power plants begin operation [5].

Australia

Australian recoverable coal reserves total over 90,000 million short tons, which is nearly equally divided between hard coal and lignite deposits (*i.e.*, 46,900 and 43,600 million short tons, respectively) with the hard coals being of Carboniferous and Permian age and the lignite of Tertiary age. Coal is mined in all of the states except for the Northern Territory. New South Wales and Queensland produce both steam and metallurgical coal for export, while production in Victoria, South Australia, and Western Australia is used for thermal electricity generation [5]. Hard coal is mined in New South Wales, Queensland, and Western Australia, while subbituminous and brown coal is mined in South Australia and Victoria. The major coal reserves are found in eastern Australia, with the Bowen, Sydney, and Gippsland basins being the most important.

The Bowen basin is located in Queensland and developed during early Permian times. The rank varies in this basin, increasing from west to east, with the higher rank coals ranging from low-volatile bituminous coal in the west to semi-anthracites and anthracites in the east. The coals have a low sulfur content (typically 0.3–0.8%) and ash contents of 8 to 10% and 8 to 16% for coking and thermal coals, respectively.

The Sydney basin is located in New South Wales, is of Permian age, and consists of several coalfields. In general, the Sydney basin coals are medium- to high-volatile bituminous coal, with the highest rank being found in the northern portion of the basin. The coals in this basin have a low sulfur content (<1%), and ash contents typically range from 6 to 24%, although one coalfield exceeds 40% ash.

The brown coal resources found in the Gippsland basin lie within the Latrobe Valley in Victoria and are of Tertiary age. This area is noted for its thick coal seams ranging from 330 to 460 feet in thickness. The brown coals have low heating values (3400–5200 Btu/lb) due to high and very variable moisture contents, which range from 49 to 70%. Ash contents, on the other hand, are low and range from 0.5 to 2%.

Western Europe

Western Europe contains approximately 101,000 million short tons of recoverable coal reserves, with 72,800 million short tons in Germany alone. Fifteen other countries (see Table 1-4) comprise the balance; Yugoslavia contains by far the largest recoverable reserves of these countries (18,000 million short tons).

Germany Germany contains nearly 73,000 million short tons of recoverable coal reserves, of which more than 25,000 and 47,000 million short tons are hard coal and lignite, respectively. Germany is Europe's largest individual lignite producer; the three main areas of lignite resources are the Rhineland, Lusatian, and Central German basins, which are of Tertiary age. In addition, Germany has a substantial hard-coal capacity, which is of Carboniferous age and located in the Ruhr and Saar basins. Of the three main lignite basins, the Rhineland deposits are now the most important and are located between the River Rhine and the German/Dutch/Belgian border. The Central German and Lusatian basins are located in eastern Germany. The lignites have heating values of 3350 to 5400 Btu/lb, and moisture contents that vary from 40 to 60%. Ash and sulfur contents vary from 1.5 to 8.5% and 0.2 to 2.1%, respectively, with Rhineland basin lignite having sulfur contents of less than 0.5%. These coals are used for producing electricity in generating stations. Because of restructuring of the hard coal mining sector, which began in 1999, the Ruhr coalfield has greater economic significance than the Saar coalfield as mines continue to close and overall production declines [5]. The Ruhr coalfield primarily consists of bituminous coal, much of which is coking coal. There are two small areas of anthracite in the basin. The ash and sulfur contents of the coals in this basin are 4 to 9% and less than 1%, respectively. The coals are used primarily for electricity generation along with some industrial applications.

Africa

Africa contains 61,000 million short tons of recoverable coal, with approximately 55,000 million short tons of those reserves being located in South Africa. The balance is found in 13 other countries, with 11 of those countries containing less than about 200 million short tons each of recoverable reserves (see Table 1-4).

South Africa South Africa's recoverable coal reserves of 55,000 million short tons consist entirely of hard coal. These coals are of Carboniferous and Permian age with significant deposits in the Great Karoo basin. This basin extends about 300 miles from west to east across northern Free State Province and south and east Mpumalanga, and about 700 miles from southern Mpumalanga in the north to the center of Kwazulu–Natal in the south [5]. Although the Great Karoo basin is the largest, several other basins and a total of 19 coalfields are located throughout South Africa. The hard coal consists of bituminous coals, anthracite, and semi-anthracite. The ash content ranges from 7% for some anthracites to over 30% for bituminous coals. Sulfur contents range from less than 1% to nearly 3%. Domestically, the coal is used for electricity generation and conversion into synthetic liquid fuels and chemical feedstocks. South Africa exports significant quantities of steam coal with minor amounts of coking coal and anthracite.

Central and South America

Central and South America contain approximately 24,000 million short tons of recoverable coal reserves, or 2.2% of the world's total. Coal is found in several countries, including Argentina, Bolivia, Brazil, Chile, Colombia, Ecuador, Peru, and Venezuela; however, two of the countries contain the majority of these reserves: Brazil (13,150 million short tons) and Colombia (7200 million short tons). Brazil's coals are subbituminous and lignitic in rank, while Colombia's coals are primarily high-volatile bituminous with a small amount of subbituminous coals. These coals formed during late Cretaceous to Tertiary times.

References

1. Van Krevelen, D. W., *Coal: Typology–Physics–Chemistry–Constitution*, Third ed. (Elsevier Science, Amsterdam, 1993).
2. EIA, *International Energy Annual 2001* (U.S. Department of Energy, Energy Information Administration, Washington, D.C., March 2003), pp. 114–115.
3. Grimston, M. C., *Coal as an Energy Source* (IEA Coal Research, London, 1999), p. 6.
4. Moore, E. S., *Coal: Its Properties, Analysis, Classification, Geology, Extraction, Uses, and Distribution* (John Wiley & Sons, New York, 1922), p. 124.
5. Walker, S., *Major Coalfields of the World* (IEA Coal Research, London, 2000).
6. Tatsch, J. H., *Coal Deposits: Origin, Evolution, and Present Characteristics* (Tatsch Associates, Sudbury, MA, 1980), p. 5.
7. Schobert, H. H., *Coal: The Energy Source of the Past and Future* (American Chemical Society, Washington, D.C., 1987).
8. Mitchell, G., *Basics of Coal and Coal Characteristics* (Selecting Coals for Quality Coke Short Course, Iron and Steel Society, Warrendale, PA, 1997).
9. Elliott, M. A. (editor), *Chemistry of Coal Utilization*, Second Suppl. Vol. (John Wiley & Sons, New York, 1981).
10. Berkowitz, N., *An Introduction to Coal Technology* (Academic Press, New York, 1979).
11. Singer, J. G. (editor), *Combustion: Fossil Power Systems* (Combustion Engineering, Windsor, CT, 1981).
12. Miller, B. G., S. Falcone Miller, R. Cooper, J. Gaudlip, M. Lapinsky, R. McLaren, W. Serencsits, N. Raskin, and T. Steitz, *Feasibility Analysis for Installing a Circulating Fluidized Bed Boiler for Cofiring Multiple Biofuels and Other Wastes with Coal at Penn State University* (U.S. Department of Energy, National Energy Technology Laboratory, DE-FG26-00NT40809, 2003), Appendix J.
13. Ward, C. R. (editor), *Coal Geology and Coal Technology* (Blackwell Scientific, Melbourne, 1984), p. 66.
14. Averitt, P., Coal resources of the U.S., January 1, 1974, *U.S. Geological Survey Bulletin*, No. 1412, 1975 (reprinted 1976), 131 pp.

15. Mackowsky, M. T., *Mineral Matter in Coal: In Coal and Coal-Bearing Strata*, D. Murchson and T. S. Westoll, editors (Oliver & Boyd, Ltd., London, 1968), pp. 309–321.
16. EIA, *U.S. Coal Reserves: 1997 Update* (U.S. Department of Energy, Energy Information Administration, Washington, D.C., February 1999), p. 5, Appendix A.
17. EIA, *International Energy Outlook 2002* (U.S. Department of Energy, Energy Information Administration, Washington, D.C., March 2002), p. 72.
18. *Keystone Coal Industry Manual* (Intertec Publishing, Chicago, IL, 1997), p. 687.

CHAPTER 2

Past, Present, and Future Role of Coal

A brief history of the use of coal is presented in this chapter along with a comparison to other energy sources. While the emphasis is on coal usage in the United States, a global perspective is also presented, especially with regard to comparing overall energy consumption. Types of technologies used in the past and developed as part of the Industrial Revolution are briefly mentioned as part of the history of coal use; however, a more in-depth discussion of major coal technologies is provided in Chapter 5 (Technologies for Coal Utilization). Similarly, the role of coal in the future energy mix of the United States, which is crucial to the U.S. economy and energy security, is introduced in this chapter but discussed in greater detail in Chapter 8 (Coal's Role in Providing U.S. Energy Security).

The Use of Coal in the Pre-Industrial Revolution Era

The use of coal as an energy source has been known from ancient times, although it was a minor resource until the Industrial Revolution. The first mention of coal in European literature dates from the fourth century B.C. [1]; however, scholars are certain that coal was first used in China as early as 1000 B.C. [2]. By 1000 A.D., coal was a primary fuel source in China, and its use was reported by the Venetian traveler Marco Polo in the thirteenth century [3,4].

The first documented use of coal in Western civilization is attributed to the Greek philosophers Pliny, Aristotle, and Theophrastus, who was Aristotle's pupil [1]. The first definitive record of the use of coal is found in Aristotle's *Meteorology*, where he writes of combustible bodies [1]. Theophrastus, in his fourth-century *Treatise on Stones*, describes a fossil substance used as a fuel [5]. Theophrastus and Pliney both mention the use of coal by smiths [1,6]. The coal mentioned in these writings was apparently brown coal from Thrace in northern Greece and from Ligurai in northwestern Italy. This coal was not normally used in iron-smelting furnaces because of its impurities and, hence, inability to produce the required high temperatures, although Pliney does mention its use in copper casting, which can be done at considerably lower temperatures [6].

Although the Greeks and Romans knew of coal around 400 B.C., they did not have much use for it because wood was plentiful. When wood is abundant, there is little incentive to mine coal. Coal was used as a domestic heating fuel in some parts of the Roman Empire, particularly in Britain, but it never made more than a marginal contribution as a fuel resource [6]. As the Romans invaded northward, they encountered the mining and use of coal in the vicinity of St. Etienne in Gaul (France) as well as in Britain, where coal cinders in Roman ruins indicated that coal was used during the Roman occupation, from approximately 50 to 450 A.D. [5].

In the middle ages, coal had to be rediscovered in Europe, and for some time coal remained of very restricted local importance. Coal had been used on a small scale in western Europe for thousands of years, as evidence shows from the discoveries of a Bronze Age corpse that was cremated with coal in South Wales, as well as remains of Roman coal-fueled fires on their northern English frontier along Hadrian's Wall; however, there is no evidence of European coal use for hundreds of years after the fall of the Roman Empire [7]. During the middle ages, records show that coal had been given to the monks in the Abbey of Peterborough in 852 A.D. as an offering or in a settlement of a claim [2]. Coal was mined in Germany as early as the 900s and was mentioned in the charter (dated 1025) of the French priory of St. Sauveuren-Rue [8].

While peasants probably continued to use surface coal for domestic heating fuel, evidence suggests that it was not until around 1200 that industrial uses were found. Coal was discovered to be a very good fuel for iron forges and metalworking, as it burned almost as slowly as charcoal, which is produced from wood and was the primary fuel of choice for village smiths. It is at this time that shipping records can be found for coal being marketed in western Europe because the smiths preferred coal over charcoal if they could get it at a reasonable price [7]. Liège in Belgium, Lyonnais in France, and Newcastle in England became important mining centers [7,8].

At first coal was used close to the areas where it was mined because it was competing against wood and charcoal, which were less expensive than coal to transport. Coal was bulky and therefore subject to the high land transport costs common for bulky materials; however, as water transportation on rivers and the sea increased and wood became scarce particularly in the cities, coal became the fuel of choice. By the mid-1200s, the term *sea-coal* was used to distinguish coal from charcoal, and sea-coal was being transported to London by sea from Newcastle. By the 1370s, 84 coal boats were traveling down the east coast of England to ten different ports along the European coast between France and Denmark, returning with iron, salt, cloth, and tiles [7].

Although coal became the fuel of choice for blacksmiths during medieval times, initially it had limited use as a heat source because of its fumes; however, as wood became increasingly scarce and coal became less expensive in the cities, coal use began to increase. With the increased use of coal (mainly in fireplaces designed to burn wood) came increased pollution problems, mainly black smoke and fumes. In 1257, Queen Eleanor was driven from Nottingham Castle by the smoke and fumes rising from the city below. In 1283 and 1288, London's citizens complained about the air quality as a result of coal being used in lime kilns. In 1307, a Royal Proclamation forbade the use of coal in lime burners in parts of South London [7]. This proclamation did not work, however, and a later commission had instructions to punish offenders with fines and ransoms for a first offense and to demolish their furnaces for a second offense. Eventually, economics and a change in government policy won out over the populace's comfort and London was to remain polluted by coal fumes for another 600 years [7]. As the price of firewood increased, it became more profitable to transport coal over longer distances. In addition, late sixteenth- and seventeenth-century England faced the dilemma of conserving its remaining forests and using the only available substitute: coal. In 1615, the English government encouraged the substitution of coal for wood whenever possible. A fundamental change in English domestic building followed with more brick chimneys constructed to accommodate the fumes from coal [7].

Early History of U.S. Coal Mining and Use

Coal was reportedly used by the Indians of the Southwest prior to the arrival of the early explorers in America [9]. The first record of coal in the United States is on a map prepared in 1673–1674 by the Frenchman Louis Joliet. It shows *charbon de terra* along the Illinois River in northern Illinois. In 1701, coal was discovered near Richmond, Virginia, and a map drawn in 1736 shows the location of several "cole mines" on the upper Potomac River, near what is now the border of Maryland and West Virginia. By the mid-1700s, coal was also reported in Pennsylvania, Ohio, and Kentucky, and the first commercial U.S. coal production was initiated near Richmond, Virginia [9]. Blacksmiths in colonial days used small amounts of coal to supplement the charcoal normally burned in their forges. Farmers dug coal from beds exposed at the surface and sold it. Although most of the coal for the larger cities along the eastern seaboard was imported from England and Nova Scotia, some of it came from Virginia [9].

The Use of Coal during the Industrial Revolution

Several developments in the eighteenth century led to an expanded use of coal in England, culminating with the Industrial Revolution, which occurred over the century of 1750 to 1850. These developments included the transport revolution, the iron industry revolution, and the demise of the forests [10]. In eighteenth-century England, no fuel other than coal was available because wood had become an exhausted resource in the populated areas. There was a continuing demand for coal to supply fuel for domestic needs and the few industries that were in place in the pre-industrial community—namely, bakeries, smithies, tanneries, sugar refineries, and breweries [10]. Transporting coal by sea could not meet the demands for coal, as this mode of transportation was not always reliable, plus sea transport was not able to satisfy the demand for coal inland. Canals, however, could meet this need. More than half of the Navigation Acts passed between 1758 and 1802 to establish a canal or river-improvement company were for companies whose primary aim was to carry coal [10]. Establishing a transport system was crucial for the success of the Industrial Revolution because the Industrial Revolution was ultimately driven by the coal and iron industries, and it was necessary to move bulky raw materials and the finished products quickly and inexpensively across England [10].

The iron industry (and the resulting increased need for coal to produce coke for iron smelting) was an important contributing factor to bringing about the Industrial Revolution. Abraham Darby successfully smelted iron with coke as early as 1709, and this technological innovation became very important in the 1750s as the price of charcoal rose and the price of coal declined [10]. When the switch from wood to coal was complete, an ironmaster's constraint on his output was not his fuel supply but his power supply to provide an adequate blast in his furnace. The invention that provided an unlimited source of power, which up to this time had primarily been water and to a lesser extent wind power, was Boulton and Watt's steam engine. The steam engine, introduced around 1775 and fueled by coal, removed any restrictions on the ironmasters with regard to the size or location of iron works. The ironmasters could now move into areas rich in coal and iron resources and reap the economies of scale of a modern industry [10]. With the invention of the steam engine came the locomotive, another means for mass transportation of raw materials and products and which also consumed coal as a fuel source.

The uses for coal did not stop with coking and solid fuel combustion for transportation. It was discovered that gases released from the coal during the coking process could be burned. This in turn led to the establishment of the manufactured gas industry to exploit the illuminating power of coal gas. In 1810, an Act of Parliament was obtained for forming a company to supply coal gas to London [2].

The Industrial Revolution began in England but spread to continental Europe, mainly France and Germany, and to the United States. These countries were able to benefit from the discoveries driving the Industrial Revolution because they too had ample supplies of coal. In the United States during the 1800s, coal became the principal fuel used by locomotives and, as the railroads branched into the coal fields, they became a vital link between the mines and markets. Coal also found growing markets as fuel for homes and steamboats and in the production of illuminating oil and gas. In fact, shortly after London began using coal gas, Baltimore, Maryland, became the first city in the United States to light streets with coal gas, in 1816. And in England, coke soon replaced charcoal as the fuel for iron blast furnaces in the latter half of the 1800s.

Post-Industrial Revolution Use of Coal

The major coal-utilization technologies—combustion, gasification, liquefaction, and carbonization (*i.e.*, the production of coke)—either got their start during the 100-year-plus period of the Industrial Revolution/post-Industrial Revolution era or, in the case of coke production, made major strides in technology development and usage. These technologies are discussed in detail in Chapter 5, but a brief discussion is provided here.

The demand for coal increased as additional fuel chemistry and engineering technologies were developed. In 1855, R. W. Bunsen invented the atmospheric gas burner, which led to a wide range of heating applications.

Electric lighting, which had its start in the late 1800s, led to widespread coal combustion for the generation of steam for power generation. Utilizing coal for generating electricity has been continually passing through various stages of technology development ever since the late 1800s; examples include such areas as advanced combustion technologies, new materials of construction, innovative system designs, and new developments in steam production, electricity generation, and pollution control (discussed in detail in later chapters).

As the use of coal gas developed, the production of gas for heating purposes was also developing. The first gas producer to make low-Btu gas was built in 1832 and at the turn of the century was an important method for heating furnaces [2]. Another development in the field of gas production was the discovery of the carbon–steam reaction, where steam is reacted with carbon to produce carbon monoxide and hydrogen. This gasification technology had its major start in the mid- to late 1800s and increased until less expensive natural gas replaced manufactured gas [2].

Coal hydrogenation began in 1913 with the Bergius concept of direct hydrogenation of coal under hydrogen pressure at an elevated temperature [2]. The production of liquid hydrocarbons via indirect liquefaction (*i.e.*, Fischer-Tropsch synthesis) was also conceived at this time.

Overview of Energy in the United States

The United States has always been a resource-rich area, but in the colonial days nearly all energy was supplied by muscle power (both human and

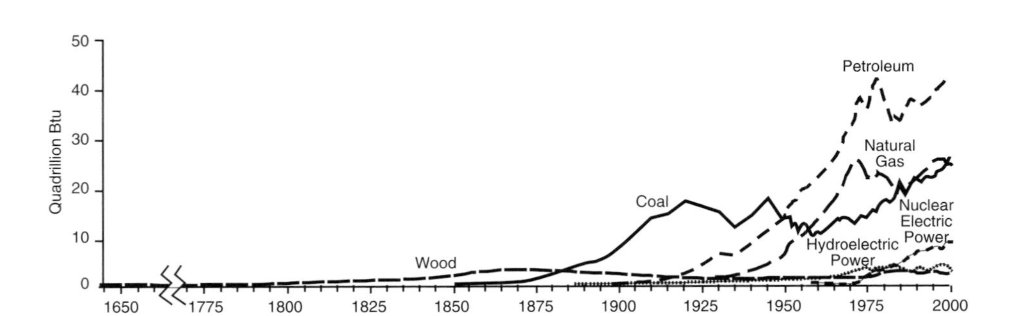

FIGURE 2-1. U.S. energy consumption by source, 1635 to 2001. (From EIA, *Annual Energy Review 2001*, U.S. Department of Energy, Energy Information Administration, Washington, D.C., November 2002.)

animal), waterpower, wind, and wood. The history of energy use in the United States begins with wood being the dominant energy source from the founding of the earliest colonies until late last century, as shown in Figure 2-1 [11], in which consumption is illustrated in quadrillion (*i.e.*, 10^{15}) Btu. Although wood use continued to expand along with the nation's economic growth, energy shortages led to the search for other energy sources; hence, coal began to be used in blast furnaces for coke production and in the making of coal gas for illumination in the early 1800s. Natural gas found limited application in lighting. It was still not until well after mid-century that the total work output from engines exceeded that of work animals.

Manifest Destiny, or the westward expansion from the seacoast to the heart of the nation, was a major factor in the expanded use of coal. As railroads drove west to the plains and mountains, they left behind the plentiful wood resources along the east coast. Coal became more attractive as deposits were found along the railroad right-of-way, and it had a higher energy content than wood. This meant more train-miles traveled per pound of fuel. Demand for coal in coke production also rose because the railroads were laying thousands of miles of new track, and iron and steel were needed for the rails and spikes. The rapid growth of the transportation and industrial sectors was fueled by coal.

Coal ended the long dominance of wood in the United States about 1885, only to be surpassed in 1951 by petroleum and then by natural gas a few years later. Hydroelectric power and nuclear electric power appeared about 1890 and 1957, respectively. Solar photovoltaic, advanced solar thermal, and geothermal technologies represent further recent developments in energy sources.

Petroleum was initially used as an illuminant and an ingredient in medicines but was not used as a fuel for many years. At the end of World War I, coal still accounted for approximately 75% of the total U.S. energy use.

This situation changed, however, after World War II. Coal relinquished its place as the premier fuel in the United States as railroads lost business to trucks that operated on gasoline and diesel fuel. The railroads themselves began switching to diesel locomotives. Natural gas also started replacing coal in home stoves and furnaces. The coal industry survived, however, mainly because nationwide electrification created new demand for coal among electric utilities.

Most of the energy produced today in the United States comes from fossil fuels—coal, natural gas, crude oil, and natural gas plant liquids (see Figure 2-2) [11]. Although U.S. energy production takes many forms, the use of all fossil fuels far exceeds that of all other forms of energy. In 2000, fossil fuels accounted for 80% of total energy production and were valued at an estimated $148 billion [11].

For most of its history, the United States was self-sufficient in energy, although small amounts of coal were imported from Britain and Nova Scotia during colonial times. Through the late 1950s, production and consumption of energy were nearly in balance; however, beginning in the 1960s and continuing today, consumption outpaces domestic production (see Figure 2-3). This imbalance is further illustrated in Figure 2-4; in 2001, the United States produced approximately 72 quadrillion Btu but consumed nearly 97 quadrillion Btu, and crude oil imports totaled nearly 25 quadrillion Btu [11]. Because of its insatiable demand for petroleum, U.S. petroleum imports reached a

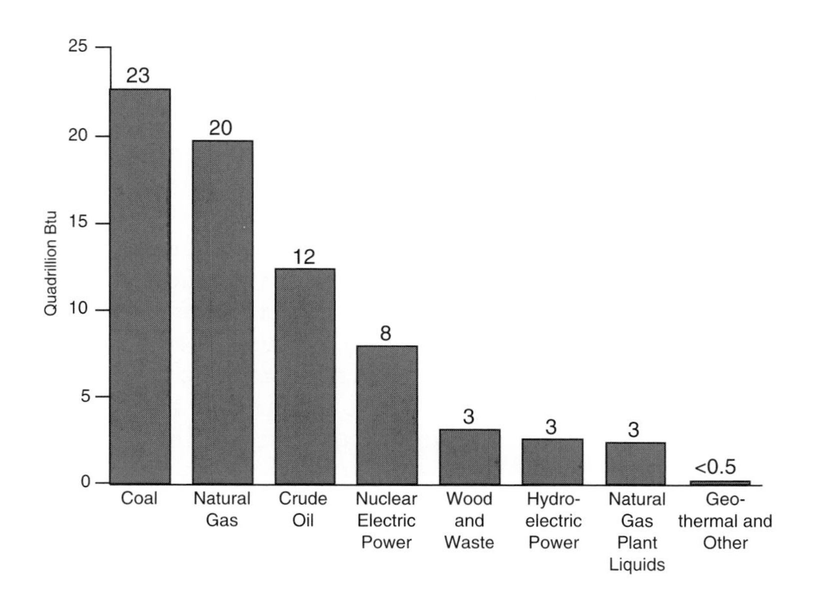

FIGURE 2-2. U.S. energy production by source for 2000. (From EIA, *Annual Energy Review 2001*, U.S. Department of Energy, Energy Information Administration, Washington, D.C., November 2002.)

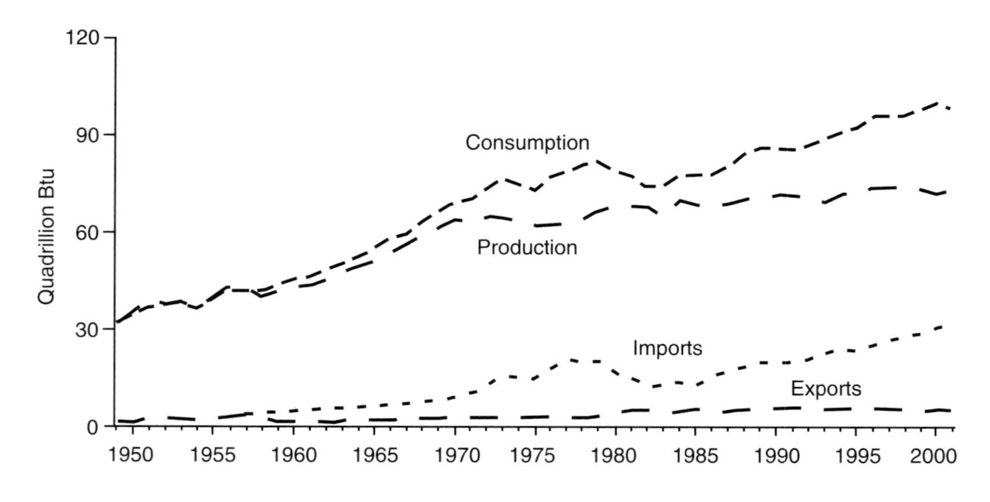

FIGURE 2-3. Energy overview of United States. (From EIA, *Annual Energy Review 2001*, U.S. Department of Energy, Energy Information Administration, Washington, D.C., November 2002.)

record level of 11 million barrels per day in 2000. This finding is disturbing, especially when compared to the U.S. petroleum imports in 1973, which totaled 6.3 million barrels per day. In October 1973, the Arab members of the Organization of Petroleum Exporting Countries (OPEC) embargoed the sale of oil to the United States, causing prices to rise sharply and sending the country into a recession. Although petroleum imports declined for two years, they increased again until prices rose dramatically from about 1979 through 1981, which suppressed imports. The increasing import trend resumed in 1986 and, except for some slight dips, has continued ever since. This dependency on foreign energy affects the security of the United States and must be addressed by its political leaders.

Energy is crucial in the operation of the industrialized U.S. economy, and energy spending is high. In recent years, American consumers have spent over a trillion dollars a year on energy [11]. Energy is consumed in four major sectors: residential, commercial, industrial, and transportation. Industry is historically the largest user of energy and the most vulnerable to fluctuating prices and consequently shows the greatest volatility, as shown in Figure 2-5. In particular, steep drops occurred in 1975, from 1980 to 1983, and again in 2001 (not shown) in response to high oil prices. Transportation was the next largest energy-consuming sector, followed by residential and commercial use.

Energy sources have changed over time for the various sectors. In the residential and commercial sectors, coal was the leading source as late as 1951 but then decreased, as illustrated in Figure 2-6. Coal was replaced by other forms of energy. Meanwhile electricity's use and related losses during generation, transmission, and distribution increased dramatically.

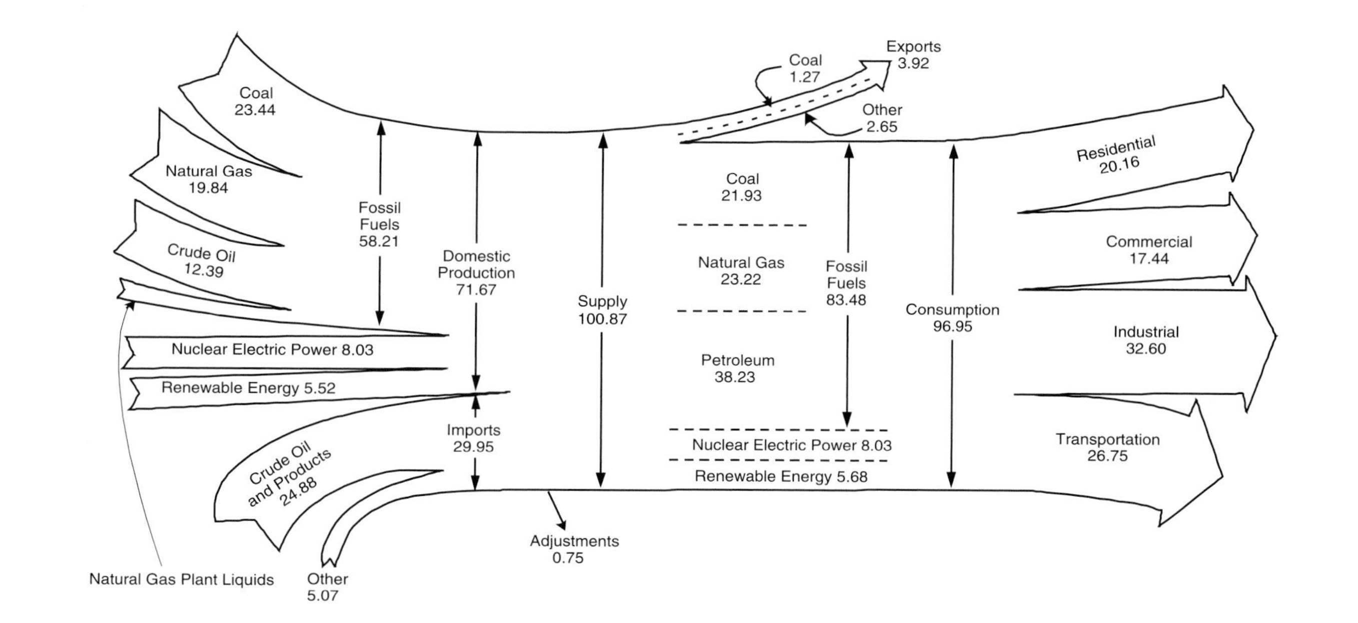

FIGURE 2-4. Energy flow of United States in 2001 (quadrillion Btu). (From EIA, *Annual Energy Review 2001*, U.S. Department of Energy, Energy Information Administration, Washington, D.C., November 2002.)

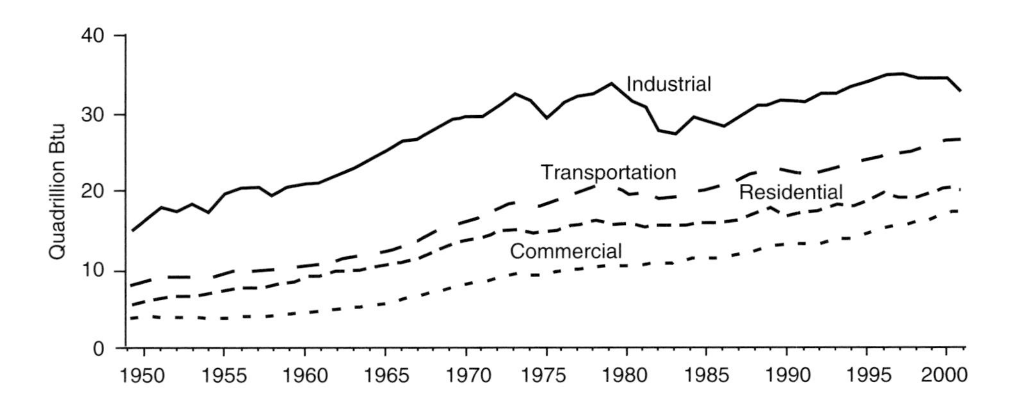

FIGURE 2-5. Energy consumption by sector in the United States. (From EIA, *Annual Energy Review 2001*, U.S. Department of Energy, Energy Information Administration, Washington, D.C., November 2002.)

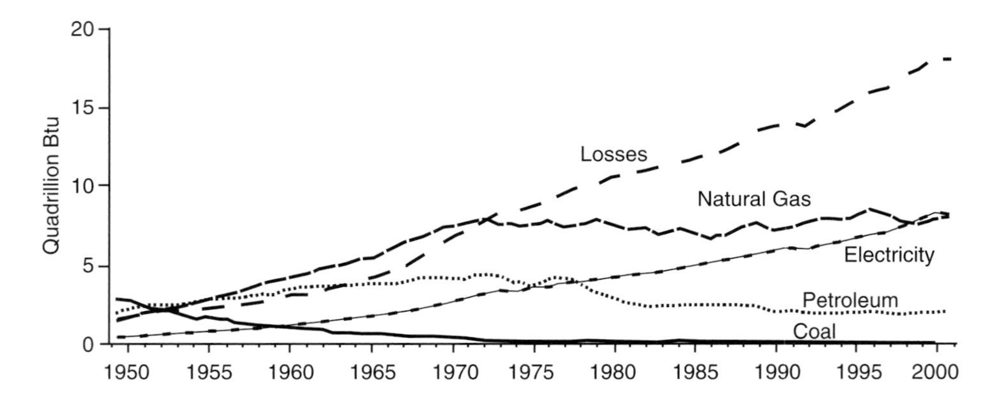

FIGURE 2-6. Residential and commercial energy consumption in the United States. (From EIA, *Annual Energy Review 2001*, U.S. Department of Energy, Energy Information Administration, Washington, D.C., November 2002.)

The expansion of electricity reflects the increased electrification of American households, which typically rely on a wide range of electrical appliances and systems. Home heating in the United States also underwent a significant change. Over a third of all housing units were heated by coal in 1950 but only 0.2% were coal heated in 1999. Similarly, distillate fuel oil lost a significant share of the home heating market, dropping from 22 to 10% over the same period. Home heating by natural gas and electricity, on the other hand, rose from 25 to more than 50% and from 0.6 to 30%, respectively [11].

In the industrial sector, the use of coal, once the leading energy source, decreased as the consumption of both natural gas and petroleum rose (see

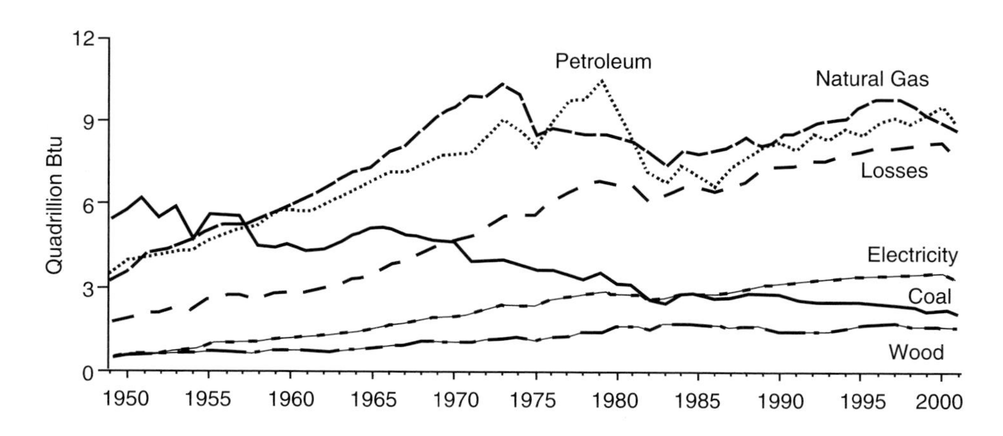

FIGURE 2-7. Industrial energy consumption in the United States. (From EIA, *Annual Energy Review 2001*, U.S. Department of Energy, Energy Information Administration, Washington, D.C., November 2002.)

Figure 2-7). Electricity and its associated losses also grew steadily in this sector. Approximately 60% of the energy consumed in the industrial sector is used for manufacturing. The remainder goes to mining, construction, agriculture, fisheries, and forestry [11]. The large consumers of energy in the manufacturing industries, for which the fuel of choice is primarily natural gas, include petroleum and coal products, chemicals and associated products, paper and associated products, and metal industries. Nearly 7% of all energy consumed in the United States is used for nonfuel purposes such as asphalt and road oil for road construction and road conditioning; roofing products; liquefied petroleum gases for feedstocks at petrochemical plants; waxes for packaging, cosmetics, pharmaceuticals, inks, and adhesives; and gases for chemical and rubber manufacture [11].

The transportation sector's use of energy, which is mainly petroleum, has more than tripled over the last 50 years, as shown in Figure 2-8. Gasoline accounts for about two-thirds of the petroleum consumed in this sector, with distillate fuel oil and jet fuel being the other main petroleum products used in this sector.

Coal Production in the United States

The total amount of coal consumed in the United States in all the years before 1800 was an estimated 108,000 short tons, much of it imported [11]; however, production and consumption began to increase as a result of the Industrial Revolution and the development of the railroads. From 1881 through 1951, coal was the leading energy source produced in the United States [11]. Coal was surpassed by crude oil and natural gas until 1982/1984, at which time coal regained its position as the top energy resource (see Figure 2-9).

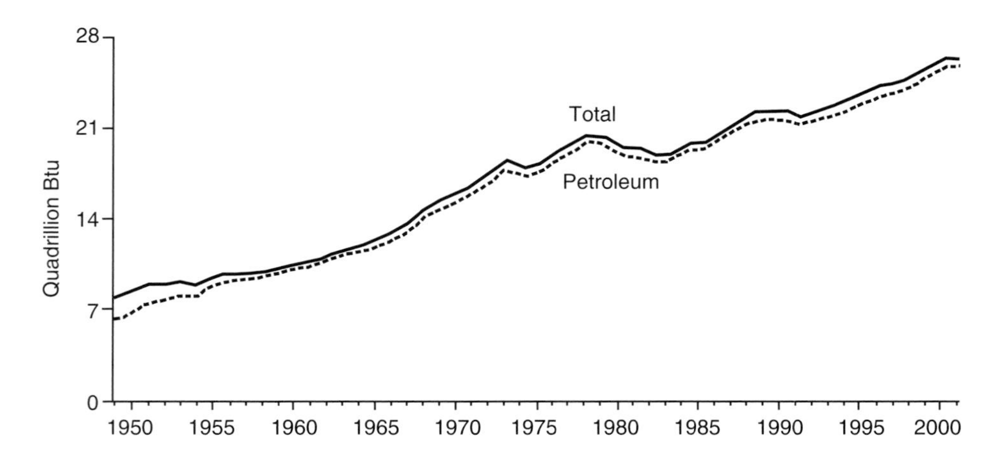

FIGURE 2-8. Transportation energy consumption in the United States. (From EIA, *Annual Energy Review 2001*, U.S. Department of Energy, Energy Information Administration, Washington, D.C., November 2002.)

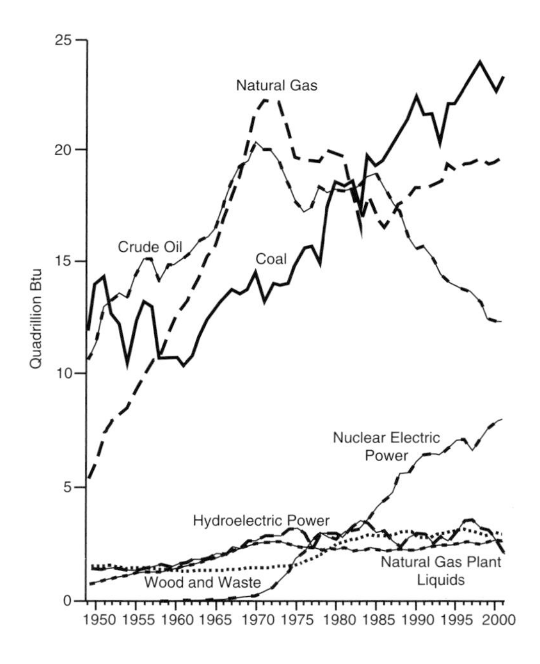

FIGURE 2-9. Energy production in the United States from 1949 to 2001. (Adapted from EIA, *Annual Energy Review 2001*, U.S. Department of Energy, Energy Information Administration, Washington, D.C., November 2002.)

Enormous quantities of coal are mined in the United States, and 1.09 billion short tons of coal were mined in 2002, which was slightly less than the record level of 1.12 billion short tons of coal mined in 2001 [12]. Coal flow in the United States for 2002 is summarized in Figure 2-10. The trend in U.S. coal production for the last 112 years is illustrated in Figure 2-11 [9]. Coal production by region and state is provided in Table 2-1 for 2002 [12]. Coal production has shifted from mainly underground mines to surface mines, as illustrated in Figure 2-12. Also, coal resources west of the Mississippi River, especially those in Wyoming, have undergone tremendous development (see Figure 2-13). Coal was produced in 26 states in 2002, with 15 of these states located in the western United States.

The shift toward surface-mined coal, especially west of the Mississippi River, came about because of the technological improvements in mining and the geological nature of the deposits (*i.e.*, thick coal seams that are located near the surface). Coal in the eastern United States generally occurs in seams that tend to be less than 15 feet thick. Thicker coalbeds are common in the western United States, particularly Wyoming, where coal seams average 65 feet, as illustrated in Figure 2-14 [9]. Coal seams that are more than 200 feet under the surface are mined by underground methods. Most underground mines are less than 1000 feet deep, although several reach depths of about 2000 feet [9]. The largest coal-producing western mines are surface mines.

Individual coalbeds commonly cover large geographical regions. For example, the heavily mined Pittsburgh coalbed is found in parts of Pennsylvania, West Virginia, Ohio, and Maryland. Similarly, the Wyodak coalbed, which is the leading source of coal in the United States, is estimated to cover at least 10,000 square miles in the Powder River Basin of Wyoming and Montana [9]. Consequently, although about 300 coalbeds were mined in the United States in 1993, nearly half of the coal produced that year was from only the ten seams listed in Table 2-2.

The most important coal deposits in the eastern United States are in the Appalachian Region, an area that encompasses more than 72,000 square miles and parts of nine states (see Figures 1-8 and 2-15). This region contains large deposits of low- and medium-volatile bituminous coal and the principal deposits of anthracite. Historically, this region has been the major source of U.S. coal, accounting for approximately 75% of the total annual production as recently as 1970 [9]; today the region produces less than 50% of the United States' total, with 396 million short tons mined in 2002, with the reduction being due to increased coal production in the western United States (see Figures 2-15 and 2-16). This region, however, is still the principal source of bituminous coal and anthracite, and three of the top four coal-producing states (West Virginia, eastern Kentucky, and Pennsylvania) are found in this region. (Note that eastern Kentucky is in the Appalachian Region, while western Kentucky is in the Interior Region.) These states, listed in Table 2-3, are among the top 10 producing states (of the 26 producing coal in 2003), which as a group produced over 85% of the coal in the United States.

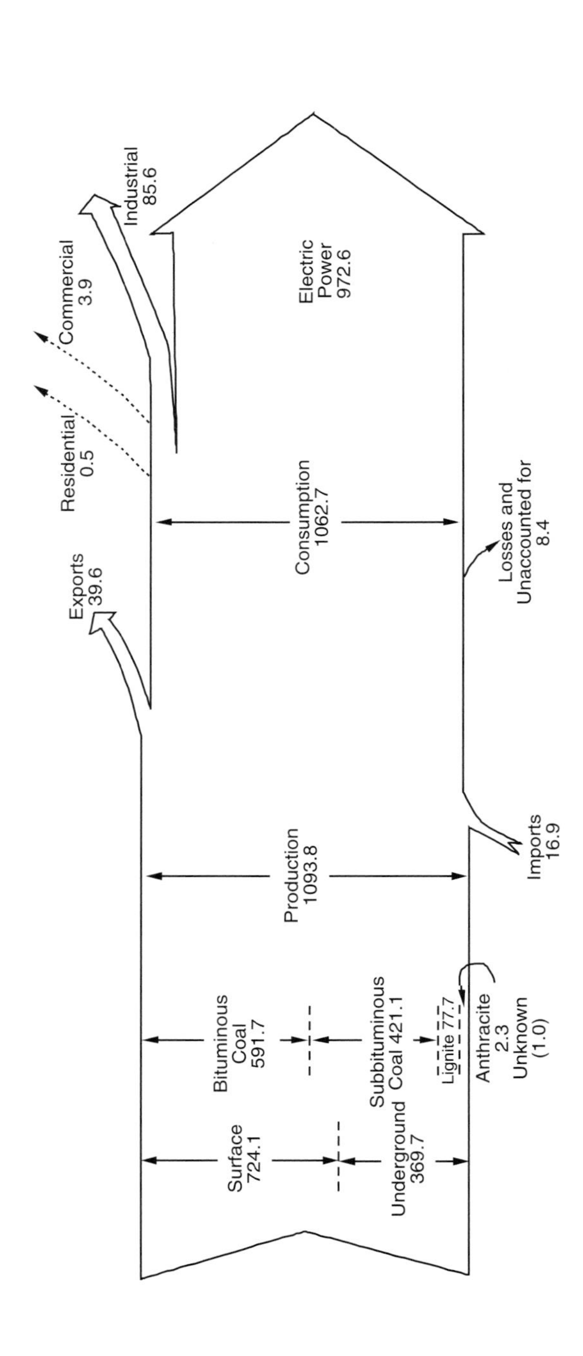

FIGURE 2-10. Coal flow in the United States in 2002 (million short tons). (Data from Freme, 2002; National Mining Association, 2003.)

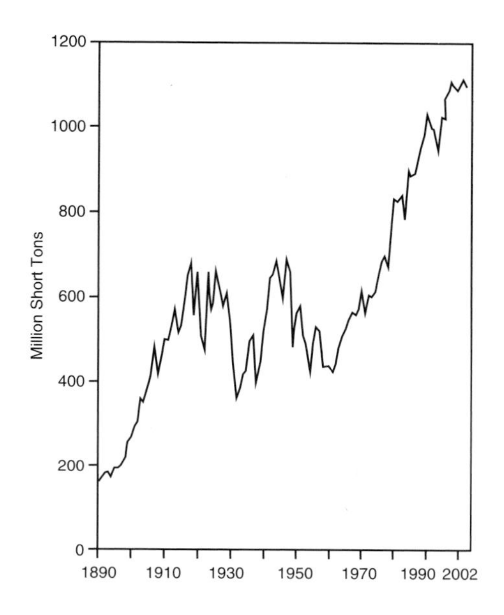

FIGURE 2-11. Coal production in the United States for the period 1890 to 2002. (Modified from EIA, *Coal Data: A Reference* U.S. Department of Energy, Energy Information Administration, Washington, D.C., February 1995.)

TABLE 2-1
U.S. Coal Production in 2002 by Region and State

Coal-Producing Region and State	*Production (million short tons)*
Appalachian total	**396.0**
Alabama	18.8
Kentucky, Eastern	98.9
Maryland	4.7
Ohio	21.3
Pennsylvania	68.7
Anthracite	1.2
Bituminous	67.4
Tennessee	3.2
Virginia	29.9
West Virginia	150.6
Interior total	**146.2**
Arkansas	<0.5
Illinois	33.3
Indiana	35.5
Kansas	0.2
Kentucky, Western	24.5
Louisiana	3.5

(continued)

TABLE 2-1
U.S. Coal Production in 2002 by Region and State *(continued)*

Coal-Producing Region and State	*Production (million short tons)*
Mississippi	2.3
Missouri	0.2
Oklahoma	1.4
Texas	45.2
Western total	**550.8**
Alaska	1.1
Arizona	12.8
Colorado	35.1
Montana	37.4
New Mexico	28.9
North Dakota	30.8
Utah	25.3
Washington	5.8
Wyoming	373.5
Refuse recovery	0.8
U.S. total	**1093.8**

Source: Freme, F., *U.S. Coal Supply and Demand: 2002 Review*, U.S. Department of Energy, Washington, D.C., 2002.

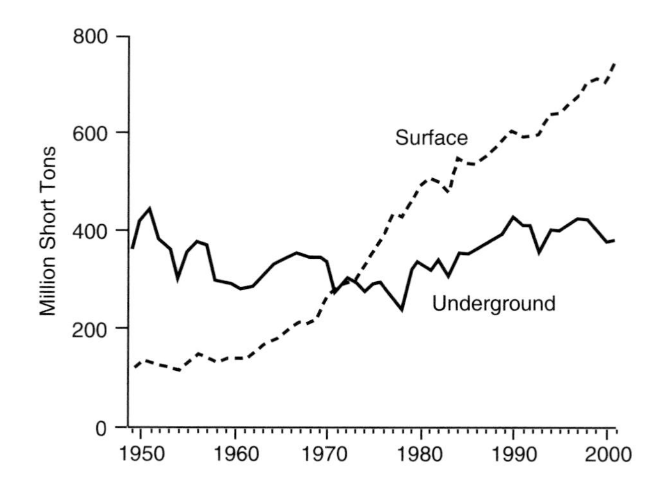

FIGURE 2-12. Coal production by mining method in the United States. (From EIA, *Annual Energy Review 2001*, U.S. Department of Energy, Energy Information Administration, Washington, D.C., November 2002.)

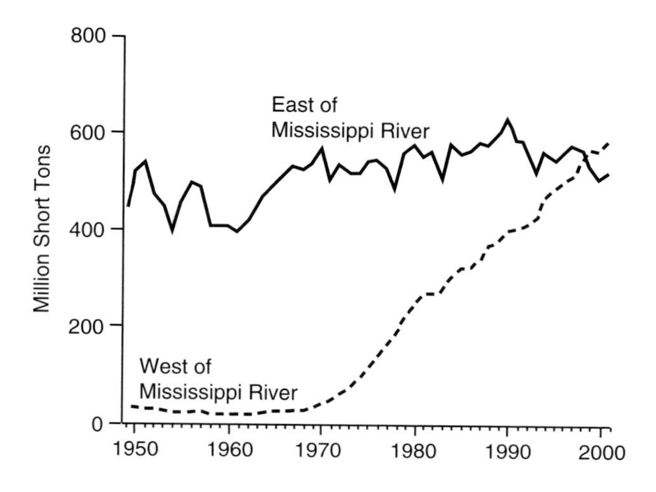

FIGURE 2-13. Coal production by location in the United States. (From EIA, *Annual Energy Review 2001*, U.S. Department of Energy, Energy Information Administration, Washington, D.C., November 2002.)

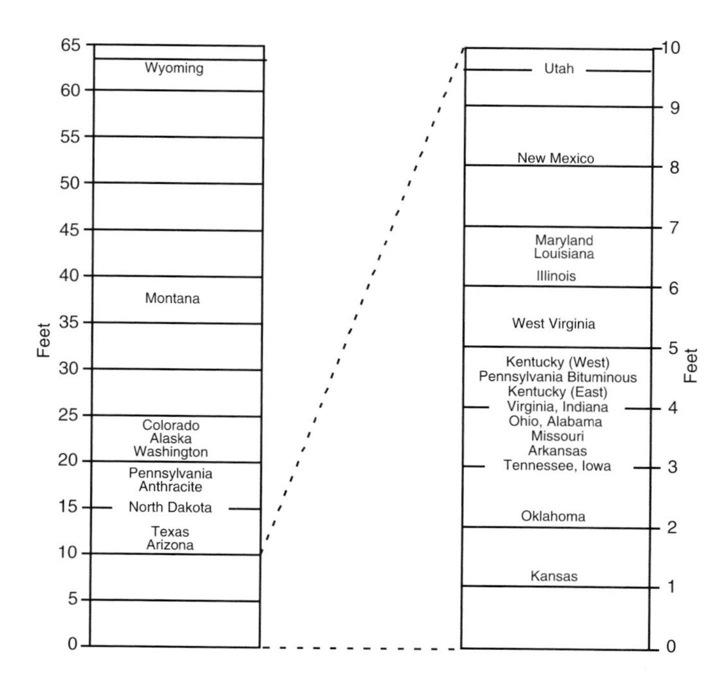

FIGURE 2-14. Average coalbed thickness in the United States. (From EIA, *Coal Data: A Reference*, U.S. Department of Energy, Energy Information Administration, Washington, D.C., February 1995.)

TABLE 2-2
U.S. Coal Production from the Ten Leading Coalbeds in 1993

Coalbed Name	*Production (million short tons)*	*State with the Largest Production in the Coalbed*
Wyodak	185.7	Wyoming
Pittsburgh	49.4	West Virginia
No. 9	34.8	Kentucky, Western
Hazard No. 5-A	32.4	Kentucky, Eastern
No. 6	30.7	Illinois
Beulah-Zap	27.7	North Dakota
Hazard No. 4	24.5	Kentucky, Eastern
Lower Kittanning	22.6	West Virginia
Lower Elkhorn	18.0	Kentucky, Eastern
Rosebud	16.2	Montana
Total	442.0	—
Percentage of U.S. total	46.8%	—

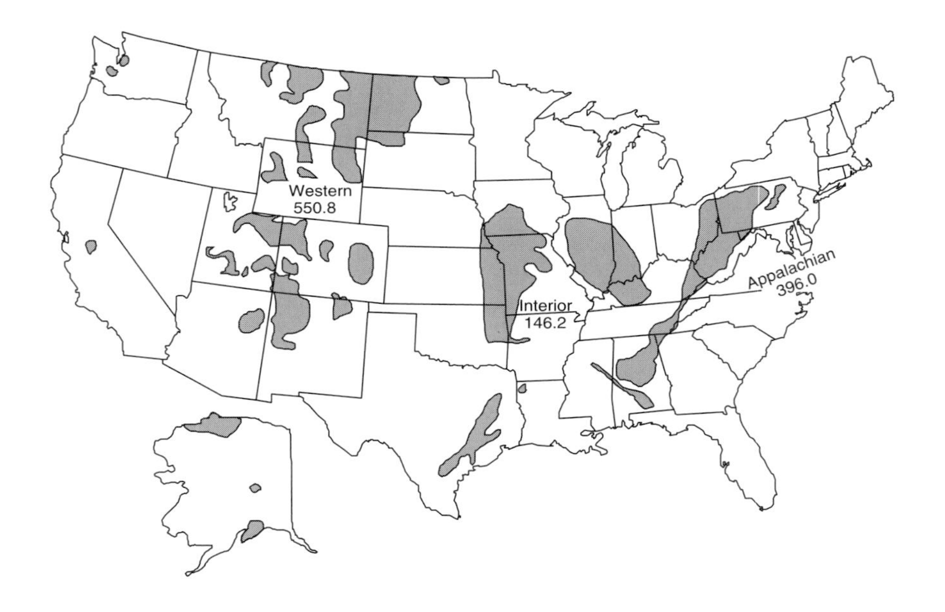

FIGURE 2-15. Coal production in 2001 by coal-producing region (million short tons). (From Freme, F., *U.S. Coal Supply and Demand: 2002 Review*, U.S. Department of Energy, Washington, D.C., 2002.)

The Interior Region, comprised of several separate basins located from Michigan to Texas, produced approximately 146 million short tons in 2002. Four states—Texas, Indiana, Illinois, and western Kentucky—produce the majority of the coal in this region. These coals range in rank from

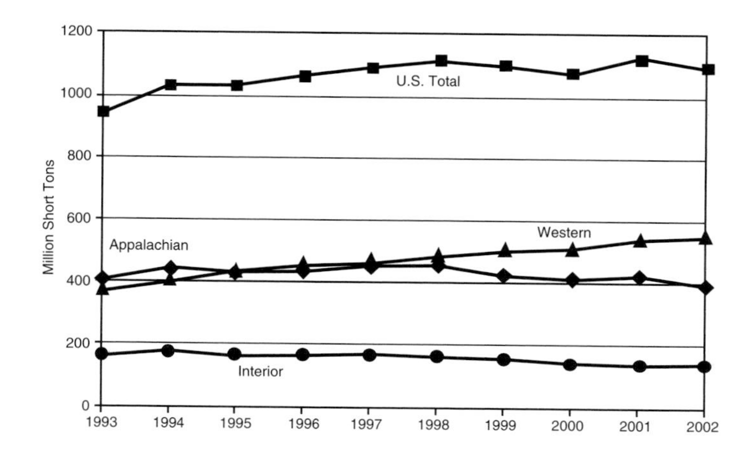

FIGURE 2-16. Coal production by region (1993–2002). (From Freme, F., *U.S. Coal Supply and Demand: 2002 Review*, U.S. Department of Energy, Washington, D.C., 2002.)

TABLE 2-3
Top Ten Coal-Producing States in 2002

State	*Production (thousand short tons)*
Wyoming	373,528
West Virginia	150,605
Kentucky	123,333
Eastern	98,866
Western	24,467
Pennsylvania	68,667
Texas	45,176
Montana	37,386
Indiana	35,513
Colorado	35,103
Illinois	33,314
North Dakota	30,799
Total	933,424
Percentage of U.S. total	85.3%

high-volatile bituminous coal in the northern part of the region to lignite in Texas.

The Western Region of the United States has several coal basins that contain all ranks of coal. Over half of the coal produced in the United States comes from this region; four of the states listed in Table 2-3 (Wyoming, Montana, Colorado, and North Dakota) produce approximately 44% of the total coal. Of these, Wyoming is by far the largest coal mining state in the

United States, as it produces approximately 34% of the country's total coal. Lignite is mined in North Dakota and Montana, subbituminous coal is mined in southeastern Montana and northeastern Wyoming, and the principal bituminous coal mining production is in Utah, Colorado, and Arizona.

Synthetic Coal

Synthetic coal, or synfuel, plants have recently assumed a larger share of the industry picture. The birth of the synfuel industry resulted from the enactment of Section 29 of the U.S. Internal Revenue Code of 1986 [14–18]. This legislation was enacted as a result of the upheaval in U.S. energy markets by the Arab oil embargo in 1973. Section 29 provides an income tax credit for fuels produced from nonconventional energy sources until 2007, when the credit is scheduled to end. In 2002, nearly 83 million tons of coal waste and run-of-mine coal were produced at more than 40 coal synfuel plants for the utility and industrial markets [12]. This volume is higher than early predictions, and it is projected that the quantity of synfuel produced will continue to increase even more, although these credits remain controversial [19–22]. The original intent of the legislation, with respect to coal, was to utilize waste coal and produce a nonconventional fuel such as a pellet. Many synfuel operators, however, are processing run-of-mine coal by simply spraying it with petroleum mixtures or emulsions, thus qualifying for the tax credit [19]. The ensuing controversy is centered on allegations of unfair competition with coal producers; consequently, while some form of tax credit program encouraging the development of alternative fuels will likely be instituted after 2007, standards for qualifying for the tax credit are expected to be more rigorous than the current ones [18].

Coal Consumption in the United States

Coal is used in all 50 states and the District of Columbia. In 2000, ten states consumed approximately 45% of the total coal produced in the United States (see Table 2-4). In 2002, most of the coal was consumed in the electric power sector, as shown in Figures 2-10 and 2-17. Nearly 92% of all coal consumed in 2002 was in the electric power sector, which includes both the electric utilities and independent power producers. Following is a breakdown, by sector, for coal consumption in 2002 (in 1000 short tons):

- Electric utilities—777,263;
- Other power producers—195,415;
- Coking—22,537;
- Other industrial—63,077;
- Residential/commercial—4369;
- Total—1,062,662.

TABLE 2-4
Top Ten Coal-Consuming States in 2000

State	*Consumption (thousand short tons)*
Texas	101,579
Indiana	70,583
Ohio	59,348
Alabama	39,797
West Virginia	39,000
Missouri	38,301
Kentucky	37,586
Michigan	36,298
Georgia	35,150
North Dakota	31,902
Total	489,544
Percentage of U.S. total	45.3%

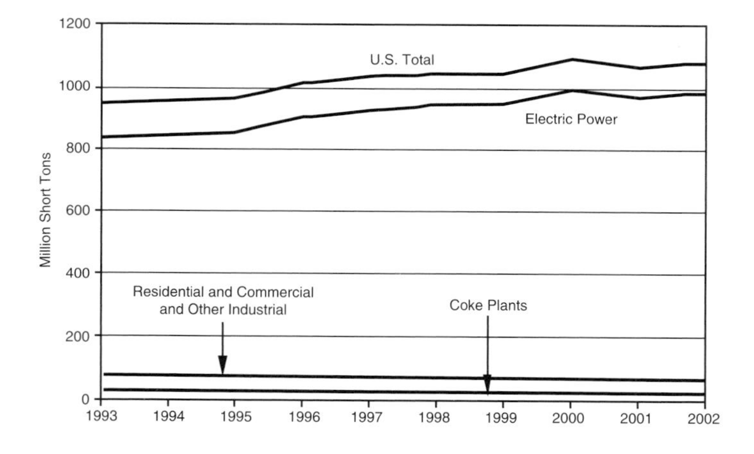

FIGURE 2-17. Coal consumption by sector (1993–2002). (From Freme, F., *U.S. Coal Supply and Demand: 2002 Review*, U.S. Department of Energy, Washington, D.C., 2002.)

The resurgence in coal consumption, after decreasing around 1950 (see Figures 2-1 and 2-9), was driven by the Arab oil embargo in the first half of the 1970s, which caused significant price increases for petroleum, and by a natural gas shortage in the second half of the 1970s. Virtually all of this growth was due to the increasing amounts of coal used to generate electricity, and coal became the dominant energy source in the electrical power industry. Nearly 50% of the electricity generated in the United States in 2002 was from coal, as illustrated in Figure 2-18; however, in 2002, coal's share of electricity generation dropped below 50% for the first time since

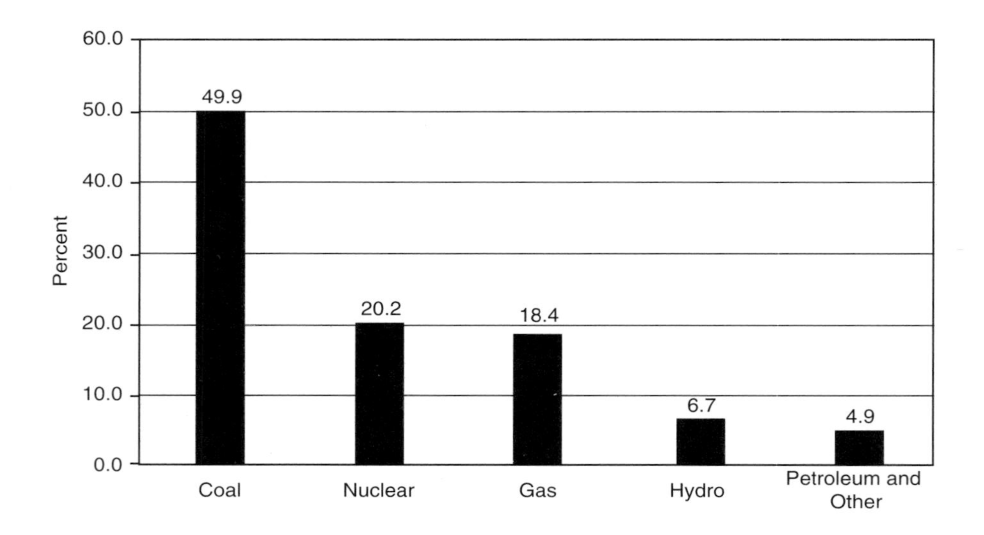

FIGURE 2-18. Production of electricity in the United States in 2002 by fuel source. (From Freme, F., *U.S. Coal Supply and Demand: 2002 Review*, U.S. Department of Energy, Washington, D.C., 2002.)

1979 due to gains by natural gas and hydroelectric plants. While a balanced energy portfolio makes sense from an energy security issue, the increasing use of natural gas for baseloaded electricity generation is questionable and is discussed in detail in Chapter 8. Coal consumption in the nonelectric power sectors is low, relative to the electric power sector. Although consumption has fluctuated slightly, total coal usage for coke production and the industrial/commercial/residential sectors has remained relatively unchanged over the last 10 years.

U.S. Coal Exports and Imports

The U.S. coal export and import markets are relatively small (see Figure 2-10), and the export market has been on a decline for many years, as illustrated in Figure 2-19 [12]. In 2002, the United States exported approximately 40 million short tons, which was nearly evenly divided between metallurgical coal and steam coal. Although this market has been declining, the EIA projects no further erosion in coal exports [12].

The United States has become only a marginal supplier in the international coal trade, particularly in the steam coal market. Canada has been the largest market for U.S. steam coal, accounting for two-thirds of the exported steam coal [12]. Europe, which traditionally had been a mainstay for U.S. steam coal exports, has been importing less coal from the United States and has contributed to the significant decrease observed in Figure 2-19.

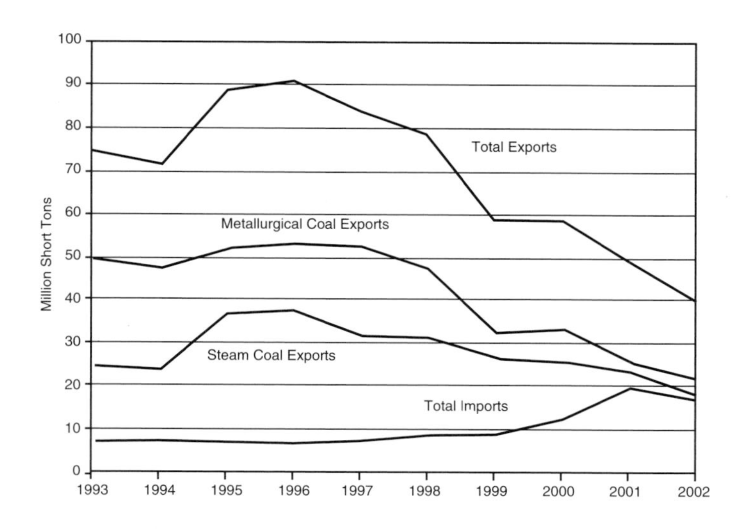

FIGURE 2-19. U.S. coal imports and exports. (From Freme, F., *U.S. Coal Supply and Demand: 2002 Review*, U.S. Department of Energy, Washington, D.C., 2002.)

Italy, Portugal, and the Netherlands all saw large decreases in imported U.S. coal [12].

As with steam coal, Canada has been the major market for U.S. metallurgical coal in 2002, accounting for nearly 22% of all metallurgical coal exports. Exports to other countries are mainly on a decline, including those to Brazil (the second largest market), Italy, the United Kingdom, Belgium, and Luxembourg. In Europe, only Spain increased imports of U.S. metallurgical coal. The Asian market (*i.e.*, Japan and Korea), which had been importing several million short tons of U.S. metallurgical coal, vanished in 2002 as this market now imports their coal from other countries, primarily Australia [12]. The loss of the Asian markets was due to increased competition and the costs associated with transporting U.S. metallurgical coal, which is mainly mined in the eastern United States, over such long distances.

The United States imports only a small amount of coal, relative to total U.S. coal consumption; less than 20 million short tons were imported out of over one billion short tons consumed. Colombia dominates the U.S. import market, accounting for approximately 55% of all coal imported in 2002, followed by Venezuela (3.3 million short tons) and Canada [12].

World Primary Energy Production and Consumption

Overviews of world energy production and consumption are provided in Tables 2-5 and 2-6, respectively. These tables contain information on the

TABLE 2-5
World Energy Production in 2001 (quadrillion Btu)

Region/ Country	*Total Primary Energy*	*Crude Oil*	*Natural Gas Plant Liquids*	*Dry Natural Gas*	*Coal*	*Net Hydroelectric Power*	*Net Nuclear Power*	*Other*[a]
North America								
Canada	18.20	4.30	1.03	6.71	1.85	3.41	0.82	0.07
Mexico	9.59	6.93	0.57	1.38	0.22	0.29	0.08	0.12
United States	71.57	12.28	2.55	19.84	23.44	2.13	8.03	1.01
Total	*99.36*	*23.51*	*4.15*	*27.93*	*25.51*	*5.83*	*8.93*	*1.21*
Central and South America								
Argentina	3.64	1.71	0.07	1.37	<0.005	0.41	0.08	<0.005
Bolivia	0.26	0.07	0.02	0.15	—	0.02	—	<0.005
Brazil	6.20	2.79	0.06	0.22	0.07	2.76	0.15	0.15
Chile	0.31	0.01	0.01	0.04	0.02	0.22	—	0.01
Colombia	3.03	1.32	0.01	0.19	1.18	0.33	—	0.01
Ecuador	0.99	—[b]	0.01	0.01	—	0.07	—	—
Paraguay	0.47	—	—	—	—	0.47	—	<0.005
Peru	0.40	—	<0.005	0.01	<0.005	0.18	—	<0.005
Trinidad and Tobago	0.83	0.25	0.02	0.56	—	—	—	<0.005
Venezuela	8.94	6.45	0.31	1.33	0.23	0.62	—	—
Other	0.53	1.30	0.00	0.03	0.00	0.26	0.00	0.07
Total	*25.61*	*13.90*	*0.51*	*3.91*	*1.49*	*5.34*	*0.22*	*0.24*
Western Europe								
Austria	0.55	0.04	<0.005	0.06	0.01	0.41	—	0.02
Belgium	0.48	—	—	0.00	<0.005	<0.005	0.46	0.01
Denmark	1.12	0.72	—	0.33	—	—	—	0.06
Finland	0.45	—	—	—	—	0.14	0.22	0.09
France	5.14	0.06	0.01	0.07	0.06	0.76	4.14	0.04
Germany	5.21	0.14	—	0.70	2.26	0.24	1.64	0.23
Greece	0.42	0.01	—	<0.005	0.38	0.02	—	0.01
Italy	1.36	0.18	0.00	0.56	<0.005	0.50	—	0.13
Netherlands	2.64	0.06	0.03	2.45	—	—	0.04	0.05
Norway	10.22	6.39	0.46	2.08	0.04	1.24	—	<0.005
Spain	1.45	0.01	0.00	0.02	0.28	0.42	0.61	0.10
Sweden	1.51	0.00	—	—	0.00	0.81	0.66	0.04
Switzerland	0.70	—	—	0.00	—	0.43	0.26	0.01
Turkey	0.91	0.10	—	0.01	0.54	0.25	—	<0.005
United Kingdom	11.16	4.83	0.42	3.95	0.79	0.03	1.07	0.06
Bosnia and Herzegovina	0.18	—	—	0.00	0.13	0.05	—	—
Croatia	0.21	0.05	0.01	0.07	0.00	0.08	—	<0.005
Yugoslavia	0.49	0.03	—	0.02	0.32	0.12	—	—
Other	0.52	0.01	0.01	0.04	0.12	0.26	0.05	0.07
Total	*44.70*	*12.63*	*0.94*	*10.36*	*4.93*	*5.76*	*9.15*	*0.92*

(continued)

TABLE 2-5
(continued)

Region/ Country	*Total Primary Energy*	*Crude Oil*	*Natural Gas Plant Liquids*	*Dry Natural Gas*	*Coal*	*Net Hydroelectric Power*	*Net Nuclear Power*	*Other*[a]
Eastern Europe and Former USSR								
Bulgaria	0.49	<0.005	—	<0.005	0.25	0.03	0.20	<0.005
Czech Republic	1.07	0.02	0.00	0.01	0.86	0.02	0.16	0.01
Slovakia	0.30	<0.005	—	0.01	0.04	0.05	0.19	—
Hungary	0.46	0.06	0.02	0.11	0.12	<0.005	0.14	<0.005
Poland	3.08	0.04	0.00	0.15	2.86	0.02	—	0.01
Romania	1.21	0.25	0.01	0.50	0.24	0.15	0.06	0.00
Azerbaijan	0.89	0.65	0.01	0.21	—	0.02	—	—
Kazakhstan	3.28	1.51	0.14	0.37	1.74	0.09	0.00	—
Lithuania	0.14	0.01	—	—	—	0.01	0.12	—
Russia	44.88	15.13	0.36	20.70	5.48	1.80	1.38	0.03
Tajikistan	0.15	<0.005	0.00	<0.005	<0.005	0.14	—	—
Turkmenistan	2.12	0.32	0.02	1.78	—	—	—	—
Ukraine	3.52	0.16	0.02	0.67	1.75	0.13	0.78	—
Uzbekistan	2.62	0.16	0.10	2.27	0.04	0.05	—	—
Other	0.44	0.11	0.00	0.02	0.01	0.31	0.03	0.00
Total	*64.65*	*18.42*	*0.68*	*26.80*	*12.82*	*2.82*	*3.06*	*0.05*
Middle East								
Bahrain	0.42	0.08	0.01	0.33	—	—	—	—
Iran	10.50	8.00	0.13	2.29	0.04	0.04	—	—
Iraq	5.31	5.17	0.03	0.10	—	0.01	—	—
Israel	<0.005	<0.005	—	<0.005	—	<0.005	—	—
Kuwait	4.85	4.32	0.18	0.35	—	—	—	—
Oman	2.57	2.06	0.01	0.51	—	—	—	—
Qatar	2.91	1.51	0.21	1.20	—	—	—	—
Saudi Arabia	20.37	17.32	1.06	1.99	—	—	—	—
Syria	1.47	1.16	0.01	0.20	—	0.10	—	—
United Arab Emirates	6.95	4.81	0.48	1.66	—	—	—	—
Yemen	0.92	0.92	—	—	—	—	—	—
Other	0.01	0.00	0.00	0.01	—	0.01	—	—
Total	*56.28*	*45.34*	*2.11*	*8.64*	*0.04*	*0.15*	*0.00*	*0.00*
Africa								
Algeria	6.24	2.57	0.46	3.20	<0.005	<0.005	—	—
Angola	1.61	1.58	—	0.02	—	0.01	—	—
Cameroon	0.20	0.17	—	—	<0.005	0.04	—	—
Congo (Brazzaville)	0.58	0.58	—	—	—	<0.005	—	—
Congo (Kinshasa)	0.11	0.05	—	—	<0.005	0.05	—	—
Egypt	2.66	1.51	0.21	0.78	—	0.15	—	—
Gabon	0.66	0.65	—	<0.005	—	0.01	—	—
Libya	3.21	2.88	0.10	0.23	—	—	—	—

(continued)

TABLE 2-5
(continued)

Region/ Country	*Total Primary Energy*	*Crude Oil*	*Natural Gas Plant Liquids*	*Dry Natural Gas*	*Coal*	*Net Hydroelectric Power*	*Net Nuclear Power*	*Other*[a]
Nigeria	5.49	4.84	—	0.58	<0.005	0.06	—	—
South Africa	5.59	0.05	0.01	0.07	5.33	0.02	0.11	—
Tunisia	0.24	0.14	<0.005	0.09	—	—	—	—
Zambia	0.08	—	—	—	<0.005	0.04	—	—
Other	1.19	0.80	0.01	0.05	0.17	0.38	0.00	0.01
Total	*28.01*	*15.82*	*0.79*	*5.02*	*5.50*	*0.76*	*0.11*	*0.01*
Asia and Oceania								
Australia	10.02	1.34	0.12	1.27	7.11	0.17	—	0.02
Bangladesh	0.36	0.01	<0.005	0.34	—	0.01	—	—
Brunei	0.88	0.42	0.04	0.42	—	—	—	—
Burma	0.35	0.03	<0.005	0.27	0.01	0.04	—	—
China	38.26	7.08	—	1.24	27.01	2.74	0.17	0.01
India	9.37	1.34	0.15	0.83	6.00	0.81	0.22	0.02
Indonesia	8.12	2.87	0.12	2.66	2.32	0.10	—	0.05
Japan	4.49	0.01	0.01	0.10	0.07	0.90	3.16	0.23
Korea, North	2.67	—	—	—	2.45	0.22	—	—
Korea, South	1.17	—	—	—	0.07	0.02	1.07	<0.005
Malaysia	3.56	1.37	0.11	2.00	0.01	0.07	—	—
Mongolia	0.05	—	—	—	0.05	—	—	—
New Zealand	0.69	0.07	0.01	0.24	0.08	0.23	—	0.07
Pakistan	1.18	0.13	<0.005	0.77	0.06	0.20	0.02	—
Papua New Guinea	0.15	0.14	—	<0.005	—	0.01	—	—
Philippines	0.38	0.02	—	<0.005	0.03	0.08	—	0.26
Taiwan	0.46	<0.005	<0.005	0.03	0.00	0.09	0.34	—
Thailand	1.33	0.26	0.10	0.65	0.24	0.06	—	0.02
Vietnam	1.24	0.78	—	0.05	0.23	0.17	—	—
Other	0.10	0.00	0.00	0.01	0.00	0.11	0.00	0.01
Total	*84.83*	*15.87*	*0.66*	*10.88*	*45.73*	*6.03*	*4.98*	*0.69*
World total	**403.44**	**145.48**	**9.86**	**93.53**	**96.02**	**26.70**	**26.45**	**3.11**

[a]Electric power generation from geothermal, solar, wind, wood, and wastes.
[b]Not applicable.

world's primary energy—petroleum, natural gas, coal, and electric power (hydro, nuclear, geothermal, solar, wind, wood, and waste)—production and consumption by world region and country for 2001. These data were tabulated from information that is published yearly by the EIA [23].

World Primary Energy Production

The world's output of primary energy totaled more than 403 quadrillion Btu in 2001. Petroleum (including both crude oil and natural gas plant liquids) continued to be the world's most important primary energy source,

TABLE 2-6
World Energy Consumption in 2001 (quadrillion Btu)

Region/ Country	*Total Primary Energy*	*Petroleum*	*Dry Natural Gas*	*Coal*	*Net Hydroelectric Power*	*Net Nuclear Power*	*Other*[a]
North America							
Canada	12.51	3.79	2.95	1.69	3.41	0.82	0.07
Mexico	6.00	3.77	1.45	0.27	0.29	0.08	0.12
United States	97.05	38.33	23.22	21.97	2.29	8.03	1.01
Other	0.02	0.02	—[b]	—	—	—	—
Total	*115.58*	*45.92*	*27.63*	*23.93*	*5.99*	*8.93*	*1.21*
Central and South America							
Argentina	2.66	0.99	1.15	0.02	0.41	0.08	<0.005
Brazil	8.78	4.46	0.35	0.52	2.76	0.15	0.15
Chile	1.06	0.49	0.24	0.09	0.22	—	0.01
Colombia	1.13	0.50	0.19	0.11	0.33	—	0.01
Cuba	0.39	0.36	0.02	<0.005	—	—	0.01
Venezuela	2.95	0.13	1.33	<0.005	0.62	—	—
Other	3.93	3.59	0.52	0.06	1.00	0.00	0.06
Total	*20.92*	*10.52*	*3.80*	*0.80*	*5.34*	*0.22*	*0.24*
Western Europe							
Austria	1.42	0.55	0.29	0.14	0.41	—	0.02
Belgium	2.77	1.26	0.58	0.36	<0.005	0.46	0.01
Denmark	0.90	0.45	0.21	0.17	—	—	0.06
Finland	1.33	0.44	0.16	0.17	0.14	0.22	0.09
France	10.52	4.20	1.61	0.48	0.76	4.14	0.04
Germany	14.35	5.82	3.27	3.15	0.24	1.64	0.23
Greece	1.39	0.85	0.07	0.41	0.02	—	0.01
Ireland	0.61	0.36	0.16	0.08	0.01	—	<0.005
Italy	8.11	3.88	2.57	0.54	0.50	—	0.13
Netherlands	4.23	1.88	1.57	0.51	—	0.04	0.05
Norway	1.91	0.40	0.16	0.07	1.24	—	<0.005
Portugal	1.09	0.70	0.10	0.12	0.14	—	0.02
Spain	5.70	0.70	0.72	0.68	0.42	0.61	0.10
Sweden	2.22	0.67	0.03	0.09	0.81	0.66	0.04
Switzerland	1.30	0.59	0.11	0.01	0.43	0.26	0.01
Turkey	2.89	1.25	0.59	0.76	0.25	—	<0.005
United Kingdom	9.81	3.45	3.47	1.63	0.03	1.07	0.06
Croatia	0.43	0.19	0.11	0.02	0.08	—	—
Yugoslavia	0.63	0.13	0.02	0.32	0.12	—	—
Other	1.15	2.93	0.10	0.27	0.16	0.05	0.05
Total	*72.76*	*30.70*	*15.90*	*9.98*	*5.76*	*9.15*	*0.92*

(continued)

TABLE 2-6
(continued)

Region/ Country	*Total Primary Energy*	*Petroleum*	*Dry Natural Gas*	*Coal*	*Net Hydroelectric Power*	*Net Nuclear Power*	*Other*[a]
Eastern Europe and Former USSR							
Bulgaria	0.93	0.20	0.20	0.35	0.03	0.20	<0.005
Czech Republic	1.53	0.36	0.35	0.73	0.02	0.16	0.01
Slovakia	0.83	0.17	0.28	0.17	0.05	0.19	—
Hungary	1.09	0.30	0.47	0.15	<0.005	0.14	<0.005
Poland	3.54	0.86	0.46	2.26	0.02	—	0.01
Romania	1.64	0.45	0.69	0.30	0.15	0.06	0.00
Azerbaijan	0.57	0.30	0.25	0.00	0.02	—	—
Belarus	1.21	0.49	0.66	0.02	—	—	<0.005
Kazakhstan	1.73	0.41	0.53	0.71	0.09	0.00	—
Lithuania	0.33	0.15	0.10	<0.005	0.01	0.12	—
Russia	28.20	5.42	14.54	5.16	1.80	1.38	0.03
Turkmenistan	0.48	0.13	0.35	0.00	—	—	—
Ukraine	6.08	0.59	2.74	1.84	0.13	0.78	—
Uzbekistan	2.08	0.30	1.62	0.04	0.05	—	—
Other	1.32	0.10	0.42	0.05	0.45	0.03	0.00
Total	*51.54*	*10.53*	*23.66*	*11.78*	*2.82*	*3.06*	*0.05*
Middle East							
Bahrain	0.39	0.06	0.33	—	—	—	—
Iran	5.18	2.65	2.45	0.05	0.04	—	—
Iraq	1.08	0.97	0.10	—	0.01	—	—
Israel	0.79	0.55	<0.005	0.26	<0.005	—	—
Kuwait	0.92	0.57	0.35	—	—	—	—
Oman	0.34	0.11	0.23	—	—	—	—
Qatar	0.64	0.06	0.59	—	—	—	—
Saudi Arabia	4.91	2.93	1.99	—	—	—	—
Syria	0.86	0.56	0.20	—	0.10	—	—
United Arab Emirates	2.06	0.66	1.40	—	—	—	—
Yemen	0.15	0.15	—	—	—	—	—
Other	0.58	0.53	0.00	0.01	0.00	—	—
Total	*17.92*	*9.80*	*7.64*	*0.32*	*0.15*	*0.00*	*0.00*
Africa							
Algeria	1.31	0.41	0.89	0.02	<0.005	—	—
Angola	0.09	0.06	0.02	—	0.01	—	—
Egypt	2.13	1.17	0.78	0.03	0.15	—	—
Gabon	0.04	0.03	<0.005	—	0.01	—	—
Libya	0.65	0.45	0.20	<0.005	—	—	—
Morocco	0.48	0.34	<0.005	0.11	0.01	—	—
Nigeria	0.92	0.56	0.29	<0.005	0.06	—	—

(continued)

TABLE 2-6
(continued)

Region/ Country	*Total Primary Energy*	*Petroleum*	*Dry Natural Gas*	*Coal*	*Net Hydroelectric Power*	*Net Nuclear Power*	*Other[a]*
South Africa	4.60	0.94	0.07	3.47	0.02	0.11	—
Zimbabwe	0.24	0.05	—	0.12	0.04	—	—
Other	1.99	1.29	0.21	0.06	0.21	0.00	0.01
Total	*12.45*	*5.30*	*2.46*	*3.81*	*0.76*	*0.11*	*0.01*
Asia and Oceania							
Australia	4.97	1.71	0.88	2.19	0.17	—	0.02
Bangladesh	0.51	0.15	0.34	0.01	0.01	—	—
Brunei	0.08	0.03	0.06	—	—	—	—
China	39.67	10.22	1.24	25.37	2.74	0.17	0.01
Hong Kong	0.87	0.55	0.03	0.21	—	—	—
India	12.80	4.40	0.83	6.51	0.81	0.22	0.02
Indonesia	4.63	2.17	1.39	0.91	0.10	—	0.05
Japan	21.92	10.97	2.97	3.69	0.90	3.16	0.23
Korea, North	2.84	0.18	—	2.45	0.22	—	—
Korea, South	8.06	4.44	0.83	1.70	0.02	1.07	<0.005
Malaysia	2.27	0.96	1.16	0.08	0.07	—	—
New Zealand	0.84	0.28	0.24	0.03	0.23	—	0.07
Pakistan	1.87	0.79	0.77	0.09	0.20	0.02	—
Philippines	1.25	0.71	<0.005	0.20	0.01	—	0.26
Singapore	1.65	1.56	0.09	0.00	—	—	—
Taiwan	4.07	2.09	0.23	1.31	0.09	0.34	—
Thailand	2.90	1.62	0.83	0.37	0.06	—	0.02
Vietnam	0.76	0.38	0.05	0.15	0.17	—	—
Other	0.79	0.50	0.09	0.05	0.23	0.00	0.01
Total	*112.75*	*43.71*	*12.03*	*45.32*	*6.03*	*4.98*	*0.69*
World total	**403.92**	**156.48**	**93.11**	**95.94**	**26.85**	**25.52**	**3.11**

[a]Geothermal, solar, wind, wood, and wastes.
[b]Not applicable.

accounting for more than 155 quadrillion Btu, or 38.5% of world primary energy production. Petroleum production in 2001 was 74.7 million barrels per day. The Middle East, as a region, produced about a third of this petroleum (47 quadrillion Btu), followed by North America (28 quadrillion Btu, of which the United States produced nearly 15 quadrillion Btu). Eastern Europe and the States of the FSU produced over 19 quadrillion Btu of petroleum (of which Russia produced 15 quadrillion Btu), with the other regions of the world producing from approximately 14 to 17 quadrillion Btu each.

Coal ranked second as a primary energy source in 2001, accounting for 23.8% of world primary energy production; more than 96 quadrillion Btu of coal (5.3 billion short tons) were produced. As a region, Asia/Oceania was the largest producer, at nearly 46 quadrillion Btu. In this region, China was the largest producer, at 27 quadrillion Btu (1.5 billion short tons), which also made it the largest coal producer in the world. Australia and India ranked a very distant second and third in coal production in this region, at approximately seven and six quadrillion Btu, respectively. This is equivalent to 357 and 339 million short tons, respectively. The second largest quantity of coal produced was on the North American continent—over 25 quadrillion Btu, of which more than 23 quadrillion Btu (or 1.1 billion short tons) were produced by the United States. The United States produced the second largest quantity of coal in 2001, second only to China.

Dry natural gas ranked third as a primary energy source, accounting for 23.2% of world primary energy production in 2001, with over 93 quadrillion Btu (90.7 trillion cubic feet) produced. North America was the largest dry natural gas producer, at about 28 quadrillion Btu, followed closely by eastern Europe and the States of the FSU, where nearly 27 quadrillion Btu were produced. All other regions of the world produced between four and 11 quadrillion Btu in 2001. Russia, the United States, and Canada were the three largest producing countries, at 20.7, 19.8, and 6.7 quadrillion Btu of dry natural gas, respectively. All other countries produced, individually, from less than 5 trillion Btu up to about four quadrillion Btu.

The remaining primary energy sources listed in Table 2-5—hydroelectric, nuclear, and other (geothermal, solar, wind, wood, and waste) electric power generation—ranked fourth, fifth, and sixth, respectively. They accounted for 6.62, 6.56, and 0.8%, respectively, of total primary energy sources in 2001. Combined, they accounted for a total of 56 quadrillion Btu (5.3 trillion kilowatt hours).

The United States produced 2.3 quadrillion Btu of renewable energy that was not used for electricity generation [23]. This included ethanol blended into motor gasoline and geothermal, solar, wood, and waste energy not used for electricity generation. This renewable energy accounted for 0.6% of world primary energy production and ranked seventh as a primary energy source.

Three countries were the leading energy producers in 2001: The United States, Russia, and China together produced 38% of the world's total energy; individually, they produced 71.6, 44.9, and 36.3 quadrillion Btu, respectively. When Saudi Arabia and Canada are included, these five countries produced almost half (*i.e.*, 47.9%) of the world's total energy. The next five leading producers of primary energy were the United Kingdom, Iran, Norway, Australia, and Mexico, which together supplied an additional 12.8% of the world's total energy [23].

World Primary Energy Consumption

As with primary energy consumption, the three largest consumers of world energy in 2001 were the United States, Russia, and China. These countries consumed approximately 97, 28, and 40 quadrillion Btu, respectively, in 2001, accounting for about 41% of the world's total energy consumed. When Japan and Germany are included, the five largest consumers of primary energy in 2001 accounted for nearly 50% of the world's total. The next five leading consumers were India, Canada, France, the United Kingdom, and Brazil, which together accounted for an additional 13.5% of world energy consumption [23].

The United States consumed over 97 quadrillion Btu, by far the most of any country; however, on a per capita basis, Kuwait was the largest consumer in the world, with 467 million Btu consumed per person, followed by Canada (403 million Btu/person), the United States (342 million Btu/person),

TABLE 2-7
Per Capita Primary Energy Consumption in 2001

Region/ Country	*Energy Consumption (Quadrillion Btu)*	*Population (millions)*	*Per Capita Energy Consumption (million Btu/person)*
North America	**115.58**	**416.93**	**277.2**
United States	97.05	283.97	341.8
Canada	12.51	31.08	402.5
Mexico	6.00	101.75	59.0
Central and South America	**20.92**	**426.20**	**49.1**
Brazil	8.78	172.39	50.9
Venezuela	2.95	24.63	119.8
Cuba	0.39	11.22	34.8
Western Europe	**72.76**	**482.42**	**150.8**
France	10.52	59.19	177.7
Germany	14.35	82.36	174.2
United Kingdom	9.81	59.54	164.8
Italy	8.11	57.95	139.9
Netherlands	4.23	16.04	263.7
Portugal	1.09	10.02	108.8
Denmark	0.90	5.33	168.9
Croatia	0.63	4.66	135.2
Eastern Europe and Former USSR	**51.54**	**386.25**	**133.4**
Russia	28.20	144.40	195.3
Ukraine	6.08	49.11	123.8
Poland	3.54	38.64	91.6
Hungary	1.09	9.92	109.9
Turkmenistan	0.48	4.88	98.4
Lithuania	0.33	3.49	94.6

(continued)

TABLE 2-7
Per Capita Primary Energy Consumption in 2001 *(continued)*

Region/ Country	*Energy Consumption (Quadrillion Btu)*	*Population (millions)*	*Per Capita Energy Consumption (million Btu/person)*
Middle East	**17.92**	**171.21**	**104.7**
Iran	5.18	64.53	80.3
Saudi Arabia	4.91	21.03	233.5
Syria	2.06	16.72	123.2
Iraq	1.08	23.58	45.8
Kuwait	0.92	1.97	467.0
Israel	0.79	6.45	122.5
Oman	0.34	2.62	129.8
United Arab Emirates	0.15	2.65	56.6
Africa	**12.45**	**811.69**	**15.3**
South Africa	4.60	44.33	103.8
Egypt	2.13	67.89	31.4
Nigeria	0.92	116.93	7.9
Zimbabwe	0.24	13.96	17.2
Asia and Oceania	**112.76**	**3450.11**	**32.7**
China	39.67	1285.00	30.9
Japan	21.92	127.34	172.1
India	12.80	1017.54	12.6
South Korea	8.06	47.34	170.3
Australia	4.97	19.49	255.0
Indonesia	4.63	214.84	26.2
North Korea	2.84	22.30	127.4
Pakistan	1.87	144.97	12.9
New Zealand	0.84	3.85	218.2
Vietnam	0.79	79.18	10.0
Bangladesh	0.51	140.37	3.6

and Australia (255 million Btu/person). Table 2-7 lists per capita energy consumption for the regions of the world along with selected countries. By region, the per capita energy consumption ranges from about 277 million Btu/person for North America to 15 million Btu/person for Africa. As expected, the per capita consumption is highest for the industrialized nations.

China's use of 1.4 billion short tons made it the largest consumer of coal in 2001. The United States consumed 1.1 billion short tons, followed by India (360 million short tons), Russia (284 million short tons), and Germany (265 million short tons). These five countries accounted for 64% of the world coal consumption in 2001.

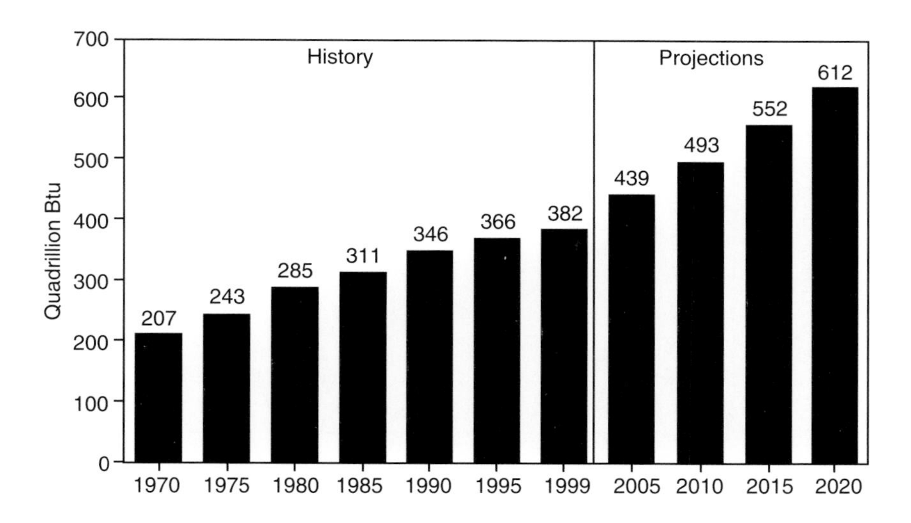

FIGURE 2-20. Projected total world energy consumption. (From EIA, *International Energy Outlook 2002*, U.S. Department of Energy, Energy Information Administration, Washington, D.C., March 2002.)

Future Projections of Energy Use and Coal's Contribution to the Energy Mix

World energy consumption is projected by the EIA to increase 60% from 382 quadrillion Btu in 1999 to 612 quadrillion Btu in 2020 [24]. This increase is illustrated in Figure 2-20. This projection appears to be on track, as world primary energy consumption was about 403 quadrillion Btu in 2001 [24], as shown in Table 2-6. The developing countries are expected to experience the greatest rate of growth, as shown in Figure 2-21, with the industrialized nations exhibiting modest increases and eastern Europe and States of the FSU showing minor increases [24].

World Energy Consumption of Oil

Oil has been the world's dominant source of primary energy for several decades, and it is expected to remain in that position for EIA's forecast period to 2020, as shown in Figure 2-22 [24]. Although oil is projected to be the dominant fuel source, its share of the world energy consumption remains relatively unchanged for the 20-year period, as illustrated in Figure 2-23 [24]. This is predicted to occur because many countries are expected to switch from oil to natural gas and other fuels, especially for electricity generation.

The largest increases in oil consumption will occur in developing countries, which will begin to utilize nearly as much oil as the industrialized nations, although they currently use only about 58% of the amount

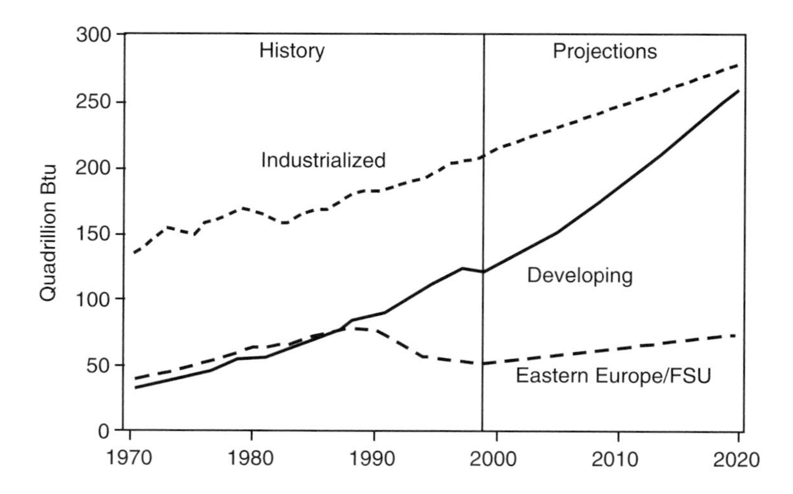

FIGURE 2-21. Projected world energy consumption by region. (From EIA, *International Energy Outlook 2002*, U.S. Department of Energy, Energy Information Administration, Washington, D.C., March 2002.)

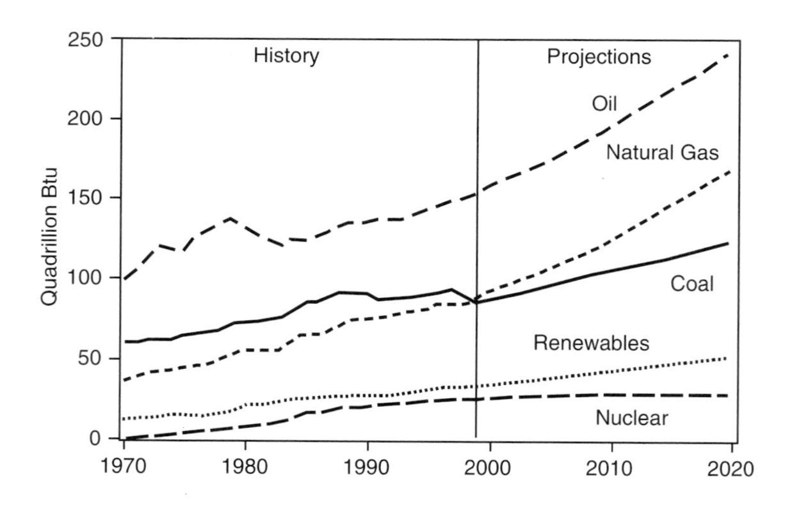

FIGURE 2-22. Projected world energy consumption by fuel type in quadrillion Btu. (From EIA, *International Energy Outlook 2002*, U.S. Department of Energy, Energy Information Administration, Washington, D.C., March 2002.)

consumed by these industrialized countries. In the industrialized world, increased use of petroleum is primarily found in the transportation sector; in the developing world, oil demand is expected in all end-use sectors. People are switching from traditional fuels, such as wood for cooking and heating, to electricity as the infrastructure of emerging technologies improves [24].

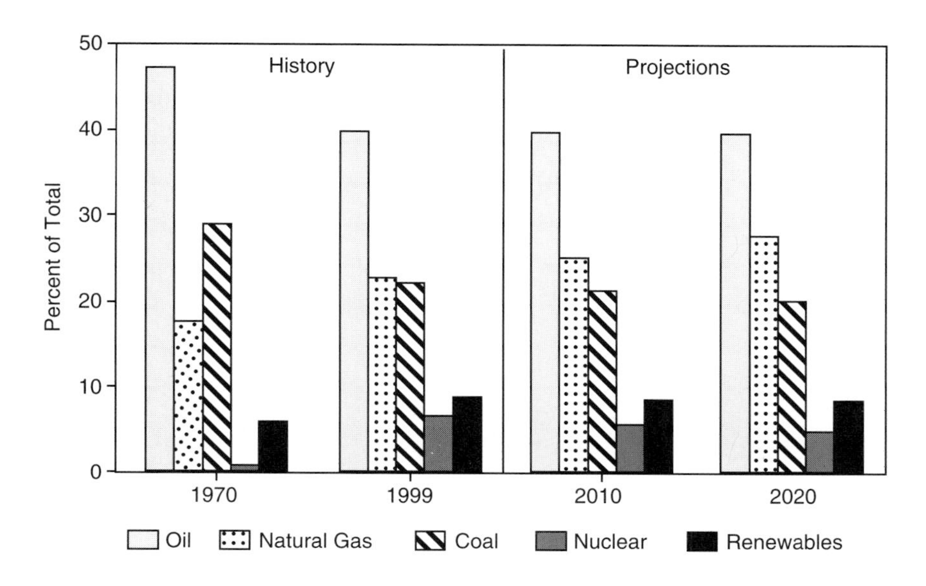

FIGURE 2-23. Projected world energy consumption by fuel type as percent of total energy consumed. (From EIA, *International Energy Outlook 2002*, U.S. Department of Energy, Energy Information Administration, Washington, D.C., March 2002.)

World Energy Consumption of Natural Gas

Natural gas is projected to be the fastest growing source of energy consumption for the EIA's forecast period and is predicted to double over 20 years. As shown in Figures 2-22 and 2-23, natural gas consumption is predicted to surpass the use of coal in both total energy (*i.e.*, on a Btu basis) and market share, respectively. Much of the demand for natural gas consumption throughout the world is due to rising demand for natural gas to fuel new gas turbine power plants along with economic advantages over fuels such as coal in the power generation and industrial markets (due to the lower capital and maintenance costs), environmental concerns, fuel diversification and energy security issues, market deregulation, and overall economic growth [24].

World Energy Consumption of Coal

Although coal use is expected to be displaced by natural gas in some parts of the world, only a slight drop in its share of total energy consumption is projected for the forecast period [24]. Coal use is expected to decline in Europe and the States of the FSU as a result of growing use of natural gas in western Europe, increased use of nuclear power in France, and the economic collapse of eastern Europe and the FSU. Increases, however, are expected in the United States, Japan, and developing Asia. Figure 2-24 shows the projected coal consumption by region for the forecast period [24]. In fact,

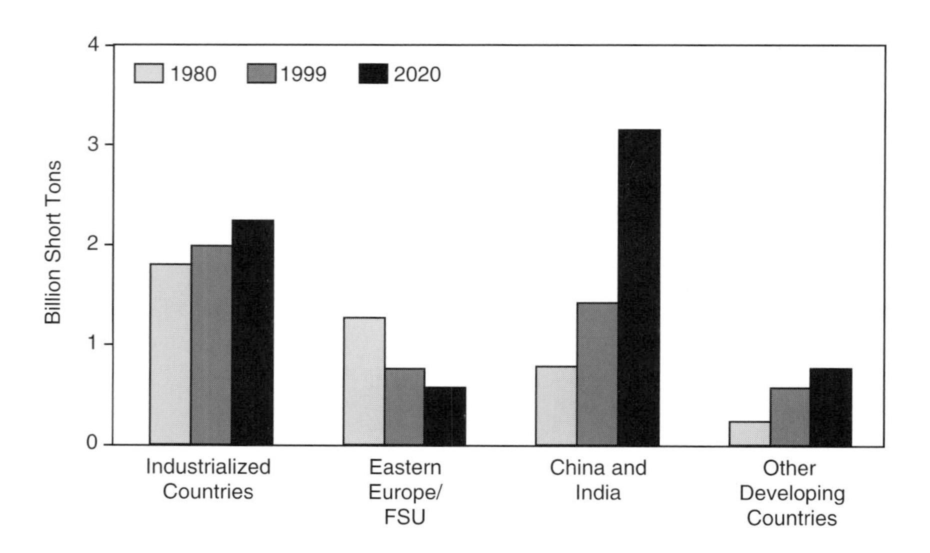

FIGURE 2-24. Projected world coal consumption by region. (From EIA, *International Energy Outlook 2002*, U.S. Department of Energy, Energy Information Administration, Washington, D.C., March 2002.)

coal's share of total energy use (see Figure 2-23) would decline even further if it were not for the large increases projected in developing Asia, where coal continues to dominate many fuel markets.

Coal consumption is concentrated in the electricity generation sector, with about 65% of the coal consumed worldwide used for producing electricity, while significant amounts of coal are also used for steel production [24]. Power generation accounts for almost all of the projected growth in coal consumption worldwide. Natural gas is expected to gain market share in the industrial, residential, and commercial sectors, except in China, where coal continues to be the main fuel in the industrial sector. Consumption of coking coal is expected to decline slightly in most regions of the world.

Overall, world coal consumption is projected to increase by approximately 2 billion short tons to a level of 6.8 billion short tons in 2020. This increase will be primarily observed in developing Asia, where an increase of 1.8 billion short tons is projected. Together, China and India are projected to account for 29% of the total increase in energy consumption worldwide and 83% of the projected increase in coal use [24]. China and India are projected to add an estimated 100 and 65 gigawatts, respectively, of new coal-fired generating capacity by 2020.

Projected Coal Consumption in North America

Coal use in North America is dominated by U.S. consumption. In 2001, the United States consumed 1060 million short tons of coal, an amount that is expected to increase to 1365 million short tons by 2020 [24]. Coal provides nearly 50% of total electricity generation in the United States, which is projected to decrease slightly to 46%. Coal use for electrical generation is expected to continue as coal has a fuel cost advantage over oil and natural gas for power production. Smaller quantities of coal are consumed in other sectors such as Independent Power Producers, metallurgical coal consumption for coke production, and residential/commercial and other industrial users; however, each of these sectors uses only about 2 to 7% per category [25]. Although environmental issues and legislation will impact the use of coal (topics discussed in detail in subsequent chapters), the overall projection for coal production and consumption is that it will continue to increase through 2020. Key features of the U.S. coal market over the next 20 years include [26]:

- Increased production of low-sulfur coal from the western coalfields;
- Continuing decline in coal prices;
- Steady improvements in productivity;
- Decreasing transportation costs.

In 2001, coal provided over 13% of Canada's primary energy consumption, most of which was used for generating electricity. Projections are that coal consumption in Canada will remain around this level through 2020 as demand for electricity in western Canada will result in the need for more coal-fired generation capacity, but the restart of six of Canada's nuclear generating units in eastern Canada will reduce the need for coal in eastern Canada [24]. Canada is a net coal exporter, the fourth largest in the world, and exports coal from western Canada to the far East and western Europe [25]. Predictions for Canada's export market vary, though, ranging from forecasts for increased exports to the far East [25] to decreased exports to the far East (due to increased competition from Australia) and western Europe (due to increased shipping costs) [24,26].

Projected Coal Consumption in Eastern Europe and the States of the Former Soviet Union

In eastern Europe and the States of the FSU, the process of economic reform continues as the transition to a market-orientated economy replaces centrally planned economic systems. While this is occurring, coal production and consumption have declined significantly. In many parts of the former Communist world, energy supplies and electricity are not paid for; hence, there is little incentive for end users to employ techniques to improve efficiency [25]. The three main coal-producing countries of the FSU—Russia, Ukraine, and Kazakhstan—are facing similar problems and have developed national programs for restructuring and privatizing their coal industries, but they are struggling with related technical and social problems [24]. For example, the Ukraine coal industry is plagued by poor safety, inefficiency, low productivity, shortages of spare parts, corruption, failure of customers to pay for coal supplied, unpaid wages, huge debts, and low morale among the workers [26]. In addition, Kazakhstan faces overcapacity, overmanning, and debt issues among both the coal mining sector and its customers [26]; consequently, projections for the FSU are that coal consumption will decrease for the forecast period while oil and natural gas are expected to fuel most of the increase in energy consumption for the region.

In eastern Europe, Poland is the largest producer and consumer of coal. Poland, in fact, is the second leading producer and consumer of coal in all of Europe, second only to Germany. Most of Poland's electricity is produced by coal; however, natural gas is being increasingly substituted for coal in the industrial and electrical power sectors due to the rising costs for coal, which make it less competitive with imported fuels [25]. Poland has also been restructuring its hard coal industry to eliminate government subsidies and achieve positive earnings [24,26]. This initiative has led to mine closures and layoffs and ultimately lower production from local mines; consequently, projections of coal consumption and production for the forecast period show a decline in coal production.

Projected Coal Consumption in Asia

The large increases in coal consumption predicted for China and India are based on anticipated strong economic growth. It is expected that much of the increased demand for energy will be met by coal, especially in the electrical and industrial sectors [24]. Coal remains the primary source of energy in China's industrial sector, primarily because China has limited reserves of oil and natural gas. China has a huge coal-fired industrial boiler market, with more than 500,000 boilers in operation producing 1.26 million short tons of steam/hour, of which 85% are coal fired [27]. For comparison, in the United States, 42,000 industrial boilers produce an average 100,000 pounds steam/hour, of which approximately 5% are coal fired [28]. Most of the projected increase in China's demand for oil and natural gas, which will be used for transportation and space heating, respectively, will be met by imports; consequently, China is planning on constructing several large coal-to-liquids plants over the next decade. In the electrical generation sector, coal use is projected to increase substantially from about 10 quadrillion Btu in 1999 to over 16 quadrillion Btu in 2020. China will require approximately 300 gigawatts of coal-fired capacity in 2020 compared to 201 gigawatts in 1999 [24]. The environmental impact of coal use in China is also of great concern. The World Bank has imposed constraints on new coal-fired power plants that it finances; hence, all new power plants are being designed to meet current emissions standards [26].

In India, the growth in coal demand is projected to occur primarily in the electricity sector. Coal consumption in this sector is rising rapidly; 50% of the coal was used to produce electricity in 1995 [25] but its share grew to more than 70% in 1999 [24]. Coal use for generating electricity is expected to increase from current levels of about five quadrillion Btu to more than 8 quadrillion Btu over the forecast period, with coal-fired capacity increasing from 59 to 125 gigawatts of coal-fired capacity over the same period.

Japan is the third largest user of coal in Asia (eighth in the world) and is the world's leading coal importer [24]. Japan imports most of its coal from Australia, is the largest customer for Australian coal exports, and is the second largest customer for the United States [25]. Coal accounts for about 20% of Japan's primary energy needs; about one-third of the coal consumed is used for power generation and more than half of the consumed coal is used in the steel industry. Japanese power companies plan to construct an additional 16 gigawatts of new coal-fired capacity between 2001 and 2020 [24].

Projected Coal Consumption in Australia

Australia is the world's leading coal exporter. In 1999, Australian coal producers shipped 189 million short tons of coal to international customers, while another 141 million short tons of Australian hard coal and lignite were used domestically, primarily for power generation [24]. Coal-fired power plants accounted for 78% of Australia's total electricity generating needs in 1999 [24], a statistic that is not expected to increase very much over the forecast period as natural gas is becoming increasingly important [25].

Projected Coal Consumption in Western Europe

The projected coal use in western Europe is uncertain as conflicting data are available. Coal use in western Europe fell for many years, as coal was first replaced by oil and then later by nuclear power and natural gas, especially in the generation of electricity [25]. Data from several sources show decreases in coal consumption, both in tons and Btu, for many years [23–26]. Environmental concerns have been a major factor in competition among coal, natural gas, and nuclear power in western Europe, and forecasts prepared around 1999 all predict a continued decrease in coal usage [24–26]. However, primary world energy consumption data collected after these dates show a reversal in the trend, with slight but progressive increases in coal consumption in 2000 and 2001: 9.55 quadrillion Btu in 1999 to 9.79 and 9.98 quadrillion Btu in 2000 and 2001, respectively [23]. Although these increases are small, they are significant in that coal consumption had been declining for years (*e.g.*, 11.25 quadrillion Btu in 1992 to 9.55 quadrillion Btu in 1999). This trend reversal is due to significant increases in coal consumption in the United Kingdom, which is the second largest consumer of coal in western Europe, and The Netherlands. Combined with smaller increases over the last two years by countries such as Italy, Belgium, Greece, and Yugoslavia, overall consumption of coal has reversed a declining trend observed for many years. The increases in the United Kingdom's coal consumption may be the result of the government reinstating coal production subsidies for 2000 through 2002 in an effort to protect the country's remaining coal operations, as many of them have shut down [24].

Coal consumption in Germany, the largest consumer of coal in western Europe, has been declining over the entire reporting period. Much of this reduction is in low-Btu lignite. Coal consumption has also been declining in several other countries including France. A modernization, rationalization, and restructuring plan submitted by the French government to the European Commission at the end of 1994 forecast the closure of all coal mines in France by 2005 [24].

Projected Coal Consumption in Africa

Africa's coal production and consumption are concentrated heavily in South Africa. In 1999, South Africa produced 248 million short tons of coal, with 70% of it going to domestic markets and the remainder to the export market [24]. South Africa is a major coal exporter but is experiencing competition from South America and Australia, as these countries are building more production capacity [26]. South Africa is the world's largest producer of coal-based synthetic liquid fuels. In 1999, about 17% of the coal consumed in South Africa, on a Btu basis, was used to produce coal liquids, which in turn accounted for more than 25% of all liquid fuels consumed in South Africa [24]. Coal consumption is projected to increase by 35 million short tons by 2020, primarily to meet increased demand for electricity. Some of this increase is expected outside of South Africa in Kenya, Nigeria, Tanzania, and Morocco.

Projected Coal Consumption in Central and South America

Historically, coal has not been a major source of energy in Central and South America and today accounts for only about 5 to 6% of the total energy consumption [24]. Hydroelectric power has provided much of the region's demand for electricity, and the use of natural gas is expected to account for most of the projected increase in electricity generation over the period to 2020. Brazil, with the eighth largest steel industry in the world in 1999, accounts for more than 66% of the region's coal demand on a tonnage basis and for more than 75% of the country's total coal consumption [24].

Projected Coal Consumption in the Middle East

Turkey (which is sometimes included in European statistics) accounts for almost 90% of the coal consumed in the Middle East. In 1999, Turkish coal consumption was 84 million short tons, most of it lignite [24]. Coal consumption, both lignite and hard coal, is projected to increase by 20 million short tons by 2020, primarily for coal-fired power production. Further increases in the use of lignite for power generation in Turkey are planned as part of an effort to reduce its dependence on foreign oil [26]. Most of the remaining coal use in the Middle East is consumed by Israel, which consumed 10 million short tons in 1999 [24]. Israel's coal consumption is expected to rise only slightly in the near future. Some environmental groups and government officials are opposed to expanding new coal-fired generating capacity and instead support the use of natural gas from local and Egyptian sources for future electricity generation [24].

World Energy Consumption of Nuclear Energy

World nuclear energy consumption will remain relatively unchanged over the forecast period. The highest growth is predicted for the developing world with Asia, in particular, expected to experience the greatest expansion in nuclear generating capacity. This region currently has 16 reactors under construction, half of the world total [24].

World Energy Consumption of Renewable Energy

Renewable energy use is projected to increase by 53% over the forecast period; however, its share of total energy consumption is expected to remain relatively unchanged at 8 to 9% [24]. Renewable energy consumption is expected to be driven by new, large-scale hydroelectric projects, particularly in China, India, Malaysia, and other developing Asian countries, but even these projects have had their share of bad publicity. Projects in China and Malaysia have continued amidst criticism of their environmental impacts and concerns about the welfare of people being relocated to accommodate the projects [24]; similarly, a project under development in Iceland has been criticized for its environmental impact [29].

Renewable energy use will be constrained by its higher fuel and capital costs and moderate fossil fuel prices. Emissions legislation, renewable energy portfolios, and tax incentives/subsidies in the industrialized countries will increase the use of renewable energy for power generation and industrial applications for chemicals and fuel production; however, renewable energy will make only a minor contribution to the overall energy mix due to its higher fuel and equipment costs. As the hydrogen economy grows, it too has the potential to increase the use of renewable energy. This technology, however, is still in its infancy, and the production of hydrogen, which is currently made predominantly from fossil fuels (*i.e.*, mainly natural gas), would need to rely on other sources to impact fossil fuel consumption. In fact, as the hydrogen economy grows, gasification of coal may become the leading source of hydrogen. Details of the gasification technology are provided in Chapter 5 (Technologies for Coal Utilization), and its potential to become a major fuel source is discussed in Chapter 7 (Future Power Generation), where future co-production technologies are presented.

Energy Outlook for the United States

Projections of U.S. energy consumption and production to 2025 made by the EIA [30] are shown in Figures 2-25 and 2-26, respectively. These projections have been developed by focusing on long-term issues such as availability of energy resources, developments in U.S. electricity markets, technology

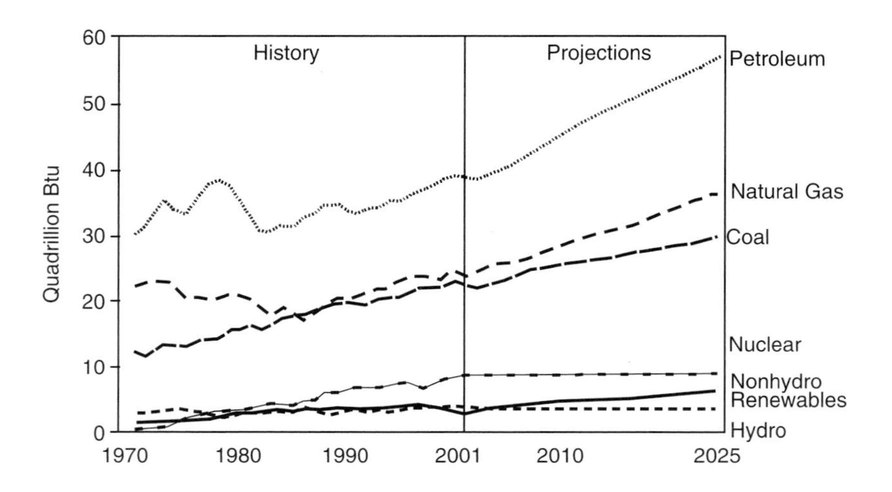

FIGURE 2-25. U.S. energy consumption projected to 2025. (From EIA, *International Energy Annual 2001*, U.S. Department of Energy, Energy Information Administration, Washington, D.C., March 2003.)

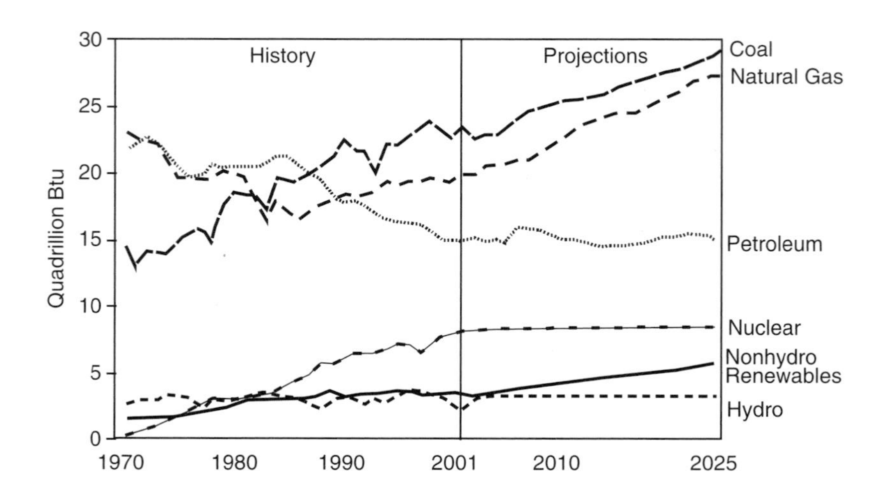

FIGURE 2-26. U.S. energy production projected to 2025. (From EIA, *International Energy Annual 2001*, U.S. Department of Energy, Energy Information Administration, Washington, D.C., March 2003.)

improvement, and the impact of economic growth on projected energy demand and prices. Less emphasis has been placed on near-term influences such as supply disruptions or political actions.

A major consideration for U.S. energy markets through 2025 will be the availability of adequate natural gas supplies at competitive prices to meet growth in demand. EIA foresees growing dependence on new, large-volume natural gas supply projects, including deepwater offshore wells, new and expanded liquefied natural gas facilities, and major pipelines in Canada and the United States to meet future demand levels [23].

Projected Energy Consumption in the United States

Energy consumption in the United States is projected to increase in all end-use sectors, with the demand for transportation energy expected to increase the most [23]. Residential energy demand is expected to increase to more than 25 quadrillion Btu/year in 2025 from current levels of about 20 quadrillion Btu/year (*i.e.*, an increase of 1.0% per year), with the most rapid growth expected for computers, electronic equipment, and appliances. Commercial energy demand is expected to grow at an average annual rate of 1.6% per year, increasing from about 17 to more than 25 quadrillion Btu/year in 2025, with the most rapid increases in demand being found in the use of computers, office equipment, telecommunications, and miscellaneous small appliances. Industrial energy demand is projected to increase by 1.3% per year and will increase from about 33 to more than 44 quadrillion Btu/year in 2025. Transportation energy demand is projected to grow at an average annual rate of 2.0% per year, increasing from about 27 to nearly 44 quadrillion Btu/year by 2025 due to increased vehicle miles traveled and a lower level of vehicle efficiency. Demand for electricity is projected to increase by 1.8% per year to 2025 due to rapid growth in electricity use for computers, office equipment, and electrical appliances in the residential and commercial sectors.

Projected energy consumption, by fuel, is shown in Figure 2-25 [23]. Increases in all fuel types are projected over the period ending in 2025, although increases for nuclear energy and hydroelectric generation are anticipated to be small. Demand for natural gas is expected to increase by 1.8% per year to 2025, increasing from 22.7 to 34.9 trillion cubic feet primarily because of rapid growth in demand for electricity generation. Total petroleum demand is projected to grow at an annual rate of 1.7% through 2025 due to demands from the transportation sector. Total renewable fuel consumption, which includes ethanol for gasoline blending, is projected to grow at 2.2% per year to 2025.

Coal consumption is projected to increase by 1.3% per year from about 1063 million short tons to 1444 million short tons in 2025 mainly due to projected growth in the electric power sector. Total demand for industrial steam coal is projected to rise slightly (by 8 million short tons in 2025), coal demand for the residential and commercial sectors is expected to remain relatively constant, and demand for coking coal is projected to decline by 8 million short tons in 2025.

Coal is projected to maintain its fuel cost advantage over both oil and natural gas; however, natural-gas-fired power generation is expected to be more economical in the near term when capital, operating, and fuel costs are considered. The EIA [23], however, projects rising natural gas costs starting in 2005, increasing demand for electricity, and retirements of some existing fossil-fuel-fired steam capacity, which will lead to increased demand for coal-fired, baseloaded capacity.

Projected Energy Production in the United States

Total energy consumption in the United States is expected to increase more rapidly than domestic energy production through 2025, increasing to approximately 139 and 90 quadrillion Btu, respectively [23]. The shortfall of about 49 quadrillion Btu in 2025 is projected to be met primarily through imports of petroleum (~41 quadrillion Btu) and natural gas (~8 quadrillion Btu). Crude oil production is projected to decrease by 0.4% per year to 2025, as shown in Figure 2-26. By 2025, net petroleum imports, which include both crude oil and refined products, are expected to account for 68% of demand, up from about 55% in 2001 [23]. Domestic natural gas production is projected to increase by 1.3% per year to 2025 from approximately 20 to 25 trillion cubic feet. This increase, however, will not meet the total demand for natural gas in the United States, and the balance will be met through imports via pipeline from Canada and Mexico as well as shipments of liquefied natural gas.

Nuclear power generation is expected to increase modestly through 2014 and then level off [23]. Renewable energy production, including hydroelectric generation, is projected to increase to over 9 quadrillion Btu in 2025 from current levels of about 6 quadrillion Btu. U.S. coal production is projected to increase to 1444 million short tons in 2025, an increase of 0.9% per year, to meet the projected demand [23]. Higher electricity demand and lower prices will be the primary reasons for the coal demand, as net coal exports are expected to decline throughout the forecast period due to declining demand in some countries and competition from other international producers. Eastern coal production will remain relatively stable, while western coal production will continue to increase because of its lower sulfur content and lower minemouth price. The mines of the Northern Great Plains, with thick seams and low overburden ratios, have higher labor productivity than other coalfields, and this advantage is expected to be maintained throughout the EIA's forecast period.

Role of Coal in the United States' 2001 Energy Policy

In May 2001, the National Energy Policy Development (NEPD) Group, chaired by Vice President Dick Cheney, unveiled a National Energy Policy for President George W. Bush. This report, titled *Reliable, Affordable, and Environmentally Sound Energy for America's Future,* is the first detailed energy policy that the United States has developed since President Jimmy Carter's administration. President Bush directed the NEPD Group to "develop a national energy policy designed to help bring together business, government, local communities and citizens to promote dependable, affordable and environmentally sound energy for the future" [31]. In the report's cover letter to President Bush, Vice President Cheney states that the report "envisions a comprehensive long-term strategy that uses leading edge technology to produce an integrated energy, environmental and economic policy. To achieve a 21st century quality of life—enhanced by reliable energy and a clean environment—we must modernize conservation, modernize our infrastructure, increase our energy supplies (including renewables), accelerate the protection and improvement of our environment, and increase our energy security" [31]. The NEPD Group recognized the importance of coal in the U.S. energy policy and security. The role of coal in the U.S. energy policy is introduced in this section, while its contribution to the energy security of the United States is discussed in detail in Chapter 8.

The United States' energy strength lies in the abundance and diversity of its energy resources and in its technological leadership in developing and efficiently using these resources. The United States has significant domestic energy resources—coal, oil, and natural gas—and remains a major producer; however, as outlined earlier in this chapter, the United States will need more energy supply than it produces. This shortfall can be made up in only three ways: import more energy, improve energy efficiency even more than expected, or increase domestic energy supply. Coal, an abundant resource in the United States, can be used to increase the domestic energy supply even more than projected through the use of advanced utilization technologies (which are discussed in detail in Chapter 7). The NEPD Group recognized this; thus coal has been given a significant role in the United States' energy policy.

The demand for electricity in the United States is projected to increase substantially, with 428 gigawatts of new generating capacity required for the period from 2001 to 2025 to meet this demand and replace retiring units [23]. Coal, nuclear energy, natural gas, and hydropower account for about 95% of total electricity generation, with oil and renewable energy contributing the remainder. Coal is used almost exclusively to generate electricity. Coal-fired power plants account for about 50% of all U.S. electricity generation (see Figure 2-18). Coal is attractive because electricity generation costs are low and coal prices have proved stable.

Although coal is the most abundant fossil energy source in the United States, production and market issues affect the adequacy of supply. As highlighted in the energy policy, issues such as protection of public health, safety, property, and the environment can limit production and utilization of some coal resources and has led, over the last several years, to using natural gas for power generation due to higher efficiencies, lower capital and operating costs, and fewer emissions; however, technological advances in cleaner coal technology have allowed for significant progress toward reducing these barriers. Coal, with its more stable prices than natural gas, must play a significant role in meeting the rising electricity demand in the United States.

Technology has been and will continue to be a key for the United States to achieve its energy, economic, and environmental goals. In recent years, technological advancements have led to substantial reductions in the cost of controlling sulfur dioxide and nitrogen oxide emissions [31]. The U.S. Department of Energy (DOE), through its Clean Coal Technology Program, has worked to provide effective control technologies. Clean Coal Technology is a category of technologies that allow for the use of coal to generate electricity while meeting environmental regulations at low cost. In the short term, the goal of the program is to meet existing and emerging environmental regulations, which will dramatically reduce compliance costs for controlling sulfur dioxide, nitrogen oxides, fine particulate matter, and mercury at new and existing coal-fired power plants. In the mid-term, the goal of the program is to develop low-cost, superclean, coal-fired power plants with efficiencies 50% higher than today's average [31]. The higher efficiencies will reduce emissions at minimal costs. In the long term, the goal of the program is to develop low-cost, zero-emission power plants with efficiencies close to double that of today's fleet (see Chapter 7). The long-term goals also include co-production of fuels and chemicals along with electricity, leading to the development of technologies that may result in coal being a major source of hydrogen for the much publicized "hydrogen economy" the United States is moving toward.

The NEPD Group recognized the importance of looking toward technology to help meet the goals of increasing electricity generation while protecting the environment. As a result, the NEPD Group recommended that President Bush direct the DOE to continue to develop advanced clean coal technology by [31]:

- Investing $2 billion over 10 years to fund research in clean coal technologies;
- Supporting a permanent extension of the existing research and development tax credit;
- Directing agencies to explore regulatory approaches that will encourage advancements in environmental technology.

In addition, the NEPD Group recommended that the President direct federal agencies to provide greater regulatory certainty relating to coal electricity generation through clearer policies that are easily applied to business decisions. President Bush has started implementing some of these recommendations, which are addressed in later chapters, including discussions on the first round of the clean coal technologies being demonstrated and regulatory/environmental legislation being proposed and implemented.

References

1. Moore, E. S., *Coal: Its Properties, Analysis, Classification, Geology, Extraction, Uses, and Distribution* (John Wiley & Sons, New York, 1922).
2. Elliot, M. A. (editor), *Chemistry of Coal Utilization*, Second Suppl. Vol. (John Wiley & Sons, New York, 1981).
3. World Book Encyclopedia, *Coal* (World Book, Chicago, IL, 2001a), Vol. 4, pp. 716–733.
4. World Book Encyclopedia, *Marco Polo* (World Book, Chicago, IL, 2001b), Vol. 15, pp. 648–649.
5. Environmental Literacy Council, www.enviroliteracy.org/article.php/18.html (accessed May 2003).
6. Landels, J. G., *Engineering in the Ancient World* (University of California Press, Berkley, 1978), p. 32.
7. U.C. Davis (University of California at Davis), www-geology.ucdavis.edu/-GEL115/115CH11coal.html (accessed May 2003).
8. Schobert, H. H., *Coal: The Energy Source of the Past and Future* (American Chemical Society, Washington, D.C., 1987).
9. EIA, *Coal Data: A Reference* (U.S. Department of Energy, Energy Information Administration, Washington, D.C., February 1995).
10. Deane, P., *The First Industrial Revolution*, Second ed. (Cambridge University Press, Cambridge, U.K., 1979), pp. 78, 103–105.
11. EIA, *Annual Energy Review 2001* (U.S. Department of Energy, Energy Information Administration, Washington, D.C., November 2002).
12. Freme, F., *U.S. Coal Supply and Demand: 2002 Review* (U.S. Department of Energy, Washington, D.C., 2002).
13. National Mining Association, *Most Requested Statistics: U.S. Coal Industry* (National Mining Association, Washington, D.C., April 2003).
14. RDI, Inc., *Synthetic Fuel Tax Credits: U.S. Coal Industry Impacts*, June 2000.
15. Sanderson, G., Section 29 Credit for Synthetic Fuels, paper presented at Coke Summit 2001, October 15–17, 2001.
16. Kalb, G. W., Tax Credit Plants Emerge, *Coal Age*, Vol. 105, No. 4, April 2000, pp. 43–44.
17. Morey, M. and C. Leshock, Tax Credit Synfuels Influence Coal Markets, *Coal Age*, Vol. 105, No. 5, May 2000, pp. 35–36.
18. Morey, M., Coal-Based Synfuel Continues to Grow, *Coal Age*, Vol. 107, No. 11, November 2002, pp. 23–24.
19. Anon., Sprayed Coal Tax Credits Under Fire, *Coal Age*, Vol. 106, No. 10, October 2000, p.8.
20. Anon., Feds Move to Probe Synfuel Tax Credits, *Coal Age*, Vol. 105, No. 11, November 2000, p. 16.
21. Anon., IRS Defines Synfuel Qualifications, *Coal Age*, Vol. 106, No. 6, June 2001.
22. Coal Outlook: IRS, Treasury Reexamine Sect. 29 Policy, Congressman Says, *Financial Times Energy*, September 18, 2000, p.11.
23. EIA, *Annual Energy Outlook 2003* (U.S. Department of Energy, Energy Information Administration, Washington, D.C., January 2003).

24. EIA, *International Energy Outlook 2002* (U.S. Department of Energy, Energy Information Administration, Washington, D.C., March 2002).
25. Grimston, M. C., *Coal as an Energy Source* (IEA Coal Research, London, 1999).
26. Walker, S., *Major Coalfields of the World* (IEA Coal Research, London, 2000).
27. CACETC, *Coal-Fired Industrial Boiler in China and the Related Applicable Advanced Technology* (Cleaner Air and Cleaner Energy Technology Cooperation, China, June 2002).
28. EPA, *Report to Congress: Wastes from the Combustion of Fossil Fuels*, Vol. 1. *Executive Summary* (U.S. Environmental Protection Agency, U.S. Government Printing Office, Washington, D.C., March, 1999), pp. 4–6.
29. Swan, J., Iceland Be Damned, *Smithsonian*, Vol. 23, No. 3, June 2002, pp. 90–97.
30. EIA, *International Energy Annual 2001* (U.S. Department of Energy, Energy Information Administration, Washington, D.C., March 2003).
31. NEPD, *National Energy Policy* (National Energy Policy Development Group, U.S. Government Printing Office, Washington, D.C., May 2001).

CHAPTER 3

The Effect of Coal Usage on Human Health and the Environment

Coal has played a significant role in the advancement of civilization and will continue to be a major fuel source for at least the next quarter century, as discussed in Chapter 2 (Past, Present, and Future Role of Coal). The value of coal is partially offset by the environmental issues it raises. Some of these environmental issues also have impacts on human health.

Coal mining has a direct impact on the environment, as it disturbs large areas of land and has the potential to affect surface water and groundwater in the case of surface mining. In some surface mines, the generation of acid mine drainage (AMD) is a major problem. Other significant impacts include fugitive dust emissions and disposal of overburden and waste rock.

In underground mining, the surface disturbance is less obvious but the effect of subsidence can be large. The generation and release of methane and other gases can be a problem. As with surface mining, groundwater can also be disturbed and AMD can become an issue. In addition, underground miners have a history of respiratory ailments and, until the last few decades, have experienced high injury and death rates from accidents.

Coal beneficiation is primarily based on wet physical processes that produce waste streams that must be dealt with. These include fine materials that are discharged as a slurry to a tailings impoundment and a coarse material that is hauled away as a solid waste. Storage, handling, and transportation of coal produce fugitive dust.

Coal utilization, specifically combustion, which is the focus of this chapter, produces several types of emissions that adversely affect the environment, particularly ground-level air quality. In addition, the generation of coal combustion by-products must be addressed, whether they are disposed of or reused/recycled.

Concern for the environment has in the past and will in the future contribute to policies that affect the consumption of coal. The main emissions from coal combustion are sulfur dioxide (SO_2), nitrogen oxides (NO_x), particulate matter (PM), and carbon dioxide (CO_2). Recent studies on the health effects of mercury have raised concerns about mercury emissions from coal-fired power plants. The environmental and health effects of these pollutants, along with other pollutants such as carbon monoxide (CO), lead, and organic emissions, are discussed in this chapter, as is the effect of coal utilization on human health and the environment. The impacts of mining, storage, handling, transportation, beneficiation, combustion by-products, and emissions from coal-fired power plants are presented with a focus on activities in the United States.

Coal Mining

The negative aspects of mining operations can lead to confrontations among citizen groups, governmental agencies, and the mining industry. The conflicts tend to be centered on the following issues [1]:

- Destruction of the landscape;
- Degradation of the visual environment;
- Disturbance of surface water and groundwater;
- Destruction of agricultural and forest lands;
- Damage to recreational lands;
- Noise pollution;
- Dust;
- Truck traffic;
- Sedimentation and erosion;
- Land subsidence;
- Vibration from blasting.

These issues, along with past operating procedures—unsafe working conditions leading to injuries, deaths, and high incidences of respiratory diseases; use of child labor; scarred landscapes; poor miner living conditions; contentious and sometimes extremely violent labor relations—resulted in coal mining being viewed negatively; consequently, a major effort has gone into addressing these issues so that today coal mining is a highly regulated industry that has seen a number of significant changes in the approach to mining and resource development along with improving miner safety. These developments include:

- Environmental impact assessment and public inquiries;
- Conditions for mining permit approval;
- Resource management and land-use planning;
- Land reclamation and rehabilitation;
- Regulations specifically addressing miner safety and training.

Coal mining falls into two categories: underground mining and surface mining. Surface mining is used when the coal seam to be mined lies near the surface, typically less than 200 feet from the surface. Each mining technique has its own set of technical and economic advantages and disadvantages, and each mining technique has its own set of health and environmental impacts. In 2002, nearly 370 million short tons of coal were mined by underground methods, while more than 724 million short tons of coal were removed by surface mining [2].

Underground Mining

Underground mining is used for deep seams, and the mining methods vary according to the site conditions. Underground mines can be classified by the types of access used to reach the coal (*i.e.*, shaft mines, slope mines, and drift mines) but are primarily classified by the coal removal system: room and pillar mining, pitch mining, or longwall mining [3]. A shaft mine uses a vertical hole dug straight down from the surface to the coal seam. A slope mine provides access at a slant and is used to follow a seam along its pitch or to cut through a mountain to reach the coal. A drift mine accesses coal outcrops on a mountainside. Room and pillar mining is used when the coal seam lies relatively level, and it is carried out by using one of two processes: conventional mining or continuous mining [3]. In 2002, over four million short tons of the coal produced in underground mines were obtained by conventional mining, but it is being replaced by other mining methods [2]. In conventional mining, explosives are used to shatter the face of the seam, and the broken pieces are manually loaded onto tram cars or conveyors and hauled out of the mine. In continuous mining, used to produce more than 175 million short tons of coal in 2002 [2], a machine that moves along on caterpillar tracks cuts the coal from the face of the seam and automatically loads the coal onto tram cars or a conveyor. Pitch mining is a technique used when the coal seams are inclined and is frequently used in anthracite mining [4]. In pitch mining, the bottom of the seam is accessed, and the coal is dropped into chutes that are gravity-fed into tram cars. Longwall mining is the removal of coal from one, long continuous face rather than removal from a number of short faces as occurs in room and pillar or pitch mining. Annual longwall mining production is similar to that of continuous mining and accounted for nearly 190 million short tons of coal produced in the United States in 2002 [2]. Regardless of the mining system used, the health and environmental impacts are common to every underground mine: land subsidence, generation of methane and other gases, liquid effluents, dust, solid waste, and miner safety.

Subsidence

Subsidence can have a major effect on the topography of the land surface. Following the removal of the coal from an underground mine, the roof materials may cave, causing collapse of the overlying rock strata and resulting in subsidence of the surface. The degree of collapse of the overlying rock strata can vary from practically no collapse with no resulting surface impacts to total collapse with more pronounced changes at the surface [5]. In general, mine subsidence problems develop where post-mining pillar support systems and coal barriers ultimately fail. Many interrelated factors control when, where, and how failure will occur, including [1]:

- Thickness of coal removed;
- Size, shape, and distribution of pillars and rooms;
- Depth of mining;
- Percent extraction of coal;
- Thickness and physical characteristics (*e.g.*, strength) of the overburden;
- Method of mining, such as longwall, shortwall (which is a slight modification of longwall mining), room and pillar, room and pillar with full or partial retreat;
- Conditions in the mine (*i.e.*, dry or flooded);
- Actual or potential level and degree of fracturing in the overburden;
- Mineralogy of the overburden (*e.g.*, clay minerals that swell when water is added, sulfide minerals that chemically and physically change in the presence of oxygen and moisture, minerals that react with water to form new minerals).

Over areas that have been longwall mined, the subsidence is often a shallow trough. In flat terrain, this trough is usually quite visible and can cause local changes in drainage. Subsidence from active coal mining has the largest effect on the land surface in terms of area undermined, although the effects are often small in terms of overall topography [5]. Subsidence impacts on surface structures and subsurface hydrologic resources are generally of more importance than the impacts on topography or surface features.

Subsidence from shallow, abandoned coal mines often results in abrupt but localized changes in topography that reflect the collapse of individual rooms or voids. This type of subsidence can be an isolated, single collapse or can involve a larger area with many individual subsidence pits.

Generation of Gases

Methane (CH_4) is produced during coalification, and only a fraction of this gas remains trapped under pressure in the coal seam and surrounding rock strata. This trapped methane is released during mining when the coal seam is fractured. The amount of methane released during coal mining depends on a number of factors, including coal rank, coal seam depth, and method of mining [6]. As coal rank increases, the amount of methane produced also increases. The adsorption capacity of coal increases with pressure, and pressure increases with depth of the coal seam; consequently, deeper coal generally seems to contain more methane than shallow seams of the same rank. Underground coal mining releases more methane than surface mining because of the higher gas content of deeper seams. The methane that is contained in the coal seams is referred to as coalbed methane (CBM). The CBM that is released from the coal during coal mining is referred to as coal mine methane (CMM) and is a subset of CBM.

Methane is highly explosive in air in concentrations between 5 and 15%, and operators have developed two types of systems for removing methane from underground mines: ventilation systems and degasification systems [6]. At present, almost all ventilation air is emitted into the atmosphere. Although the methane concentrations are low, the amount of methane released into the atmosphere each year is significant. Emissions factors for underground mining range from 10 to 25 cubic meters emitted per ton of coal mined, compared to 0.3 to 2 cubic meters per ton of coal that is surface mined. As a greenhouse gas, methane is more than 21 times as potent as carbon dioxide (CO_2). Table 3-1 lists global estimates of methane emissions from coal mining in 1990 along with coal production that year [7].

Some of the gassiest mines in the United States (as well as Russia, Australia, and other countries) have installed desgasification systems to extract the methane from the coal seams in advance of mining, during coal recovery, and during post-mining operations [6]. Depending on the quality of this gas, mine operators can use recovered methane for on-site electricity generation or can sell it to a local pipeline.

TABLE 3-1
Estimate of Global Methane Emissions from Coal Mining in 1990

Country	*Coal Production (million short tons)*			*CH_4 Emissions (Tg)*[a]	
	Underground	*Surface*	*Total*	*Low*	*High*
China	1024	43	1066	9.5	16.6
United States	385	548	934	3.6	5.7
Former USSR	393	309	701	4.8	6.0
Germany	77	359	436	1.0	1.2
India	109	129	238	0.4	0.4
Poland	154	58	212	0.6	1.5
Australia	52	154	206	0.5	0.8
South Africa	112	63	175	0.8	2.3
Czechoslovakia	22	85	107	0.3	0.5
United Kingdom	75	14	89	0.6	0.9
Subtotal (top 10 countries)	2043	1762	4164	22.1	35.9
World total	—	—	4740	24.4	39.6

[a]Tg = teragram = 1×10^9 kg; 1 Tg CH_4 = 52 billion cubic feet (Bcf).

Source: EPA, *International Anthropogenic Methane Emissions in the United States* (Office of Policy, Planning and Evaluation, U.S. Environmental Protection Agency, U.S. Government Printing Office, Washington, D.C., 1994).

The total volume of CMM liberated in the United States in 2000 was 196 billion cubic feet (Bcf); of that, underground mining activities liberated 142 Bcf [8]. CMM emissions account for approximately 10% of total U.S. methane emissions. Globally, coal mines account for 8% of all methane emissions. Underground mines are the largest source of CMM and account for 72% of the total CMM liberated.

The U.S. coal industry has made substantial progress in recovering and using CMM through drainage systems. Of the 142 Bcf of CMM liberated from underground mines in 2000, about 42 Bcf were emitted through drainage systems, with the remainder being emitted through ventilation systems [8]. Coal mines in the United States recovered 86%, or 36 Bcf, of the gas liberated through the drainage systems, which is nearly a threefold increase over 1990, when only 14 Bcf of gas were recovered. Approximately 100 Bcf of CMM are emitted through ventilation systems each year. Ventilation air represents a considerable source of greenhouse gas emissions and is a major focus area for recovery and use.

Some coal seams, particularly in Australia but also in France and Poland, contain high carbon dioxide concentrations, which can comprise as much as 100% of the gases in the coal seam [6]. This gas is thought to have a magmatic origin. Although carbon dioxide is not toxic, it can cause asphyxiation by displacing breathable oxygen.

Liquid Effluents/Acid Mine Drainage

The generation of liquid effluents from major mining techniques tends to be higher for underground mining than for surface mining. For underground mining, liquid effluent rates of 1.0 and 1.6 tons per 1000 short tons of coal produced have been documented for conventional and longwall mining, respectively [9]. If groundwater systems are disturbed, the possibility then exists for serious pollution from highly saline or highly acidic water.

The highly acidic water, commonly known as acid mine drainage, is produced by the exposure of sulfide minerals (most commonly pyrite) to air and water, resulting in the oxidation of sulfur and the production of acidity and elevated concentrations of iron, sulfate, and other metals [1]. Pyrite and other sulfide minerals are generally contained in the coal, overburden, and coal processing wastes.

Historically, coal extraction in the northern Appalachian coalfields has resulted in serious problems related to contaminated mine drainage [10]. Acid drainage from closed and abandoned mines (both underground and surface) has far-ranging effects on water quality and, therefore, on fish and wildlife. Drainage from closed mines is particularly acidic in Pennsylvania, Ohio, northern West Virginia, and Maryland.

The formation of AMD is primarily a function of the geology, hydrology, and mining technology employed for the mine site [11,12]. AMD is formed by a series of complex geochemical and microbial reactions that occur when water comes in contact with pyrite (iron disulfide minerals) in coal, refuse, or the overburden of a mine operation. The resulting water is usually high in acidity and dissolved metals. The metals stay dissolved in solution until the pH raises to a level where precipitation occurs.

Four commonly accepted chemical reactions represent the chemistry of pyrite weathering to form AMD. An overall summary is:

$$4FeS_2 + 15O_2 + 14H_2O \longrightarrow 4Fe(OH)_3^- + 8H_2SO_4 \quad (3\text{-}1)$$

Pyrite + Oxygen + Water ⟶ "Yellowboy" + Sulfuric Acid

The first reaction in the weathering of pyrite includes the oxidation of pyrite by oxygen. Sulfur is oxidized to sulfate and ferrous iron is released. This reaction generates 2 moles of acidity for each mole of pyrite oxidized:

$$2FeS_2 + 7O_2 + 2H_2O \longrightarrow 2Fe^{2+} + 4SO_4^{2-} + 4H^+ \quad (3\text{-}2)$$

Pyrite + Oxygen + Water ⟶ Ferrous Iron + Sulfate + Acidity

The second reaction involves the conversion of ferrous iron to ferric iron. The conversion of ferrous iron to ferric iron consumes one mole of acidity. Certain bacteria increase the rate of oxidation from ferrous iron to ferric iron. This reaction rate is pH dependent, with the reaction proceeding slowly under acidic conditions (pH of 2 to 3) with no bacteria present and several orders of magnitude faster at pH values near 5. This reaction is often referred to as the *rate-determining step* in the overall acid-generating sequence:

$$4Fe^{2+} + O_2 + 4H^+ \longrightarrow 4Fe^{3+} + 2H_2O \quad (3\text{-}3)$$

Ferrous Iron + Oxygen + Acidity ⟶ Ferric Iron + Water

The third reaction which may occur is the hydrolysis of iron. Hydrolysis is a reaction which splits the water molecule. Three moles of acidity are generated as a by-product. Many metals are capable of undergoing hydrolysis. The formation of ferric hydroxide precipitate (*i.e.*, a solid product) is pH dependent. Solids form if the pH is above about 3.5, but below pH 2.5 few or no solids will precipitate. The third reaction is:

$$4Fe^{3+} + 12H_2O \longrightarrow 4Fe(OH)_3^- + 12H^+ \quad (3\text{-}4)$$

Ferric Iron + Water ⟶ Ferric Hydroxide ("Yellowboy") + Acidity

The fourth reaction is the oxidation of the additional pyrite by ferric iron. The ferric iron is generated in Reactions (3-2) and (3-3). This is the cyclic and self-propagating part of the overall reaction that takes place very rapidly and continues until either ferric iron or pyrite is depleted. Note that in this reaction ferric iron is the oxidizing agent, not oxygen:

$$FeS_2 + 14Fe^{3+} + 8H_2O \longrightarrow 15Fe^{2+} + 2SO_4^{2-} + 16H^+ \quad (3\text{-}5)$$

Pyrite + Ferric Iron + Water ⟶ Ferrous Iron + Sulfate + Acidity

Treatment of AMD includes both chemical and passive techniques. In Pennsylvania, for example, strict effluent discharge limitations were placed on mine operations in 1968 [11]. Many companies used chemical treatment methods to meet these new effluent limits. In these systems, the acidity is buffered by the addition of alkaline chemicals such as calcium carbonate, sodium hydroxide, sodium bicarbonate, or anhydrous ammonia. These chemicals raise the pH to acceptable levels and decrease the solubility of dissolved metals. Precipitates form from the solution. These chemicals are expensive, however, and the treatment system requires additional costs associated with operation and maintenance as well as the disposal of metal-laden sludges.

Many variations of AMD passive treatment systems were studied as early as 1978 [11]. During the last 15 years, passive treatment systems have been implemented on full-scale sites throughout the United States with promising results. The concept behind passive treatment is to allow the naturally occurring chemical and biological reactions that aid in AMD treatment to take place in the controlled environment of the treatment system and not in the receiving water body. Passive systems do not require the expensive chemicals necessary for chemical treatment systems, and operation and maintenance requirements are considerably less. Passive AMD treatment technologies being implemented include:

- Aerobic wetland;
- Compost or anaerobic wetland;
- Open limestone channels;
- Diversion wells;
- Anoxic limestone drains;
- Vertical flow reactors;
- Pyrolusite process.

Hydrologic Impact

With underground mining, subsidence and fracturing of overlying strata may cause surface runoff to be diverted underground and may disrupt aquifers, causing local water level declines and changing the direction of groundwater flow near the mine [1]. Dewatering required by mining operations affects groundwater quantity by depleting aquifers when mine features extend below the water table and become a drain [5]. Although groundwater depletion is the most obvious and immediate effect of mining on groundwater, longer term effects on the environment may be equally important. Redistribution and/or change in groundwater recharge rates may affect the time and degree to which aquifers will recover to a static condition. Conscientious management practices minimize water-related environmental impacts [13]. Coal mining activities are highly regulated, requiring extensive surface and groundwater sampling and monitoring to ensure compliance with federal, state, and local statutes. Also, hydrological impacts must be taken into consideration as part of the permitting process; therefore, coal company hydrologists study and monitor the quality of surface and underground water resources before, during, and after mining activity to ensure minimal hydrological impacts.

Health Effects/Miner Safety

There is no argument that mining is a dangerous occupation and that historically there has been a tremendous social cost associated with coal mining. Mine accidents and diseases have taken their toll on miners. In the United States, more than 100,000 miners died over the period of about 1885 to 1985 [4]. In 1900, the annual death rate among underground miners was 3.5 deaths per thousand miners [4]; however, a commitment to enhancing safety training and the development of new technologies have yielded remarkable improvements in the mine as a workplace. In addition, coal mines are subject to regular, comprehensive inspections by the federal Mine Safety and Health Administration (MSHA), and safety and health reporting requirements are much more stringent than those required by the Occupational Safety and Health Administration (OSHA), which regulates most other U.S. industries. According to MSHA, the twentieth century saw remarkable improvements in safety and health for U.S. miners, and the rate of fatal injuries for underground miners declined by 92% over the period from 1960 to 1999 [13]. Data from the U.S. Department of Labor show that in 2002 there were 12 fatalities in underground mining, 6 in surface mining, and 2 in preparation plants [14]. This is an annual death rate of 0.2 per thousand mining personnel, nearly a 20-fold decrease since 1900. The annual death rate for underground miners, which experienced the largest incidences of fatalities with 12, was 0.3 per thousand miners. According to the U.S. Department of labor, mining has a lower rate of injuries and illness per 100 employees than the agricultural, construction, or retail trades [13]. The accident and injury rates for miners today are comparable to those of grocery store workers.

Dust issues in mines, particularly underground mines, have had a negative impact on the coal industry. Many miners, particularly underground miners, have contracted respiratory diseases. These diseases—pneumoconiosis (black lung) and silicosis—have taken years to manifest themselves, and the industry has been working to minimize dust in mines. In an underground mine, the walls of the tunnels or shafts are covered with pulverized limestone to help settle coal dust [13]. Water sprayers on the mechanical equipment, such as in continuous and longwall mines, help reduce dust concentrations in the mines. The ventilation fans that remove methane also remove the lingering dust, and a continuous supply of fresh air is brought into the mine. Plus, miners now wear air-purifying systems. Dust production in underground mines is 0.0006 and 0.01 short tons per 1000 short tons of coal produced for conventional and longwall mining, respectively [9].

Surface Mining

Surface mining of coal is an alternative to underground mining and is practiced as strip mining, the most common form, and auger mining. Some coal is also produced through dredging and culm bank (*i.e.*, anthracite waste piles) recovery; however, these quantities are relatively small compared to underground, strip, and auger mining. The quantities of coal produced in 2002 by the various methods, as reported by MSHA, are listed in Table 3-2 [14]. Note that the production totals reported by MSHA and those of the National Mining Association/Energy Information Administration (see Figure 2-10) differ slightly.

Strip mining is favored when the overburden (the overlying rock strata) is typically 200 feet or less in thickness but can be economical with overburden thickness up to 600 feet [15]. Another factor that is considered when determining the economics of strip mining is the stripping ratio, which is the ratio of overburden thickness to coal seam thickness. Typically, the maximum stripping ratio that is economical is 20 to 1 [4]. The two general methods of strip mining are area mining and contour mining. In area mining, a trench is dug, the overburden is piled to one side, and the coal is removed. As mining progresses, the overburden from the new trench is dumped into the first trench. In contour mining, which is used when coal lies beneath hilly terrain or outcrops on a hillside, the overburden is removed and dumped on the downhill side of the mining operation [3].

Auger mining can be used for coal outcrops on a hillside or in strip mines where the stripping ratio has become too high to be economical for strip mining [4]. In auger mining, an auger drills it way into the face of the coal seam, and the coal is removed via conveyor and loaded directly onto a truck.

Many adverse environmental impacts could potentially result from area surface mining of coal if no reclamation practices are used. Such measures are used with varying degrees of effectiveness in the United States [1]. Air quality can be affected in several ways. Fugitive dust from coal

TABLE 3-2
Coal Production by Type in the United States in 2002

Mining Type	*Production (short tons)*
Total underground	353,273,486
Total surface	736,546,576
Strip	730,630,458
Auger	5,071,011
Culm bank recovery	398,536
Dredge	446,571
Total coal	1,089,820,062

haul roads, unvegetated spoil surfaces, topsoil stockpiles, and coal stockpiles is a potential problem. Overburden blasting can produce troublesome noise, air shock, and ground vibration. The quality and quantity of surface water and groundwater can be affected if effective reclamation practices are not used. Sedimentation of surface waters may occur. Erosion of reclaimed slopes is often experienced due to the unconsolidated nature of the reclaimed materials. Aesthetically, the disturbed land is very unsightly prior to reclamation.

Surface Disturbance

Surface mining leads to large-scale disturbances on the surface of the Earth. The natural land surface is drastically changed by the mining activities through the removal of soil, rock, and coal. Depending on mining conditions and equipment, widespread changes in the locations of materials will occur [5]. For example, an inadequate amount of material may be available to fill the final pit of a surface coal mine. As a result, the areas will usually be graded to a topography that includes a lake or basins with at least a portion of the area having relatively steep slopes. Reclamation of contour mines and sometimes mountain-top removal mines often results in some very steep slopes [5]. These areas are prone to erosion and mass failure due to the steepness of the slopes and the loose, nonhomogeneous nature of the materials present. Unfortunately, erosion and mass wasting are often found to be the major problems associated with surface mining because these reclaimed land forms are often in states of disequilibrium relative to the natural environment where they were created [5]. The primary role of reclamation, therefore, is to achieve a landscape that approximates pre-mining conditions assumed to be near equilibrium with the local environmental factors.

Generation of Gases

Although a surface mine releases less methane than an underground mine, as discussed earlier, the amount emitted into the atmosphere is still significant due to the large amount of surface-mined coal. Measurements of actual emissions of methane from surface mining are technically difficult, costly, and generally not available for inventory purposes; however, emission factors have been developed from a number of country-specific studies [6]. Irving and Tailakov [6] estimate that surface mining releases 0.3 to 2.0 cubic meters of methane per metric ton of coal mined.

Liquid Effluents/Acid Mine Drainage

Surface mining operations raise some of the same issues that underground mining does. Contour and area mining generates 0.24 and 1.2 tons of liquid effluents per 1000 tons of coal produced, respectively, as compared to 1.0 and 1.6 tons, respectively, for conventional and longwall mining operations [9]. The presence of soluble salts such as sodium in discarded overburden can cause saline and caustic conditions in topsoils if conditions allow the upward migration of these salts. Also, oxidation processes result in significant changes in chemistry. As discussed previously, when sulfides such as pyrite are present, acid is produced, and the solubility of elements tends to increase; hence, acid mine drainage is produced.

Hydrologic Impact

Surface mining affects surface stream runoff. The runoff may increase and subsequent channel erosion may occur as a result of reduced infiltration rates [1]. Conversely, streams may also be affected by decreased surface runoff where more permeable rock strata become exposed by the surface mining. Modifications of the local or regional recharge zones involve changing the infiltration rates by removal of vegetative cover, alteration of soil profiles, and compaction. Reduced infiltration rates decrease groundwater storage and reduce water availability. These disruptions are of particular concern in the semi-arid western region of the United States. Shallow and coal seam aquifers can be drained by mining activity, causing temporary or permanent loss of existing wells near mined areas [1].

The disturbance of the overburden during surface mining also causes significant changes in the chemical nature of the system [5]. Such changes are due to the influence of water on the now-available soluble salts and to the changing redox conditions resulting from the influx of oxygen into the system that was previously oxygen depleted. The movement of high concentrations of salts and/or elements into existing or reestablished groundwater aquifer systems can occur due to the disruption of the consolidated overburden and increased water penetration into reclaimed land.

Solid Waste/Dust

Waste rock is a product of the mining process that influences the post-mining land surface. In the case of mountain-top removal and contour mining methods, waste materials are often used to fill adjacent canyons or hollow areas. When these materials are used for canyon fill, steep slopes are formed that tend to be very erosive. Surface mining produces more solid waste than underground mining techniques, with 10 tons of solid waste produced per 1000 tons of coal removed for both contour and area mining [9]. Conventional and longwall underground mining, on the other hand, produces 3 and 5 tons of solid waste, respectively, per 1000 tons of coal removed. Similarly, dustiness associated with surface mining is significantly greater than that of underground mining. The World Bank Group [9] reports dust generation of 0.1 and 0.06 ton per 1000 tons of coal produced for contour and area mining, respectively, while only 0.0006 and 0.01 ton of dust is generated during conventional and longwall underground mining, respectively.

Health Effects/Miner Safety
Surface miners experience different safety issues from underground miners; in 2002, the annual death rate for surface miners was 0.2 per thousand miners [14]. Historically, underground miners suffered a greater rate of respiratory diseases than did surface miners until regulations improved dust control practices and mandated the use of personal air-purifying systems. Now the fatality rates are similar for surface and underground miners.

Legislation/Reclamation

Because coal mining can have a significant impact on the environment and health of miners, coal producers are required to go through a complicated process to obtain local, state, and federal permits to mine. Coal mining is one of the most extensively regulated industries in the United States [16,17]. A company must comply with many laws and regulations, and meeting all the requirements is a long and difficult process [16,18]. Up to 10 years can pass between the start of planning a mine and mining the first ton of coal. A coal company must provide detailed information on how the coal will be mined and how the land will be reclaimed, on the quality and quantity of surface and underground water sources and how the mining operations will affect them, and on the method to transport the coal from the mine and how the area will be affected by the transportation. The coal company has to return the land to approximately the same physical contour and to a state of productivity equal to or better than pre-mining conditions. Wildlife habitat cannot be permanently disrupted, and archeological sites must be protected. Companies must post bonds as high as $10,000/acre to ensure that the sites will be reclaimed [16].

Unfortunately, concern for the environment was not always a high priority in the past and as a result there are many abandoned mines. A tax on mined coal is earmarked for the Federal Abandoned Mines Land Fund to finance reclamation projects of these orphaned mines. Some states had reclamation laws on their books since the 1930s; however, it was not until 1977 that Congress enacted the Surface Mining Control and Reclamation Act (SMCRA), which mandated strict regulation of surface mining [16]. This act, along with other federal laws (specifically, the Clean Air Act, the Clean Water Act, and the National Environmental Policy Act), has had a significant impact on surface mining. In addition, many other legislative acts affect surface mining in the United States and, as noted from the names listed below, cover a host of subject areas [16]:

- American Indian Religious Freedom Act of 1978;
- Antiquities Act of 1906;
- Archeological and Historical Preservation Act of 1974;
- Archeological Salvage Act;
- Bald Eagle Protection Act of 1969;
- Endangered Species Act of 1963;

- Fish and Wildlife Coordination Act of 1934;
- Forest and Rangeland Resources Planning Act of 1974;
- Historic Preservation Act of 1966;
- Migratory Bird Treaty Act of 1918;
- Mining and Minerals Policy Act of 1970;
- Multiple Use–Sustained Yield Act of 1960;
- National Forests Management Act of 1976;
- National Trails System Act;
- Noise Control Act of 1976;
- Resource Conservation and Recovery Act;
- Safe Drinking Water Act of 1974;
- Soil and Water Resources Conservation Act of 1977;
- Wild and Scenic Rivers Act;
- Wilderness Act of 1964.

Although the mining industry must be cognizant of many laws and regulations, and past mining operations were not environmentally conscientious, mine reclamation has become a success story in the United States. Mine operators are addressing the issues discussed earlier in this chapter, and as coal continues to be a major energy source for the United States the mining industry; federal, state, and local governments; and the general public will have to continue working closely together to minimize the environmental and health impacts of coal mining.

Surface mining of coal should be regarded as a temporary land use. The enactment of stringent laws requiring that the land be returned to its original condition or better has meant that restoration of the land surface and rehabilitation of the soil materials have now become normal parts of the planning, approval, and operation of most surface mines. Many examples of high-quality restored land can now be found, and the Office of Surface Mining annually recognizes companies and individuals whose efforts are exemplary. These companies and individuals are honored for not only doing the reclamation required of them but also for going beyond the requirements to achieve outstanding landscape restoration [19]. Internationally, many countries have legislation governing the restoration and rehabilitation of the land [17]. Major coal-producing countries, such as Canada, Germany, the United Kingdom, Australia, and South Africa, practice reclamation. Unfortunately, for economic reasons, in less-developed countries such as China, India, and Indonesia less reclamation is being practiced.

Coal Preparation

The purpose of coal preparation is to improve the quality of the coal to make it suitable for a specific purpose by [20,21]:

- Crushing;
- Screening;

- Conventional cleaning;
- Deep cleaning;
- Blending;
- Dedusting.

Run-of-mine (ROM) coal generally falls into two major groups: that from underground mining and that from strip mining. Underground mining tends to produce a finer product than strip mining; however, both products are further crushed to produce the desired size for a coal cleaning process or directly for utilization (*e.g.*, combustion or gasification) [20]. In addition, increased mechanization in the underground mining industry has decreased selectivity and increased the volume of refuse [21]. Equipment such as continuous miners or longwall shearers often takes roof and floor rock in addition to the coal. Also, equipment currently used to mine and transport coal produces more fine coal particles than did earlier equipment [21].

End-use facilities such as power plants are designed for optimal combustion to burn a coal with a specific composition of ash, sulfur, energy, and, sometimes, volatile matter content. These requirements are becoming more difficult to meet with ROM coal or coal from one specific source because of several factors: coal variability (*e.g.*, ash and sulfur contents) within the coal seam, considerable variation in underground mined coal quality due to the inclusion of roof and floor layers, the need for reduced sulfur content due to limitations on sulfur dioxide emissions from power plants, and the declining quality of coal being mined in the eastern United States as higher quality reserves have been depleted. Consequently, techniques have been implemented to upgrade the coal quality, and the need for advanced coal cleaning processes for both coarse coal and fine coal has grown [21].

In the past, coal was predominantly cleaned by dry methods, which have been abandoned in recent years in favor of wet cleaning processes due to several factors: particle size requirements (grinding the coal to finer sizes liberates more ash from the coal); dust emissions (which can be controlled by the use of water in underground mines); transportation issues; health, safety, and noise impacts; and the better performance of wet processes for coal cleaning [21]. A coal preparation plant separates the material it receives into a product stream and a reject stream, which may be further divided into coarse and fine refuse streams. Depending on the source, 20 to 50% of the material delivered to a coal preparation plant may be rejected [20]. One of the reject streams is a slurry, a blend of water, coal fines, silt, sand, and clay particles, which is most commonly disposed of in an impoundment [21].

The coal cleaning processes predominantly in use now include dense medium separation, hydraulic separation, froth flotation, and agglomeration [20]. Dense medium separations include those coal preparation processes that clean raw coal, coarse or fine, by immersing it in a fluid having a density intermediate between clean coal and the rejects. Because there is a general correlation between ash content and specific gravity, it is possible to achieve the required degree of removal of impurities from raw coal by regulating the specific gravity of the separating fluid, which can be organic liquids, dissolved salts in water, aerated solids, or suspensions of fine solids suspended in water [20]. Hydraulic separation jigging is a process of coarse particle stratification in which the particle rearrangement results from an alternative expansion and compaction of a bed of particles by pulsating flow. The rearrangement results in layers of particles that are arranged by increasing density from top to bottom of the bed, with the coal near the top. Hydraulic concentration of fine coal is performed using wet concentrating tables, cyclones, launders, feldspar jigs, and hydrorotators. These processes depend on the physical characteristics—size, shape, and density—of particles suspended in a liquid medium to effect a concentration of desired quality [20]. Froth flotation is a chemical process that depends on the selective adhesion of some solids (*i.e.*, fine coal) to air and the simultaneous adhesion of other particles (*i.e.*, refuse) to water. A separation of coal from coal waste then occurs as finely disseminated air bubbles are passed through a feed coal slurry. Agglomeration works on a similar principal as froth flotation (*i.e.*, differences in the surface properties of coal and inorganic matter), and fine coal particles in a suspension can be readily be agglomerated by the addition of a bridging liquid (many different oils), under agitation, and then recovered while the inorganic constituents remain in the aqueous suspension and are rejected.

These processes, briefly described here, impact the environment in several ways, including making it necessary to deal with the contamination aspects of fine coal cleaning and "blackwater" disposal, air contamination, refuse disposal and control, and operator health and safety.

Water Contamination from Preparation Plants

The effluents from coal preparation plants and waters draining from plant site surfaces contain fine coal and coal refuse materials in suspension. At older plants, the disposal of effluent continues to present a serious problem as it is becoming increasingly difficult to comply with the standards required by many water authorities and pollution control agencies [20]. New plants, however, are installing more complex clarification facilities, and new plants are designed to operate on a closed circuit to satisfy pollution requirements.

Air Contamination from Preparation Plants

Preparation of fine coal can cause air pollution if proper dust and gas removal equipment is not installed. The air effluent from a fine coal preparation plant consists of entrained dust, both coal and ash, and various gases, primarily consisting of products of coal combustion from thermal dryers [20]. Sources of particulates include thermal dryers, pneumatic coal-cleaning equipment, coal processing and conveying equipment, screening equipment, coal storage, coal transfer points, and coal handling facilities. The coal dust and ash can be controlled by a combination of mechanical separators, wet scrubbers, electrical precipitators, and filters.

Refuse Contaminants from Preparation Plants

Coal refuse, varying in composition and size, is a function of the seam mined, the mining system, and the preparation system. Coal refuse consists mainly of unsalable coal, shale, bone, calcite, gypsum, clay, pyrite, or marcasite [20]. Disposal methods for coarse and fine coal refuse developed differently. Prior to modern coal preparation plants, coarse refuse was handpicked from the coal and discarded either back into the mine or deposited on the surface. When fine coal cleaning came into widespread use, it became more of an issue to deal with. Early practices were to discharge the "blackwater" or liquid effluent containing solids into the nation's stream systems [20]. This is no longer acceptable, and the effluent must be filtered to remove the fine solids. Embankments are constructed using compacted coarse refuse material to impound fine coal slurry, and the impoundments serve as a filter to clarify this effluent as it flows through the permeable structure. These impoundments are also used as settlement basins and water-supply reservoirs.

The environmental impacts for coarse coal refuse sites include the nonproductive use of land, the loss of aesthetic value, water pollution, and air pollution. Many burning coal refuse sites in the United States are sources of local air pollution and safety hazards. Coal refuse disposal impoundments are constructed for the permanent disposal of any coal, rock, and related material removed from a coal mine in the process of mining. The environmental impacts of impoundments include the nonproductive use of land, the loss of aesthetic value, the danger of slides, dam failure, and water pollution [22]. Most coal waste impoundments in the United States, of which there are approximately 700, are found in the eastern United States, predominantly in West Virginia, Pennsylvania, Kentucky, and Virginia [21]. Most of the coal mined in the western United States is shipped without extensive processing; therefore, coal waste impoundments are rarely used there. The majority of coal from underground mines is processed before sale. Of the over 1 billion short tons of coal mined each year in the United States, about 600 to 650 million short tons are processed to some degree, 350 to 400 million short tons are handled in wet-processing systems, and 70 to 90 million short tons of fine refuse are produced [21]. It is estimated that 2 billion short tons of refuse are contained in impoundments in the United States [22].

Health and Safety Issues

In 2002, the incidences of injuries and fatalities at coal preparation facilities were similar to coal mining, ~8 and 0.2 per 1000 workers, respectively [14]; however, the largest impact of coal cleaning facilities is on the environment.

Coal waste facilities have been involved in several accidents or incidents over the last 30 years. The dramatic failure on Buffalo Creek in West Virginia in 1972 revealed the hazards associated with embankments. Prior to this accident, little consideration was given to the control of water entering an impoundment from a preparation plant or as runoff or to the discharge of contaminated effluent to the stream system. The accident in West Virginia, which occurred when a coal waste impounding structure collapsed on a Buffalo Creek tributary, resulted in a flood that killed 125 people; injured 1100 people; left more than 4000 people homeless; demolished 1000 cars and trucks, 502 houses, and 44 mobile homes; damaged 943 houses and mobile homes; and caused $50 million in property damages [21]. At the time of the accident, no federal standards required either impoundments or hazardous refuse piles to be constructed and maintained in an approved manner. This situation changed, however, as a result of this accident, and numerous federal and state statutes and regulations now apply to the disposal of coal waste impoundments. The Mining Enforcement and Safety Administration developed standards for impoundments and refuse piles, and nearly every coal waste facility is subject to regulatory requirements imposed by MSHA, OSHA, or the state with a regulatory program approved under the Surface Mining and Reclamation Act of 1977 [21].

Although the industry has become more regulated, there still have been some embankment failures since 1972 [21]. These failures have been less significant than the Buffalo Creek incident in that the quantity of slurry released, damage occurring, injuries, and loss of life (1 person) have not been as great. A notable exception is the October 2000 impoundment failure near Inez, Kentucky, in which a 7-acre surface impoundment failed and approximately 250 million gallons of slurry were released into a nearby underground coal mine. The slurry flowed through the mine and into nearby creeks and rivers, flooding stream banks to a depth of 5 feet [21]. No loss of life was experienced; however, the environmental impact was significant, and local water supplies that were taken from the rivers were disrupted for days. This incident resulted in Congress requesting the National Research Council to examine ways to reduce the potential for similar accidents in the future [21]. The National Research Council appointed the Committee on Coal Waste Impoundments to:

- Examine engineering practices and standards currently being applied to coal waste impoundments;
- Evaluate the accuracy of mine maps and explore ways to improve surveying and mapping of underground mines to determine how underground mines relate to current or planned slurry impoundments;
- Evaluate alternative technologies that could reduce the amount of coal waste generated or allow productive use of the waste;
- Examine alternative disposal options for coal slurry.

The committee offered many conclusions and recommendations. The implementation of these recommendations should substantially reduce the potential for uncontrolled release of coal slurry from impoundments [21].

Coal Transportation

Environmental impacts result from the transport of coal. Coal transportation is accomplished through rail, truck, water, slurry pipeline, or conveyor; however, most is performed by rail. Environmental impacts occur during loading, en route, or during unloading and affect natural systems, manmade buildings and installations, and people (*e.g.*, due to injuries or deaths) [23]. All forms of coal transportation have certain common environmental impacts, which include use of land, structural damage to facilities such as buildings or highways, air pollution from engines that power the transportation systems, and injuries and deaths related to accidents involving workers and the general public (*e.g.*, railway crossing accidents). In addition, fugitive dust emissions are experienced with all forms of coal transport, although precautionary measures are increasingly being taken [23]. It is estimated that 0.02% of the coal loaded is lost as fugitive dust with a similar percentage lost when unloading. Coal losses during transit are estimated to range from 0.05 to 1.0%. The amount is dependent upon mode of transportation and length of trip but can be a sizeable amount, especially for unit train coal transit across the country.

Coal Combustion By-Products (CCB)

More than 100 million tons of coal-related residues are generated annually by coal-burning plants [24,25]. These materials have many names—they are referred to as fossil fuel combustion wastes (FFCWs) by the U.S. Environmental Protection Agency (EPA); as coal combustion products (CCPs) by the utility industry, ash marketers, and ash users; and as coal combustion by-products (CCBs) by the U.S. Department of Energy and other federal agencies. These residues become products when utilized and wastes when disposed of [24]. They include fly ash, bottom ash, boiler slag, and flue gas desulfurization (FGD) material (*i.e.*, synthetic gypsum). The fly ash is the fine fraction of the CCBs that is entrained in the flue gas exiting a boiler and is captured by particulate control devices. Bottom ash is the large ash particles that accumulate at the bottom of a boiler. Boiler slag is the molten inorganic material that is collected at the bottom of some boilers and discharged into a water-filled pit where it is quenched and removed as glassy particles. FGD units, which remove sulfur dioxide using calcium-based reagents, generate large quantities of synthetic gypsum, which is a mixture of mainly gypsum ($CaSO_4$) and calcium sulfite ($CaSO_3$) but which can also contain fly ash and unreacted lime or limestone [24,25]. In 2000, CCB production in the United States was 108,050,000 short tons and was comprised of [25]:

- Fly ash—62,810,000 short tons, or 58.1% of the total generated;
- Bottom ash—16,940,000 short tons, or 15.7% of the total generated;
- Boiler slag—2,670,000 short tons, or 2.5% of the total generated;
- FGD material—25,630,000 short tons, or 23.7% of the total generated.

In the United States, approximately 30% of the CCBs are used in a variety of applications, with the remainder being disposed of [25]. The components of the CCBs have different uses because they have distinct chemical and physical properties that make them suitable for specific applications. CCBs are used in cement and concrete; mine backfill, agriculture, blasting grit, and roofing applications; waste stabilization; wallboard production; acid mine drainage control; and as road base/subbase, anti-skid material, fillers, and extenders [24,25].

Globally, CCB use varies significantly. In Europe, more CCBs are used than in the United States; for example, in 1999, 56% of the CCBs were profitably used in Europe compared to about 30% in the United States [25]. The CCBs are used in a number of applications, primarily in concrete, portland cement manufacture, and road construction. Raw materials shortages and favorable state regulations account for higher usage in Europe than in the United States. Countries such as Canada, India, and Japan utilize 27, 13, and 84% of their CCBs, respectively [25]. Canada's usage is similar to that of the United States, CCB usage in India is low due to the relatively large amount of CCBs produced because of the coal's high ash content, and Japan utilizes most of its CCBs due to the high cost of disposal in Japan.

Coal combustion by-products primarily contain elements such as iron, aluminum, magnesium, manganese, calcium, potassium, sodium, and silica, which for the most part are innocuous. CCBs also contain small amounts of trace elements such as arsenic, barium, beryllium, cadmium, cobalt, chromium, copper, nickel, lead, selenium, zinc, and mercury. These elements can be classified as essential nutrients, toxic elements, or priority pollutants and are considered to have some environmental or public health impacts [24]. The risks include potential groundwater contamination of trace elements and above-ground human health impacts through inhalation and ingestion of contaminants released through wind erosion and surface water erosion and runoff [26].

The Resource Conservation and Recovery Act (RCRA) has been the primary statute governing the management and use of CCBs. The EPA was considering some form of Subtitle C regulation (*i.e.*, classify CCBs as hazardous wastes) under the RCRA for CCBs used in mine backfill or for agricultural applications [26]; however, the agency investigated the dangers of CCBs to human health and the environment and concluded that CCBs do not pose sufficient danger to the environment to warrant regulations under RCRA Subtitle C. The EPA does intend to develop national regulations under RCRA Subtitle D (nonhazardous solid waste) or to work with the U.S. Department of the Interior toward modifying existing regulations under the SMCRA when CCBs are placed in landfills or surface impoundments or are used as fill in surface or underground mines [24].

Emissions from Coal Combustion

Coal combustion produces large quantities of products that may be released into the atmosphere. The emissions are largely steam (*i.e.*, water vapor), which is what is most often observed coming from the stack of a power plant, carbon dioxide, and nitrogen from the air and do not present any direct health hazard. However, the emissions do contain small concentrations of atmospheric pollutants, which translate into large quantities emitted due to the vast amount of coal consumed. The principal pollutants that can cause health problems are sulfur and nitrogen oxides, particulate matter, trace elements (including arsenic, lead, mercury, fluorine, selenium, and radionuclides), and organic compounds. The environmental impacts and health effects of these pollutants, along with carbon monoxide and carbon dioxide, are discussed here.

The definition of air pollution is the addition to the atmosphere of any material that may have a deleterious effect on life [27]. Air pollution is produced by natural processes or by anthropogenic, or manmade, actions. The legal definition of an air pollutant is any air pollution agent or combination of such agents, including any physical, chemical, biological, or radioactive substance, or matter that is emitted into or otherwise enters the ambient air [27]. These agents include primary and secondary pollutants, which are classifications that indicate how the various pollutants are formed. Primary pollutants have the same state and chemical composition in the ambient atmosphere as when emitted from sources, but secondary pollutants have changed in form after leaving the source due to oxidation, decay, or reaction with other primary pollutants.

Sulfur Oxides

Gaseous emissions of sulfur oxides from coal combustion are mainly sulfur dioxide (SO_2) and, to a much lesser extent, sulfur trioxide (SO_3) and gaseous sulfates. The sulfur in the coal reacts with oxygen to form the sulfur oxides:

$$S + O_2 \longrightarrow SO_2 \qquad (3\text{-}6)$$

$$S + 1.5O_2 \longrightarrow SO_3 \qquad (3\text{-}7)$$

Sulfur dioxide is a nonflammable, nonexplosive, colorless gas that causes a taste sensation at concentrations from 0.1 to 1.0 part per million by volume (ppmv) in air [28]. At concentrations greater than 3.0 ppm, the gas has a pungent, irritating odor. Sulfur dioxide is partly converted to sulfur trioxide or sulfuric acid (H_2SO_4) and its salts by photochemical or catalytic processes in the atmosphere. Sulfur trioxide and water form sulfuric acid.

Environmental Effects

Environmental effects of sulfur compounds include impaired visibility, damage to materials, damage to vegetation, and deposition as acid rain. Fine particles in the atmosphere reduce the visual range by scattering and absorbing light [28]. Aerosols of sulfuric acid and other sulfates comprise from 5 to 20% of the total suspended particulate matter in urban air, thus contributing to the reduction in visibility.

Sulfur compounds are responsible for major damage to materials. Sulfur oxides generally accelerate metal corrosion by first forming sulfuric acid either in the atmosphere or on the metal surface. Sulfur dioxide is the most detrimental pollutant with regard to metal corrosion [28]. Temperature and relative humidity also significantly influence the rate of corrosion. Sulfurous and sulfuric acids are capable of damaging a wide variety of building materials including limestone, marble, roofing slate, and mortar. Textiles made of nylon are also susceptible to pollutants in the atmosphere.

In general, the damage to plants from air pollution usually occurs in the leaf structure, as the leaf contains the building blocks for the entire plant [28]. Sulfur dioxide enters the leaf, and the plant cells convert it to sulfite and then into sulfate. Apparently, when excessive sulfur dioxide is present, the cells are unable to oxidize sulfite to sulfate fast enough and disruption of the cell structure begins. Spinach, lettuce, and other leafy vegetables are most sensitive, as are cotton and alfalfa. Pine needles are also affected, with either the needle tip or the entire needle becoming brown and brittle.

Acidic deposition or acid rain occurs when emissions of sulfur dioxide and oxides of nitrogen in the atmosphere react with water, oxygen, and oxidants to form acidic compounds. These compounds fall to the ground in either dry form (*i.e.*, gas and particles) or wet form (*i.e.*, rain, snow, and fog). The acidity of water is reported in terms of the pH, where pH is the logarithm (base 10) of the molar concentration of hydrogen ions:

$$pH = -\log_{10}[H^+] \tag{3-8}$$

Pure water contains a hydrogen ion concentration that is approximately 10^{-7} molar or pH = 7, which is referred to as neutral pH. Water droplets formed in the atmosphere, however, normally have a pH of ~5.6 because atmospheric carbon dioxide is dissolved in the rain and forms carbonic acid (H_2CO_3). When sulfur dioxide or nitrogen oxides are also dissolved in the water, the pH drops; average yearly pH values of 4.0 to 4.5 are reported in the eastern United States [28]. Sulfur dioxide can be absorbed from the gas phase into an aqueous droplet, creating acidic conditions as follows [28]:

$$SO_2(g) \Longleftrightarrow SO_2(aq) \tag{3-9}$$

$$SO_2(aq) + H_2O \Longleftrightarrow HSO_3^- + H^+ \tag{3-10}$$

$$HSO_3^- \Longleftrightarrow SO_3^{2-} + H^+ \tag{3-11}$$

$$SO_3^{2+} + H_2O \Longleftrightarrow SO_3^{2-} + 2H^+ \tag{3-12}$$

Several effects of acid rain are of concern. Acidification of natural water sources can have a devastating effect on fish; trout and salmon are particularly sensitive to a low pH [28], and reproduction in many fish fails to occur at a pH less than 5.5. A decrease in plankton and bottom fauna is also observed as the pH drops which reduces the food supply for the fish. Leaching of nutrients occurs from the soil, and this demineralization can lead to loss in productivity of crops and forests or a change in natural vegetation. Vegetation itself can be directly damaged. An increase in corrosion to materials is also observed.

Health Effects

Sulfur dioxide and other oxides of sulfur have been studied extensively; however, many questions concerning the effects of sulfur dioxide upon health remain unanswered [28]. Few epidemiological studies have been able to differentiate adequately the effects of individual pollutants because sulfur oxides tend to occur in the same kinds of atmospheres as particulate matter and high humidity. High concentrations of sulfur dioxide can result in temporary breathing impairment in asthmatic children and adults who are active outdoors [29]. Short-term exposures of asthmatic individuals at moderate exertion to elevated sulfur dioxide levels may result in reduced lung function that may be accompanied by such symptoms as wheezing, chest tightness, or shortness of breath. Other effects that have been associated with longer term exposures to high concentrations of sulfur dioxide in conjunction with high levels of particulate matter include respiratory illness, alterations in the lung's defenses, and aggravation of existing cardiovascular disease. Those who may be affected under these conditions include individuals with cardiovascular disease or chronic lung disease, as well as children and the elderly.

The relationship between human health and the concentration of sulfur dioxide in the atmosphere is a very complex one to decipher, especially because compounding factors such as other pollutants (*e.g.*, particulate matter) and high humidity often occur simultaneously. Consequently, the Electric Power Research Institute (EPRI) has designed a program called the Aerosol Research and Inhalation Epidemiology Study (ARIES) to address the issue of air pollution components by coupling an extensive air quality monitoring effort with five health studies: daily mortality, emergency room visits, heart rate variability, arrhythmic events, and unscheduled physician visits [30]. ARIES started in 1998 and Atlanta, Georgia, was selected by the EPA to be the first site. In Atlanta, the health effects of suspended particulate matter (PM)—fine particulate matter less than 2.5 μm ($PM_{2.5}$)—were studied, and preliminary results suggest that adverse health effects appear to be associated with carbon-containing PM and not the sulfate and nitrate components primarily derived from coal combustion. The study has been expanded to include Chicago and will be followed by studies in four additional cities.

Nitrogen Oxides (NO_x)

Seven oxides of nitrogen are present in ambient air [28]. These include nitric oxide (NO), nitrogen dioxide (NO_2), nitrous oxide (N_2O), NO_3, N_2O_3, N_2O_4, and N_2O_5. Nitric oxide and nitrogen dioxide are collectively referred to as NO_x due to their interconvertibility in photochemical smog reactions. The term NO_y is often used to represent the sum of the reactive oxides of nitrogen and all other compounds that are atmospheric products of NO_x. NO_y includes compounds such as nitric acid (HNO_3), nitrous acid (HNO_2), nitrate radical (NO_3), dinitrogen pentoxide (N_2O_5), and peroxyacetyl nitrate (PAN). It excludes N_2O and ammonia (NH_3) because they are not normally the products of NO_x reactions [28].

Nitrogen oxide emissions from coal combustion are produced from three sources: thermal NO_x, fuel NO_x, and prompt NO_x. Nitrogen oxides are primarily produced as a result of the fixation of atmospheric nitrogen at high temperatures (thermal NO_x) and the oxidation of coal nitrogen compounds (fuel NO_x). Prompt NO_x is formed when hydrocarbon radical fragments in the flame zone react with nitrogen to form nitrogen atoms, which then form NO. The majority of the oxide species produced is NO, with NO_2 accounting for less than 5% of the total [31].

The production of thermal NO is a function of the combustion temperature and fuel-to-air ratio and increases exponentially at temperatures above 2650°F. Thermal NO can be predicted by the following equation [32]:

$$[NO] = K_1 e^{-k_2/T}[N_2][O_2]^{1/2}t \qquad (3\text{-}13)$$

where T is temperature, t is time, K_1 and K_2 are constants, and $[N_2]$ and $[O_2]$ are concentrations in moles. Accordingly, thermal NO can be decreased by reducing the time, temperature, and concentration of N_2 and O_2. The principal reactions in the formation of thermal NO, which are referred to as the extended Zeldovich mechanism, are:

$$N_2 + O_2 \Longleftrightarrow NO + N \qquad (3\text{-}14)$$

$$N + O_2 \Longleftrightarrow NO + O \qquad (3\text{-}15)$$

$$N + OH \Longleftrightarrow NO + H \qquad (3\text{-}16)$$

Reaction (3-14) is assumed to be the rate-determining step due to the high activation energy required to break the triple bond in the nitrogen molecule. Reaction (3-16) has been found to contribute under fuel-rich conditions. The general conclusion is that very little thermal NO is formed in the combustion zone and that the majority is formed in the post-flame region, where the residence time is longer.

Prompt NO is produced by the reaction of hydrocarbon fragments and molecular nitrogen in the flame front. Prompt NO is most significant in fuel-rich flames, where the concentration of radicals such as O and OH can exceed equilibrium values, thereby enhancing the rate of NO formation. Prompt NO occurs due to the collision of hydrocarbons with molecular nitrogen in the fuel-rich flames to form HCN (hydrogen cyanide) and N. The HCN is then converted to NO by a series of reactions among NCO, H, O, OH, NH, and N. The amount of prompt NO generated is proportional to the concentration of N_2 and the number of carbon atoms present in the gas phase, but the total amount produced is low in comparison to the total thermal and fuel NO in coal combustion. The two reactions believed to be the most significant with regard to the mechanism for formation of prompt NO are [33]:

$$CH + N_2 \Longleftrightarrow HCN + N \tag{3-17}$$

$$C + N_2 \Longleftrightarrow CN + N \tag{3-18}$$

Reaction (3-17) was originally proposed, and Reaction (3-18) was added as a minor, but non-negligible, contributor to prompt NO; its importance grows with increasing temperature.

Fuel NO is the primary source of NO_x in flue gas from coal combustion and is formed from the gas-phase oxidation of devolatilized nitrogen-containing species and the heterogeneous combustion of nitrogen-containing char in the tail of the flame [31]. At temperatures below 2650°F, fuel NO can account for more than 75% of the measured NO in coal flames and can be as high as 95%. The reason for the dominance of fuel NO in coal systems is because of the moderate temperatures (2240–3140°F) and the locally fuel-rich nature of most coal flames. Fuel NO is produced more readily than thermal NO because the N–H and N–C bonds common in fuel-bound nitrogen are weaker than the triple bond in molecular nitrogen in the air, which must be dissociated to produce thermal NO. Combustion conditions and the nitrogen content of a coal affect the quantity of NO emissions. During devolatilization, a portion of the coal nitrogen is released as HCN and to a lesser extent as NH_3. HCN readily reacts with oxygen to form NO, but some of this NO can be converted to N_2 by reaction with hydrocarbon radicals in fuel-rich zones:

$$\text{Fuel N}\begin{pmatrix}NH_3\\HCN\end{pmatrix}\xrightarrow[NO/O_x]{}\begin{pmatrix}N_2\\NO\end{pmatrix} \tag{3-19}$$

Nitrogen retained in the char is also oxidized to NO, which may react with the char surface or hydrocarbon radicals to form N_2. Much of the coal nitrogen is initially converted to NO. The final NO emissions, however, are determined largely by the extent of conversion of NO to N_2 in the various regions of the combustion unit. Unlike thermal NO, the production of fuel NO is relatively insensitive to temperature over the range found in pulverized coal flames and more sensitive to the air-to-fuel ratio [32,34].

Environmental Effects

Both NO_x and NO_y (*i.e.*, HNO_3) have been shown to accelerate damage to materials in the ambient air. NO_x affects dyes and fabrics, resulting in fading, discoloration of archival and artistic materials and textile fibers, and loss of textile fabric strength [28]. NO_2 absorbs visible light and at a concentration of 0.25 ppmv will cause appreciable reduction in visibility. NO_2 affects vegetation, as studies have shown suppressed growth of pinto beans and tomatoes and reduced yields of oranges. In the presence of sunlight, nitrogen oxides react with unburned hydrocarbons—volatile organic compounds (VOCs) that are emitted primarily from motor vehicles but also from chemical plants, refineries, factories, consumer and commercial products, and other industrial sources—to form photochemical smog.

Nitrogen oxides also contribute to the formation of acid rain. NO and NO_2 in the ambient air can react with moisture to form NO_3^- and H^+ in the aqueous phase (*i.e.*, nitric acid), which can cause considerable corrosion of metal surfaces. The kinetics of nitric acid formation are not as well understood as those for the formation of sulfuric acid discussed earlier. Nitrogen oxides contribute to changes in the composition and competition of some species of vegetation in wetland and terrestrial systems, acidification of freshwater bodies, eutrophication (*i.e.*, explosive algae growth leading to depletion of oxygen in the water) of estuarine and coastal waters, and increases in the levels of toxins harmful to fish and other aquatic life [29].

Health Effects

Nitrogen dioxide acts as an acute irritant and in equal concentrations is more injurious than NO; however, at concentrations found in the atmosphere, NO_2 is only potentially irritating and related to chronic obstructive pulmonary disease [28]. EPRI has shown from their early results from the ARIES study that the nitrate components of coal combustion do not have adverse health effects [30]. The EPA reports that short-term exposures (*e.g.*, less than three hours) to current NO_2 concentrations may lead to changes in airway responsiveness and lung function in individuals with preexisting respiratory illnesses and increases in respiratory illnesses in children from 5 to 12 years in age [29]. The EPA also reports that long-term exposures to NO_2 may lead to increased susceptibility to respiratory infection and may cause alterations in the lung. Atmospheric transformation of NO_x can lead to the formation of ozone and nitrogen-bearing particles (most notably in some western United States urban areas), which are associated with adverse health effects [29].

Particulate Matter (PM)

Particulate matter is the general term used for a mixture of solid particles and liquid droplets found in the air. Some particles are large or dark enough to be seen as soot or smoke, while others are so small they cannot be seen with the naked eye. These small particles, which come in a wide range of sizes, originate from many different stationary and mobile sources as well as natural sources [29]. Fine particles, those less than 2.5 μm ($PM_{2.5}$), result from fuel combustion from motor vehicles, power generation, industrial facilities, and residential fireplaces and woodstoves. Coarse particles, those larger than 2.5 μm but classified as less than 10 μm (PM_{10}), are generally emitted from sources such as vehicles traveling on unpaved roads, materials handling, crushing and grinding operations, and windblown dust [29]. Some particles are emitted directly from their sources, such as smokestacks and cars. In other cases, gases such as SO_2, NO_x, and VOCs react with other compounds in the air to form fine particles.

Coal generally contains from 5 to 20 weight percent (wt.%) mineral matter (*i.e.*, ash content per a proximate analysis) [5]. During combustion, most of the minerals are transformed into dust-sized glassy particles and, along with some unaltered mineral grains and unburned carbon, are emitted from smokestacks. Particle composition and emission levels are complex functions of firing configuration, boiler operation, and coal properties [27]. In dry-bottom, pulverized coal-fired systems (various types of combustion systems are described in Chapter 5), combustion is very good, and the particles are largely composed of inorganic ash residue. In wet-bottom, pulverized coal-fired units and cyclone-fired boilers, the amount of fly ash is less than in dry-bottom units because some of the ash melts and is removed from the system as slag. Spreader stokers, which fire a mixture of fine and coarse coal, tend to have a significant quantity of unburned carbon in the fly ash. Overfed and underfed stokers emit considerably less particulate than pulverized coal-fired units or spreader stokers because combustion takes place on a relatively undisturbed bed [27]. Fly ash reinjection for increased consumption of unburned carbon or load changes can also affect particulate emissions.

Environmental Effects

Particulate matter is responsible for reduction in visibility. Visibility is principally affected by fine particles that are formed in the atmosphere from gas-phase reactions. Although these particles are not directly visible, carbon dioxide, water vapor, and ozone in increased concentrations change the absorption and transmission characteristics of the atmosphere [28]. Particulate matter can cause damage to materials depending upon its chemical composition and physical state [28]. Particles will soil painted surfaces, clothing, and curtains merely by settling on them. Particulate matter can cause corrosive damage to metals either by intrinsic corrosiveness or by the action of corrosive chemicals absorbed or adsorbed by inert particles. Little is known of the effects of particulate matter in general on vegetation [28]. The combination of particulate matter and other pollutants such as sulfur dioxide may affect plant growth. Coarse particles, such as dust, may be deposited directly onto leaf surfaces and reduce gas exchange, increase leaf surface temperature, and decrease photosynthesis. Toxic particles containing elements such as arsenic or fluorine can fall onto agricultural soils or plants that are ingested by animals and thus can affect the animal's health.

Health Effects

Particulate matter alone or in combination with other pollutants constitutes a very serious health hazard. The pollutants enter the human body mainly via the respiratory system. Inhalable particulate matter includes both fine and coarse particles. These particles can accumulate in the respiratory system and are associated with numerous health effects [29]. Exposure to coarse particles is primarily associated with the aggravation of respiratory conditions such as asthma. Fine particles are most closely associated with such health effects as increased hospital admissions and emergency room visits for heart and lung disease, increased respiratory symptoms and disease, decreased lung function, and even premature death. Sensitive groups that appear to be at greatest risk include the elderly, individuals with cardiopulmonary disease such as asthma, and children [35].

As previously mentioned, because particulate matter has been linked with adverse health effects at levels currently observed in the United States, EPRI has initiated the epidemiological study ARIES and comprehensive air quality monitoring to identify many of the components that might be associated with a particular health endpoint [30]. This is a critical undertaking, as knowledge of the true causative agents allows for better protection of public health through regulation of those sources that produce harmful pollutants, especially in light of the fact that health effects drive the regulatory agenda for proposed multipollutant legislation and air quality standards (see Chapter 4 for a discussion of such legislation).

ARIES, initiated in 1998 in Atlanta, was designed to address the issue of air pollution components by coupling an extensive air quality monitoring effort with five health studies focused on distinct endpoints to cover a wide range of possible health effects: a daily mortality study conducted by Klemm Analysis Group; an emergency room visit study conducted by Emory University; a heart rate variability study conducted by Harvard University; a cardiac arrhythmic events study conducted by Emory University; and an unscheduled physician visit study conducted by Kaiser Permanente [30]. Atlanta is the first of six cities that will be studied, with Chicago currently in the planning stages to be followed (tentatively) by Houston, Texas; Pittsburgh, Pennsylvania; a northeast city; and a western city. The major findings from the Atlanta study include the following [30]:

- Current levels of air pollution in the United States appear to be causing health effects;
- Different pollutants appear to impact different health endpoints (*e.g.*, for respiratory disease, the links are stronger for large particles and gases);
- Particulate matter components are important, but gases should not be overlooked;
- The most toxic fraction of fine particulate matter appears to be the carbon-containing fractions (organic and elemental carbon);
- There are no statistically significant associations between health impacts and sulfates and nitrates;
- Controlling the wrong pollutants will not yield health benefits.

It is generally recognized that coal-fired power plants can be important contributors to ambient fine particulate matter ($PM_{2.5}$) mass concentrations and regional haze. In 1999, coal-fired power plants emitted 1.5% of the total primary $PM_{2.5}$ in the United States [36]. In response to growing concerns over $PM_{2.5}$ being emitted into the atmosphere from coal-fired power plants, the U.S. Department of Energy, National Energy Technology Laboratory (DOE-NETL) is pursuing a major research effort to provide the applied science necessary to quantitatively relate the emissions from energy production to ambient $PM_{2.5}$ concentrations and composition at downwind receptors and to inform decision makers about management options applicable to coal-fired power generation to achieve the national standards for $PM_{2.5}$ and regional haze. Special emphasis has been given to Pittsburgh, Pennsylvania, and the surrounding upper Ohio River valley region because, on many occasions throughout the year, this region is downwind of major coal-fired power plants and other industrial sources of air emissions and upwind of the Boston–Washington corridor, the largest regional complex of urban areas in the United States.

Organic Compounds

Organic compounds are unburned gaseous combustibles that are emitted from coal-fired boilers but generally in very small amounts [37]; however, for brief periods, unburned combustible emissions may increase significantly such as during system startup or upsets. The organic emissions from pulverized coal-fired or cyclone-fired units are lower than from smaller stoker-fired boilers, where operating conditions are not as well controlled [28].

Organic emissions are due to constituents present in the coal or are formed as products of incomplete combustion. Polycyclic organic matter (POM) has also been referred to as polynuclear or polycyclic aromatic compounds (PACs). Nine major categories of POM have been identified by the EPA [38]. The most common organic compounds in the flue gas of coal-fired boilers are polycyclic aromatic hydrocarbons (PAHs). Hydrocarbons as a class are not listed as criteria pollutants, although a large number of specific hydrocarbon compounds are listed among the 188 hazardous air pollutants under Title III of the Clean Air Act Amendments of 1990 (see Chapter 4 for a discussion of regulations). Furthermore, the EPA has identified 16 PAH compounds as priority pollutants: naphthalene, acenaphthylene, acenaphthene, fluorene, phenanthrene, anthracene, fluoranthene, pyrene, chrysene, benzo[*a*]anthracene, benzo[*b*]fluoranthene, benzo[*k*]fluoranthene, benzo[*a*]pyrene, indeno[1,2,3-*c,d*]pyrene, benzo[*g,h,i*]perylene, and dibenzo[*a,h*]anthracene.

Environmental Effects

Gaseous hydrocarbons as a broad class do not appear to cause any appreciable corrosive damage to materials [28]. Of all the hydrocarbons, only ethylene has adverse effects on plants at known ambient concentrations, including inhibition of plant growth and injury to orchids and cotton.

Health Effects

Studies of the effects of ambient air concentrations of many of the gaseous hydrocarbons have not demonstrated direct adverse effects upon human health [28]. Certain airborne PAHs, however, are known carcinogens. Also, studies of the carcinogenicity of certain classes of hydrocarbons indicate that some cancers appear to be caused by exposure to aromatic hydrocarbons found in soot and tars. An extreme example is the occurrence of highly elevated incidences of lung cancer in China from PAH exposure [39]. PAHs are released during unvented coal combustion of "smoky" coal in homes, resulting in lung cancer mortality that is five times the national average in China.

Carbon Monoxide

Carbon monoxide (CO) is a colorless, odorless gas that is very stable and has a life of 2 to 4 months in the atmosphere [28]. Similar to organic compounds, it is formed when fuel is not burned completely. It is a component of motor vehicle exhaust, which contributes about 60% of all CO emissions nationwide [29]. High concentrations of CO occur in areas with heavy traffic congestion where as much as 95% of all CO emissions may come from automobile exhaust. Other sources include industrial processes, non-transportation fuel combustion, and natural sources such as wildfires. Carbon monoxide emissions from coal-fired boilers are generally low. Like organic hydrocarbon emissions, CO can be formed during system startup or upset. Also, systems with good combustion control, which is typical of power generation plants, produce little CO.

Environmental Effects

Carbon monoxide appears to have no detrimental effects on material surfaces [28]. Experiments have not shown that CO at ambient concentrations produces harmful effects on plant life. Carbon monoxide has been found to be a minor participant in photochemical reactions leading to ozone formation.

Health Effects

High concentrations of CO can cause physiological and pathological changes and ultimately death. Carbon monoxide enters the bloodstream through the lungs and reduces the delivery of oxygen to the body's organs and tissues [29]. The health threat from lower levels of CO is most serious for those who suffer from cardiovascular disease, such as angina pectoris. At much higher levels of exposure, CO can be poisonous and even healthy individuals can be affected. Visual impairment, reduced work capacity, reduced manual dexterity, poor learning ability, and difficulty in performing complex tasks are all associated with exposure to elevated CO levels.

Trace Elements

All coals contain small concentrations of trace elements. Trace elements enter the atmosphere through natural processes, and sources of trace elements include soil, seawater, and volcanic eruptions. Human activities, such as power generation and industrial and commercial sectors, also lead to emissions of some elements. Although these elements are present in small concentrations in the coal (*i.e.*, parts per million, ppm, by weight), the large amount of coal burned annually mobilizes tons of these pollutants as particles or gases.

Title III of the U.S. Clean Air Act Amendments (CAAA) of 1990 designates 188 hazardous air pollutants (HAPs). Included in the list are eleven trace elements: antimony (Sb), arsenic (As), beryllium (Be), cadmium (Cd), chromium (Cr), cobalt (Co), lead (Pb), manganese (Mn), mercury (Hg), nickel (Ni), and selenium (Se). In addition, barium (Ba) is regulated by the Resources Conservation and Recovery Act, and boron (B) and molybdenum (Mo) are regulated by Irrigation Water Standards [37]. Vanadium (V) is regulated based on its oxidation state, and vanadium pentoxide (V_2O_5) is a highly toxic regulated compound. Other elements, such as fluorine (F) and chlorine (Cl), which produce acid gases (*i.e.*, HF and HCl) upon combustion, and radionuclides such as radon (Rn), thorium (Th), and uranium (U) are also of interest.

The distribution of trace elements in the bottom ash, ash collected in the air pollution control device, and fly ash and gaseous constituents emitted into the atmosphere depends on many factors, including the volatility of the elements, temperature profiles across the system, pollution control devices, and operating conditions [27,40]. Numerous studies have shown that trace elements can be classified into three broad categories based on

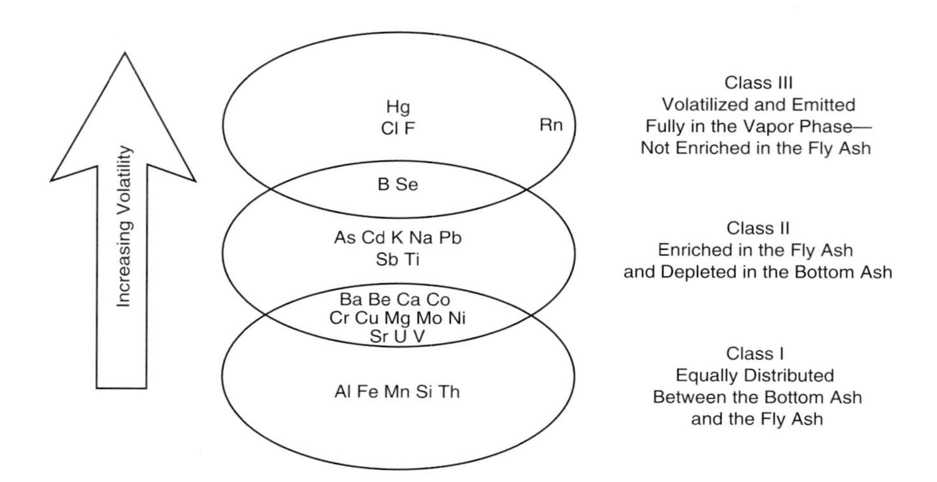

FIGURE 3-1. Classification scheme for selected trace elements relative to their volatility and partitioning in power plants. (Adapted from Miller *et al.* [37] and Clarke and Sloss [40].)

their partitioning during coal combustion. A summary of these studies is presented by Clarke and Sloss [40], and Figure 3-1 illustrates the classification scheme for selected elements.

Class I elements are the least volatile and are concentrated in the coarse residues (*i.e.*, bottom ash) or are equally divided between coarse residues and finer particles (*i.e.*, fly ash). Class II elements will volatilize in the boiler but condense downstream and are concentrated in the finer-sized particles. Class III elements are the most volatile and exist entirely in the vapor phase. Overlap between classifications exists and is a function of fuel, combustion system design, and operating conditions, especially temperature [40].

Environmental Effects

The environmental effects of trace elements are a function of the chemical and physical form in which they are found [40]. Environmental effects may occur due to the element itself or as a result of a combination of the element and other compounds. Linking a specific environmental effect to an individual element is difficult as is determining the contribution from human activities because the trace elements also occur naturally.

Some trace elements may have an immediate effect in the atmosphere. Trace element metals such as Mn(II) and Fe(III) may contribute to acid rain by promoting oxidation of sulfur dioxide to sulfate in water droplets [40]. Trace elements may also be involved in the complex atmospheric chemistry that forms photochemical smog and may affect cloud formation.

Soils may contain high concentrations of certain trace elements due to natural minerals and ores. In addition, deposition of trace elements downwind from power plants can lead to high concentrations in the soils and uptake by plants. Some elements found in coal are major plant nutrients—specifically, calcium, magnesium, and potassium [40]. These elements are not considered trace elements because they occur in quantities greater than 1000 ppm (>0.1% by weight). Other elements (both major and trace) are considered minor plant nutrients, such as iron, manganese, copper, zinc, molybdenum, cobalt, and selenium. Elements such as aluminum, sodium, and vanadium are considered essential for some species while others are potentially toxic including chromium, nickel, lead, arsenic, and cadmium.

Cadmium Cadmium (Cd) is a silvery metal and is both toxic and carcinogenic [41]. The correlation between soil concentration of cadmium and plant uptake is not clear; consequently, there is concern about cadmium concentrations in the environment and ingestion of plant-based foods by the general population.

Mercury Mercury (Hg) is a liquid, silvery metal that is considered toxic; in the form of methyl mercury, it is extremely toxic [35,41]. The mercury directly emitted from power plants is measured as three forms: elemental (Hg°), oxidized (Hg^{+2}), and condensed on ash particles (Hg_p). In the natural environment, mercury can go through a series of chemical transformations to convert it to a highly toxic form, methylmercury (CH_3Hg), which is concentrated in fish and birds [42]. Methylation rates in the ecosystems are a function of mercury availability, bacterial population, nutrient load, acidity and oxidizing conditions, sediment load, and sedimentation rates. Methylmercury enters the food chain, particularly in aquatic organisms, and bioaccumulates. Volatile elements that are emitted from power plants, such as mercury, are mostly found either in gaseous form or enriched on the surface of fine particles and physically should be available for uptake by plants [40]. There is evidence, however, that almost no mercury from the soil is taken into the shoots of the plants; hence, plants appear to be an important barrier against entry of mercury into the above-ground ecosystem, even if accumulation in the soil has occurred.

Lead Lead (Pb) is the only metal currently listed as a criteria pollutant. Lead is a gray metal with a low melting point. It is soft, malleable, ductile, resistant to corrosion, and a relatively poor electrical conductor [28]. For these reasons, lead has been used for over 4000 years for plates and cups, food storage vessels, paints, piping, roofing, storage containers for corrosive materials, radiation shields, lead-acid batteries, and as an organolead additive in gasoline. As a result, lead can be found throughout the world, including trace amounts in Antarctica and the Arctic [28]. In the past, automotive sources were the major contributor of lead emissions to the atmosphere [29]. Due to the EPA's regulatory efforts to remove lead from gasoline, along with banning lead from paint pigments and solder, a decline in lead emissions has been observed. The highest concentrations of lead are found in the vicinity of nonferrous and ferrous smelters and battery manufacturers. Lead can be deposited on the leaves of plants, presenting a hazard to grazing animals.

Selenium Crops, including animal forages are sensitive to the addition of small amounts of selenium (Se) in the soil [40]. Selenium is a silvery metallic allotrope or red amorphous powder [41] and is toxic to plants at low concentrations. It has been shown to cause stunting and brown spots in some varieties of beans and to reduce germination and cause stunting in cereals and cotton.

Other Trace Elements Clarke and Sloss [40] listed examples where trace elements from coal combustion may have beneficial effects in some areas. Boron is a micronutrient that is required in trace amounts by many plants and animals, and the amount of boron released from coal-fired power plants is likely to be beneficial to local agriculture. Copper, iron, manganese, and zinc are necessary for normal growth of plants. It is recognized, however, that excessive concentrations of copper and zinc lead to damage to root formation and growth, but the quantities being deposited around coal-fired power stations are likely to be beneficial to local soils. Also, manganese can be harmful in large doses, especially in acidic soils.

Health Effects

Trace element emissions have the potential to cause a number of harmful effects on human health. While there is no evidence that most trace elements from coal-fired power plants are causing health effects at their low ambient air concentration, there is concern that pollutants may accumulate throughout the food chain. This is especially true of mercury, which is discussed in more detail later in this section.

Arsenic The combustion of most coals is unlikely to contribute toxic amounts of arsenic (As) to air [43]. There is some concern, however, that arsenic in fly ash disposal and coal cleaning wastes may be leached into the groundwater. Arsenic, which is gray, metallic, soft, and brittle, may be considered essential; however, it is toxic in small doses [41]. Arsenic can cause anemia, gastric disturbance, renal symptoms, ulceration, and skin and lung cancer [40]. In addition, arsenic can damage peripheral nerves and blood vessels and is a suspected teratogen (*i.e.*, causes damage to embryos and fetuses). The chemical form of arsenic can affect its toxicity, with organic forms of arsenic being more toxic than elemental arsenic. An extreme example of chronic arsenic poisoning is occurring in the Guizhou Province of China [39,44]. In this province, the villagers bring their chili pepper harvest indoors in the autumn to dry. They hang their peppers over open-burning stoves, where arsenic-rich coal (up to 35,000 ppm) is used to heat and cook. These chili peppers, which normally contain <1 ppm arsenic, can contain as much as 500 ppm arsenic after drying. About 3000 people are exhibiting typical symptoms of arsenic poisoning, including hyperpigmentation (flushed appearance, freckles), hyperkeratosis (scaly lesions on the skin, generally concentrated on the hands and feet), Bowen's disease (dark, horny, precancerous lesions of the skin), and squamous cell carcinoma.

Boron Boron (B) is similar to arsenic in that the concentration emitted to the atmosphere is small and is unlikely to cause problems as an airborne pollutant [43]; however, boron in the fly ash can become soluble in ash disposal sites. Boron, a dark powder, is essential for plants but can be toxic in excess [41].

Beryllium Beryllium (Be), a silvery, lustrous, relatively soft metal, is toxic and carcinogenic [41]. It can cause respiratory disease and lymphatic, liver, spleen, and kidney effects [40].

Cadmium Cadmium (Cd) has no known biological function and is therefore not a nutritional requirement [40]. Cadmium is toxic, carcinogenic, and teratogenic [41] and can cause emphysema, fibrosis of the lung, renal injury, and possibly cardiovascular disease.

Chromium Chromium (Cr), a hard, blue-white metal, is an essential trace element; however, its chromates are toxic and carcinogenic [41]. Chromium, which is ingested by humans through food and drink, can be toxic when it accumulates in the liver and spleen [40]. The oxidation state of chromium affects its mobility and toxicity. Chromium (III) is nontoxic and has a tendency to absorb to clays, sediments, and organic matter and therefore is not very mobile. Chromium (IV), however, is more mobile and toxic and may account for ~5% of the total chromium particles emitted from power plants.

Fluorine Fluorine (F) is a pale yellow gas and is the most reactive of all elements [41]. It is an essential element and is commonly used for protecting the enamel of teeth, but excess fluoride is toxic. Fluorosis includes mottling of tooth enamel (dental fluorosis) and various forms of skeletal damage, including osteosclerosis, limited movement of the joints, and outward manifestations such as knock-knees, bow legs, and spinal curvature [39]. Fluorosis, combined with nutritional deficiencies in children, can result in severe bone deformation. An extreme example of this is exhibited in China, where the health problems caused by fluorine volatilized during domestic coal use are far more extensive than those caused by arsenic [39,44].

More than 10 million people in the Guizhou Province and surrounding areas suffer from various forms of fluorosis caused by corn being dried over unvented ovens burning high-fluorine (>200 ppm) coal.

Mercury Mercury (Hg) exists in trace amounts in fossil fuels, including coal, vegetation, crustal material, and waste products. Through combustion or natural processes, mercury vapor can be released to the atmosphere, where it can drift for a year or more, spreading with air currents over vast regions of the world [45]. Research indicates that mercury poses adverse human health effects, and fish consumption is the primary pathway for human and wildlife exposure. Mercury bioaccumulates in fish as methylmercury (CH_3Hg) and poses a serious health hazard for humans. Other research suggests that other forms of mercury may be harmful as well [46,47]. Ingested mercury in elemental, organic, and inorganic form is converted to mercuric mercury, which is slowly eliminated from the kidneys but remains fixed in the brain indefinitely [47]. Exposure to high levels of metallic, inorganic, or organic mercury can permanently damage the brain, kidneys, and developing fetus [46]. Documented associations have been reported between low-dose, prenatal exposure to methylmercury and neurodevelopmental effects on attention, motor function, language, visual-spatial, and verbal abilities [48]. Loss of sight has been associated with cases of extreme mercury ingestion [39]. Chronic thallium poisoning has been reported in the Guizhou Province in China, where vegetables are grown on mercury-/thallium-rich mining slag. Most symptoms that have been reported, such as hair loss, are typical of thallium poisoning; however, many patients from this region have lost their vision, which is being attributed to mercury poisoning as the mercury concentration of this coal is 55 ppm, or about 200 times the average mercury concentration in U.S. coals.

Manganese Manganese (Mn) is unlikely to cause health problems as an airborne pollutant from the combustion of most coals, but leaching of ash may be a concern [43]. Manganese, a hard, brittle, silvery metal, is considered an essential nutrient, is nontoxic, and is a suspected carcinogen [41]. It is also reported to cause respiratory problems [40].

Molybdenum Molybdenum (Mo), a lustrous, silvery, and fairly soft metal, is an essential nutrient, is moderately toxic, and is a teratogen [41]. Under some conditions, molybdenosis may occur in animals, notably ruminants, due to consumption of vegetation with relatively high concentrations of molybdenum [43].

Nickel Nickel (Ni), a silvery, lustrous, malleable, and ductile metal, has no known biological role, and nickel and nickel oxide are carcinogenic [41]. Nickel can cause dermatitis and intestinal disorders [40].

Lead Exposure to lead (Pb) occurs mainly through inhalation of air or ingestion in food, water, soil, or dust. It accumulates in the blood, bones, and soft tissues [29]. Lead can adversely affect the kidneys, liver, nervous system, and other organs. Excessive exposure to lead may cause neurological impairments, such as seizures, mental retardation, and behavioral disorders. Even at low doses, lead exposure is associated with damage to the nervous system of fetuses and young children. Lead may be a factor in high blood pressure and heart disease.

Selenium Selenium (Se) is considered an essential element but is toxic in excess of dietary requirements and is also a carcinogen [41]. Livestock consuming plants with excessive amounts of selenium can suffer two diseases, alkali disease or blind staggers, and can experience infertility and cirrhosis of the liver, as well as death in extreme cases [40]. In humans, selenium can cause gastrointestinal disturbance, liver and spleen damage, and anemia and is a suspected teratogen. Symptoms of selenium poisoning include hair and nail loss. Selenosis has been reported in southwest China, where selenium-rich carbonaceous shales, known locally as "stone coals," are used for home heating and cooking [39]. The ash from this selenium-rich coal (as much as 8400 ppm) is then used as a soil amendment, thereby introducing high concentrations of selenium into the soil that is subsequently taken up by crops.

Vanadium Vanadium (V), a shiny, silvery, soft metal, is an essential trace element, although some compounds, specifically vanadium pentoxide (V_2O_5), are quite toxic [41]. Health effects associated with vanadium include acute and chronic respiratory dysfunction [40].

Radionuclides Radionuclides are listed generically as 1990 CAAA HAPs. Radioactivity arises mainly from isotopes of lead, radium, radon, thorium, and uranium [37,40,43,49]. Health effects from radiation are well documented and include various forms of cancer; however, radionuclide emissions from power plants are quite low [37,40]. During coal combustion, most of the uranium and thorium and their decay products are released from the original coal matrix and are distributed between the gas phase and solid combustion products [49]. Virtually 100% of the radon gas present in the coal feed is transferred to the gas phase and is lost in stack emissions. In contrast, less volatile elements, such as thorium and uranium, and the majority of their decay products are retained in the solid combustion wastes [37,49]. Fly ash is commonly used as an additive to concrete building products, but the radioactivity of typical fly ash is not significantly different from that of more conventional concrete additives or other building materials such as granite or red brick [49].

Greenhouse Gases: Carbon Dioxide

The Earth naturally absorbs and reflects incoming solar radiation and emits longer wavelength terrestrial (thermal) radiation back into space [50]. On average, the absorbed solar radiation is balanced by the outgoing terrestrial radiation emitted to space (see Figure 3-2). A portion of this terrestrial radiation, though, is absorbed by gases in the atmosphere. These gases, known as greenhouse gases, have molecules that have the right size and shape to absorb and retain heat. These gases include water vapor (H_2O), carbon dioxide (CO_2), methane (CH_4), nitrous oxide (N_2O), and, to a lesser extent, halocarbons consisting of hydrochlorofluorocarbons, perfluorocarbons, and sulfur hexafluoride (SF_6). The energy from this absorbed terrestrial radiation warms the Earth's surface and atmosphere, creating what is known as the natural greenhouse effect, which makes the Earth inhabitable. Without the natural heat-trapping properties of these atmospheric gases, the average surface temperature of the Earth would be about 60°F lower.

Although the Earth's atmosphere consists of mainly oxygen and nitrogen (*i.e.*, over 99%), neither plays a significant role in enhancing the greenhouse effect because both are essentially transparent to terrestrial radiation [50]. The greenhouse gases comprise the remaining ~1% of the atmosphere, of which over 97% is water vapor. Methane, carbon dioxide,

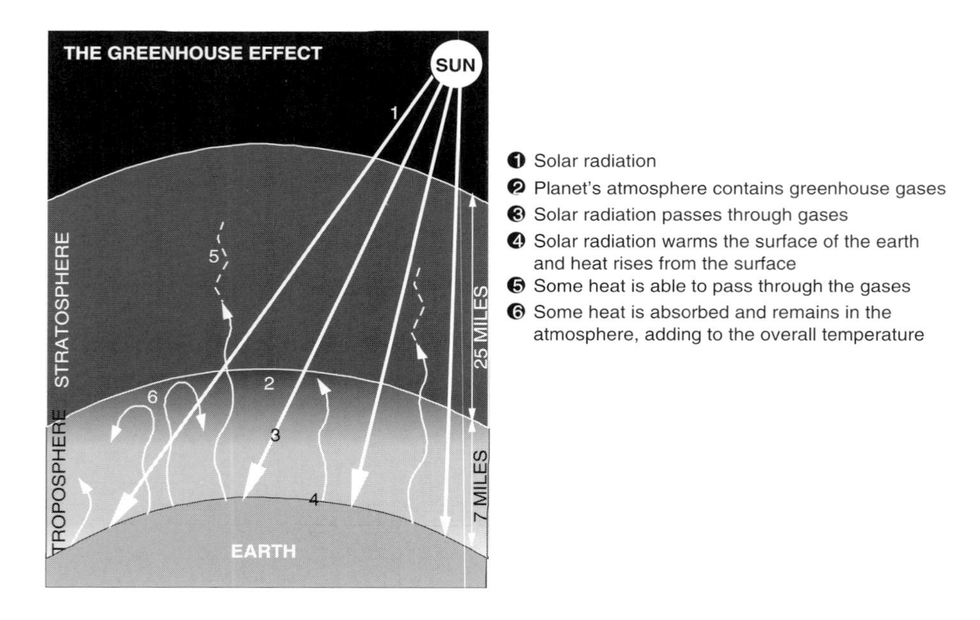

FIGURE 3-2. The greenhouse effect. (Adapted from DOE [51] and Peabody Holding Co. [52].)

nitrous oxide, and other gases comprise less than 3% of the greenhouse gases [52]. The composition of the greenhouse gases is [51]:

- Energy-related CO_2—81%;
- Methane (CH_4)—9%;
- Nitrous oxide (N_2O)—6%;
- Other CO_2—2%;
- Other gases—2%.

Carbon dioxide, methane, and nitrous oxide are continuously emitted to and removed from the atmosphere by natural processes on Earth. Anthropogenic activities, however, can cause additional quantities of these and other greenhouse gases to be emitted or sequestered, thereby changing their global average atmospheric concentrations. Natural activities such as respiration by plants or animals and seasonal cycles of plant growth and decay generally do not alter average atmospheric greenhouse gas concentrations over decadal time frames [50]. Climatic changes, which are long-term fluctuations in temperature, precipitation, wind, and other elements of the Earth's climate system, that result from anthropogenic activities can have positive or negative feedback effects on these natural systems.

Overall, the most abundant and dominant greenhouse gas in the atmosphere is water vapor. Human activities, however, are not believed to directly affect the average global concentration of water vapor, although this issue is currently being debated [50,53]. In nature, carbon dioxide is cycled between various atmospheric, oceanic, land biotic, marine biotic, and mineral reservoirs. Of all the greenhouse gases, human activity has the largest influence on carbon dioxide, which is a product of the combustion of fossil fuels. Carbon dioxide concentrations in the atmosphere increased from approximately 280 ppmv in pre-industrial times to 367 ppmv in 1991 [50].

Methane is primarily produced through anaerobic decomposition of organic matter in biological systems. Agricultural processes, such as wetland rice cultivation, enteric fermentation in animals, and the decomposition of animal wastes, emit methane, as does the decomposition of municipal solid wastes [50]. Methane is also emitted during the production and distribution of natural gas and petroleum. Methane is released as a by-product of coal mining and, to a lesser extent, incomplete fossil fuel combustion.

Anthropogenic sources of nitrous oxide include agricultural soils, due, particularly, to the use of fertilizers; fossil fuel combustion, especially from mobile sources; nylon and nitric acid production; wastewater treatment; waste combustion; and biomass burning [50]. Halocarbons that contain chlorine (*e.g.*, chlorofluorocarbons, hydrofluorocarbons, methyl chloroform, and carbon tetrachloride) and bromine (*e.g.*, halons, methyl bromide, and hydrobromofluorocarbons), perfluorocarbons, and sulfur hexafluoride (SF_6) are manmade chemicals and not products of combustion. They are, however, powerful greenhouse gases.

The concept of global warming potentials (GWPs) has been developed to evaluate the relative effects of emissions over a given time period in the future [54]. GWPs take into account the differing times that gases remain in the atmosphere, their greenhouse effect while in the atmosphere, and the time period over which climatic changes are of concern. GWPs are intended as a quantified measure of the globally averaged relative radiative forcing impacts of a particular greenhouse gas [50]. It is defined as the cumulative radiative forcing—both direct and indirect effects—integrated over a period of time from the emission of a unit mass of gas relative to some reference gas. Table 3-3 summarizes the greenhouse gases, their major anthropogenic sources, and their GWPs [50]. Carbon dioxide has been chosen as the reference gas; as an example, methane's GWP is 21, which means that methane is 21 times better at trapping heat in the atmosphere than carbon dioxide. GWPs are typically reported on a 100-year time horizon. The EPA uses a time period of 100 years for policy making and reporting purposes.

TABLE 3-3
Greenhouse Gases, Major Anthropogenic Sources, and Global Warming Potentials

Greenhouse Gas	*Major Anthropogenic Sources*	*Global Warming Potential*
Carbon dioxide (CO_2)	Fossil fuel combustion, iron and steel production, cement manufacture	1
Methane (CH_4)	Landfills, enteric fermentation, natural gas systems, coal mining, manure management	21
Nitrous oxide (N_2O)	Agriculture soil management, mobile sources, nitric and adipic acid production, manure management, stationary sources, human sewage	310
HFC-23	Substitution of ozone-depleting substances, semiconductor manufacture, mobile air conditioners, hydrochlorofluorocarbon (HCFC)-22 production	11,700
HFC-125		2800
HFC-134a		1300
HFC-143a		3800
HFC-152a		140
HFC-227		2900
HFC-236fa		6300
HFC-431mee		1300
CF_4	Substitution of ozone-depleting substances, semiconductor manufacture, aluminum production	6500
C_2F_6		9200
C_4F_{10}		7000
C_6F_{14}		7400
SF_6	Electrical transmission and distribution systems, magnesium casting	23,900

The greenhouse gas emissions of direct greenhouse gases in the U.S. inventory are reported in terms of equivalent emissions of carbon dioxide, using teragrams of carbon dioxide equivalents (Tg CO_2 Eq.). The relationship between gigagrams (Gg) of a gas and Tg CO_2 Eq. can expressed as follows:

$$\text{Tg } CO_2 \text{ Eq.} = (\text{Gg of gas}) \times (\text{GWP}) \times \left(\frac{\text{Tg}}{1000 \text{ Gg}}\right) \qquad (3\text{-}20)$$

Quantities of emissions are provided in Chapter 4.

Environmental Effects

Much uncertainty surrounds global climate changes. The global climate is a massive and highly complex system with many interrelated subsystems. Evidence indicates that there have been times in Earth's history when the concentrations of greenhouse gases in the atmosphere have been higher and lower than today, but it is difficult to determine the causes and effects of those situations, and the data from those time periods are limited and imprecise [53].

One important cause for uncertainty in the area of global science lies in feedback loops. Complex climate change models have been developed, but the various modelers consider different feedbacks to be more important, so markedly different predictions regarding the Earth's climate are obtained [53]; however, many of the undisputed facts have not changed over the last decade. For example, based on samples of air trapped in arctic ice, scientists have determined that, prior to the Industrial Revolution, the concentration of carbon dioxide in the atmosphere had been stable at a level of around 280 to 290 ppmv. When people started to burn fossil fuels, the concentration of carbon dioxide began to increase and is now at approximately 370 ppmv [53]. This correlation indicates that increased concentrations of greenhouse gases in the atmosphere have likely increased the amount of heat from the sun that stays within the Earth's ecosystem, thereby contributing to increased global temperatures. The Goddard Institute for Space Studies has reported that the average temperature at the Earth's surface has risen approximately 2°F from 1870 to 1998 [55]. Studies project that globally averaged surface temperatures will increase by 2.5 to 10.4°F between 1990 and 2100 at current rates of increase under a business-as-usual scenario [53].

Potential environmental impacts of global warming include effects on agriculture production, forests, water resources, coastal areas, species, and natural areas [56,57]. Warmer temperatures can lead to more intense rainfall and flooding in some areas (*e.g.*, U.S. Pacific Northwest and Midwest), with more frequent drought-like conditions in other areas such as the western United States. Predictions of the agricultural effects of climate change remain uncertain, but models indicate potential changes in cereal grain production and irrigation demands in the United States. Models also predict that rising temperatures could affect current land use such as the reduction of coffee growing areas in countries such as Uganda. Predictions indicate forest areas could be affected. As temperatures increase, the forest composition could change, with warmer climate varieties moving into traditionally colder climate areas. Forest health and productivity could be impacted.

The effect of global warming on freshwater resources is uncertain. Some studies indicate that global water conditions will worsen, while others suggest that climate change could have a net positive impact on global water resources. Impacts on water supply, water quality, and competition for water are predicted. Coastal areas are predicted to be subject to erosion of beaches and inundation of coastal lands, as well as increased costs to protect coastal communities. Current rates of sea-level rise are expected to increase 2 to 5 times due to both thermal expansion of the oceans and the partial melting of mountain glaciers and polar ice caps. Low-lying areas along the U.S. Gulf of Mexico and estuaries such as Chesapeake Bay are especially vulnerable. Bangladesh is the country most vulnerable to sea-level rise, while the Nile Delta, Egypt's only suitable agriculture area, would be flooded.

Changes in temperature and water availability could decimate non-intensively managed ecosystems, such as forests, rivers, and wetlands. For example, global warming could dry out the wetlands that support over 50% of North American waterfowl. Arctic and northern latitudes are likely to experience above-average warming and are especially vulnerable to its effects, including thinning of the arctic sea ice, and this disruption may affect fisheries, human structures built on permafrost, and northern ecosystems.

Health Effects

Human health is also predicted to be impacted by global climate change [56]. This includes increases in weather-related mortality, infectious diseases, and air-quality respiratory illnesses. Small increases in average temperatures can increase the spread of diseases, such as malaria and dengue fever, and lead to a significant rise in the number of extreme heat waves. Elderly people are particularly vulnerable to heat stress. Heat waves could also aggravate local air quality problems, which pose threats to young children and individuals with asthma.

References

1. Sengupta, M., *Environmental Impacts of Mining: Monitoring, Restoration, and Control* (Lewis Publishers, Boca Raton, FL, 1993).
2. National Mining Association, *Most Requested Statistics: U.S. Coal Industry* (National Mining Association, Washington, D.C., April 2003).
3. World Book Encyclopedia, *Coal* (World Book, Chicago, IL, 2001), Vol. 4, pp. 716–733.
4. Schobert, H. H., *Coal: The Energy Source of the Past and Future* (American Chemical Society, Washington, D.C., 1987).

5. Marcus, J. J (editor), *Mining Environmental Handbook: Effects of Mining on the Environment and American Environmental Controls on Mining* (Imperial College Press, London, 1997).
6. Irving, W. and O. Tailakov, CH_4 Emissions: Coal Mining and Handling, Good Practice Guidance and Uncertainty Management in National Greenhouse Gas Inventories, 1999, accessed from www.ipcc-nggip.iges.or.jp/public/gp/bgp/2_7_Coal_Mining_Handling.pdf, June 2003.
7. EPA, *International Anthropogenic Methane Emissions in the United States* (Office of Policy, Planning and Evaluation, U.S. Environmental Protection Agency, U.S. Government Printing Office, Washington, D.C., 1994).
8. EPA, *Coalbed Methane Outreach Program* (U.S. Environmental Protection Agency, Washington, D.C.), www.epa.gov/coalbed/about.htm#section2 (accessed March 3, 2003).
9. World Bank Group, *Coal Mining and Production, Pollution Prevention, and Abatement Handbook*, www.natural-resources.org.minerals/CD/docs/twb/PPAH/52.coal.pdf, pp. 282–285 (accessed July 1998).
10. USGS, *Coal-Extraction: Environmental Prediction* (USGS Fact Sheet FS-073-02, U.S. Geological Survey, U.S. Government Printing Office, Washington, D.C., August 2002).
11. PADEP, *The Science of Acid Mine Drainage and Passive Treatment* (Pennsylvania Department of Environmental Protection, Harrisburg), www.dep.state.pa.us/dep/deputate/minres/bamr/amd/science_of_amd.htm (accessed April 5, 2002).
12. OSM, *Factors Controlling Acid Mine Drainage Formation* (U.S. Office of Surface Mining, Washington, D.C.), www.osmre.gov/amdform.htm (accessed March 5, 2002).
13. Coalition for Affordable and Reliable Energy (CARE), www.careenergy.com, 2003.
14. Mine Safety and Health Administration (MSHA), www.msha.gov, 2003.
15. EIA, *Coal Data: A Reference* (Energy Information Administration, U.S. Department of Energy, Washington, D.C., February 1995).
16. ACF, *Strict Regulations Govern Coal Mining* (American Coal Foundation, Washington, D.C.), www.ket.org/trips/Coal/AGSMM/agsmmregs.html, 2001.
17. Jackson, L. J., *Surface Coal Mines: Restoration and Rehabilitation* (IEA Coal Research, London, 1991).
18. OSM, *Chronology of the Office of Surface Mining and the Surface Mining Law Implementation* (U.S. Office of Surface Mining, Washington, D.C.), www.osmre.gov (accessed February 13, 2003).
19. OSM, *Regulating Surface Coal Mining: Restoring Mining Landscapes* (U.S. Office of Surface Mining), www.osmre.gov/news/101702.txt (accessed June 2003).
20. Leonard, J. W (editor), *Coal Preparation*, Fourth ed. (The American Institute of Mining, Metallurgical, and Petroleum Engineers, New York, 1979).
21. National Research Council, *Coal Waste Impoundments: Risks, Responses, and Alternatives* (National Academy Press, Washington, D.C., 2002).

22. Falcone Miller, S., J. L. Morrison, and A. W. Scaroni, The Utilization of Coal Pond Fines as Feedstock for Coal-Water Slurry Fuels, in *Proc. of the 20th International Technical Conference on Coal Utilization & Fuel Systems* (Coal and Slurry Technology Association, Washington, D.C., 1995), pp. 535–558.
23. Chadwick, M. J., N. H. Highton, and N. Lindman, *Environmental Impacts of Coal Mining and Utilization* (Pergamon Press/The Beijer Institute, Elmsford, NY, 1987), pp. 73–80.
24. Kim, A. G., W. Aljoe, and S. Renninger, Wastes from the Combustion of Fossils Fuels: Research Perspective on the Regulatory Determination, in *Proc. of the 16th International Conference on Fluidized Bed Combustion* (Council of Industrial Boiler Owners, Washington, D.C., 2001).
25. Kalyoncu, R., Coal Combustion Products, in *Proc. of the 19th Annual International Pittsburgh Coal Conference: Coal Energy and the Environment* (University of Pittsburgh, 2002).
26. EPA, *Report to Congress: Wastes from the Combustion of Fossil Fuels*, Vol. 2, *Methods, Findings, and Recommendations* (U.S. Environmental Protection Agency, U.S. Government Printing Office, Washington, D.C., March 1999), chap. 3.
27. Davis, W. T. (editor), *Air Pollution Engineering Manual*, Second ed. (John Wiley & Sons, New York, 2000), p. 9.
28. Wark, K., C. F. Warner, and W. T. Davis, *Air Pollution: Its Origin and Control*, Third ed. (Addison-Wesley Longman, Menlo Park, CA, 1998).
29. EPA, *Latest Findings on National Air Quality: 1997 Status and Trends* (U.S. Environmental Protection Agency, Office of Air Quality Planning and Standards, Washington, D.C., 1998).
30. EPRI, *Air Pollution and Health Effects Research at EPRI: The ARIES Program, Strategic Overview Fact Sheet* (Electric Power Research Institute, Hillview, CA, December 2002).
31. Wall, T. F., *Principles of Combustion Engineering for Boilers* (Harcourt Brace Javanovich, London, 1987), pp. 197–294.
32. Singer, J. G. (editor), *Combustion: Fossil Power Systems* (Combustion Engineering, Windsor, CT, 1981).
33. Miller, J. A., and C. T. Bowman, Mechanism and Modeling of Nitrogen Chemistry in Combustion, *Progress Energy Combustion Science*, Vol. 15, 1989, pp. 287–338.
34. Wood, S. C., Select the Right NO_X Control Technology, *Chemical Engineering Progress*, Vol. 90, No. 1, 1994, pp. 32–38.
35. EPA, *Mercury Study Report to Congress* (U.S. Environmental Protection Agency, Office of Air Quality Planning and Standards, U.S. Government Printing Office, Washington, D.C., December 1997).
36. Aljoe, W. W. and T. J. Grahame, The DOE-NETL Air Quality Research Program: Airborne Fine Particulate Matter (PM2.5), in *Proc. of the Conference on Air Quality III: Mercury, Trace Elements, and Particulate Matter* (University of North Dakota, Grand Forks, 2002).
37. Miller, S. J., S. R. Ness, G. F. Weber, T. A. Erickson, D. J. Hassett, S. B. Hawthorne, K. A. Katrinak, and P. K. K. Louie, *A Comprehensive Assessment*

of Toxic Emissions from Coal-Fired Power Plants: Phase I Results from the U.S. Department of Energy Study Final Report, 1996, Contract No. DE-FC2I-93MC30097 (Subtask 2.3.3).

38. Brooks, A. W., *Estimating Air Toxic Emissions from Coal and Oil Combustion Sources* (Report No. EPA-450/2-89-001, U.S. Environmental Protection Agency, Research Triangle Park, NC, 1989).
39. Finkelman, R. B., W. Orem, V. Castranova, C. A. Tatu, H. E. Belkin, B. Zheng, H. E. Lerch, S. V. Maharaj, and A. L. Bates, Health Impacts of Coal and Coal Use: Possible Solution, *International Journal of Coal Geology*, Vol. 50, 2002, pp. 425–443.
40. Clarke, L. E. and L. L. Sloss, *Trace Elements: Emissions from Coal Combustion and Gasification* (IEA Coal Research, London, 1992).
41. Emsley, J., *The Elements* (Clarendon Press, Oxford, 1989).
42. USGS, *Mercury in U.S. Coal: Abundance, Distribution, and Modes of Occurrence* (USGS Fact Sheet FS-095-01, U.S. Geological Survey, U.S. Government Printing Office, Washington, D.C., September 2001).
43. Swaine, D. J. and F. Goodarzi (editors), *Environmental Aspects of Trace Elements in Coal* (Kluwer Academic Publishers, Dordrecht, The Netherlands, 1995).
44. Finkelman, R. B., H. C. W. Skinner, G. S. Plumlee, and J. E. Bunnell, *Medical Geology* (Geosciences and Human Health series, American Geological Institute, Alexandria, VA, 2001), pp. 20–23.
45. Feeley, T. J., J. Murphy, J. Hoffmann, and S. A. Renninger, *A Review of DOE/NETL's Mercury Control Technology R&D Program for Coal-Fired Power Plants* (U.S. Department of Energy, Pittsburgh, PA, 2003).
46. CERHR, *Mercury* (Center for the Evaluation of Risks to Human Reproduction, Research Triangle Park, NC) http://cerhr.niehs.nih.gov/genpub/topics/mercury2-ccae.html, May 16, 2002.
47. Aposhian, H. V. and M. M. Aposhian, Elemental, Mercuric, and methylmercury: Biological Interactions and Dilemmas, in *Proc. of the Air Quality II: Mercury, Trace Elements, and Particulate Matter Conference* (University of North Dakota, Grand Forks, 2000).
48. National Research Council, *Toxicological Effects of Methyl Mercury* (National Academy Press, Washington, D.C., 2000).
49. USGS, *Radioactive Elements in Coal and Fly Ash: Abundance, Forms, and Environmental Significance* (USGS Fact Sheet FS-163-97, U.S. Geological Survey, U.S. Government Printing Office, Washington, D.C., 1997).
50. EPA, *Greenhouse Gases and Global Warming Potential Values* (U.S. Greenhouse Gas Inventory Program Office of Atmospheric Programs, U.S. Environmental Protection Agency, U.S. Government Printing Office, Washington, D.C., April 2002).
51. DOE, *Flare for Fuel: Environmental Solutions* (U.S. Department of Energy, U.S. Government Printing Office, Washington, D.C., Fall 2000).
52. Peabody Holding Co., *Clearing the Air about the Greenhouse Effects* (Peabody Holding Co., St. Louis, MO), www.peabodyenergy.com/index-ie.html (accessed November 1997).

53. DOE, *Greenhouse Gases and Global Climate Change: Frequently Asked Questions* (U.S. Department of Energy, Washington, D.C.), www.netl.doe.gov/coalpower/sequestration/index.html (accessed June 27, 2003).
54. Smith, I. M., C. Nilsson, and D. M. B. Adams, *Greenhouse Gases: Perspectives on Coal* (IEA Coal Research, London, 1994), p. 10.
55. Anon., Technology to Cool Down Global Warming, *Chemical Engineering*, Vol. 106, No. 1, January 1999.
56. NRDC, *Global Warming: In Depth Testimony* (National Resources Defense Council, New York), www.nrdc.org/globalwarming/tdl0300.asp (accessed March 20, 2000).
57. Climateark, *Potential Climate Changes Impact* (National Energy Technology Laboratory, Fall 2000), www.climateark.org/vital/20.htm (accessed June 27, 2003).

CHAPTER 4

Coal-Fired Emissions and Legislative Action in the United States

The use of coal has a long history of emitting a variety of pollutants into the atmosphere and has contributed to several known health episodes in the past. Regulations on coal usage date back to medieval times when nobility became inconvenienced, but major incidents in the 1940s and 1950s in the United States and England that had severe impacts on human health were the impetus for legislation in these two countries to protect the health and welfare of the general public. In the United States, the major development of legislative and regulatory acts occurred from 1955 to 1970, and by the mid-1970s the basis for national regulation of air pollution was well developed. The regulations are continually changing, as more information on the effect of emissions on health and the environment is obtained, new control technologies are developed, and society demands a safe living environment. The use of coal for power generation is a highly regulated industry with more regulations soon to be implemented. This chapter briefly discusses past health episodes, primarily in light of their role in the passage of regulations on the use of coal. The emphasis of this chapter is on federal legislation and regulatory trends in the United States. A history of legislative action in the United States as it pertains to coal-fired power plants is presented, and impending legislation regarding emissions not currently regulated is also discussed. The types and quantities of emissions, with an emphasis on coal-fired power plants, are presented, along with their trends over time as a consequence of legislative action. A brief discussion of emissions and legislation from other countries and how they compare to the United States is also provided.

Major Coal-Related Health Episodes

Air pollution is not a modern phenomenon, although legislative action to successfully regulate emissions and protect human health is. This section discusses the major health episodes caused or contributed to by the use of coal, especially those episodes that have raised public awareness of the ill effects of air pollution on health and have led to air pollution regulations, specifically with respect to the use of coal.

Pre-Industrial Revolution

Documentation of air pollution begins as early as ancient Rome when the statesman Seneca complained about the "stink, soot and heavy air" in the city [1]. Coal usage increased in England during medieval times because coal became less expensive and wood became scarce, and air pollution problems intensified, particularly black smoke and fumes. It became so bad that in 1257 Queen Eleanor was driven from Nottingham Castle by the smoke and fumes rising from the city below [2]. In 1283 and 1288, there were complaints about air quality in London because coal, which was used by the various smithies and the general public for home heating, was now being used in lime kilns. By 1285, London's air was so polluted that King Edward I established the world's first air pollution commission, and 22 years later the king made it illegal to burn coal [1]. This Royal Proclamation in 1307 forbade the use of coal in lime burners in parts of South London; however, this proclamation did not work, and a later commission had instructions to punish offenders with fines and ransoms for a first offense and to demolish their furnaces for a second offense [2]. Eventually, economics (high-priced wood versus low-priced coal) and a change in government policy (laws to save the few remaining forests) won out over the populace's comfort, and London was to remain polluted by coal fumes for another 600 years [2], leading the poet Shelley to write in the early 1800s: "Hell must be much like London, a smoky and populous city" [1].

Post-Industrial Revolution

Two major air pollution health episodes raised awareness of the effect of pollution on human health and were instrumental in passage of the English Clean Air Act in the United Kingdom in 1956 and the Clean Air Act in the United States in 1970 (although federal air pollution control acts began to be passed in 1955, as discussed later in this chapter) [3,4]. These episodes occurred in Donora, Pennsylvania, in 1948 and in London, England, in 1952 and illustrated the fact that people will largely ignore pollution until it begins to kill in a dramatic way. In fact, local officials in Donora and London initially placed blame on weather inversions, frail health, and flu epidemics [3].

Donora, Pennsylvania, which is located south of Pittsburgh in the Monongahela Valley, was a town of approximately 26,000 people in the 1940s and home to a steel mill and zinc works [3]. The inhabitants were accustomed to dreary days, dirty buildings, and barren ground where no vegetation would grow, but they ignored the effect that the steel and zinc mills had on the population and environment as about two-thirds of the workers were employed in the steel and zinc mills. This continued until the "killer smog" of October 1948. Nearly half of the town's inhabitants were sick by the end of the second day of the thick smog, 20 people died in 3 days, 50 more deaths than would be expected from other causes occurred in the month following the episode, and many people experienced breathing difficulties for the rest of their lives [3]. It was discovered later that the cause for many of the fatalities was not sulfur dioxide from firing coal in the steel and zinc mills, as initially thought and often reported [5], but was fluoride poisoning from the fluorspar used in the zinc mill [3]. Regardless of the cause, this episode brought increased public awareness of air pollution and helped in the passage of air pollution laws.

A similar and even more severe episode occurred in London. London has been renowned for centuries for its thick fog, the infamous London smog. Throughout the eighteenth century, London experienced about 20 foggy days per year, but by the end of the nineteenth century, this had increased to 60 days [1]. People were beginning to become aware of a connection between pollution and certain sicknesses, and bronchitis was initially known as the "British disease." A severe pollution episode occurred in London in December 1952 with substantial loss of life and was followed by a similar one in January 1956 [4]. The more severe episode, which occurred in 1952, happened at a time when London was economically depressed and still recovering from World War II [3]. The government was selling its cleaner burning anthracite reserves to help pay its war debt while nearly all households were burning cheap, brown coal for heat. This situation was exacerbated in December 1952 when a cool spell (and temperature inversion) settled over London. London's 8 million residents burned the brown coal for warmth and the week-long inversion kept the smoke at ground level. In 7 days, an estimated 2800 people died [3].

History of Legislative Action for Coal-Fired Power Plants

The major development of air pollution legislative and regulatory acts occurred from 1955 to 1970; however, the early acts were narrow in scope as the U.S. Congress was hesitant to grant the federal government a high degree of control because air pollution problems were viewed as local or regional. This approach was found to be impractical as some states were hesitant to regulate industry and atmospheric transport of pollutants is not bounded by geographic lines. By the mid-1970s, the basis for national regulation of air pollution was developed and the actual regulations are continually changing. Regulations on coal-fired emissions essentially started in 1970 with the passing of the Clean Air Act Amendments of 1970. There were a few regulatory changes in the 1980s, but the Clean Air Act Amendments of 1990 resulted in significant regulatory changes. In December 2003, the U.S. Environmental Protection Agency (EPA) proposed a rule to permanently cap and reduce mercury emissions from power plants and is expected to promulgate legislation by December 2004, with full compliance expected by 2007 [6]. Currently, legislation is under consideration that would further reduce levels of sulfur dioxide and nitrogen oxides. Legislation for controlling carbon dioxide is currently being debated.

Pre-1970 Legislation

The history of federally enacted air pollution legislation begins in 1955 with the Air Pollution Control Act of 1955. The act was narrow in scope because of the federal government's hesitation to encroach on states' rights; however, it was the first step toward identifying air pollution sources and its effects and laid the groundwork for the effective legislation and enforcement by regulatory agencies developed over the next 15 years. The act initiated the following [4]:

- Research by the U.S. Public Health Service on the effects of air pollution;
- Provision for technical assistance to the states by the federal government;
- Training of individuals in the area of air pollution;
- Research on air pollution control.

The Air Pollution Control Act of 1955 was amended in 1960 and 1962 (*i.e.*, Air Pollution Control Act Amendments of 1960 and 1962) because of worsening conditions in urban areas due to mobile sources. Through these acts, Congress directed the Surgeon General to study the effect of motor vehicle exhausts on human health. A more formal process for the continual review of the motor vehicle pollution problem was included in the Clean Air Act of 1963, which provided, for the first time, federal financial aid for air pollution research and technical assistance [4]. The act supported state, regional, and local programs for the control and abatement of air pollution while reserving federal authority to intervene in interstate conflicts, thereby preserving the classical three-tier system of government. The act provided for [4]:

- Acceleration in the research and training program;
- Matching grants to state and local agencies for air pollution regulatory control programs;
- Developing air quality criteria to be used as guides in setting air quality standards and emissions standards;
- Initiating efforts to control air pollution from all federal facilities;

- Federal authority to abate interstate air pollution;
- Encouraging efforts by automotive companies and the fuel industries to prevent pollution.

The Clean Air Act of 1963 also provided for research authority to develop standards for sulfur removal from fuels, and a formal process for reviewing the status of the motor vehicle pollution problem. This, in turn, led to the Motor Vehicle Air Pollution Control Act of 1965, which formally recognized the technical and economic feasibility of setting automotive emission standards. The act also gave the secretary of the Department of Health, Education, and Welfare (HEW) the authority to intervene in intrastate air pollution problems of "substantial significance."

National Air Quality Control Act of 1967

The first federal legislation to impact stationary combustion sources was the National Air Quality Control Act of 1967. The act provided for a 2-year study on the concept of national emissions standards for stationary sources and was the basis for the 1970 legislative action. The provisions of the National Air Quality Control Act of 1967 included [4]:

- Establishment of eight specific areas in the United States on the basis of common meteorology, topography, and climate;
- Designation of air quality control regions (AQCRs) within the United States where evaluations were to be conducted to determine the nature and extent of the air pollution problem;
- Development and issuance of air quality criteria (AQC) for specific pollutants that have identifiable effects on human health and welfare;
- Development and issuance of information on recommended air pollution control techniques, which would lead to recommended technologies to achieve the levels of air quality suggested in the AQC reports;
- Requirement of a fixed time schedule for state and local agencies to establish air quality standards consistent with air quality criteria.

The states were allowed to set higher standards than recommended in the AQC reports; however, if a state did not act, the secretary of HEW had the authority to establish air quality standards for each air quality region. The states were given primary responsibility for action, but a very strong federal fallback authority was provided. Unfortunately, the federal program was not implemented according to the required time schedule because federal surveillance of the overall program was understaffed and the process to set up the AQCRs proved to be too complex; consequently, both President Nixon and Congress proposed new legislation in 1970.

Clean Air Act Amendments of 1970

The Clean Air Act Amendments of 1970 extended the geographical coverage of the federal program aimed at the prevention, control, and abatement of air pollution from stationary and mobile sources. The act transferred administrative functions assigned to the secretary of HEW to the newly created Environmental Protection Agency. The act provided for the first time national ambient air quality standards and national emission standards for new stationary sources. It initiated the study of aircraft emissions and imposed carbon monoxide, hydrocarbons, and nitrogen oxide emissions control on automobiles.

The major goal of the act was the achievement of clean air throughout the United States by the middle of the decade. Two types of pollutants were to be regulated: criteria air pollutants and hazardous air pollutants (HAPs). The criteria air pollutants were to be regulated to achieve attainment of National Ambient Air Quality Standards by establishing emission standards (developed by state and local agencies) for existing sources and national emission standards for new sources through promulgation of New Source Performance Standards. HAPs were to be regulated under National Emission Standards for Hazardous Air Pollutants. The major provisions of the act included [4]:

- The EPA was to establish National Ambient Air Quality Standards (NAAQSs), including primary standards for the protection of public health and secondary standards for the protection of public welfare;
- New Source Performance Standards (NSPSs) were to be required, with each state implementing and enforcing the standard of performance. Before a new stationary source could begin operation, state or federal inspectors were required to certify that the controls would function, and the new stationary sources had to remain in compliance throughout the lifetime of the plant;
- National Emission Standards for Hazardous Air Pollutants (NESHAPs) were to be established and would apply to existing as well as new plants;
- Funding was provided for fundamental air pollution studies, research on health and welfare effects of air pollutants, research on cause and effects of noise pollution, and research on fuels at stationary sources, including methods of cleaning fuels prior to combustion rather than flue gas cleaning techniques, improved combustion techniques, and methods for producing new or synthetic fuels with lower potential for creating polluted emissions;
- State and regional grant programs were authorized and matching grants established for implementing standards;
- The designation of AQCRs was to be completed;
- Establishment of statewide plans to be implemented (*i.e.*, state implementation plans, or SIPs) and designed to achieve primary or

public health standards within 3 years was required, and the overall plan had to be convincing as to its ability to meet and maintain the standards;

- Industry was required to monitor and maintain emission records and to make these records available to EPA officials, and the EPA was given the right of entry to examine records;
- Fines and criminal penalties were imposed for violation of implementation plans, emissions standards, and performance standards that were stricter than those under the earlier law;
- New automobile emission standards were set;
- Aircraft emission standards were to be developed by the EPA;
- Citizen's suits were permitted against those alleged to be in violation of emission standards, including the United States, and suits could be brought against the EPA administrator if he or she failed to act in cases where the law specified that action must be taken.

Air Quality Criteria and National Ambient Air Quality Standards

The Air Quality Act of 1967 addressed the development and issuance of air quality criteria (AQC), and the need for such criteria was reaffirmed in the 1970 amendments. AQC indicate qualitatively and quantitatively the relationship between various levels of exposure to pollutants and the short- and long-term effects on health and welfare [4]. AQC describe effects that can be expected to occur when pollutant levels reach or exceed specific values over a given time period and delineate the effects from combinations of contaminants as well as from individual pollutants. Economic and technical considerations are not relevant to the establishment of AQC.

The development of AQC is essential in providing a quantitative basis for air quality standards. Standards prescribe the pollutant levels that cannot be legally exceeded during a specific time period in a specific geographical region. The Clean Air Act Amendments of 1970 required federal promulgation of national primary and secondary standards that are to be established equitably in terms of the social, political, technological, and economic aspects of the problem. Standards are subject to revision as aspects change over time.

The purpose of the primary standards is immediate protection of the public health, including the health of sensitive populations such as people with asthma, children, and the elderly. Primary standards are to be achieved regardless of cost and within a specified time limit. Secondary standards are intended to protect the public welfare from known or anticipated adverse effects, including protection against decreased visibility, damage to animals, crops, vegetation, and buildings. Both standards have to be consistent with AQC, and, in addition, the standards have to prevent the continuing deterioration of air quality in any portion of an air quality control region.

The Clean Air Act Amendments of 1970 defined the first six criteria pollutants as carbon monoxide, nitrogen dioxide, sulfur dioxide, total particulate matter, hydrocarbons, and photochemical oxidants, and NAAQSs were established for these; subsequently, the list has been revised with the following actions [4]:

- Lead was added to the list in 1976;
- The photochemical oxidant standard was revised and restated as ozone in 1979;
- The hydrocarbon standard was withdrawn in 1983;
- The total suspended particulate matter standard was revised in 1987 to include only particles with an aerodynamic particle size of $<10\ \mu m$ and referred to as the PM_{10} standard;
- The $PM_{2.5}$ (*i.e.*, particles with an aerodynamic particle size of $<2.5\ \mu m$) standard was added in 1997;
- In 1997, the EPA reviewed the air quality standard for ground-level ozone and established the primary and secondary 8-hour NAAQS for ozone at 0.08 ppm and the primary and secondary 1-hour NAAQS for ozone at 0.12 ppm.

The current list of NAAQSs is provided in Table 4-1 and includes carbon monoxide, nitrogen dioxide, ozone, lead, PM_{10}, $PM_{2.5}$, and sulfur dioxide [7].

National Emissions Standards

Emissions standards place a limit on the amount or concentration of a pollutant that may be emitted from a source. It is often necessary for certain industries to be regulated by emission standards promulgated by the federal or state government in order to maintain or improve ambient air quality within a region to comply with national or state air quality standards. A number of factors must be considered when establishing emission standards [4]:

- The availability of technology appropriate for cleanup of a given type of industry should be determined;
- Monitoring stations must be available to measure the actual industrial emissions for which control is considered, as well as the ambient air quality so that the effectiveness of the standards can be determined;
- Regulatory agencies must be organized to cope with the measurement and enforcement of the standards;
- The synergistic effects of various pollutants must be determined;
- Models must be developed that reasonably predict the effects of reducing various emissions on the ambient air quality;
- Reasonable estimates of future emissions must be made based on the growth or decline of industry and population within a region.

TABLE 4-1
National Ambient Air Quality Standards

Pollutant	Standard Value		Standard Type
Carbon monoxide (CO)			
8-hour average	10 mg/m^3	9 ppm	Primary
1-hour average	40 mg/m^3	35 ppm	Primary
Nitrogen dioxide (NO_2)			
Annual arithmetic mean	100 $\mu g/m^3$	0.053 ppm	Primary and secondary
Ozone (O_3)			
1-hour average	235 $\mu g/m^3$	0.12 ppm	Primary and secondary
8-hour average	157 $\mu g/m^3$	0.08 ppm	Primary and secondary
Lead (Pb)			
Quarterly average	1.5 $\mu g/m^3$	—	Primary and secondary
Particulate (PM_{10})			
Annual arithmetic mean	50 $\mu g/m^3$	—	Primary and secondary
24-hour average	150 $\mu g/m^3$	—	Primary and secondary
Particulate ($PM_{2.5}$)			
Annual arithmetic mean	15 $\mu g/m^3$	—	Primary and secondary
24-hour average	65 $\mu g/m^3$	—	Primary and secondary
Sulfur dioxide (SO_2)			
Annual arithmetic mean	80 $\mu g/m^3$	0.030 ppm	Primary
24-hour average	365 $\mu g/m^3$	0.14 ppm	Primary
3-hour average	1300 $\mu g/m^3$	0.50 ppm	Primary and secondary

Emission standards can be categorized into the following general types of standards [4]:

- *Visible emissions standards*—The opacity of the plume from a stack or a point of fugitive emissions is not to equal to or does not exceed a specified opacity;
- *Particulate concentration standards*—The maximum allowable emission rate is specified in mass/volume (grams per dry standard cubic meter [g/dscm] or grains per dry standard cubic foot [gr/dscf]), and for combustion processes it is common to specify the concentration at a fixed oxygen (O_2) or carbon dioxide (CO_2) level so as to prevent dilution, thereby lowering the concentration;
- *Particulate process weight (or mass) standards*—The maximum allowable particulate emissions are tied to the actual mass of material being processed or used, and in combustion systems the standards are commonly reported in pounds of particulate matter per million Btu of fuel burned (lb/10^6 Btu or lb/MM Btu);
- *Gas concentration standards*—Gas standards are typically reported in mass per volume or volume per volume (g/dscm or ppm);

- *Prohibition of emissions*—Processes are banned outright;
- *Fuel regulations*—Fuel standards may be specified for various fuel-burning equipment such as limiting sulfur concentration in a fuel;
- *Zoning restrictions*—Emissions may be limited by passing zoning ordinances that dictate facilities that can be constructed;
- *Dispersion-based standards*—These standards limit the allowable emission of pollutants based on their contribution to the ambient air quality.

National emission or performance standards have been set for a number of industries, including fossil fuel-fired electric utility steam-generating units. National standards are necessary for industries that are spread geographically across the country and provide a basic commodity essential for the development of the country. Unfair economic advantages might be gained if the standards were set by individual states, thus allowing a state to relax the standards to attract industries. These national emissions standards are referred to as New Source Performance Standards (NSPSs) and apply to construction of new sources as well as sources that undergo operational and physical changes, which either increase emission rates or initiate new emissions from the plant [4].

40 CFR, Part 60, Subpart D On December 23, 1971, the first five final standards were published. Since December 23, 1971, the EPA administrator has promulgated nearly 75 standards. The complete text for each NSPS is available in the *Code of Federal Register* at Title 40 (Protection of the Environment), Part 60 (Standards of Performance for New Stationary Sources), with steam electric plants found in Subpart D [8]. The 1971 regulations addressed coal usage in utility and industrial steam generation units, and the regulations were amended for lignite on March 7, 1978. The maximum SO_2 emissions allowable from electric utility steam-generating units of more than 73 MW (megawatts) or 250 million (MM) Btu/hr of heat input and under construction or modification after August 17, 1971, was 1.2 lb SO_2 per MM Btu for solid fuels. Particulate matter was limited to 0.10 lb/MM Btu heat input, and opacity was not to exceed 20% for one 6-minute period per hour. NO_X standards were 0.70 lb NO_X per MM Btu heat input for solid fossil fuel or solid fossil fuel and wood residue (except for lignite); 0.60 lb NO_X per MM Btu heat input for lignite or lignite and wood residue; and 0.80 lb NO_X per MM Btu for lignite mined in North Dakota, South Dakota, or Montana and burned in a cyclone-fired unit (various combustion technologies are discussed in Chapter 5, Technologies for Coal Utilization).

40 CFR, Part 60, Subpart D(a) The original regulations were significantly revised as of February 6, 1980. The EPA promulgated new regulations for the control of SO_2, NO_X, and particulate matter from steam-generating units of more than 73 MW or 250 MM Btu/hr of heat input (40 CFR, Part 60, Subpart

D(a)) and under construction after September 18, 1978. For a coal-fired unit, the 1980 standard requires at least a 90% reduction of potential SO_2 emissions and limits the rate to 1.2 lb SO_2 per MM Btu heat input, or requires at least a 70% reduction and limits the emission rate to 0.6 lb SO_2 per MM Btu heat input. In addition, the standard specifies a unique maximum allowable emission rate and unique minimum reduction of potential emission based on the sulfur content and heating value of the coal [9]. If the uncontrolled emission rate (UER) is determined, then the required efficiency is as follows:

UER of SO_2	Required efficiency or maximum allowable efficiency rate
<2 lb SO_2 per MM Btu	70%
2–6 lb SO_2 per MM Btu	0.6 lb SO_2 per MM Btu; efficiency (%) = [(UER − 0.6)/UER] × 100%
6–12 lb SO_2 per MM Btu	90%
>12 lb SO_2 per MM Btu	1.2 lb SO_2 per MM Btu; efficiency (%) = [(UER − 1.2)/UER] × 100%

The NO_X standard varies according to fuel type and, with respect to coal, is 0.50 lb NO_X per MM Btu heat input from subbituminous coal, shale oil, or any solid, liquid, or gaseous fuel derived from coal; 0.80 lb NO_X per MM Btu heat input from the combustion in a slag tap furnace of any fuel containing more than 25%, by weight, lignite, which has been mined in North Dakota, South Dakota, or Montana; and 0.60 lb NO_X per MM Btu heat input from anthracite or bituminous coal. The NO_X standard is based on a 30-day rolling average.

Particulate emissions are limited to 0.03 lb/MM Btu heat input. In addition, opacity is limited to 20% for a 6-minute average. The NO_X standards for Subparts D(a) and D(b) (see below for Subpart D(b)) were revised on September 16, 1998 [10]. Only those electric utility steam generating units for which construction, modification, or reconstruction is commenced after July 9, 1997, would be affected by these revisions. The revisions changed the existing standards for NO_X emission limits to reflect the performance of best-demonstrated technology. The revisions also changed the format of the revised NO_X emission limit for new electric utility steam-generating units to an output-based format to promote energy efficiency and pollution prevention. The NO_X emission limit in Subpart D(a) is 1.6 lb NO_X per megawatt-hour (MWh) gross energy output regardless of fuel type for new utility boilers. For existing utility boilers that would become subject to the standards due to a modification or reconstruction, the EPA revised the NO_X limit to be consistent with the requirements for new units but expressed the emission limits in an equivalent input-based format: 0.15 lb NO_X per MM Btu. This provision was withdrawn by the EPA on August 7, 2001, however, after industry groups filed petitions for review and a motion to vacate the standards as applied to modified boilers in the U.S. Court of

Appeals [11]. On September 21, 1999, the court issued an order granting the petitioner's motion and, as a result, owners and operators of electric utility boilers on which modification is commenced after July 9, 1997, are required to comply with the applicable nitrogen oxides emission limits specified in the pre-existing NSPS, which is 0.50 lb NO_X per MM Btu.

40 CFR, Part 60, Subparts D(b) and D(c) Although the acid rain provisions of the 1990 Clean Air Act Amendments (discussed further below) place additional requirements on the electric utility industry, the NSPS, as revised in 1980, is still applicable for new sources. Smaller sources have been addressed over time, including industrial/commercial/institutional steam generators constructed after June 19, 1984, with heat inputs of 29 to 73 MW (40 CFR, Part 60, Subpart D(b)) and small industrial/commercial/institutional steam-generating units with 2.9 to 29 MW of heat input constructed after June 9, 1989 (40 CFR, Part 60, Subpart D(c)). On November 25, 1986, standards of performance for industrial/commercial/institutional steam-generation units were promulgated [12]. Coal-fired facilities having heat input capacity between 29 and 73 MW (100 and 250 million Btu/hr) are subject to the particulate matter and NO_X standards under 40 CFR, Part 60, Subpart D(b). Particulate matter (PM) standards for coal are more complicated than previous regulations:

- 0.05 lb PM/MM Btu heat input if the facility combusts only coal or combusts coal and other fuels and has an annual capacity factor for the other fuels of 10% or less;
- 0.10 lb PM/MM Btu heat input if the facility combusts coal and other fuels and has an annual capacity factor for the other fuels greater than 10% and is subject to a federally-enforceable requirement limiting operation of the affected facility to an annual capacity factor greater than 10% for fuels other than coal;
- 0.20 lb PM/MM Btu heat input if the affected facility combusts coal or coal and other fuels and has an annual capacity factor for coal or coal and other fuels of 30% or less, has a maximum heat input capacity of 73 MW (250 million Btu/hr) or less, had a federally-enforceable requirement limiting operation of the affected facility to an annual capacity factor of 30% or less for coal or coal and other solid fuels, and construction of the facility commenced after June 19, 1984, and before November 25, 1986.

The NO_X standards for coal-fired facilities identified in 1986 in Subpart D(b) are:

- 0.50 lb NO_X (expressed as NO_2) per MM Btu for mass-feed stokers;
- 0.60 lb NO_X per MM Btu for spreader stoker and fluidized-bed combustors;

- 0.70 lb NO_x per MM Btu for pulverized coal-fired units;
- 0.60 lb NO_x per MM Btu for lignite units except for lignite mined in North Dakota, South Dakota, or Montana and combusted in a slag tap furnace for which the emissions limits are 0.8 lb NO_x per MM Btu;
- 0.50 lb NO_x per MM Btu for coal-derived synthetic fuels.

As previously mentioned, the NO_x standard for Subpart D(b) was revised on September 16, 1998 [10]. Only those industrial steam-generating units for which construction, modification, or reconstruction is commenced after July 9, 1997, would be affected by these revisions. For coal-fired Subpart D(b) units, the NO_x emission limit promulgated was 0.20 lb/MM Btu heat input; however, this provision was withdrawn on August 7, 2001, and owners and operators of industrial/commercial/institutional boilers on which modification is commenced after July 9, 1997, are required to comply with the applicable nitrogen oxides emission limits specified in the pre-existing NSPS (*i.e.*, 0.50 to 0.80 lb NO_x per MM Btu, depending on fuel type and boiler configuration as listed above) [11].

On September 12, 1990, standards of performance for small industrial/commercial/institutional steam-generation units were promulgated (40 CFR, Part 60, Subpart D(c)) [13]. Under Subpart D(c), coal-fired facilities having heat input capacity between 2.9 and 29 MW (10 and 100 million Btu/hr) are not subject to NO_x standards nor are they subject to SO_2 or particulate matter emissions limits during periods of combustion research. For nonresearch operations, SO_2 standards include the following:

- When firing only coal, SO_2 emissions are limited to no more than 10% of the potential SO_2 emissions rate (*i.e.*, 90% reduction) and less than 1.2 lb SO_2 per MM Btu heat input;
- If coal is combusted with other fuels, the affected facility is subject to the 90% SO_2 reduction requirement, and the emission limit is determined by the equation:

$$E_{SO_2} = \frac{K_a H_a + K_b H_b + K_c H_c}{H_a + H_b + H_c} \qquad (4\text{-}1)$$

where:

E_{SO_2} is the SO_2 emission limit, expressed in lb/million Btu heat input;

K_a is 1.2 lb/million Btu;

K_b is 0.60 lb/million Btu;

K_c is 0.50 lb/million Btu;

H_a is the heat input from the combustion of coal, except coal combusted in a facility that uses an emerging technology for SO_2 control, in million Btu;

H_b is the heat input from the combustion of coal in a facility that uses an emerging technology for SO_2 control, in million Btu;
H_c is the heat input from the combustion of oil in million Btu.

- When firing only coal in a facility that uses an emerging technology for SO_2 control, SO_2 emissions are limited to no more than 50% of the potential SO_2 emissions rate (*i.e.*, 50% reduction) and less than 0.60 lb SO_2 per MM Btu heat input;
- If coal is combusted with other fuels, a 50% SO_2 reduction is required, and the emission limit is determined using Equation (4-1);
- Percent reduction requirements are not applicable for affected facilities that have input capacity of 22 MW (75 million Btu/hr) or less; affected facilities that have an annual capacity for coal of 55% or less; affected facilities located in a noncontinental area; and affected facilities that combust coal in a duct burner as part of a combined cycle system where 30% or less of the heat entering the steam-generating unit is from combustion of coal in the duct burner and 70% or more of the heat entering the steam generating unit is from exhaust gases entering the duct burner;
- Reduction of the potential SO_2 emission rate through fuel pretreatment is not credited toward the percent reduction requirement unless fuel pretreatment results in a 50% or greater reduction in the potential SO_2 emission rate and emissions from the pretreated fuel (without either combustion or post-combustion SO_2 control) are equal or less than 0.60 lb/MM Btu.

The particulate matter standards in Subpart D(c) state that a facility that combusts coal or mixtures of coal with other fuels and has a heat input capacity of 8.7 MW (30 million Btu/hr) or greater is subject to the following emission limits:

- 0.05 lb PM per MM Btu heat input if the facility combusts only coal, or combusts coal with other fuels and has an annual capacity factor for the other fuels of 10% or less;
- 0.10 lb PM per MM Btu heat input if the facility combusts coal with other fuels, has an annual capacity factor for the other fuels greater than 10%, and is subject to a federally enforceable requirement limiting operation of the facility to an annual capacity factor greater than 10% for fuels other than coal.

Emission Factors Once an NSPS has been established, it is necessary for new sources constructed after a defined date to meet the standards. It is not possible to sample new sources to determine required collection or removal efficiencies. In these cases, knowledge of the emission factors for the specific regulated pollutant from these sources is used to estimate the approximate level of control required to meet the NSPS. The EPA has published a document, *Compilation of Air Pollutant Emission Factors* (referred to as AP-42) since 1972. Supplements to AP-42 have been routinely published to add new emissions source categories and to update existing emission factors. This document is also provided on the EPA website at their CHIEF (Clearinghouse for Inventories and Emissions Factors; www.epa.gov/ttn/chief/ap42) bulletin board.

The EPA routinely updates AP-42 in order to respond to new emission factor needs of state and local air pollution control programs, industry, and the agency itself. The current emission factors for bituminous and subbituminous coal, lignite, and anthracite firing are provided in Appendix A and are from Sections 1.1 (Bituminous and Subbituminous Coal Combustion), 1.2 (Anthracite Coal Combustion), and 1.7 (Lignite Combustion) of AP-42, fifth ed., vol. I, suppls. A through G. The emission factors have been developed for (not inclusive):

- Various fuel firing configurations;
- Uncontrolled and controlled emissions;
- Criteria gaseous pollutants (SO_x, NO_x, CO);
- Filterable particulate matter and condensable particulate matter;
- Trace elements;
- Various types of polynuclear organic matter (POM), polynuclear aromatic hydrocarbons (PAHs), and organic compounds;
- Acid gases (HCl and HF);
- Other gaseous pollutants such as CO_2, N_2O, and CH_4;
- Cumulative ash particle size distribution and size-specific emissions.

Emissions factors and emissions inventories have long been fundamental tools for air quality management. Emission estimates are important for developing emission control strategies, determining applicability of permitting and control programs, and ascertaining the effects of sources and appropriate mitigation strategies. Users include federal, state, and local agencies; consultants; and industry. Data from source-specific emission tests or continuous emission monitors are usually preferred for estimating a source's emissions because those data provide the best representation of the tested source's emissions; however, test data from individual sources are not always available and they may not reflect the variability of actual emissions over time. Consequently, emission factors are often the best or only method available for estimating emissions.

The passage of the Clean Air Act Amendments of 1990 and the Emergency Planning and Community Right-To-Know Act of 1986 has increased the need for both criteria and hazardous air pollutant emission factors and inventories. The Emission Factor and Inventory Group (EFIG) of the EPA's Office of Air Quality Planning and Standards develops and maintains emission-estimating tools. The AP-42 series is the principal means by which the EFIG can document its emission factors.

Emission factors may be appropriate to use in a number of situations such as source-specific emission estimates for area-wide inventories. These inventories have many purposes, including ambient dispersion modeling and analysis, control strategy development, as screening sources for compliance investigations, and in some permitting applications. Emission factors in AP-42 are neither EPA-recommended emission limits—for example, best available control technology (BACT) or lowest achievable emission rate (LAER)—nor standards (*e.g.*, NSPSs or NESHAPs).

Figure 4-1 depicts various approaches to emission estimation in a hierarchy of requirements and levels of sophistication that need to be considered when analyzing the tradeoffs between cost of the estimates and the quality of the resulting estimates. More sophisticated and more costly emission determination methods may be necessary where risks of either adverse environmental effects or adverse regulatory outcomes are high. Less expensive estimation techniques such as emission factors and emission models may be appropriate and satisfactory where risks of using a poor estimate are low. Note that the reliability of the AP-42 emission factors are rated from A through E, which is a general indication of the robustness of that factor. This rating is assigned based on the estimated reliability of the tests used to develop the factor. In general, factors based on many observations, or on

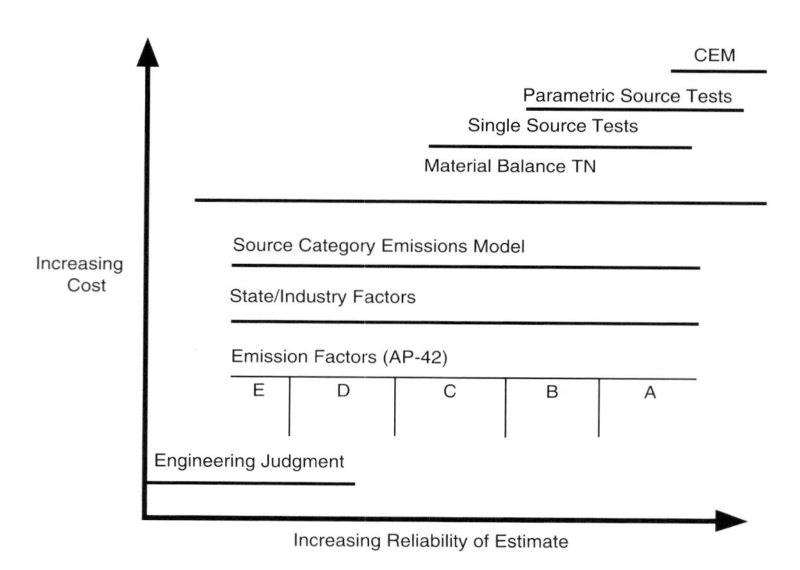

FIGURE 4-1. Approach to emission estimation. (From AP-42, External Combustion Sources, in *Emission Factors*, Fifth ed., Office of Air Quality Planning and Standards and Office of Air and Radiation, U.S. Environmental Protection Agency, U.S. Government Printing Office, Washington, D.C., 1993 [latest revisions in 1998].)

more widely accepted test procedures, are assigned higher rankings, with A being the best. The emission factors rating is provided with the emission factors contained in Appendix A.

National Emission Standards for Hazardous Air Pollutants

The 1970 Clean Air Act Amendments also provided for national emission standards for hazardous air pollutants (NESHAPs). Only seven standards were established between 1970 and 1990, as the NESHAPs were the subject of numerous suits and court decisions regarding how to address emission limits on carcinogenic pollutants [4]. None of the original seven NESHAPs—asbestos, beryllium, mercury, vinyl chloride, benzene, radionuclides, and arsenic—specifically addressed coal-fired steam-generating units. The Clean Air Act Amendments of 1990 made major changes in the approach taken to address hazardous air pollutants (HAPs) with additional amendments being added as recently as May 30, 2003 [15].

Clean Air Act Amendments of 1977 and Prevention of Significant Deterioration

By 1977, most areas of the country had still not attained the NAAQSs for at least one pollutant [4]. For those areas that had not attained a NAAQS (*i.e.*, nonattainment areas), states were required to submit and have an approved state implementation plan (SIP) revision by July 1, 1979, that demonstrated how attainment would be achieved by December 31, 1982. This requirement was a precondition for the construction or modification of major emission sources in nonattainment areas after June 30, 1979. If a state could not attain primary standards for carbon monoxide or photochemical oxidants after implementation of all reasonably available measures, it was required to submit a second SIP revision by December 31, 1982, that would demonstrate how attainment would be achieved by December 31, 1987.

Prevention of Significant Deterioration

The concern over nonattainment areas and the controversy generated by the provision in the 1970 Clean Air Act Amendments on standards preventing continuing deterioration of air quality led to a set of guidelines issued in 1974 by the EPA to prevent the significant deterioration of air quality in areas that were cleaner than required by NAAQS (*i.e.*, had attained the NAAQSs). This was necessary because some interpreted the 1970 act to mean that a region could not backslide in air quality even though the current air quality may be superior to the national standard. This interpretation would have stifled economic growth in a region (new industrial and commercial operations in the region could not contribute zero pollution), and would have failed to force sources in the region to decrease their contaminant emissions. This concern led to the passage of regulations regarding the prevention of significant deterioration (PSD). The PSD regulations for attainment areas required that all of these areas be designated as Class I, II, or III, depending upon the degree of deterioration to be allowed, and incremental limits were placed on the amount of increase in deterioration allowed. The classifications are [4]:

- Class I—Pristine areas, including international parks, national parks, and national wilderness areas in which very little deterioration would be allowed;
- Class II—Areas where moderate change would be allowed, but where stringent air quality constraints are desirable;
- Class III—Areas where major growth and industrialization would be allowed.

Congress specified which of the areas must be protected by the most stringent Class I designation, designated all other areas within the United States as Class II areas, and provided the option for redesignation of Class II areas to Class I or Class III areas by public referendum. Congress also specified the maximum allowable incremental increases in concentration of sulfur dioxide and particulate matter and charged the EPA to determine comparable increments for hydrocarbons, carbon monoxide, photochemical oxidants, and nitrogen oxides. The PSD increments are listed in Table 4-2 [4].

A requirement was placed on major sources in the preconstruction PSD review process that specified that each major new plant must install BACT, which was defined to be at least as stringent as NSPS, to limit its emissions. Major sources subject to PSD review are those with the potential to emit 100 short tons or more per year of any regulated pollutant under the Clean Air Act Amendment of 1977 [4]. All sources emitting greater than 250 short tons

TABLE 4-2
Prevention of Significant Deterioration (PSD) Increments

Pollutant	*Maximum Allowable Increase (mg per m^3)*		
	Class I	*Class II*	*Class III*
Particulate matter			
PM_{10}, annual arithmetic mean	4	17	34
PM_{10}, 24-hour maximum	8	30	60
Sulfur dioxide			
Annual arithmetic mean	2	20	40
24-hour maximum	5	91	182
3-hour maximum	25	512	700
Nitrogen dioxide			
Annual arithmetic mean	2.5	25	50

per year are subject to PSD review. For non-NSPS sources, a BACT review document is prepared in the preconstruction review.

The 1974 Guidelines on PSD and the Clean Air Amendments of 1977 established the protocol for new sources and proposed modifications to major sources in attainment areas where they would be subjected to a new source review and would have to meet certain PSD requirements. This review requires that dispersion modeling be conducted of the proposed emissions from the sources to ensure that the emissions from the proposed facility would not exceed the increments listed in Table 4-2 or cause an exceedance of the NAAQS. In concept, sources are allowed to use some fraction of the increment, as determined by the state or local agency and based on dispersion modeling. The PSD increment can only be used to the extent that it does not cause the ambient concentration to exceed the NAAQS. Another important issue is the extent to which any single source would be allowed to use the available increment and the ramifications on long-term growth if a single source was allowed to use it all. Some state and local agencies address this on a case-by-case basis, while others have taken the approach of allowing only a certain fraction of the increment, or the remaining increment, to be used in a single PSD application [4].

Nonattainment Areas

There was concern that no industrial growth could occur in nonattainment areas, as these areas are in violation of one or more NAAQSs and, because PSD does not apply in these areas, no increments are available. This was remedied on December 21, 1976, with the Offsets Policy Interpretive Ruling, which pertained to the preconstruction review requirements for all new or modified stationary sources of air pollution in nonattainmant areas [4]. This ruling, also known as the emission offset policy, requires that three conditions be met: (1) the source must meet the lowest achievable emission rate (LAER), defined as being more stringent than BACT; (2) all existing sources owned by the applicant in the same region must be in compliance or under an approved schedule to achieve compliance; and (3) the source must provide an offset or reduction of emissions from other sources greater than the proposed emissions that the source would contribute such that there is a net improvement in air quality.

The Clean Air Act Amendments of 1977 also provided for the banking of offsets. If the offsets achieved are considerably greater than the new source's emissions, a portion of this excess emission reduction (also known as emission reduction credits) can be banked by the source for use in future growth or traded to another source, depending on each state's offset/trading policy.

Clean Air Act Amendments of 1990

In June 1989, President Bush proposed major revisions to the Clean Air Act. Both the House of Representatives and the Senate passed Clean Air bills by large votes that contained the major components of the president's proposals. After a joint conference committee met to work out the differences in the bills, Congress voted for the package recommended by the conferees, and President Bush signed the bill, the Clean Air Act Amendments of 1990 (1990 CAAAs), on November 15, 1990. The 1990 CAAAs are the most substantive regulations adopted since passage of the Clean Air Act Amendments of 1970. Specifically, the new law:

- Encourages the use of market-based principles and other innovative approaches, such as performance-based standards and emission banking and trading;
- Provides a framework from which alternative clean fuels will be used by setting standards in the fleet and a California pilot program that can be met by the most cost-effective combination of fuels and technology;
- Promotes the use of clean low-sulfur coal and natural gas, as well as innovative technologies to clean high-sulfur coal through the Acid Rain Program;
- Reduces enough energy waste and creates enough of a market for clean fuels derived from grain and natural gas to cut dependency on oil imports by 1 million barrels per day;
- Promotes energy conservation through an Acid Rain Program that gives utilities flexibility to obtain needed emission reductions through programs that encourage customers to conserve energy.

The 1990 CAAAs contains eleven major divisions, referred to as Titles I through XI, which either provided amendments to existing titles and sections of the Clean Air Act or provided new titles and sections. The titles for the 1990 CAAAs are:

- Title I—Provisions for Attainment and Maintenance of National Ambient Air Quality Standards;
- Title II—Provisions Relating to Mobile Sources;
- Title III—Air Toxics;
- Title IV—Acid Deposition Control;
- Title V—Permits;
- Title VI—Stratospheric Ozone and Global Climate Protection;
- Title VII—Provisions Relating to Enforcement;
- Title VIII—Miscellaneous Provisions;
- Title IX—Clean Air Research;
- Title X—Disadvantaged Business Concerns;
- Title XI—Clean Air Deployment Transition Assistance.

The titles that directly impact the use of coal, Titles I, III, IV, and V, are discussed in the following sections. The concepts of NAAQSs, NSPSs, and

PSD, as defined in Title I, remained virtually unchanged; however, major changes have occurred in regulations and approaches used to address nonattainment areas in Title I, hazardous air pollutants in Title III, acid rain in Title IV, and permitting in Title V.

Title I: Provisions for Attainment and Maintenance of National Ambient Air Quality Standards

Although the Clean Air Act Amendments of 1970 and 1977 brought about significant improvements in air quality, urban air pollution persisted, and many cities failed attainment for ozone, carbon monoxide, and PM_{10}. Of these, the most widespread pollution problem is ozone (*i.e.*, smog). One component of smog, hydrocarbons, comes from automobile emissions, petroleum refineries, chemical plants, dry cleaners, gasoline stations, house painting, and printing shops while another key component, nitrogen oxides, comes from the combustion of fossil fuels for transportation, utilities, and industries [16].

The 1990 CAAAs created a new, balanced strategy for the nation to address urban smog. The new law gave states more time to meet the air quality standard but it also required states to make constant progress in reducing emissions. It required the federal government to reduce emissions from cars, trucks, and buses; from consumer products such as hair spray and window-washing compounds; and from ships and barges during loading and unloading of petroleum products. The federal government must also develop guidance that states the need to control stationary sources.

The 1990 CAAAs address the urban air pollution problems of ozone, carbon monoxide, and PM_{10}. Specifically, they clarify how areas are designated and redefine the terms of attainment. The 1990 CAAAs also allow the EPA to define the boundaries of nonattainment areas and establish provisions defining when and how the federal government can impose sanctions on areas of the country that have not met certain conditions.

For the pollutant ozone, the 1990 CAAAs establish nonattainment area classifications ranked according to the severity of the area's air pollution problem. These classifications are marginal, moderate, serious, severe, and extreme and were designated based on the air quality of the nonattainment area during the period of 1987 to 1989. The EPA assigns each nonattainment area one of these categories, thus triggering varying requirements with which the area must comply in order to meet the ozone standard. Table 4-3 lists the ozone design value, which determines the classification, attainment deadlines, minimum size of a new or modified source that would be affected, and offset requirements [4].

Nonattainment areas have to implement different control measures depending upon their classification. Nonattainment areas with worse air quality problems must implement greater control measures. For example, in areas classified as extreme, boilers with emission rates greater than

TABLE 4-3
Classifications for Nonattainment Areas

Pollutant	*Classification*	*Design Value (ppm)*	*Attainment Deadline*	*Major Source*[a]	*Offset Ratio for Sources*
Ozone	Marginal	0.121–0.138	11/15/1993	100	1.1–1
	Moderate	0.138–0.160	11/15/1996	100	1.15–1
	Serious	0.160–0.180	11/15/1999	50	1.2–1
	Severe	0.180–0.190	11/15/2005	25	1.3–1
	Severe	0.190–0.280	11/15/2007	25	1.3–1
	Extreme	>0.280	11/15/2007	10	1.5–1
Carbon monoxide	Moderate	9.1–16.4	12/31/1995	—	None
	Serious	>16.4	12/31/2000	50	None

[a]Short tons VOCs per year for ozone; short tons CO per year for carbon monoxide.

25 short tons per year are required to burn clean fuels or install advanced control technologies.

The 1990 CAAAs also established similar programs for areas that do not meet federal health standards for carbon monoxide and PM_{10}. Areas exceeding the standards for these pollutants are divided into moderate and serious classifications. The classifications for nonattainment areas for carbon monoxide are shown in Table 4-3.

Areas not attaining required levels of PM_{10} at the time the 1990 CAAAs were passed were designated as moderate areas and given an attainment deadline of December 31, 1994 [4]. Nonattainment areas subsequent to passage of the 1990 CAAAs are designated moderate and are given 6 years to achieve compliance. Major sources in moderate areas are those that emit 100 short tons or more of particulate matter per year. Moderate areas require the adoption of reasonably available control measures (RACM).

Moderate areas that fail to reach attainment are redesignated as serious areas and have 10 years from the date of designation as nonattainment to achieve attainment. For serious areas, major sources include those that emit 70 short tons or more of particulate matter per year. Serious areas must also adopt best available control measures (BACMs).

Title III: Air Toxics

Hazardous air pollutants, also known as toxic air pollutants or air toxics, are those pollutants that cause or may cause cancer or other serious health effects, such as reproductive effects or birth defects, or adverse environmental and ecological effects but are not specifically covered under another portion of the Clean Air Act. Most air toxics originate from human-made sources, including mobile sources (*e.g.*, cars, trucks, buses), stationary sources (*e.g.*, factories, refineries, power plants), and indoor sources (*e.g.*, building materials and activities such as cleaning) [16]. The Clean

Air Act Amendments of 1977 failed to result in substantial reductions of the emissions of these very threatening substances. Over the history of the air toxics program, only seven pollutants had been regulated. Title III established a list of 189 (later modified to 188) HAPs associated with approximately 300 major source categories. The list of HAPs is provided in Appendix B [17]. Under Title III, a major source is defined as any new or existing source with the potential to emit, after controls, 10 short tons or more per year of any of the 188 HAPs or 25 short tons or more per year of any combination of those pollutants. These sources may release air toxics from equipment leaks, when materials are transferred from one location to another, or during discharge through emissions stacks or vents.

The EPA must then issue maximum achievable control technology (MACT) standards for each listed source category according to a prescribed schedule. These standards are based on the best demonstrated control technology or practices within the regulated industry, and the prescribed schedule dictated that the EPA must issue the standards for 40 source categories within 2 years, 25% of the source categories within 5 years, 50% of the source categories within 7 years, and 100% of the source categories within 10 years of passage of the new law. Eight years after MACT is installed on a source, the EPA must examine the risk levels remaining at the regulated facilities and determine whether additional controls are necessary to reduce unacceptable residual risk [16].

The Bhopal, India, tragedy, where an accidental release of methyl isocyanate at a pesticide-manufacturing plant in 1984 killed approximately 4000 people and injured more than 200,000, inspired the 1990 CAAAs requirement that factories and other businesses develop plans to prevent accidental releases of highly toxic chemicals. In addition, the Act established the Chemical Safety Board to investigate and report on accidental releases of HAPs from industrial plants.

Title III did not directly regulate air toxics from power plants but did state that regulation of air toxics from utility power plants would be based on scientific and engineering studies. Mercury is one pollutant that was identified for study and will be discussed in more detail later in this chapter. At power plants, compounds in the vapor phase (*e.g.*, polycyclic organic matter) and those combined with or attached to particulate matter (*e.g.*, arsenic) are subject to the Title III provisions [18].

Title IV: Acid Deposition Control

The Acid Rain Program was established under Title IV of the 1990 CAAAs. The program required major reductions of sulfur dioxide (SO_2) and nitrogen oxides (NO_X) emissions, the pollutants that cause acid rain. Using an innovative market-based or cap-and-trade approach to environmental protection, the program sets a permanent cap on the total amount of SO_2 that may be emitted by electric power plants nationwide. The cap is set at about one-half the amount of SO_2 emitted in 1980, and the trading component allows for flexibility for individual fossil-fuel-fired combustion units to select their own methods of compliance. The program also sets NO_x emission limitations for certain coal-fired electric utility boilers, representing about a 27% reduction from 1990 levels [19].

Under the Acid Rain Program, each unit must continuously measure and record its emissions of SO_2, NO_x, and CO_2, as well as volumetric flow and opacity [19]. In most cases, a continuous emissions monitoring (CEM) system must be used. Units report hourly emissions data to the EPA on a quarterly basis. These data are then recorded in the Emissions Tracking System, which serves as a repository of emission data for the utility industry. Emissions monitoring and reporting are critical to the program as they instill confidence in allowance transactions by certifying the existence and quantity of the commodity being traded and provide assurance that NO_x averaging plans are working. Monitoring also ensures, through accurate accounting, that the SO_2 and NO_x emissions reduction goals are met.

The SO_2 Program Title IV of the 1990 CAAAs called for a two-step program to reduce SO_2 emissions by 10 million short tons from 1980 levels and, when fully implemented in 2000, placed a cap of approximately 8.9 million short tons per year on SO_2 emissions, forcing all generators that burn fossil fuels after 2000 to possess an emissions allowance for each ton of SO_2 they emit. By January 1, 1995, the deadline for Phase I, half of the total SO_2 reductions were to have occurred by requiring 110 of the largest SO_2-emitting power plants (with 263 boilers or units) located in 21 eastern and Midwestern states to cut their emissions to an annual average rate of 2.5 lb SO_2 per million Btu. These stations, specifically identified in the 1990 CAAAs (see Appendix C), consisted of boilers with output greater than or equal to 100 MW and sulfur emissions of greater than 2.5 lb SO_2 per million Btu. Plants deciding to reduce SO_2 emissions by 90% were given until 1997 to meet the requirements. By the year 2000, the deadline for Phase II, virtually all power plants greater than 75 MW and discharging SO_2 at a rate more than 1.2 lb/million Btu were required to reduce emissions to that level. In 2001, 2792 units were affected by the SO_2 provision of the Acid Rain Program [19].

The Phase I reductions were accomplished by issuing the utilities that operated these units emission allowances equivalent to what their annual emission would have been at these plants in the years 1985 to 1987 based on burning coal with emissions of 2.5 lb SO_2 per million Btu. One allowance is equivalent to the emission of one short ton of SO_2 per year. Utilities were allowed the flexibility of determining which control strategies to be used on existing plants as long as the total emissions from all plants listed in Phase I and owned by the utility did not exceed the available allowances. The law was designed to let industry find the most cost-effective way to stay under the cap. This differs in approach from previous air quality regulations, such as the NSPSs and PSD, which are based on controlling emissions at their source and then monitoring to ensure compliance [20]. If a utility's emissions exceeded the available allowances, they were subject to fines assessed at \$2000/short ton of excess emissions with a requirement to offset the emissions in future years. Any emission reductions achieved that were in excess of those required could be banked by the utility for use at a later date or traded or sold to another utility.

The SO_2 component of the Acid Rain Program represents a dramatic departure from traditional regulatory approaches that establish source-specific emissions limitations; instead, the program uses an overall emissions cap for SO_2 that ensures that emission reductions are achieved and maintained and provides for a trading system that facilitates lowest-cost emissions reductions. The program features tradable SO_2 emissions allowances, where one allowance is a limited authorization to emit 1 short ton of SO_2. A fixed number of allowances is issued by the government and they may be bought, sold, or banked for future use by utilities, brokers, or anyone else interested in holding them. Existing units are allocated allowances for each year; new units do not receive allowances and must buy them. New coal-fired boilers are subject to the NSPSs that remain in effect (*i.e.*, 70–90+% reduction), with the provision that they must acquire emission allowances to emit the residual SO_2 that is not controlled [4]. At the end of the year, all participants in the program are obliged to surrender to the EPA the number of allowances that correspond to their annual SO_2 emissions [19].

The NO_x Program Title IV also required the EPA to develop a NO_x reduction program and set a goal of reducing NO_x by 2 million short tons from 1980 levels. As with the SO_2 emission reduction requirements, the NO_x program was implemented in two phases, beginning in 1996 and 2000 [21]. The NO_x program embodies many of the same principles of the SO_2 trading program in its design: a results orientation, flexibility in the method to achieve emission reductions, and program integrity achieved through measurement of the emissions. However, it does not cap NO_x emissions as the SO_2 program does nor does it utilize an allowance trading system, although NO_x trading programs have been implemented, as discussed later in this chapter.

Emission limitations for the NO_x boilers provide flexibility for utilities by focusing on the emission rate to be achieved, expressed in pounds of NO_x per million Btu of heat input. Two options for compliance with the emission limitations are provided:

- Compliance with an individual emission rate for a boiler;
- Averaging of emission rates over two or more units, that have the same owner or operator to meet an overall emission rate limitation.

These options give utilities flexibility to meet the emission limitations in the most cost-effective way and allow for the further development of

TABLE 4-4
Number of NO_X-Affected Units by Boiler Type (2001)

Coal-Fired Boiler Type[a]	*Standard Emission Limit (lb/MM Btu)*	*Number of Units*
Phase, I Group 1: tangentially fired	0.45	135
Phase I, Group 1: dry bottom wall fired	0.50	130
Phase II, Group 1: tangentially fired	0.40	304
Phase II, Group 1: dry bottom wall fired	0.46	312
Cell burners	0.68	37
Cyclones >155 MW	0.86	56
Wet bottom >65 MW	0.84	31
Vertically fired	0.80	41
Total	—	1046

[a] All coverage for boilers >25 MW unless otherwise noted.

technologies to reduce the cost of compliance. If a utility properly installs and maintains the appropriate control equipment designed to meet the emission limitation established in the regulations but is still unable to meet the limitation, the NO_X program allows the utility to apply for an alternative emission limitation (AEL), which corresponds to the level that the utility demonstrates is achievable. Phase I of the program, which was delayed a year due to litigation, began on January 1, 1996, and affected two types of boilers, which were among those already targeted for Phase I SO_2 reductions: dry-bottom, wall-fired boilers and tangentially fired boilers. The regulations to govern the Phase II portion of the program, which began in 2000, were promulgated December 19, 1996. These regulations set lower emission limits for Group 1 boilers and established NO_X limitations for Group 2 boilers which include boilers applying cell-burner technology, cyclone boilers, wet bottom boilers, and other types of coal-fired boilers. The NO_X limitations and number of units affected in 2001 are provided in Table 4-4 [19].

Title V: Permitting

The 1990 CAAAs introduced a national permitting program to ensure compliance with all applicable requirements of the Clean Air Act and to enhance the EPA's ability to enforce the Act [16]. Sources are required to submit applications for Title V operating permits through the state agencies to the EPA. Air pollution sources subject to the program must obtain an operating permit; states must develop and implement the program; and the EPA must issue permit program regulations, review each state's proposed program, and oversee the state's efforts to implement any approved program. The EPA must also develop and implement a federal permit program when a state fails to adopt and implement its own program.

The following sources are required to submit Title V permits:

- Major sources as determined under Title I—(1) ≥100 short tons per year and listed pollutants, excluding CO; and (2) ≥10 to 100 short tons year, or sources in nonattainment areas depending on the classification of marginal to extreme;
- All NSPSs, PSD review sources, and NESHAPs sources;
- Major sources as determined under Title III—(1) ≥10 short tons per year any air toxic, and (2) ≥25 short tons per year multiple air toxics;
- Sources under Title IV;
- Sources emitting ≥100 short tons per year of ozone depleting substances under Title VI;
- Other sources required to have state or federal operating permits.

Additional NO_X Regulations and Trading Programs

Many urban areas do not meet the ozone standard and are classified as nonattainment areas. To address this, along with the fact that NO_X can be transported great distances, more restrictive requirements for NO_X emissions from electric power generating plants and other large stationary boilers in 22 eastern states and the District of Columbia were established [22]. The requirements set statewide NO_X emissions budgets including budget components for the electric power industry and certain industrial sources. States are required to develop state implementation plans (SIPs) that include NO_X emission limits for certain sources in order to achieve the required statewide emission budgets.

Title I of the 1990 CAAAs includes provisions designed to address both the continued nonattainment of the existing ozone NAAQSs and the transport of air pollutants across state boundaries. These provisions allow downwind states to petition for tighter controls on upwind states that contribute to their NAAQS nonattainment status. In general, Title I nitrogen oxide provisions with regard to ozone nonattainment regions:

- Require existing major stationary sources to apply reasonably available control technology (RACT), which is the lowest emission limitation that a particular source is capable of meeting by application of control technology that is reasonably available considering technological and economic feasibility;
- Require new or modified stationary sources to offset their emissions and install controls representing the lowest achievable emission rate, which is the minimum emissions rate accepted by the EPA for major new or modified sources in nonattainment areas;
- Require each state with an ozone nonattainment region to develop an SIP that may, in some cases, include reductions in stationary source

NO_X emissions beyond those required by the RACT provisions of Title I.

Section 184 of the Clean Air Act delineated a multistate ozone transport region (OTR) in the northeast and requires specific additional nitrogen oxide and volatile organic compound controls for all areas in this region. It also established the Ozone Transport Commission (OTC) for the purpose of assessing the degree of ozone transport in the OTR and recommending strategies to mitigate the interstate transport of pollution. The OTR consists of the states of Connecticut, Delaware, Maine, Maryland, Massachusetts, New Hampshire, New Jersey, New York, Pennsylvania, Rhode Island, Vermont, the northern counties of Virginia, and the District of Columbia. The OTR states confirmed that they would implement RACT on major stationary sources of NO_X (Phase I) and agreed to a phased approach for additional controls, beyond RACT, for power plants (>25 MW) and other large fuel combustion sources (industrial boilers with a rated capacity >250 million Btu per hour input; Phases II and III). This agreement, known as the OTC Memorandum of Understanding (MOU), was approved on September 27, 1994, at which time all OTR states except for Virginia signed the MOU. The MOU establishes an emission trading system to reduce the costs of compliance with the control requirements under Phase II, which began on May 1, 1999, and Phase III, which began on May 1, 2003. The OTC program capped summer-season (May 1 through September 30) NO_X emissions at approximately 219,000 short tons in 1999 and 143,000 short tons in 2003, which represents approximately 55 and 70% reductions in NO_X, respectively, from the 1990 baseline emission level of 490,000 short tons. While there are 13 affected regions (see Table 4-5), the actual reductions during the 1999 season reflect participation by only eight of the jurisdictions: Connecticut, Delaware, Massachusetts, New Hampshire, New Jersey, New York, Pennsylvania, and Rhode Island [23].

The EPA promulgated a rule on October 27, 1998, known as the NO_X SIP call, to address long-range transport of ozone. The purpose of this rule is to limit summer-season NO_X emissions in 22 northeastern states and the District of Columbia that EPA considers significant contributors to ozone nonattainment in downwind areas (see Table 4-5 for affected states). These states were required to amend their SIPs through a procedure established in Section 110 of the Clean Air Act to further reduce NO_X emissions by taking advantage of newer, cleaner control strategies. The EPA finalized a summer-season state NO_X budget and developed a state implemented and federally enforced NO_X trading program to provide for emissions trading by certain electric and industrial stationary sources. Each affected state's NO_X budget is based on a population-wide 0.15 lb/MM Btu NO_X emission rate for large electric generating stations and a 60% reduction from uncontrolled emissions for large electric generating units. This effort is projected to reduce

TABLE 4-5
Summary of Nitrogen Oxide Reduction and Trading Programs

	Ozone Transport Commission (OTC) NO_x Budget Program	*NO_x State Implementation Plan (SIP) Call*	*Section 126 Federal NO_x Budget Trading Program*	*Acid Rain Program*
Affected regions	District of Columbia and 12 states: CT, DE, MA, ME, MD, NH, NJ, NY, PA, RI, VA, VT	District of Columbia and 22 states: AL, CT, DE, GA, IL, IN, KY, MA, MD, MI, MO, NC, NJ, NY, OH, PA, RI, SC, TN, VA, WI, WV	District of Columbia and 12 states: DE, IN, KY, MD, MI, NC, NJ, NY, OH, PA, VA, WV	Entire nation
Compliance period	May 1–September 30 of each year	May 1–September 30 of each year (in 2004 the compliance period began May 31)	May 1–September 30 of each year	Annual
Initial compliance year	1999	2004 (phase I)	2003	Phase I January 1, 1996 Phase II January 1, 2000
Emissions cap	219,000 short tons in 1999; 143,000 short tons in 2003	—	289,983 short tons	—
NO_x reductions	246,000 short tons in 1999; 322,000 short tons in 2003	1.2 million short tons in 2007	800,000 short tons in 2007	340,000 short tons per year in Phase I; 2.06 million short tons per year in Phase II
Baseline year	1990	1995	1995	1980
Baseline emissions	490,000 million short tons NO_x	—	—	—
Program owner	OTC; allowances set by OTC, program administered by the EPA	States and the EPA; states have the option of participating in the trading program and establishing unit allocations; program administered by the EPA	EPA	EPA

Source: Adapted from EPA [23].

summer-season NO_X emissions by 1.2 million short tons in the affected 22 states and the District of Columbia by 2007 [24].

In addition to promulgating the NO_X SIP call, the EPA responded to petitions filed by eight northeastern states under Section 126 of the Clean Air Act. The petitions, known as the Section 126 petitions, request that the EPA make a finding that NO_X emissions from certain major stationary sources significantly contribute to ozone nonattainment problems in the petitioning states. The final Section 126 rule requires upwind states to take action to reduce emissions of NO_X that contribute to nonattainment of ozone standards in downwind states. The findings affect large electric generating stations and both nonelectric generating boilers and turbines located in 12 northeastern states and the District of Columbia (see Table 4-5 for affected states). Like the NO_X SIP call, the EPA has finalized a federal NO_X budget trading program based on application of a population-wide emission rate of 0.15 lb NO_X per MM Btu for large electric generating units and a 60% reduction from uncontrolled emissions for large nonelectric generating units. The final Section 126 actions are projected to reduce summer season NO_X emissions by 800,000 short tons/year in the affected area [25].

New Source Review

The New Source Review (NSR) program is one of many programs created by the Clean Air Act to reduce emissions of air pollutants, particularly "criteria" pollutants that are emitted from a wide variety of sources and have an adverse impact on human health and the environment. The NSR program was established in parts C and D of Title I of the Clean Air Act to protect public health and welfare, as well as national parks and wilderness areas, as new sources of air pollution are built and when existing sources are modified in a way that significantly increases air pollutant emissions. Specifically, the purpose of the NSR is to ensure that when new sources are built or existing sources undergo major modifications the air quality improves if the change occurs where the air currently does not meet federal air quality standards and air quality is not significantly degraded where the air currently meets federal standards [26].

The original intent of the NSR program was to ensure that major new facilities that are sources of emissions or existing facilities that are modified and result in increased emissions would install state-of-the-art controls. Subsequently, the EPA provided interpretive guidance that complicated the review program and expanded it to include maintenance or improvement. While the determination of whether an activity is subject to the major NSR program is fairly straightforward for a newly constructed source, the determination of what should be classified as a modification subject to a major NSR presents a more difficult issue; consequently, installation of new technology, greater energy efficiency, and improved environmental performance at facilities were being inhibited. In addition, there was much controversy between industry and the EPA over what "triggers" applicability of the major NSR program, which led to litigation between Wisconsin Electric Power Company (WEPCO) and the EPA. In 1992, the EPA promulgated revisions to the applicability regulations, creating special rules for physical and operational changes at electric utility steam-generating units.

On July 23, 1996, a number of changes were proposed to the existing major NSR requirements as part of a larger regulatory package [27]. This was followed by a Notice of Availability published by the EPA in the *Federal Register* on July 24, 1998, requesting comment on three of the proposed changes. After public comment, on December 31, 2002, the EPA issued an NSR final rule to improve the NSR program and a proposed rule to provide a regulatory definition of routine maintenance, repair, and replacement. In summary, the final rule that became effective March 3, 2003:

- Reforms the emissions accounting system for determining when a change is triggered (now the trigger is based on actual emissions);
- Allows already-controlled "clean units" to make changes without triggering an NSR;
- Broadens the exclusion from an NSR for projects intended for pollution control;
- Sets new rules for establishing plant-wide applicability limits of specific pollutants under the program.

The EPA took action to improve the NSR program after performing a comprehensive review of the program and, in June 2002, issued a Report to the President on NSR [26]. The report concluded that the program, as administered and as it related to the energy sector, impeded or resulted in the cancellation of projects that would maintain or improve the reliability, efficiency, or safety of existing power plants. The EPA issued the final rule improvements after the culmination of a 10-year process that included pilot studies and the engagement of state and local governments, environmental groups, private sector representatives, academia, and concerned citizens in an open and far-reaching public rulemaking process. In addition, the nation's governors and environmental commissioners, on a bipartisan basis, called for NSR reform.

The final rule implements the following major improvements to the NSR program [28]:

- *Plantwide applicability limits (PALs)*—To provide facilities with greater flexibility to modernize their operations without increasing air pollution, facilities that agree to operate within strict site-wide emission caps referred to as PALs will be given the flexibility to modify their operations without undergoing an NSR, as long as the modifications do not cause emissions to violate their plant-wide cap;

- *Pollution control and prevention projects*—To maximize investments in pollution prevention, companies that undertake certain specified environmentally beneficial activities will be free to do so upon submission to their permitting authority of a notice, rather than having to wait for adjudication of a permit application. The EPA is also creating a simplified process for approving other environmentally beneficial projects;
- *Clean unit provision*—To encourage the installation of state-of-the-art air pollution controls, the EPA will give plants that attain clean unit status flexibility in the future if they continue to operate within permitted limits. This flexibility is an incentive for plants to voluntarily install the best available pollution controls. Clean units must have an NSR permit or other regulatory limit that requires the use of the best air pollution technologies;
- *Emissions calculation test methodology*—To provide facilities with a more accurate procedure for evaluating the effect of a project on future emissions, the final regulations improve how a facility calculates whether a particular change will result in a significant emissions increase and thereby trigger NSR permitting requirements.

The EPA's proposed rule would make improvements to the routine maintenance, repair, and replacement exclusion currently contained in the EPA regulations. These proposed improvements will be subject to a full and open public rulemaking process. Since 1980, EPA regulations have excluded from NSR review all repairs and maintenance activities that are routine, but a complex analysis must be made to determine what activities meet the standard. This has deterred companies from conducting repairs and replacements that are necessary for the safe, efficient, and reliable operation of facilities.

After the new NSR final and proposed rules were issued, certain environmental groups and state and local governments petitioned the EPA to reconsider specific aspects of the final NSR reform rule. The EPA announced on July 25, 2003, that it would reconsider parts of the NSR final rule [29]. The EPA's notice was a response in part to the petitions and requests comment on six limited areas. The EPA will take action on the remaining issues raised by the petitioners at a later date. The EPA has not yet acted on the remaining issues but plans to make a decision within the time period it takes to reconsider the six items EPA has agreed to address. The EPA solicited comments on the following six areas [29]:

- The EPA report titled *Supplemental Analysis of the Environmental Impact of the 2002 Final NSR Improvement Rules*, which concluded that the NSR improvement rule will likely result in greater environmental benefits than the prior program;

- The decision to allow certain sources of air emissions to maintain Clean Unit status after an area is redesignated from attainment to nonattainment for one of the six criteria air pollutants;
- The EPA's inclusion of the reasonable possibility standard as it pertains to the need to maintain records and file certain reports which project actual emissions following a physical or operational change;
- The method for assessing air emissions from process units built after the 24-month baseline period used to establish PAL limits;
- The decision to allow a PAL to supersede existing emission limits established for NSR applicability purposes. Compliance with the PAL is then used to determine if NSR requirements apply in the future;
- The method of measuring emission increases when existing emission units are replaced.

On October 27, 2003, the EPA published its final NSR Equipment Replacement Rule [30]. This final version of the rule applies only to equipment replacement. It became effective December 26, 2003, and states will have up to 3 years to revise their state implementation plans to reflect these requirements. The regulation specifies that replacement components must be functionally equivalent to existing components in that there are no changes in the basic unit design or to pollutant emitting capacity. It also sets a 20% limit on replacement cost for equipment. If these restrictions are exceeded, the replacement work is subject to the NSR process.

Impending Legislation and Pollutants under Consideration for Regulation

Developing new or modifying current regulations of air pollutant emissions from coal-fired boilers is a continual process. This section summarizes impending legislation and pollutants under consideration for regulation and complements the previous discussion of possible NSR revisions. The following are discussed in this section: impending legislation for fine particulate matter and mercury emissions, the Clear Skies Act of 2002 and multi-pollutant legislation under consideration, and global climate changes issues.

Fine Particulate Matter

Epidemiological research over the past 10 years has revealed a consistent statistical correlation between levels of airborne fine particulate matter ($PM_{2.5}$) and adverse respiratory and cardiopulmonary effects in humans [31]. This has resulted in the EPA's promulgation of NAAQSs that limit the allowable mass concentrations of $PM_{2.5}$ (see earlier discussion of current NAAQSs).

Attainment of the $PM_{2.5}$ NAAQSs requires an annual average mass concentration of less than 15 mg/m^3 and a daily maximum concentration of less than 65 mg/m^3. The EPA and the states are now in the process of identifying attainment and nonattainment areas of the $PM_{2.5}$ NAAQSs. Ambient $PM_{2.5}$ has also been found to contribute significantly to the impairment of long-range visibility (regional haze) in many areas of the United States [32]. The EPA issued a Regional Haze Rule in 1999 that established goals for reducing regional haze in areas of the United States where long-range visibility has been determined to have exceptional value (Class I areas) and has outlined methods for achieving these goals [32].

It is generally recognized that coal-fired power plants can be important contributors to ambient $PM_{2.5}$ mass concentrations and regional haze; therefore, it is very likely that the EPA and/or state and local air pollution control agencies will require additional restrictions of coal power plant emissions from 2005 to 2008 as they develop SIPs for achieving and/or maintaining compliance with the $PM_{2.5}$ NAAQSs and the Regional Haze Rule [33]. However, specific requirements as to the types of pollutants to be reduced, and the timing, magnitude, and locations of these emissions restrictions have yet to be determined, although the process is underway to do this. The EPA provided guidance to state and local air pollution control agencies and tribes for designating areas for the purpose of implementing the $PM_{2.5}$ NAAQSs and plans to issue the final designations on December 15, 2004 [34,35]. The EPA's timeline for their $PM_{2.5}$ NAAQS implementation program is:

- September 2003—EPA issues proposed $PM_{2.5}$ implementation rule;
- February 15, 2004—State and tribal recommendations are due for $PM_{2.5}$ designations (recommendations can be based on 2002–2005 data);
- July 2004—The EPA notifies states and tribes concerning any modifications to their recommendations;
- September 2004—The EPA issues final $PM_{2.5}$ implementation rule;
- December 15, 2004—The EPA issues final $PM_{2.5}$ designations;
- December 2007—State implementation plans are due for $PM_{2.5}$ nonattainment areas (3 years after designation date);
- December 2009–2014—$PM_{2.5}$ standards must be attained by this date (5 years after designation date; an extension of up to 5 years is possible with an adequate demonstration).

Mercury

The EPA prepared a Mercury Study Report, dated December 1997, which they submitted to Congress on February 24, 1998, as a requirement of Section 112(n)(1)(B) of the 1990 CAAAs [36]. The report provided an assessment of the magnitude of U.S. mercury emissions by source, the health and environmental implications of those emissions, and the availability and cost of control technologies. The report identified electric utilities as the largest remaining source of mercury emissions in the air, as the EPA has regulated mercury emissions from municipal waste combustors, medical waste incinerators, and hazardous waste combustion.

The EPA also submitted the Utility Hazardous Air Pollutant Report to Congress on February 24, 1998, in which the EPA examined 67 air toxics emitted from 52 fossil fuel-fired power plants and concluded that mercury is the air toxic of greatest concern [37]. Although not conclusive, the report finds evidence suggesting a link between utility emissions and the methylmercury found in soil, water, air, and fish from contaminated waters. The report identified the need for additional information on the amount of mercury in U.S. coals and mercury emissions from coal-fired power plants. Specifically, such data identified by the EPA included additional data on the quantity of mercury emitted from various types of generating units; the amount of mercury that is divalent versus elemental; and the effect of pollution control devices, fuel type, and plant configuration on emissions and speciation. To obtain these data, the EPA issued a three-part Information Collection Request (ICR) for calendar year 1999 [38]. Part I collected information on the size and configuration of all coal-fired utility boilers greater than 25 MW and their pollution control devices. Part II obtained data quarterly on the origin, quantity, and analysis of coal shipments delivered to the generating units (which totaled more than 1100), including a minimum of three analyses per month for mercury and chlorine contents, together with any other available analyses such as ash and sulfur contents and heating value. Part III required emission tests on 84 generating units selected at random from 36 categories representing different plant configurations and coal rank to measure total and speciated mercury concentrations in the flue gas before and after the final air pollution control device upstream of the stack.

On December 14, 2000, the EPA announced it would regulate mercury emissions from power plants [39]. On December 15, 2003, the EPA proposed a rule to permanently cap and reduce mercury emissions from power plants [6]. The schedule requires a final rule by December 2004 and implementation of controls by the end of 2007. The EPA is proposing two alternatives for controlling emissions of mercury and will take comment on the alternatives before taking final action. The first alternative requires utilities to install controls known as maximum achievable control technologies (MACTs) under Section 112 of the Clean Air Act. If implemented, this proposal will reduce nationwide emissions of mercury by 14 tons (29% reduction of current levels) by the end of 2007. The second alternative establishes standards of performance limiting mercury emissions from new and existing units. This proposal, under Section 111 of the Clean Air Act, would create a market-based cap-and-trade program that, if implemented, would reduce nationwide utility emissions in two phases. In the first phase, extending to 2010, emissions would be reduced by taking advantage of co-benefit controls (*e.g.*, mercury reductions achieved by reducing SO_2 and NO_X emissions). When fully implemented in 2018, mercury emissions will be reduced by 33 tons (69% reduction from current levels).

Multi-Pollutant Legislation

For the last 5 years, Congress has discussed and started proposing legislation for more stringent control of power plant emissions, including SO_2, NO_X, mercury, and CO_2. Recent proposals include the Clear Skies Initiative, Clean Power Act, and Clean Air Planning Act. The Clear Skies Initiative was the first of several multi-pollutant bills proposed and would create a mandatory program to reduce power plant emissions of SO_2, NO_X, and mercury by setting a national cap on each pollutant. Clear Skies was proposed in response to a growing need for an emission reduction plan that will protect human health and the environment while providing regulatory certainty to the industry. The program was submitted as proposed legislation in July 2002. The program was reintroduced in the U.S. House of Representatives (H.R. 999) and the U.S. Senate (S. 485) on February 27, 2003 [40].

The emission reductions from Clear Skies would help to alleviate air pollution-related health and environmental problems, including fine particles, ozone, mercury, acid rain, nitrogen deposition, and visibility impairment [41]. Specifically, Clear Skies would: (1) reduce SO_2 emissions by 73%, from year 2000 emissions of 11 million short tons to a cap of 4.5 million short tons in 2010 and to a cap of 3 million short tons in 2018; (2) reduce NO_X emissions by 67%, from year 2000 emissions of 5 million short tons to a cap of 2.1 million short tons in 2008 and to a cap of 1.7 million short tons in 2018; and (3) reduce mercury emissions, through a first-ever national cap on mercury emissions, by 69%, from 1999 emissions of 48 short tons to a cap of 26 short tons in 2010 and to a cap of 15 short tons in 2018. The reduction in emissions is illustrated in Figure 4-2.

Clear Skies, which is modeled on the cap-and-trade provisions of the Acid Rain Program of the 1990 CAAAs, would use an emission caps system that is predicted to [41]:

- Protect against diseases by reducing smog and fine particles, which contribute to respiratory and cardiovascular problems;
- Protect our wildlife, habitats, and ecosystem health by reducing acid rain, nitrogen, and mercury deposition;
- Deliver a rapid reduction in emissions with certain improvements in air quality;
- Enable power generators to continue to provide affordable electricity while quickly and cost-effectively improving air quality and the environment;
- Encourage the use of new and cleaner pollution control technologies that would further reduce compliance costs.

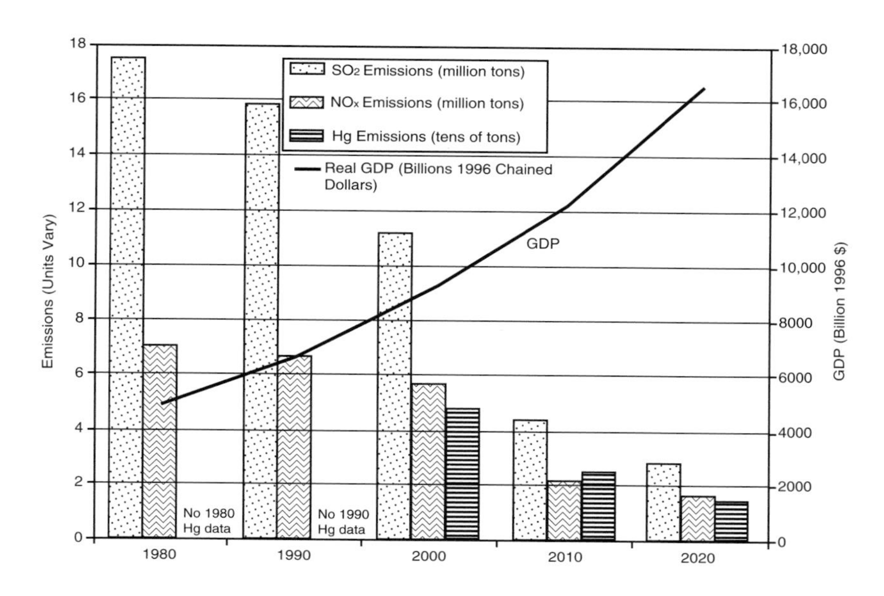

FIGURE 4-2. Emission levels of SO_2, NO_X, and mercury under the Clear Skies Act.

Clear Skies is expected to provide significant benefits to public health and the environment at a reasonable cost [41]. The EPA projects that, by 2020, the public health benefits alone would include avoiding more than 8400 to 14,000 premature deaths and saving $21 to $110 billion per year, depending on the methodology for calculating the health-related benefits. The annual cost of achieving the Clear Skies emission levels is projected at $6.3 billion. Americans would also experience approximately 30,000 fewer visits to the hospital and emergency room, 23,000 fewer nonfatal heart attacks, 1.6 million fewer work loss days, and 200,000 fewer school absences each year. Benefits of improvements in visibility in national parks and wildernesses in 2020 are projected to be $3 billion annually. In addition, by 2020, an estimated 77 counties with 26 million people would meet the fine particulate NAAQSs.

The future of the Clear Skies Act is unknown. Congressional Republican support for President Bush's planned legislation is becoming increasingly uncertain [42]. As of early July 2003, three bipartisan bills have been introduced in Congress that would reduce SO_2, NO_X, and mercury emissions to lower levels than the Clear Skies Initiative and on a faster timeline. In addition, the other bills have plans to limit CO_2 emissions. Hearings are planned to explore criticisms of the Clear Skies Initiative before Congress decides whether to move the legislation forward [42].

Two other proposals receiving considerable attention include the Clean Power Act (S. 386) sponsored by Senator Jim Jeffords (Independent–Vermont) and the Clean Air Planning Act sponsored by Senators Tom Carper

TABLE 4-6
Overview of Multipollutant Bills

	Clear Skies Act	*Clean Power Act*	*Clean Air Planning Act*
SO_2 cap	4.5 million short tons in 2008; 3.0 million short tons in 2018	2.25 million short tons by 2009 (0.28 million short tons in western region and 1.98 million short tons in eastern region)	4.5 million short tons by 2009; 3.5 million short tons by 2013; 2.25 million short tons by 2016
NO_X cap	2.1 million short tons in 2008; 1.7 million short tons in 2018	1.51 million short tons by 2009	1.87 million short tons by 2009; 1.7 million short tons by 2013
CO_2 cap	None	2.05 billion short tons by 2009	In 2009, stabilize at 2006 levels (~2.57 billion short tons) plus flexibility measures; in 2013, cap at 2001 levels (~2.47 billion short tons) plus flexibility measures
Mercury cap	26 short tons in 2010; 15 short tons in 2015	5 short tons by 2009	24 short tons by 2009; 10 short tons by 2013; 50 and 70% reductions required at each plant in 2009 and 2013, respectively
Emission trading	Trading allowed for SO_2, NO_X, and mercury	Trading allowed for SO_2, NO_X, and CO_2; no trading for mercury	Cap-and-trade for NO_X, SO_2, CO_2, and mercury, along with facility-specific mercury requirements

Source: Adapted from Tatsutani [43].

(Democrat–Delaware), Lincoln Chaffee (Republican–Rhode Island), and Judd Gregg (Republican–New Hampshire). These bills are currently being debated, and key elements of each are summarized in Table 4-6. The Clear Skies Act is included in Table 4-6 for comparison.

In addition to the proposed legislation listed in Table 4-6, the EPA announced a proposal, called the Interstate Air Quality Rule, on December 4, 2003, that will require coal-fired power plants to make the largest reductions in NO_X and SO_2 in over a decade [44]. The rule focuses on states that significantly contribute to ozone and fine particulate pollution in the eastern United States. This rule will reduce power plant emissions in two phases. Sulfur dioxide emissions will drop by 3.7 million short tons by 2010 (a reduction of ~40% from current levels) and by another 2.3 million short tons when the rules are fully implemented after 2015, resulting in a total reduction of ~70% from current levels. Nitrogen oxide emissions will be cut by 1.4 million short tons by 2010 and by a total of 1.7 million short tons by 2015, a reduction of ~50% from current levels in the 30 states covered under the rule. In addition, emissions will be permanently capped and cannot increase. The EPA is taking public comment on the rule, and a final rule is planned for 2005.

The future of these bills (Table 4-6) and the EPA's recent proposals (mercury and interstate air quality rules) is uncertain; however, industry is expecting and readying itself for some form of multi-pollutant legislation. Industry, the U.S. Department of Energy (DOE), and the Electric Power Research Institute are developing and commercializing a number of multi-pollutant control technologies to optimize removal efficiencies of various power plant emissions. These control technologies are discussed in Chapter 6 (Emissions Control Strategies for Power Plants).

Global Climate Change

Similar to the multi-pollutant proposals discussed previously, global climate change is on the agenda of the 108th Congress and the Bush Administration [45–48]. Congress is considering several legislative initiatives, with key issues being considered that include better coordination of government research, setting caps on greenhouse gas (GHG) emissions, and whether reporting systems should be voluntary or mandatory. Climate change debates are expected to occur in connection with stand-alone climate change legislation, as well as during discussions of energy policy and the Clear Skies Initiative [45]. Several proposed bills address climate change research and data management, managing risks of climate change, reporting of GHG emissions, and stabilization of and caps on GHG emissions. This is a hotly debated topic, and industry expects some form of legislation to be passed; however, it is not clear what the final form of legislation will contain or if it will be stand-alone legislation or combined with other legislation such as a modified Clear Skies Initiative.

The Bush Administration is also taking steps in addressing GHG emissions. On June 25, 2003, DOE Secretary Spencer Abraham and energy ministers from around the world signed the first international framework for research and development on the capture and storage of CO_2 emissions. This initiative is called the Carbon Sequestration Leadership Forum and includes the following participants: Australia, Brazil, Canada, China, Colombia, India, Italy, Japan, Mexico, Norway, Russian Federation, United Kingdom, United States, and the European Commission [46]. It was soon followed with the announcement on July 23, 2003, of the Bush Administration's unveiling of a long-term strategic plan to study global change [47,48]. Presented by Commerce Secretary Don Evans and DOE Secretary Spencer Abraham, the 10-year plan sets five goals, including:

- Identify the natural variability in the Earth's climate;
- Understand the forces that cause global warming;
- Reduce the uncertainties in climate forecasting;
- Improve the understanding of sensitivity and adaptability of ecosystems to climate change;
- Develop more exact methods for calculating the risks of global warming.

The Bush Administration also plans to spend $103 million performing a 2-year, high-priority research study to collect data relating to carbon pollution, aerosols, and the oceans.

Emissions Legislation in other Countries

A brief discussion of regulatory requirements for sulfur dioxide, nitrogen oxides, particulate matter, trace elements (specifically, mercury), and carbon dioxide for many countries is presented in this section. The emissions standards of other countries are compared to those from the United States.

Sulfur Dioxide

The United Nations Economic Commission for Europe's (UNECE) Convention on Long-Range Transboundary Air Pollution (LRTAP) was the first legally binding instrument to address air pollution on a broad regional context [49]. The Convention was adopted in 1979, came into force in 1983, and now has been ratified by 48 countries, as listed in Table 4-7 [49,50]. Under the convention, the countries recognize the transboundary problems of air pollution and accept general responsibility to move toward a solution to these problems. Following the LRTAP Protocol, the SO_2 Helsinki Protocol (the "30% Club") was signed in 1985 and came into force in 1987 (see Table 4-7). Under this protocol, the signatories agreed to reduce their SO_2 emissions by 30% (based on 1980 values) by 1993 [49]. The Second Sulfur Protocol, the Protocol on Further Reductions of Sulfur Emissions, was signed in June 1994 by 27 European countries, the European Community (EC), and Canada. The protocol came into force in August 1998. All signatories, which are listed in Table 4-7, were allocated targets for 2000, while some countries agreed to additional targets for 2005 and 2010.

The UNECE Gothenburg Protocol to abate acidification, eutrophication, and ground-level ozone was signed by 27 countries in December 1999. As of June 15, 2003, it had 31 signatures and 4 ratifications [50]. The protocol

TABLE 4-7
Signatories to the UNECE LRTAP and Subsequent SO_2 Protocols

1979 LRTAP Convention	*1985 First Sulfur Protocol*	*1994 Second Sulfur Protocol*
Armenia (+)[a]	—	—
Austria (+)	Austria (+)	Austria (+)
Belarus (+)	Belarus (+)	—
Belgium (+)	Belgium (+)	Belgium (+)
Bosnia and Herzegovina (Sc)[b]	—	—
Bulgaria (+)	Bulgaria (+)	Bulgaria
Canada (+)	Canada (+)	Canada (+)
Croatia (Sc)	—	Croatia (+)
Cyprus (Sc)	—	—
Czech Republic (Sc)	Czech Republic (Sc)	Czech Republic (+)
Denmark (+)	Denmark (+)	Denmark (+)
Estonia (+)	—	—
European Community (+)	—	European Community (+)
Finland (+)	Finland (+)	Finland (+)
France (+)	France (+)	France (+)
Georgia (+)	—	—
Germany (+)	Germany (+)	Germany (+)
Greece (+)	—	Greece (+)
Holy See	—	—
Hungary (+)	Hungary (+)	Hungary
Iceland (+)	—	—
Ireland (+)	—	Ireland (+)
Italy (+)	Italy (+)	Italy (+)
Kazakhstan (+)	—	—
Kyrgyzstan (+)	—	—
Latvia (+)	—	—
Liechtenstein (+)	Liechtenstein (+)	Liechtenstein (+)
Lithuania (+)	—	—
Luxembourg (+)	Luxembourg (+)	Luxembourg (+)
Malta (+)	—	—
Monaco (+)	—	—
Netherlands (+)	Netherlands (+)	Netherlands (+)
Norway (+)	Norway (+)	Norway (+)
Poland (+)	—	Poland
Portugal (+)	—	—
Republic of Moldova (+)	—	—
Romania (+)	—	—
Russian Federation (+)	Russian Federation (+)	Russian Federation
San Marino	—	—
Slovakia (Sc)	Slovakia (Sc)	Slovakia (+)
Slovenia (Sc)	—	Slovenia (+)
Spain (+)	—	Spain (+)

(continued)

TABLE 4-7
(continued)

1979 LRTAP Convention	*1985 First Sulfur Protocol*	*1994 Second Sulfur Protocol*
Sweden (+)	Sweden (+)	Sweden (+)
Switzerland (+)	Switzerland (+)	Switzerland (+)
Former Yugoslav Republic of Macedonia (Sc)	—	—
Turkey (+)	—	—
Ukraine (+)	Ukraine (+)	Ukraine
United Kingdom (+)	—	United Kingdom (+)
United States (+)	—	—
Yugoslavia (Sc)	—	—

[a]Ratification, accession, approval, or acceptance.
[b]Succession.
Sources: Soud, H. N., *Developments in FGD*, IEA Coal Research, London, 2000; Wu, Z., *NO_x Control for Pulverized Coal Fired Power Stations*, IEA Coal Research, London, 2002. With permission.

sets emission ceilings for SO_2, NO_x, volatile organic compounds, and ammonia for the year 2010.

On November 24, 1998, the EC adopted the Large Combustion Plants Directive (LCPD) and, with an amendment in December 1994, has set targets and emission limits for air pollutants, including SO_2, for plants >50 megawatts thermal (MW_t) for both exiting and new facilities. Revisions to the directive are proposing stricter SO_2 and NO_x emission ceilings to be achieved by the year 2010, and regulations on smaller-sized units are being discussed [49]. In 1996, the European Union Environment Ministers adopted a Directive on Integrated Pollution Prevention and Control (IPPC) [49]. The directive came into force in October 1996, mandated implementation by October 1999, and applies to all new installations and those undergoing a substantial change. The main purpose of the IPPC directive is to achieve integrated prevention and control of air pollution.

In most countries, the legislators target new and large facilities. More than 30 countries have adopted or are in the processing of introducing legislation limiting SO_2 emissions from their coal-fired power plants [49]. A range of national emissions standards for sulfur emissions is given in Table 4-8 [49].

Nitrogen Oxides

Nitrogen oxide emission standards have been introduced or are becoming more stringent around the world due to increased concerns about the local, regional, and transboundary effects of NO_x emissions. Generally, international legislation has been an important factor in developing national regulations in many parts of the world. Recognition of the transboundary effect of air pollution has led to a number of international agreements [50]. Those that pertain to NO_x emissions include the UNECE LRTAP (discussed in the previous section), EC directives, and World Bank environmental guidelines. In addition to the SO_2 emissions discussed earlier, the UNECE

TABLE 4-8
Range of National Emission Standards for Sulfur Emissions (mg SO_2/m^3)[a]

Country	*New Plants*	*Existing Plants*	*Coal Sulfur Limit (%)*	*Applicable Plant Size for Coal Sulfur Limit*
Austria[b]	200–550	200–2000	≤1 (Brown coal)	Plants <10 MW_t
Belgium[b]	250–2000	1700	≤1	All plants
Bulgaria	610–3335	1875–3500	—	—
Canada[c]	740	—	—	—
China	—	—	—	—
Hong Kong	~200	—	≤1	>550 MW_t
Mainland	1200–2000	Plant specific	—	—
Croatia	400–2000[d]	2000[d]	< 0.7 g/MJ	<50 MW_t
Czech Republic	500–2500	500–2500	—	—
Denmark[b]	400–2000	—[e]	0.9	Industrial plants
European Union	400–2000[f]	—[g]	—	—
Finland[b]	380–620	620	≤390 g/GJ	Plants without FGD[h]
France[b]	400–2000	400–2000	—	—
Germany[b]	400–2000	400–2000	<1	Plants <1 MW_t
Greece[b]	400–2000	—	—	—
Hungary	400–2000	—	—	—
Indonesia	820	1635[i]	—	—
Ireland[b]	400–2000	—	—	—
Italy[b]	400–2000	—	1	All plants
Japan	—[j]	—[j]	—	—
Korea, South	345	430-770	—	—
Luxembourg[b]	400–2000	—	—	—
Netherlands[b]	200–700	400	1.2 (industrial)	Plants without FGD
Philippines	175–765	—	<1	All existing plants
Poland	540–1755	675–2890	—	—
Portugal[b]	100–2000	—	—	—
Romania	400–2000	—	—	—
Slovakia	400–2000	500–2500	—	—
Slovenia	400–2000	2000	—	—
Spain[b]	400–2000	2400–9000	—	—
Sweden[b]	160–270	270–540	—	—

(continued)

TABLE 4-8
(continued)

Country	*New Plants*	*Existing Plants*	*Coal Sulfur Limit (%)*	*Applicable Plant Size for Coal Sulfur Limit*
Switzerland	430–2145	430–2145	≤1	All plants
Taiwan	570–1430	570–1430	—	—
Thailand	180	290–390[k]	—	—
Turkey	430–2500	430–2500	—	—
United Kingdom[b]	200–2000	—[e]	—	—
United States	740–1480	1480[e]	—	—

[a]Based on dry flue gas at 6% O_2, STP (0°C [275 K], 101.3 kPa).
[b]European Union (EU) country.
[c]National guidelines.
[d]Proposed standards.
[e]Based on annual quota totals.
[f]EU proposed limits scheduled to come into operation after January 1, 2000. New plants, 50–100 MW_t, 850 mg/m^3; >300 MW_t, 200 mg/m^3; 100–300 MW_t, sliding scale within upper and lower limits proposed.
[g]European Parliament proposed limits for existing plants granted a license before January 1, 2000, scheduled to become mandatory from January 1, 2005. Existing plants, 50–100 MW_t, 900 mg/m^3; >300 MW_t, 300 mg/m^3; 100–300 MW_t, sliding scale within upper and lower limits proposed.
[h]Flue gas desulfurization.
[i]Must have met new plant standards by January 1, 2000.
[j]Set on a plant-by-plant basis according to nationally defined formula.
[k]New and existing industrial plants.
Source: Soud, H. N., *Developments in FGD*, IEA Coal Research, London, 2000. With permission.

Convention on LRTAP also addressed transboundary NO_x emissions. Table 4-9 lists the signatories to the UNECE Convention on LRTAP and the status of each country [50]. The Sofia NO_x protocol was signed in 1998 by 23 countries and came into force in 1991. The protocol required that NO_x emissions be frozen at 1987 levels by the end of 1994 and that these levels be maintained in subsequent years. The protocol has 25 signatures and 28 ratifications as listed in Table 4-9 [50]. Table 4-9 also contains the signatories to the 1999 Protocol to Abate Acidification, Eutrophication, and Ground-Level Ozone. The protocol, signed in Gothenburg, Sweden, places NO_x emission limits of 400, 300, and 200 mg NO_x per m^3 for new installations with capacities of 50 to 100, 100 to 300, and >300 MW_t, respectively [50]. Existing installations are limited to 650 mg NO_x per m^3 for solid fuels in general and 1300 mg NO_x per m^3 for solid fuels with >10% volatile matter content.

The EC has adopted several directives and amendments, including the Directive on Controlling Emissions from Large Combustion Plants, the Directive on the Limitation of Emissions of Certain Pollutants into the Air

TABLE 4-9
Signatories to the UNECE Convention on LRTAP and Subsequent NO_X Protocols as of June 15, 2002

1979 LRTAP Convention	*1988 Sofia Protocol*	*1999 Gothenburg Protocol*
Armenia (Ac)	Armenia	—
Austria (R)	Austria (R)	Austria
Belarus (R)	Belarus (At)	—
Belgium (R)	Belgium (R)*	Belgium
Bosnia and Herzegovina (Sc)	—	—
Bulgaria (R)	Bulgaria (R)	Bulgaria
Canada (R)	Canada (R)	Canada
Croatia (Sc)	Croatia	—
Cyprus (Ac)	—	—
Czech Republic (Sc)	Czech Republic (Sc)	Czech Republic
Denmark (R)	Denmark (At)*	Denmark (Ap)
Estonia (Ac)	Estonia (At)	—
European Community (Ap)	European Community (Ac)	—
Finland (R)	Finland (R)*	Finland
France (Ap)	France (Ap)*	France
Georgia (Ac)	—	—
Germany (R)	Germany (R)*	Germany
Greece (R)	Greece (R)	Greece
Holy See	—	—
Hungary (R)	Hungary (Ap)	Hungary
Iceland (R)	—	—
Ireland (R)	Ireland (R)	Ireland
Italy (R)	Italy (R)*	Italy
Kazakhstan (Ac)	—	—
Kyrgyzstan (Ac)	—	—
Latvia (Ac)	Latvia	—
Liechtenstein (R)	Liechtenstein (R)*	Liechtenstein
Lithuania (Ac)	—	—
Luxembourg (R)	Luxembourg (R)	Luxembourg (R)
Malta (Ac)	—	—
Monaco (At)	—	—
Netherlands (At)	Netherlands (At)*	—
Norway (R)	Norway (R)*	Norway (R)
Poland (R)	Poland	Poland
Portugal (R)	Portugal	—
Republic of Moldova (Ac)	Republic of Moldova	—
Romania (R)	Romania	—
Russian Federation (R)	Russian Federation (At)	—
San Marino	—	—
Slovakia (Sc)	Slovakia (Sc)	Slovakia
Slovenia (Sc)	Slovenia	—

(continued)

TABLE 4-9
(continued)

1979 LRTAP Convention	*1988 Sofia Protocol*	*1999 Gothenburg Protocol*
Spain (R)	Spain (R)	Spain
Sweden (R)	Sweden (R)*	Sweden (R)
Switzerland (R)	Switzerland (R)*	Switzerland
The Former Yugoslav Republic of Macedonia (Sc)	—	—
Turkey (R)	—	—
Ukraine (R)	Ukraine (At)	—
United Kingdom (R)	United Kingdom (R)	United Kingdom
United States (At)	United States (At)	United States
Yugoslavia (Sc)	—	—

Note: R, ratification; Ac, accession; Ap, approval; At, acceptance; Sc, succession; *, committing to 30% reduction.

Source: Wu, Z., *NO_X Control for Pulverized Coal Fired Power Stations*, IEA Coal Research, London, 2002. With permission.

from Large Combustion Plants, and the Directive on National Emission Ceilings for Certain Atmospheric Pollutants [50]. NO_X emission limits for solid fuel-fired boilers in the new EC directives are 600 and 500 mg NO_X per m^3 for existing installations with capacities of 50 to 500 and >500 MW_t, respectively. These limits become stricter beginning January 1, 2016, for larger units, with the limit decreasing to 200 mg NO_X per m^3 for units >500 MW_t. Emission limits for new installations with capacities of 50 to 100, 100 to 300, and >300 MW_t are, respectively, 400, 200 (300 for biomass-fired units), and 200 mg NO_X per m^3. National ceilings for NO_X emissions for 2010, under the EC directive, are shown in Table 4-10 [50].

The World Bank has also developed environmental guidelines that are to be followed in all the projects it funds, thereby covering a host of developing countries. The World Bank has determined that environmental standards of developed countries may not be appropriate for developing countries or economies in transition; therefore, their guidelines are flexible and try to maintain and improve environmental quality on an ongoing basis [50]. The World Bank's NO_X standards are 750 mg NO_X per m^3 for all coal-fired power plants except for those firing coal with less than 10% volatile matter content, where the NO_X emission limit is 1300 mg NO_X per m^3.

National standards for NO_X emissions from coal-fired power plants have been adopted or are being introduced in more than 30 countries [50]. They vary widely between countries and are often determined by taking into account the technology available, type of plant (new or existing), size of plant, and boiler configuration. A comparison of emission standards for various countries is provided in Table 4-11 [51].

TABLE 4-10
National NO_X Emission Ceilings for 2010

Country	NO_X Emissions (metric kiloton)	
	1990[a]	2010
Austria	193	103
Belgium	339	176
Denmark	272	127
Finland	300	170
France	1865	810
Germany	2706	1051
Greece	326	344
Ireland	118	65
Italy	1935	990
Luxembourg	23	11
Netherlands	580	260
Portugal	317	250
Spain	1156	847
Sweden	338	148
United Kingdom	2756	1167
Total	13,227	6519

[a]The latest reported by each country to the LRTAP Convention.
Source: Wu, Z., *NO_X Control for Pulverized Coal Fired Power Stations*, IEA Coal Research, London, 2002. With permission.

Particulate Matter

Standards for the control of particulate emissions were first introduced in the early 1900s in Japan, the United States, and Western Europe [52]. Over the decades, they have become increasingly more stringent and widespread, with recent emphasis being placed on the control of fine particulate matter. Similar to those for NO_X and SO_2 emissions, international agreements have been signed to reduce particulate emissions from coal-fired power plants. The EC set limits for particulate emissions from coal-fired power plants in the LCPD. These set particulate emissions limits of 100 and 50 mg/m^3 for new plants with capacities of 50 to 500 and $\geq$500 MW_t, respectively [52]. Even stricter limits are under review to conserve the environment and human health. Increasingly more stringent national emission standards have been adopted in Japan, North America, and Western Europe. The growing importance of using coal in an environmentally acceptable manner for power generation as well as in the industrial and residential sectors has led to the introduction of particulate emission standards in other countries as well. Currently, 30 countries have emissions standards for particulate emissions from coal-fired power plants [52]. Examples of particulate emissions standards in some of these counties are provided in Table 4-12 [51].

TABLE 4-11
NO_X Emission Standards Applicable to New and Existing Coal-Fired Power Plants of Thermal Capacity >300 MW_t

Country	*NO_X Emissions (mg/m^3)[a]*	
	New Plant	*Existing Plant*
Austria	200	200–300
Canada[b]	490–740	—
European Community	650–1300	—
Denmark	200–650	—
Germany	200	200
Italy	200–650	200–650
Japan	410–515	410–515
Korea, South	720	720
Netherlands	200–400	650–1100
Poland	405–460	610–1335
Spain	650–1300	—
United Kingdom	650–1300	—
United States[b]	615–740	555–615

[a]Standards are given in mg/m^3 corrected to standard conditions (6% O_2, standard temperature and pressure—0°C (273 K), 101.3 kPa—on dry flue gas); when converting from lb/million Btu to mg/m^3, dry flue gas volume is assumed to be 350 m^3/GJ (based on gross heat value). Note that ranges exist because emission standards may vary according to plant type, size, location, construction/commissioning date, boiler configuration, and type of coal used.

[b]Federal standards only; state standards may be more stringent.

Source: McConville, A., *An Overview of Air Pollution Emission Standards for Coal-Fired Plants Worldwide,* Coal and Slurry Technology Association, Washington, D.C., 1997), pp. 1–12.

Trace Elements/Mercury

Concern over environmental effects of trace elements emissions, specifically cadmium, chromium, copper, mercury, nickel, lead, selenium, and zinc, from human activities has led to the introduction of legislation on emissions in many countries; however, this legislation sets limits for medical waste incinerators, municipal solid waste combustors, and hazardous waste incinerators [53]. Trace element emissions from coal combustion are not currently regulated. An overview of mercury regulations in the European Union, Japan, and the United States reveals a variety of different approaches [54]. The 1998 United Nations Protocol on Heavy Metals set emissions limits for hazardous and municipal waste incinerators and directed signatories to the protocol to set limits for medical waste incinerators. The European Union has approved the protocol, and the United States has accepted but not ratified it. Separate from the United Nations protocol, the European Council issued a directive in 1996 ordering limit values and alert thresholds for a variety of air pollutants including mercury [54]. The directive resembles

TABLE 4-12
Particulate Emission Standards Applicable to New and Existing Coal-Fired Power Plants of Thermal Capacity >300 MW_t

Country	*Particulate Emissions (mg/m^3)*[a]	
	New Plant	*Existing Plant*
Austria	50	50
Canada[b]	145	—
European Community	50–100	—
Denmark	40	40–120
Germany	50	80–125
India	150–350	150–350
Italy	50	50
Japan	50–300	50–300
Korea, South	50–100	50–100
Netherlands	50	—
Poland	190–350	460–700
Spain	50–100	200–500
United Kingdom	50–100	—
United States[b]	40–125	40–125

[a]Standards are given in mg/m^3 corrected to standard conditions (6% O_2, standard temperature and pressure—0°C (273 K), 101.3 kPa—on dry flue gas); when converting from lb/million Btu to mg/m^3, dry flue gas volume is assumed to be 350 m^3/GJ (based on gross heat value); note that ranges exist because emission standards may vary according to plant type, size, location, construction/commissioning date, boiler configuration, and type of coal used.

[b]Federal standards only; state standards may be more stringent.

Source: McConville, A., *An Overview of Air Pollution Emission Standards for Coal-Fired Plants Worldwide* (Coal and Slurry Technology Association, Washington, D.C., 1997), pp. 1–12.

aspects of the U.S. federal regulation in that it assigns to member states the responsibility to implement the limit values and all attainment programs. In response to the directive, the EC proposed an ambient air quality standard of 0.05 mg/m^3 for elemental mercury. The proposed standard is rarely exceeded in Europe. The United States and Japan do not have such a standard. The United States has regulated all significant sources of mercury emissions in a manner consistent with the United Nations protocol with the exception of power plant emissions. As discussed earlier, however, regulations are still forthcoming. The European Union also has not regulated mercury emissions from power plants.

Carbon Dioxide

Fossil fuel consumption is projected to increase over the next 20 years, as was discussed in Chapter 2 (Past, Present, and Future Role of Coal), with coal being the leading energy source in some countries, especially certain developing countries; consequently, carbon dioxide (CO_2) emissions are projected to increase. The increase in CO_2 emissions and concern about global warming have received international attention. The first major action was taken in New York on May 9, 1992, when the United Nations Framework Convention on Climate Change was adopted. The objective of the Convention is to achieve stabilization of greenhouse gas concentrations in the atmosphere at a level that would prevent dangerous interference with the climate system [55]. Stabilization must be achieved in such a time-frame as to ensure that food production is not threatened and to allow economic development to proceed in a sustainable manner. The Convention contains a legally binding framework that commits the world's governments to voluntary reductions of greenhouse gases or other actions such as enhancing greenhouse gas sinks aimed at stabilizing atmospheric concentrations of greenhouse gases at 1990 levels by the year 2000 [56].

On June 12, 1992, at the Earth Summit in Rio de Janeiro, 154 nations, including the United States, signed the United Nations Framework Convention on Climate Change. In October 1992, the United States became the first industrialized nation to ratify the treaty, which came into force on March 21, 1994. The treaty was not legally binding and, because reducing emissions would likely cause great economic damage, many nations were not expected to meet the goal [57]. The Convention has become a cornerstone of global climate policy representing a compromise among a wide range of different interests among the member countries. The concept of a common goal but different responsibilities provided for different roles for industrialized and developing countries, notably in the obligations imposed on them in connection with climate protection policy [58]. This led to a grouping of the member states of the Convention into Annex I, Annex II, and non-Annex I countries, the latter including developing countries with no commitments to reducing climate gases. Annex I countries agreed, among other issues, to adopt national policies and take corresponding measures on the mitigation of climate change, periodically provide information on its policies and measures to mitigate climate change, and calculate emissions sources and removal through sinks. The developed countries in Annex II agreed, along with additional provisions, to provide new and additional financial resources to meet the agreed full costs incurred by developing countries in complying with their obligations.

Representatives from around the world met again in December 1997 at a conference in Kyoto to sign a revised agreement. The Clinton Administration negotiators agreed to legally binding, internationally enforceable limits on the emissions of greenhouse gases as a key tenet of the treaty. The protocol called for a worldwide reduction of emissions of carbon-based gases by an average of 5.2% below 1990 levels by 2010. Different countries adopted different targets. Those countries agreeing to reduce specified amounts of climate gases within a specified time period are listed as Annex B countries, which is a subcategory of the Annex I countries. For example, the EU committed to a reduction of 8% in climate gases, the United States to 7%, and Japan to 6%, while Russia and the Ukraine agreed to stabilize at 1990 levels. Table 4-13 lists the Annex I, Annex II, and Annex B countries, along with the specified amounts of climate gases agreed upon by the Annex B countries [58,59].

Conflicts with regard to the distribution of different obligations have become apparent since the Kyoto conference. In March 2001, the United States announced that it would not support the Kyoto Protocol [60]. The United States insists that the rules pertaining to the Annex B countries (*i.e.*, voluntary commitment to reducing climate gases) be extended to at least the major developing countries and made this a precondition to ratifying the Kyoto Protocol.

In November 2001, the participating member countries of the United Nations Seventh Conference of Parties (COP-7) met in Marrakesh, Morocco, and reached final agreement for the procedures and institutions needed to make the Kyoto Protocol fully operational [60]. On March 4, 2002, the EU voted to ratify the protocol, committing its 15 member countries to reductions in greenhouse gas emissions as specified in the accord. No agreement has been reached among the EU member countries, however, with regard to the individual emission reductions that will be required. Some countries feel they have been given a disproportionate share of the EU's total reduction burden [60]. The Kyoto Protocol enters into force 90 days after it has been ratified by at least 55 parties to the United Nations Framework Climate Change Convention, including a representation of Annex I countries accounting for at least 55% of the total 1990 CO_2 emissions from the Annex I group. Although the United States had the largest share of Annex I emissions in 1990 at 35%, even without U.S. participation the Protocol could enter into force for the other signatories [60].

Air Quality and Coal-Fired Emissions

The EPA evaluates the status and trends in the nation's air quality on a yearly basis and tracks air pollution by evaluating the air quality measured from over 5200 ambient air monitors located at over 3000 sites across the nation that are operated primarily by state, local, and tribal agencies [61]. In addition, the EPA has tracked emissions from all sources for the last 30 years. In the most recent report (for the year 2002) on the latest findings on air quality in the United States, the EPA stated that aggregate emissions of the six principal (*i.e.*, criteria) air pollutants tracked nationally have been reduced by 48% since 1970 [62]. During this same period, the U.S. gross domestic product increased 164%, energy consumption increased 42%, the population increased 38%, and vehicle miles traveled increased 155%. This reduction in emissions in criteria air pollutants between 1970 and 2002 is illustrated

TABLE 4-13
Annex I, Annex II, and Annex B Countries of the United Nations Framework Convention on Climate Change and the Kyoto Protocol

Annex I Countries	*Annex II Countries*	*Annex B Countries (percent of base year or period)*
Australia	Australia	Australia (108)
Austria	Austria	Austria (92)
Belarus[a]	—	—
Belgium	Belgium	Belgium (92)
Bulgaria[a]	—	Bulgaria[a] (92)
Canada	Canada	Canada (94)
—	—	Croatia (95)
Czechoslovakia[a]	—	Czechoslovakia[a] (92)
Denmark	Denmark	Denmark (92)
Estonia[a]	—	Estonia[a] (92)
European Economic Community	European Economic Community	European Economic Community (92)
Finland	Finland	Finland (92)
France	France	France (92)
Germany	Germany	Germany (92)
Greece	Greece	Greece (92)
Hungary[a]	—	Hungary[a] (94)
Iceland	Iceland	Iceland (110)
Ireland	Ireland	Ireland (92)
Italy	Italy	Italy (92)
Japan	Japan	Japan (94)
Latvia[a]	—	Latvia[a] (92)
—	—	Liechtenstein (92)
Lithuania[a]	—	Lithuania[a] (92)
Luxembourg	Luxembourg	Luxembourg (92)
—	—	Monaco (92)
Netherlands	Netherlands	Netherlands (92)
New Zealand	New Zealand	New Zealand (100)
Norway	Norway	Norway (101)
Poland[a]	—	Poland[a] (94)
Portugal	Portugal	Portugal (92)
Romania[a]	—	Romania[a] (92)
Russian Federation[a]	—	Russian Federation[a] (100)
—	—	Slovakia[a] (92)
—	—	Slovenia[a] (92)
Spain	Spain	Spain (92)
Sweden	Sweden	Sweden (92)
Switzerland	Switzerland	Switzerland (92)
Turkey	Turkey	—
Ukraine[a]	—	Ukraine[a] (100)
United Kingdom	United Kingdom	United Kingdom (92)
United States	United States	United States (93)

[a]Countries that are undergoing the process of transition to a market economy.

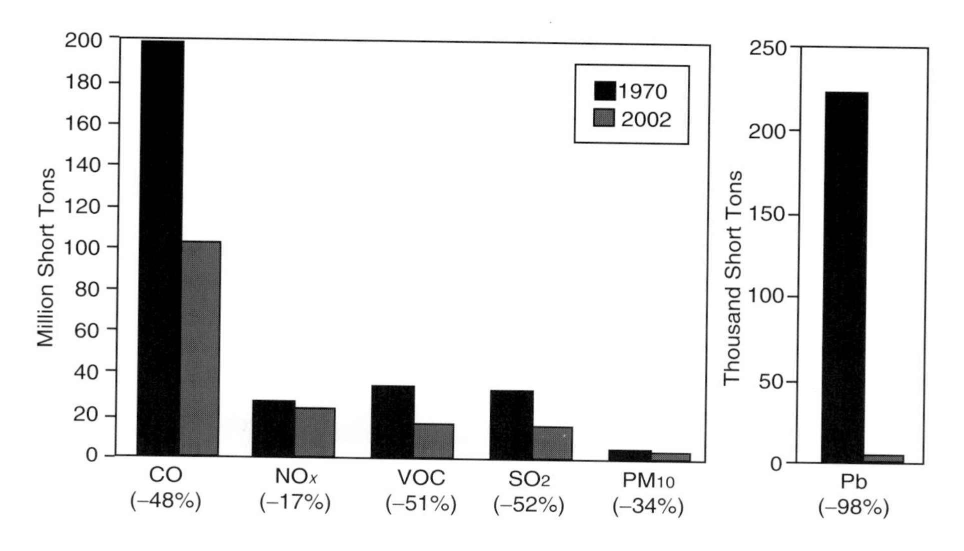

FIGURE 4-3. Comparison of 1970 and 2001 emissions of criteria air pollutants. (From EPA, *Latest Findings on National Air Quality 2002 Status and Trends*, Office of Air Quality Planning and Standards, U.S. Environmental Protection Agency, U.S. Government Printing Office, Washington, D.C., August 2003.)

in Figure 4-3 [62]. Despite this progress, about 160 million short tons of pollution are emitted into the air each year in the United States, and approximately 146 million people live in counties where air monitored in 2002 was unhealthy because of high levels of at least one of the six criteria air pollutants [62]. Most of the areas that experienced the unhealthy air did so because of particulate matter and/or ozone. This section summarizes the air quality and emissions trends in the United States for the criteria pollutants: NO_2, ozone, SO_2, particulate matter, carbon monoxide, and lead, along with acid rain, trace elements (specifically, mercury), and CO_2. Air quality is based on actual measurements of pollutant concentrations in the ambient air at monitoring sites. Trends are derived by averaging direct measurements from these monitoring stations on a yearly basis. Emissions of ambient pollutants and their precursors are estimated based on actual monitored readings or engineering calculations of the amounts and types of pollutants emitted by vehicles, factories, stationary combustion, and other sources.

Six Principal Pollutants

As previously discussed, under the Clean Air Act, the EPA established air quality standards to protect human health and public welfare. The EPA has set national air quality standards for six principal or criteria air pollutants, which include nitrogen dioxide, ozone, sulfur dioxide, particulate matter, carbon monoxide, and lead. Four of these pollutants—NO_2, SO_2, CO, and lead—result primarily from direct emissions from a variety of sources. Particulate matter results from direct emissions but is also commonly formed when emissions of nitrogen oxides, sulfur oxides, ammonia, organic compounds, and other gases react in the atmosphere. Ozone is not directly emitted but is formed when nitrogen oxides and volatile organic compounds react in the presence of sunlight.

Nitrogen Dioxide (NO_2)

Nitrogen oxides (NO_x), the term used to describe the sum of NO, NO_2, and other oxides of nitrogen, contribute to the formation of ozone, particulate matter, haze, and acid rain. While the EPA traces national emissions of NO_x, the national monitoring network measures ambient concentrations of NO_2 for comparison to national air quality standards. The major sources of anthropogenic NO_x emissions are high-temperature combustion processes, such as those that occur in vehicles and power plants. Over the past 20 years, monitored levels of NO_2 have decreased 21% [62]. All areas of the country that once violated the NAAQS for NO_2 now meet the standard. National emissions of NO_x have declined by almost 15% over the past 20 years. While overall NO_x emissions are declining, emissions from some sources such as nonroad engines have actually increased since 1983. Figures 4-4 and 4-5

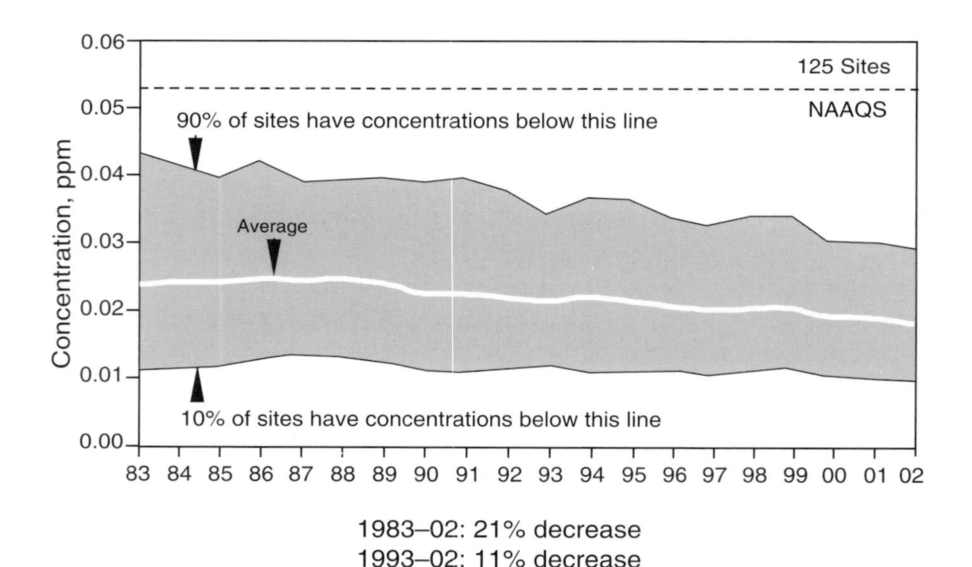

FIGURE 4-4. NO_2 air quality from 1983 to 2002. (From EPA, *Latest Findings on National Air Quality 2002 Status and Trends*, Office of Air Quality Planning and Standards, U.S. Environmental Protection Agency, U.S. Government Printing Office, Washington, D.C., August 2003.)

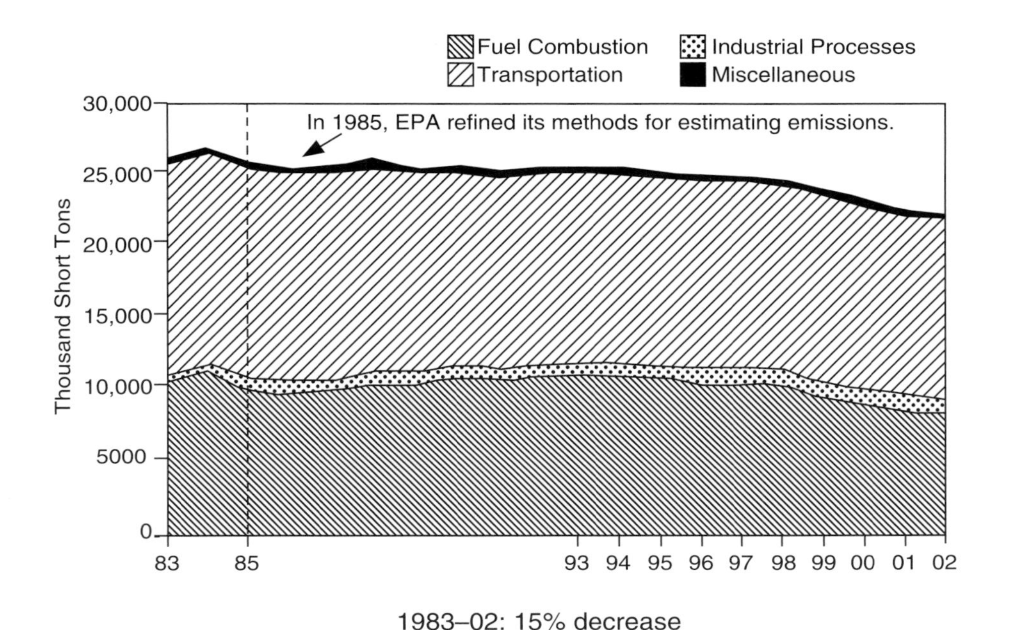

FIGURE 4-5. NO_X emissions from 1983 to 2002. (From EPA, *Latest Findings on National Air Quality 2002 Status and Trends*, Office of Air Quality Planning and Standards, U.S. Environmental Protection Agency, U.S. Government Printing Office, Washington, D.C., August 2003.)

illustrate the trends of NO_2 air quality and NO_X emissions, respectively, for the last 20 years [62]. Of the approximately 10 million short tons of NO_X emitted from fuel combustion (Figure 4-5), power plants contributed less than 4.5 million short tons [62].

Ozone

Ground-level ozone, which is the primary constituent of smog, continues to be a pollution problem throughout many areas of the United States. Ozone is not emitted directly into the air but is formed by the reaction of volatile organic compounds (VOCs) and NO_X in the presence of heat and sunlight. The trends of VOC emissions and their sources for the past 20 years are shown in Figure 4-6 [62]. Fuel combustion contributes approximately 5% of the VOC emissions, with power stations comprising less than half of the 5% [4]. Over the past 20 years, national ambient ozone levels decreased 22 and 14% based on 1-hour and 8-hour data, respectively [62]. During this period, emissions of VOCs decreased 40% (excluding wildfires and prescribed burning). Ozone air quality trends are illustrated in Figures 4-7 and 4-8 [62].

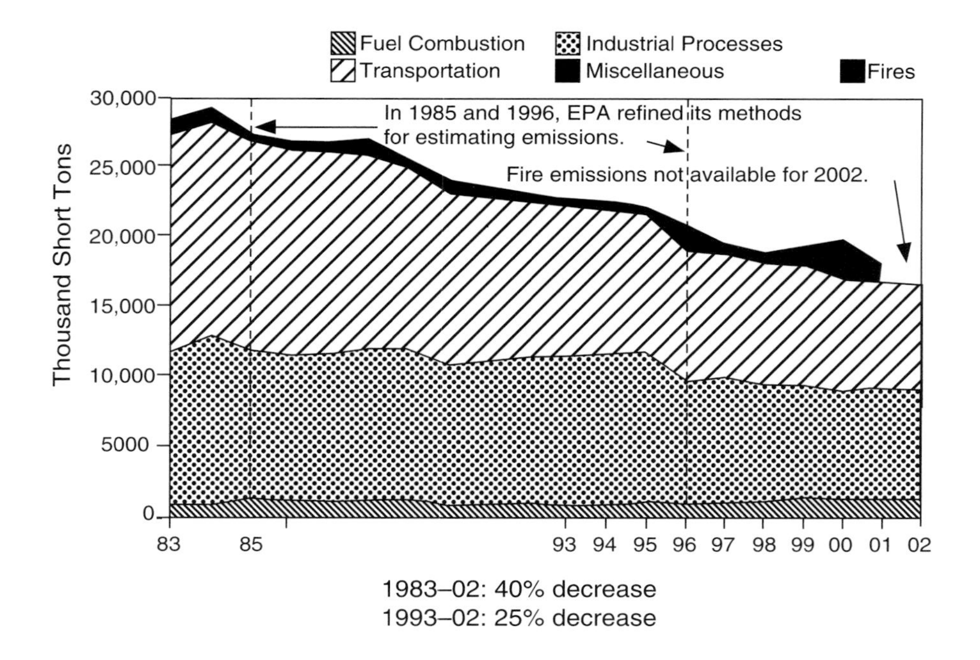

FIGURE 4-6. VOC emissions from 1983 to 2002. (From EPA, *Latest Findings on National Air Quality 2002 Status and Trends*, Office of Air Quality Planning and Standards, U.S. Environmental Protection Agency, U.S. Government Printing Office, Washington, D.C., August 2003.)

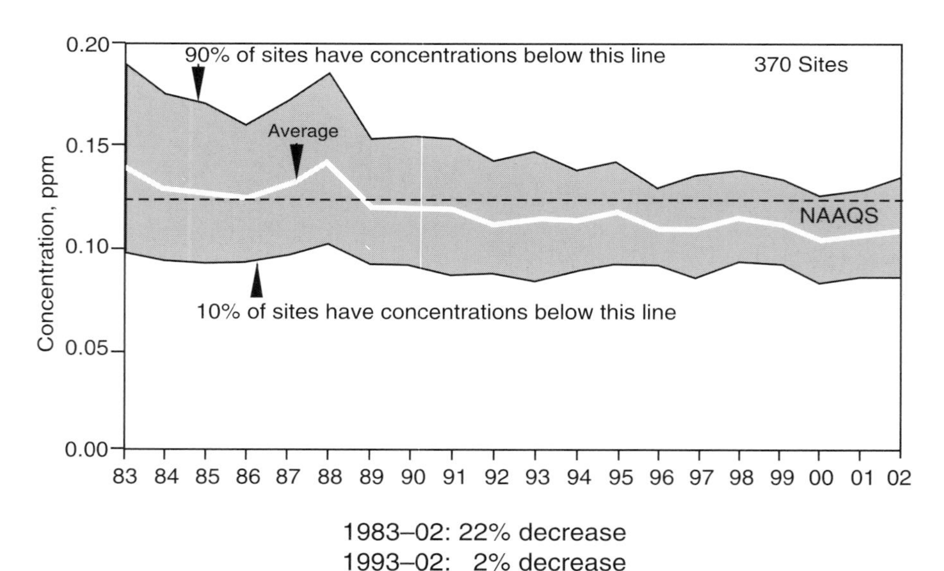

FIGURE 4-7. Ozone air quality from 1983 to 2002 based on 8-hour averages. (From EPA, *Latest Findings on National Air Quality 2002 Status and Trends*, Office of Air Quality Planning and Standards, U.S. Environmental Protection Agency, U.S. Government Printing Office, Washington, D.C., August 2003.)

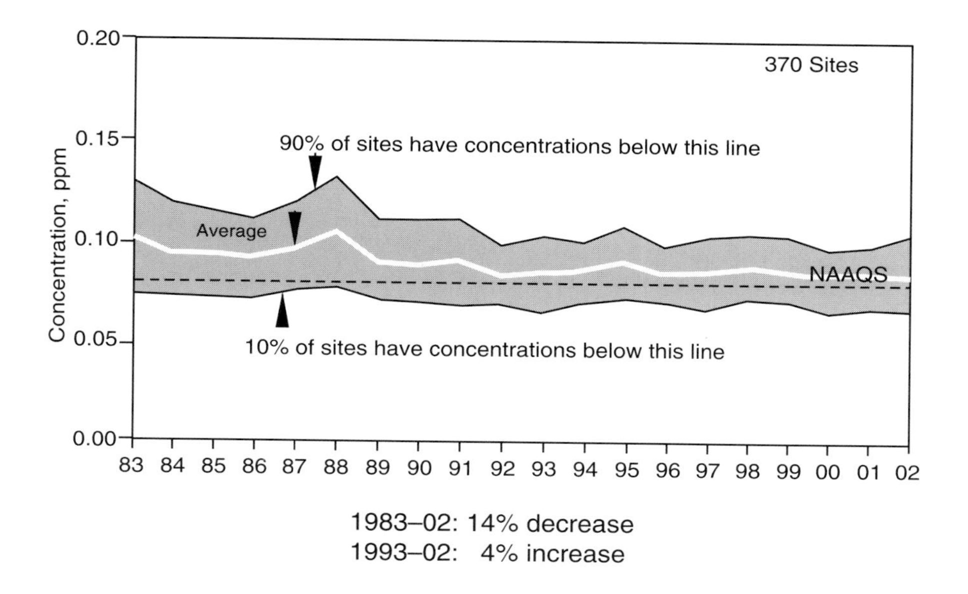

FIGURE 4-8. Ozone air quality from 1983 to 2002 based on 1-hour averages. (From EPA, *Latest Findings on National Air Quality 2002 Status and Trends*, Office of Air Quality Planning and Standards, U.S. Environmental Protection Agency, U.S. Government Printing Office, Washington, D.C., August 2003.)

Sulfur Dioxide (SO_2)

Nationally, average SO_2 ambient concentrations have decreased 54% from 1983 to 2002 and 39% over the last 10 years, as shown in Figure 4-9 [62]. SO_2 emissions decreased 33% from 1983 to 2002 and 31% from 1993 to 2002. Reductions in SO_2 concentrations and emissions since 1990 are due to controls implemented under the EPA's Acid Rain Program beginning in 1995. As shown in Figure 4-10, fuel combustion, primarily coal and oil, accounts for most of the total SO_2 emissions. Coal combustion accounts for approximately 11 of the 15 million short tons of SO_2 emitted in 2002.

Particulate Matter

Between 1993 and 2002, PM_{10} concentrations decreased 13%, while PM_{10} emissions decreased 22% [62], as illustrated in Figures 4-11 and 4-12, respectively. Fuel combustion accounts for about one-third of total particulate emissions (see Figure 4-12), while electric utilities account for approximately 5% of the total particulate matter emitted [4]. Figure 4-13 shows that direct $PM_{2.5}$ emissions from anthropogenic sources decreased 17% nationally between 1993 and 2002 [62]. Figure 4-13 tracks only directly emitted particles and does not account for secondary particles, which are primarily sulfates and nitrates formed when emissions of NO_X, SO_2, ammonia, and other gases react in the atmosphere.

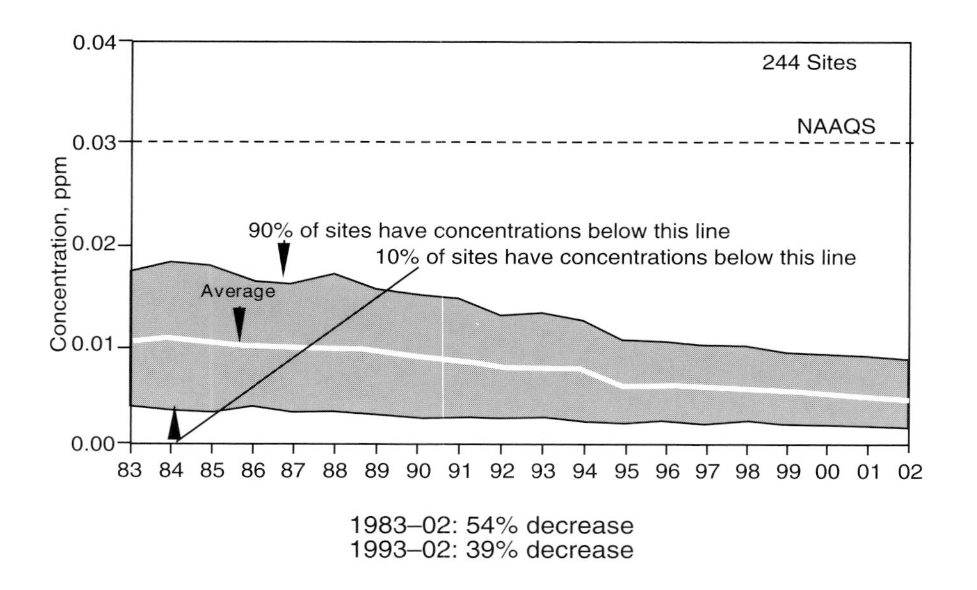

FIGURE 4-9. SO_2 air quality from 1983 to 2002. (From EPA, *Latest Findings on National Air Quality 2002 Status and Trends*, Office of Air Quality Planning and Standards, U.S. Environmental Protection Agency, U.S. Government Printing Office, Washington, D.C., August 2003.)

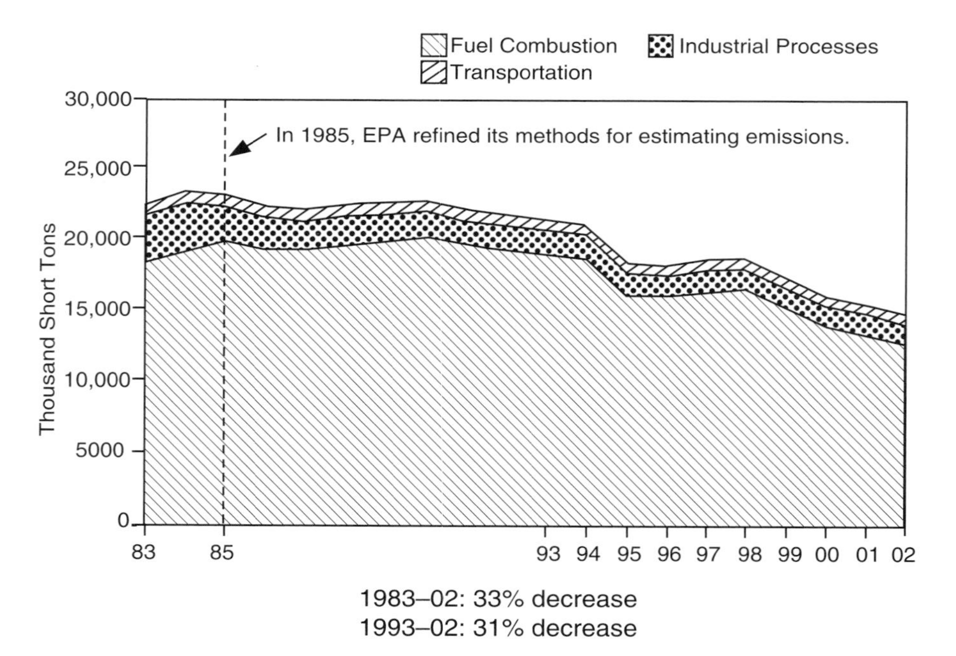

FIGURE 4-10. SO_2 emissions from 1983 to 2002. (From EPA, *Latest Findings on National Air Quality 2002 Status and Trends*, Office of Air Quality Planning and Standards, U.S. Environmental Protection Agency, U.S. Government Printing Office, Washington, D.C., August 2003.)

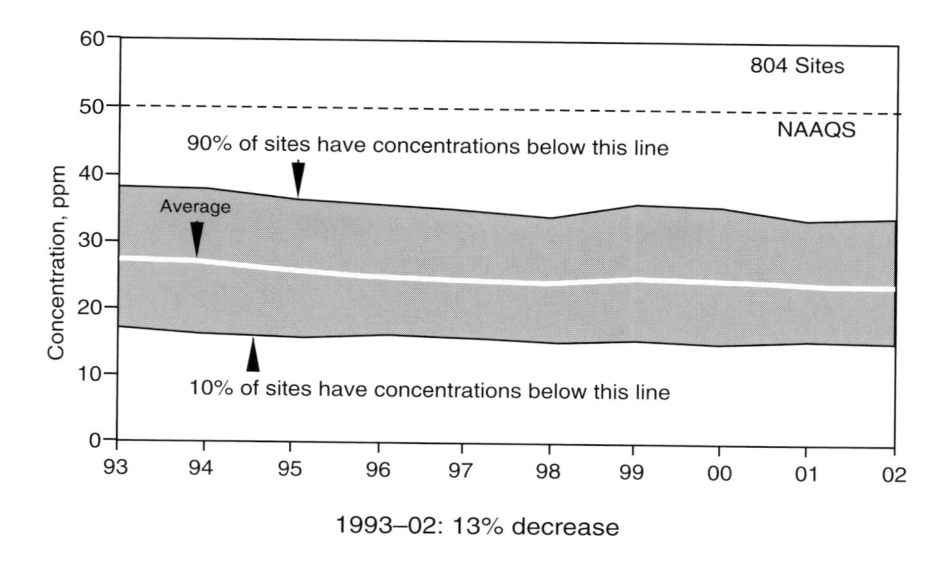

FIGURE 4-11. PM_{10} air quality from 1993 to 2002. (From EPA, *Latest Findings on National Air Quality 2002 Status and Trends*, Office of Air Quality Planning and Standards, U.S. Environmental Protection Agency, U.S. Government Printing Office, Washington, D.C., August 2003.)

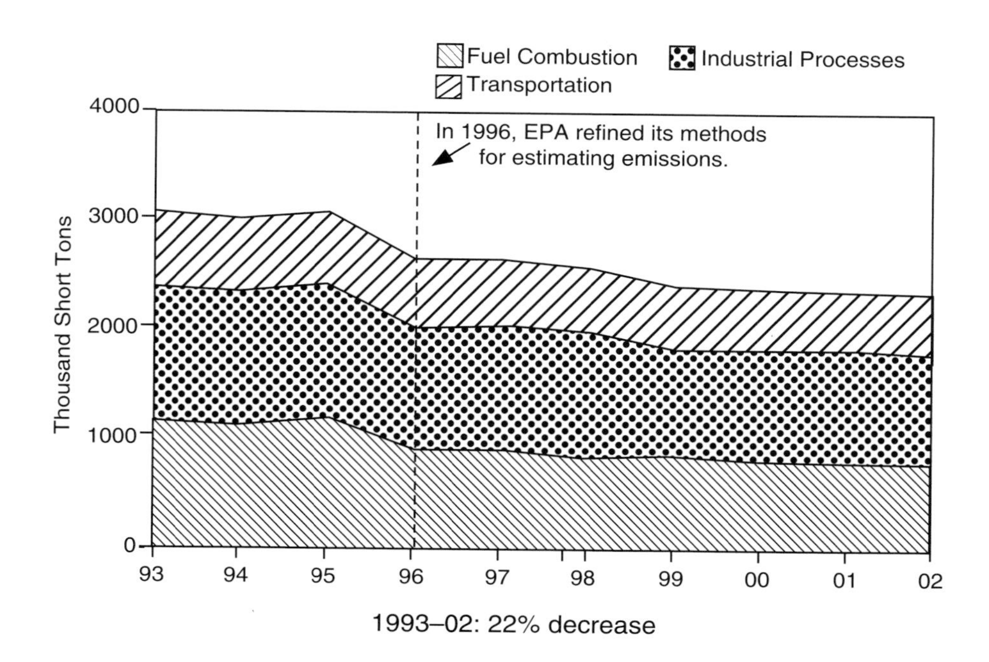

FIGURE 4-12. PM_{10} emissions from 1993 to 2002. (From EPA, *Latest Findings on National Air Quality 2002 Status and Trends*, Office of Air Quality Planning and Standards, U.S. Environmental Protection Agency, U.S. Government Printing Office, Washington, D.C., August 2003.)

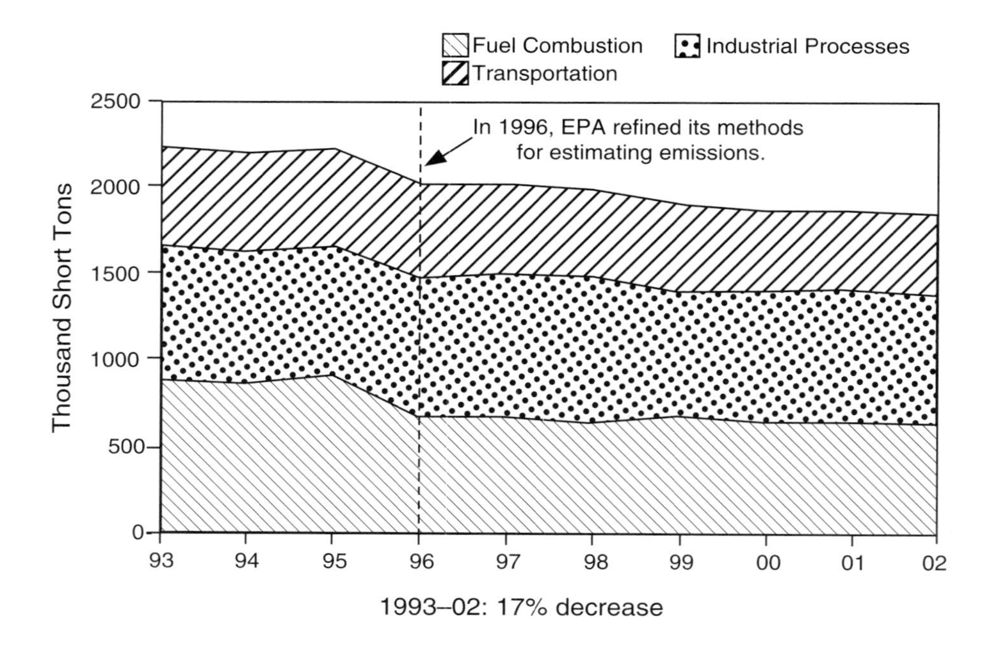

FIGURE 4-13. Direct $PM_{2.5}$ emissions from 1993 to 2002. (From EPA, *Latest Findings on National Air Quality 2002 Status and Trends*, Office of Air Quality Planning and Standards, U.S. Environmental Protection Agency, U.S. Government Printing Office, Washington, D.C., August 2003.)

Carbon Monoxide

Carbon monoxide (CO) is a component of motor vehicle exhaust, which contributes about 60% of all CO emissions nationwide. Other sources of CO emissions include industrial processes, nontransportation fuel combustion, and natural sources such as wildfires. Nationally, the 2002 ambient average CO concentration was nearly 65% lower than that for 1983, which is illustrated in Figure 4-14 [62]. CO emissions decreased about 42% over the last 10 years despite an approximately 23% increase in vehicle miles traveled. Transportation sources are the largest contributors to CO emissions, with fuel combustion accounting for about 7% of the CO emissions. Electric utilities account for less than 0.5% of the total CO emissions [4]. The trend in CO emissions is shown in Figure 4-15 [62].

Lead

In the past, automotive sources were the major contributor of lead (Pb) emissions to the atmosphere. The emissions of lead from the transportation sector have greatly declined over the last 20 years as leaded gasoline was phased out. Today, industrial processes, primarily metals processing, are the major sources of lead emissions to the atmosphere. As a result of the phase-out

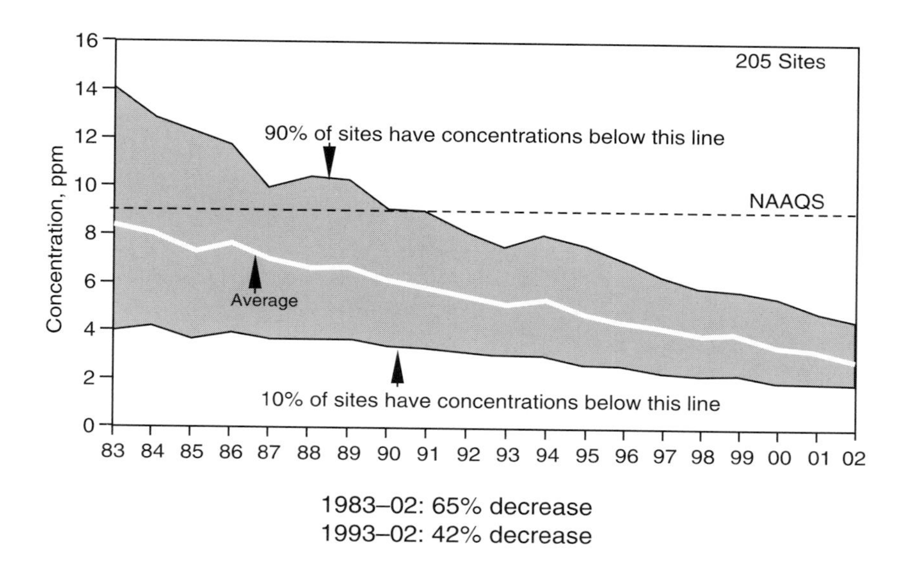

FIGURE 4-14. CO air quality from 1983 to 2002. (From EPA, *Latest Findings on National Air Quality 2002 Status and Trends*, Office of Air Quality Planning and Standards, U.S. Environmental Protection Agency, U.S. Government Printing Office, Washington, D.C., August 2003.)

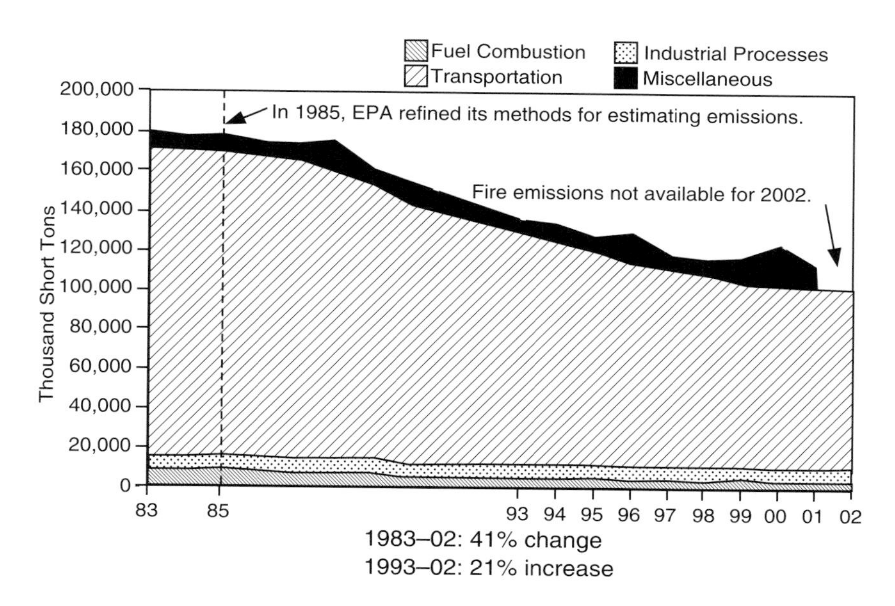

FIGURE 4-15. CO emissions from 1983 to 2002. (From EPA, *Latest Findings on National Air Quality 2002 Status and Trends*, Office of Air Quality Planning and Standards, U.S. Environmental Protection Agency, U.S. Government Printing Office, Washington, D.C., August 2003.)

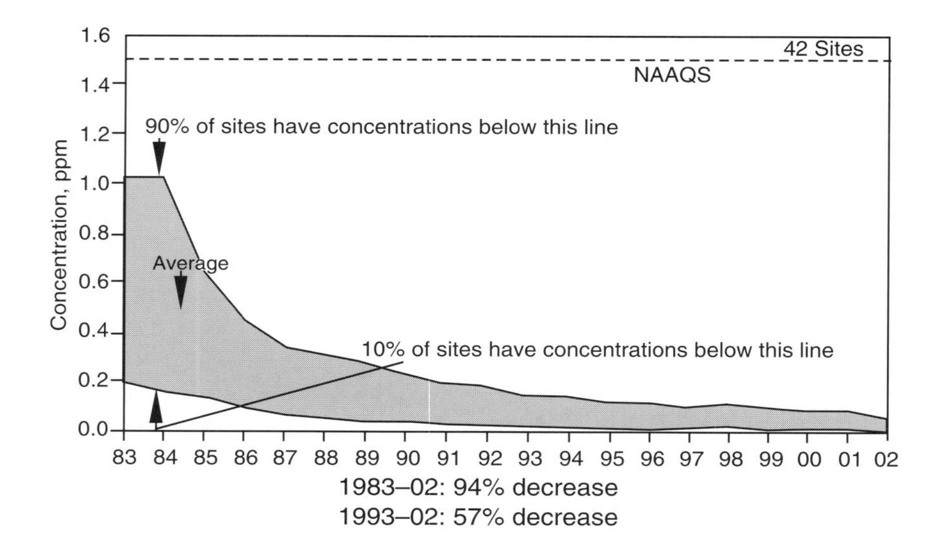

FIGURE 4-16. Lead air quality from 1983 to 2002. (From EPA, *Latest Findings on National Air Quality 2002 Status and Trends,* Office of Air Quality Planning and Standards, U.S. Environmental Protection Agency, U.S. Government Printing Office, Washington, D.C., August 2003.)

of leaded gasoline, lead concentrations and emissions have decreased significantly, as shown in Figures 4-16 and 4-17, respectively [62]. The 2002 average air quality concentration for lead is 94% lower than in 1982, and lead emissions decreased by 93% over the same period. Lead emissions from electric utilities are less than 10% of the total (*i.e.,* less than 500 short tons [4]), and the only violations of the lead NAAQS that occur today are near large industrial sources such as lead smelters and battery manufacturers [62].

Acid Rain

As discussed earlier, acid rain or acidic deposition occurs when emissions of sulfur dioxide and nitrogen oxides in the atmosphere react with water, oxygen, and oxidants to form acidic compounds. These compounds then fall to Earth in either dry form (gas and particles) or wet form (rain, snow, and fog). In the United States, about 63% of annual SO_2 emissions and 22% of NO_x emissions are produced by electric utility plants that burn fossil fuels [62]. The EPA's Acid Rain Program will reduce annual SO_2 emissions by 10 million short tons from 1980 levels by 2010. The program sets a permanent cap of 8.95 million short tons on the total amount of SO_2 that may be emitted by power plants nationwide, which is about half of that emitted in 1980 [62]. Approximately 3000 units are now affected by the Acid

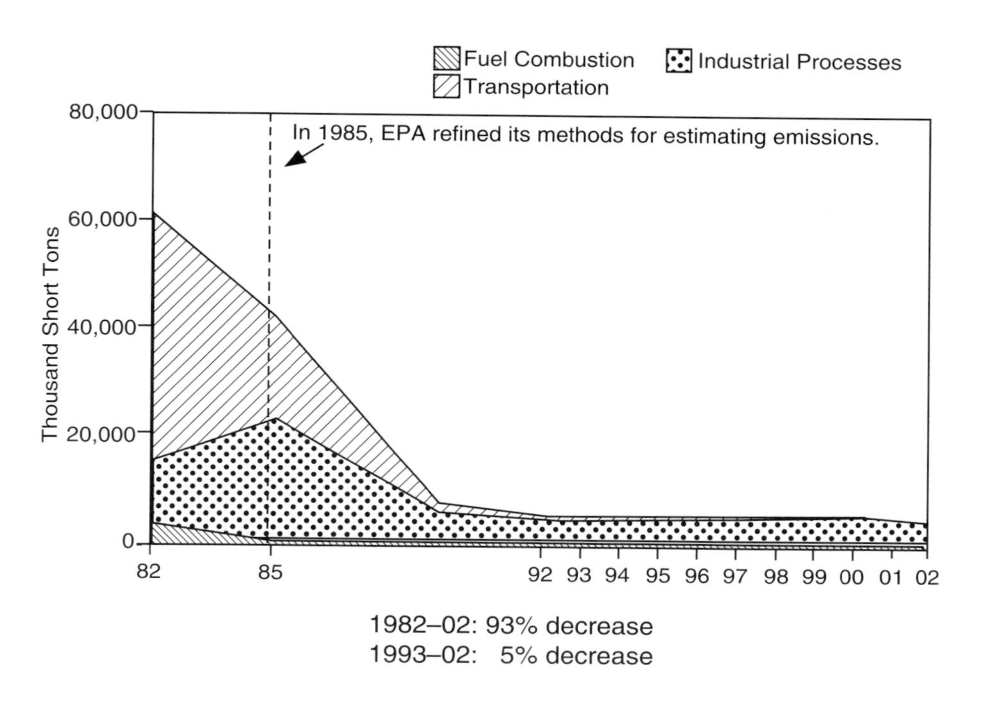

FIGURE 4-17. Lead emissions from 1982 to 2002. (From EPA, *Latest Findings on National Air Quality 2002 Status and Trends*, Office of Air Quality Planning and Standards, U.S. Environmental Protection Agency, U.S. Government Printing Office, Washington, D.C., August 2003.)

Rain Program. Figure 4-18 shows the SO_2 reductions achieved as of 2002 and illustrates that SO_2 emissions were reduced to about 10 million short tons in 2002 [62]. The NO_X component of the Acid Rain Program limits the emission rate for all affected utilities, resulting in an NO_X reduction of 2 million short ton from 1980 levels by 2000. NO_X emissions, shown in Figure 4-19, have declined since 1990, with NO_X emissions from ~1000 affected sources totaling slightly more than 4 million short tons in 2002 [62]. In the atmosphere, sulfate concentrations, which are a major component of fine particles, especially in eastern United States, have decreased since 1990 [19]. In 2001, concentrations in the Northeast and Mid-Atlantic were 8 to 12 mg/m^3, as much as 8 mg/m^3 lower than in 1990. Wet sulfate deposition, a major component of acid rain, has also decreased since 1990. In 2001, deposition in the Northeast and Midwest was 20 to 30 kg/ha/yr, as much as 12 kg/ha/yr lower than in 1990 [19]. Wet nitrate deposition has not decreased regionally because of the overall increase in NO_X emissions. Acid-neutralizing capacity, a major indicator of recovery in acidified lakes and streams, is beginning to rise in streams in the Northeast, including the Adirondacks. This is an indictor that recovery from acidification is beginning in those areas.

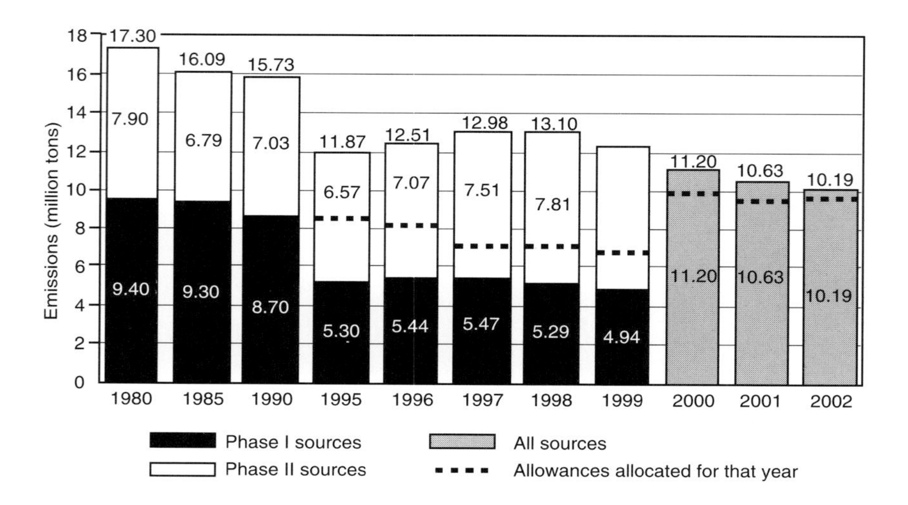

FIGURE 4-18. SO_2 emissions covered under the Acid Rain Program. (From EPA, *Latest Findings on National Air Quality 2002 Status and Trends*, Office of Air Quality Planning and Standards, U.S. Environmental Protection Agency, U.S. Government Printing Office, Washington, D.C., August 2003.)

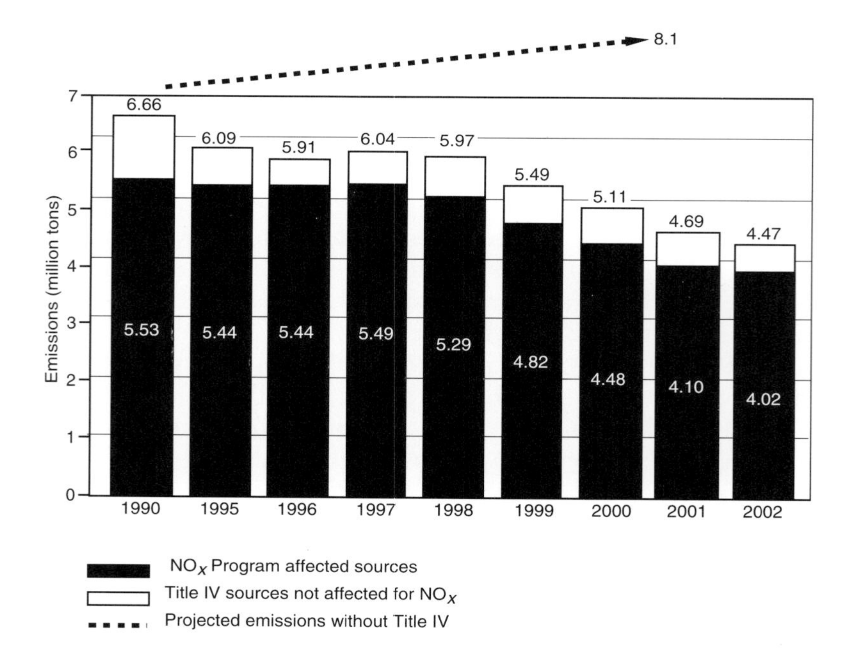

FIGURE 4-19. NO_X emissions covered under the Acid Rain Program. (From EPA, *Latest Findings on National Air Quality 2002 Status and Trends*, Office of Air Quality Planning and Standards, U.S. Environmental Protection Agency, U.S. Government Printing Office, Washington, D.C., August 2003.)

Hazardous Air Pollutants

Currently, emissions of hazardous air pollutants (HAPs) are not regulated by the EPA, although mercury regulations will be implemented by 2007. Emissions of HAPs from coal-fired power plants have been estimated by the EPA, which performed a study (*i.e.*, Study of Hazardous Air Pollutant Emissions from Electric Utility Steam Generating Units) to determine the quantity of hazardous air pollutants being emitted from fossil-fuel-fired power plants [37]. In this study, discussed earlier, HAP emissions tests data were gathered from 52 utility units (*i.e.*, boilers), including a range of coal-, oil-, and natural-gas-fired utility boilers. The emission tests data, along with facility specific information (*e.g.*, boiler type, control devices, fuel usage) were used to estimate HAP emissions from all 684 utility plants in the United States. These utilities are fueled primarily by coal (59%), oil (12%), or natural gas (29%). Many plants have two or more units, and several plants burn more than one type of fuel (*e.g.*, contain both coal- and oil-fired units). In 1990, 426 plants burned coal as one of their fuels, 137 plants burned oil, and 267 plants burned natural gas. The overall summary of the study is presented in Table 4-14,

TABLE 4-14
Nationwide Utility (Coal-Fired) Emissions for Thirteen Priority HAPs[a]

HAP	*Nationwide HAP Emission Estimates (short tons per year)*[b]		
	1990	*1994*	*2010*
Arsenic	61	56	71
Beryllium	7.1	7.9	8.2
Cadmium	3.3	3.2	3.8
Chromium	73	62	87
Lead	75	62	87
Manganese	164	168	219
Mercury	46	51	60
Nickel	58	52	69
Hydrogen chloride	143,000	134,000	155,000
Hydrogen fluoride	20,000	23,000	26,000
Acrolein	25	27	34
Dioxins[c]	0.000097	0.00012	0.00020
Formaldehyde	35	29	45

[a]Radionuclides are the one priority HAP not included on this table because radionuclide emissions are measured in different units (*i.e.*, curies per year) and, therefore, would not provide a relevant comparison to the other HAPs shown.

[b]The emission estimates in this table are derived from model projections based on a limited sample of specific boiler types and control scenarios; therefore, there are uncertainties in these numbers.

[c]These emission estimates were calculated using the toxic equivalency (TEQ) approach, which is based on the summation of the emissions of each congener after adjusting for toxicity relative to 2,3,7,8-tetrachlorodibenzo-*p*-dioxin (*i.e.*, 2,3,7,8-TCDD).

TABLE 4-15
Estimated Emissions for Nine Priority HAPs from Characteristic Utility Units (1994; short tons per year)[a]

Fuel *Unit size (MWe):*	*Coal* *325*
Arsenic	0.0050
Cadmium	0.0023
Chromium	0.11
Lead	0.021
Mercury	0.05
Hydrogen chloride	190
Hydrogen fluoride	14
Dioxins[b]	0.00000013
Nickel[c]	NC

[a]There are uncertainties in these numbers. Based on an uncertainty analysis, the EPA predicts that the emission estimates are generally within a factor of roughly three of actual emissions.
[b]These emission estimates were calculated using the toxic equivalency (TEQ) approach, which is based on the summation of the emissions of each congener after adjusting for toxicity relative to 2,3,7,8-tetrachlorodibenzo-*p*-dioxin (*i.e.*, 2,3,7,8-TCDD).
[c]Not calculated.

which lists nationwide utility emissions estimates for 13 priority HAPs [37]. Table 4-15 contains estimated emissions for 9 priority HAPs from characteristic utility units. In summary, the Utility Hazardous Air Pollutant Report to Congress analyzed 66 other air pollutants (other than mercury, which is discussed in the next section) from 684 power plants that are 25 MW or larger and burning coal, oil, or natural gas [37]. The report noted potential health concerns about utility emissions of dioxin, arsenic, hydrogen chloride, hydrogen fluoride, and nickel, although uncertainties exist about the health data and emissions for these pollutants.

Mercury

The best estimate of annual anthropogenic U.S. emissions of mercury from 1994 to 1995 is 158 short tons [36]. Approximately 87% of these emissions were from combustion sources, including waste and fossil fuel combustion, as illustrated in Figure 4-20 [63]. Contemporary anthropogenic emissions are only one part of the mercury cycle. Releases from human activities today are adding to the mercury reservoirs that already exist in land, water, and air, both naturally and as a result of previous human activities. One estimate of the total annual global input to the atmosphere from all sources, including natural, anthropogenic, and oceanic emissions, is 5500 short tons [36].

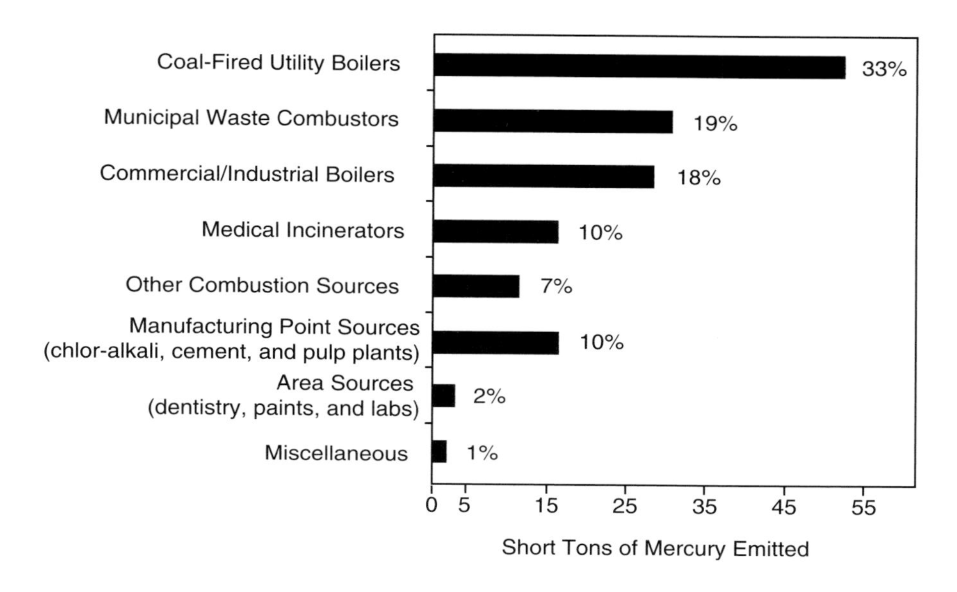

FIGURE 4-20. Annual mercury emissions in the United States. (From DOE, Quicksilver: Don't Play With It, *FETC Focus*, Issue 2, March 1999, p 25.)

Based on this, U.S. sources are estimated to have contributed about 3% of the 5500 tons in 1995. Mercury emissions from coal-fired boilers are estimated to be 48 short tons per year.

In a report released by the United Nations Environment Program (UNEP) in February 2003, coal-fired power plants were identified as the largest single anthropogenic source of mercury air emissions [64]. UNEP reported that power plants in Asia contribute the most mercury—some 860 metric tons (note that 1 metric ton = 1.1023 short tons) per year, more than a third of the 2200 metric tons of annual emissions reported coming from major anthropogenic sources. In all, 1470 metric tons of mercury were emitted in 1995 by coal-fired power plants. Other major sources of mercury include metal production (200 metric tons), cement production (130 metric tons), waste disposal (110 metric tons), and small-scale gold mining (300 metric tons).

Carbon Dioxide (CO_2)

Carbon dioxide emissions from energy use are shown in Figure 4-21, which provides CO_2 emissions by sector and fuel for 1990 and 2001 and projections up to 2025 [65]. Petroleum products are the leading source of CO_2 emissions from energy use. In 2025, petroleum is projected to account for 971 million metric tons carbon equivalent, a 43% share of the projected total. Coal is the second leading source of CO_2 emissions and is projected to produce 73 million metric tons carbon equivalent in 2025, or 34% of the total.

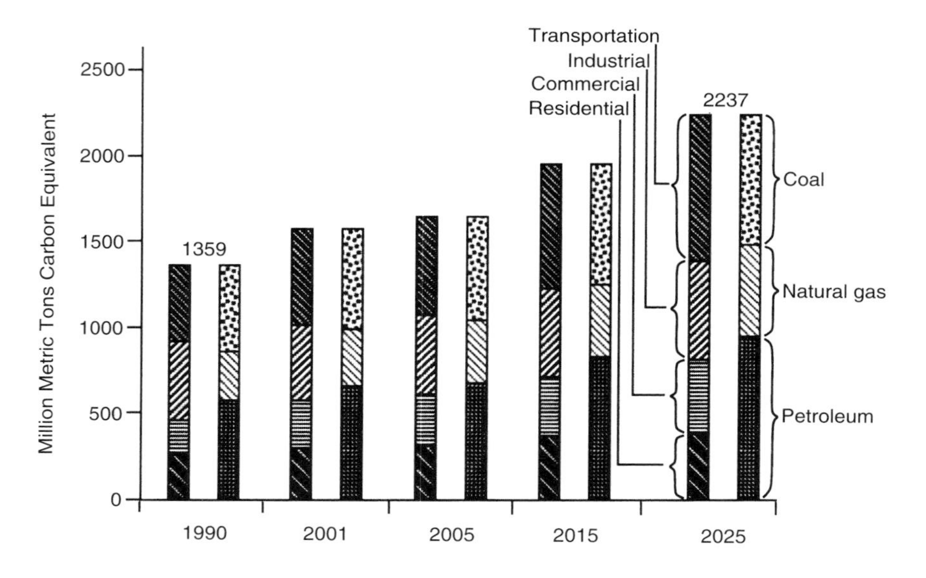

FIGURE 4-21. Current and projected carbon dioxide emissions by sector and fuel. (From EIA, *Annual Energy Outlook 2003*, Energy Information Administration, U.S. Department of Energy, Washington, D.C., January 2003.)

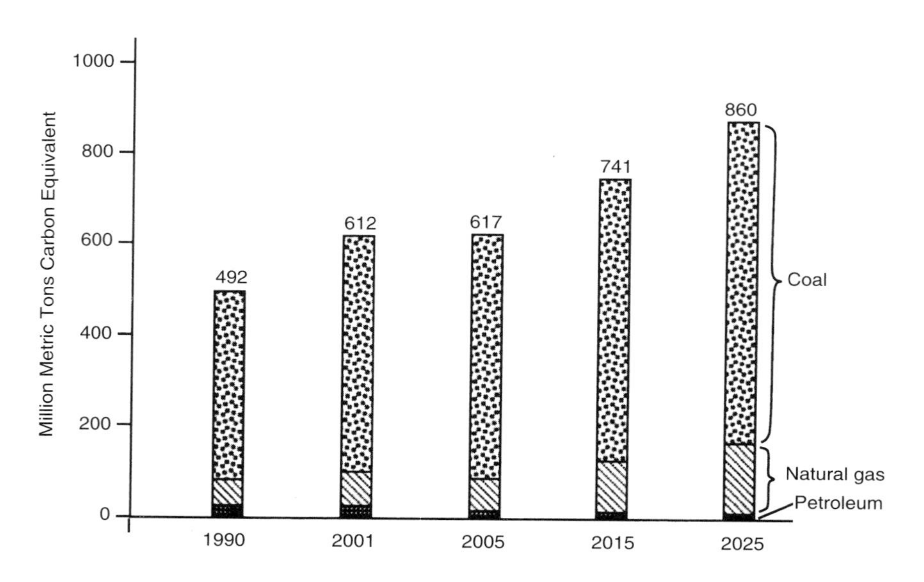

FIGURE 4-22. Current and projected carbon dioxide emissions from the electric power sector by fuel. (From EIA, *Annual Energy Outlook 2003*, Energy Information Administration, U.S. Department of Energy, Washington, D.C., January 2003.)

In 2025, natural gas use is projected to produce a 23% share of the total CO_2 emissions with 512 million metric tons carbon equivalent.

The use of fossil fuels in the electric power industry accounted for 39% of total energy-related CO_2 emissions in 2001, and the share is projected to be 38% in 2025, as shown in Figure 4-22 [65]. Coal is projected to account for 50% of the power industry's electricity generation in 2025 and to produce 81% of electricity-related CO_2 emissions. In 2025, natural gas is projected to account for 27% of electricity generation and 18% of electricity-related CO_2 emissions.

References

1. Lomborg, B., *The Skeptical Environmentalist: Measuring the Real State of the World* (Cambridge University Press, Cambridge, U.K., 2001).
2. University of California at Davis, www-geology.ucdavis.edu/~GEL115/115CH11coal.html (accessed May 2003).
3. Anon., Recognizing Pollution's Hazard's, *Chemical and Engineering News*, Vol. 81, No. 15, April 14, 2003, pp. 54–55.
4. Wark, K., C. F. Warner, and W. T. Davis, *Air Pollution: Its Origin and Control*, Third ed. (Addison-Wesley Longman, Menlo Park, CA, 1998).
5. Dunn, S., King Coal's Weakening Grip on Power, *World Watch*, September/October 1999, pp. 10–19.
6. EPA, *EPA Proposes Options for Significantly Reducing Mercury Emissions from Electric Utilities* (U.S. Environmental Protection Agency, Washington, D.C.), www.epa.gov/mercury/ (accessed December 15, 2003).
7. EPA (United States Environmental Protection Agency), National Ambient Air Quality Standards (NAAQS) (U.S. Environmental Protection Agency, Washington, D.C.), www.epa.gov/air/criteria.html (accessed November 15, 2002).
8. EPA, Subpart D(a)—Standards of Performance for Fossil Fuel-Fired Steam Generators for Which Construction is Commenced After August 17, 1971, *Federal Register*, www.epa.gov/ttn/atw/combust/boiler/boilnsps.html#rule (December 23, 1971).
9. EPA, Standards of Performance for New Stationary Sources; Electric Utility Steam Generating Units; Decision in Response to Petitions for Reconsideration, *Federal Register*, Vol. 45, No. 26, February 6, 1980, pp. 8210–8213.
10. EPA, Revision of Standards of Performance for Nitrogen Oxide Emissions from New Fossil Fuel-Fired Steam Generating Units; Revisions to Reporting Requirements for Standards of Performance for New Fossil Fuel-Fired Steam Generating Units, *Federal Register*, Vol. 63, No. 179, September 16, 1998, pp. 49442–49454.
11. EPA, *Fact Sheet: Revision of Standards of Performance for Nitrogen Oxides Emissions from Fossil Fuel-Fired Steam Generating Units* (U.S. Environmental Protection Agency, Washington, D.C.), www.epa.gov/ttn/oarpg, (accessed August 7, 2001).

12. EPA, Subpart D(b)—Standards of Performance for Industrial-Commercial-Institutional Steam Generating Units, *Federal Register*, www.epa.gov/ttn/atw/combust/boiler/boilnsps.html#rule (November 25, 1986).
13. EPA, Subpart D(c)—Standards of Performance for Small Industrial–Commercial–Institutional Steam Generating Units, *Federal Register*, www.epa.gov/ttn/atw/combust/boiler/boilnsps.html#rule (September 12, 1990).
14. AP-42, External Combustion Sources, in *Emission Factors*, Fifth ed. (Office of Air Quality Planning and Standards and Office of Air and Radiation, U.S. Environmental Protection Agency, Washington, D.C., 1993 [revisions in 1998]).
15. Anon., EPA Finalizes NESHAP Amendments, *Chemical Engineering Progress*, Vol. 99, No. 7, July 2003, p. 18.
16. EPA, *Overview: The Clean Air Act Amendments of 1990* (U.S. Environmental Protection Agency, Washington, D.C.), www.epa.gov/oar/caa/overview.txt, (accessed November 15, 2002).
17. EPA, CAAA: Original List of Hazardous Air Pollutants (U.S. Environmental Protection Agency, Washington, D.C.), www.epa.gov/ttn/atw/orig189.html, (accessed February 11 2002).
18. Makanski, J., Clean Air Act Amendments: The Engineering Response, *Power*, Vol. 135, No. 6, 1991, pp. 11–66.
19. EPA, *EPA Acid Rain Program 2001 Progress Report* (Office of Air and Radiation, U.S. Environmental Protection Agency, U.S. Government Printing Office, Washington, D.C., November 2002).
20. Leone, M., Cleaning the Air the Market-Based Way, *Power*, Vol. 129, No. 10, December 1990, pp. 9–10.
21. EPA, *Acid Rain Program: Program Overview* (U.S. Environmental Protection Agency, Washington, D.C.), www.epa.gov/airmarkets/arp/index.html (accessed April 1999).
22. Smith, D. N., H. G. McIlvried, and A. N. Mann, Understanding NO_X and How It Impacts Coal, *Coal Age*, Vol. 105, No. 11, November 2000, p. 35.
23. EPA, NO_X Trading Programs, www.epa.gov/airmarkets/progregs/noxview.html, October 29, 2002.
24. EPA, *The Regional Transport of Ozone* (Office of Air Quality Planning and Standards, U.S. Environmental Protection Agency, U.S. Government Printing Office, Washington, D.C., September 1998).
25. D'Aquino, R., The NO_X Market: Play It Like a Pro, *Chemical Engineering Progress*, Vol. 99, No. 6, June 2003, p. 9.
26. EPA, *New Source Review: Report to the President* (U.S. Environmental Protection Agency, Washington, D.C., June 2002).
27. EPA, Prevention of Significant Deterioration (PSD) and Nonattainment New Source Review (NSR): Final Rule and Proposed Rule, *Federal Register*, Vol. 67, No. 251, December 31, 2002, pp. 80186–80289.
28. EPA, *EPA Announces Improvements to New Source Review Program* (U.S. Environmental Protection Agency, Washington, D.C.), www.epa.gov/air/nsr-review/press_release.html (accessed March 13, 2003).
29. EPA, *New Source Review* (U.S. Environmental Protection Agency, Washington, D.C.), www.epa.gov/air/nsr-review/ (accessed July 25, 2003).

30. Anon., EPA Finalizes New Source Review Rule, *Chemical Engineering Progress*, Vol. 99, No. 12, December 2003, p. 24.
31. DOE, *Atmospheric Aerosol Source–Receptor Relationships: The Role of Coal-Fired Power Plants—Project Facts* (Office of Fossil Energy, National Energy Technology Laboratory, U.S. Department of Energy, Washington, D.C., January 2003).
32. EPA, *Fact Sheet: Final Regional Haze Regulations for Protection of Visibility in National Parks and Wilderness Areas* (U.S. Environmental Protection Agency, Washington, D.C.), www.epa.gov/oar/visibility/program.html (accessed June 2, 1999).
33. Aljoe, W. W. and T. J. Grahame, The DOE-NETL Air Quality Research Program: Airborne Fine Particulate ($PM_{2.5}$), in *Proc. of the Conference on Air Quality III: Mercury, Trace Elements, and Particulate Matter* (University of North Dakota, Grand Forks, 2002).
34. Anon., EPA Issues $PM_{2.5}$ Guidance, *Chemical Engineering Progress*, Vol. 99, No. 5, May 2003, p. 24.
35. EPA, *Designations for the Fine Particle National Ambient Air Quality Standards* (U.S. Environmental Protection Agency, Washington, D.C.), www.epa.gov/ttn/naaqs/pm/pm25_guide.html (accessed April 1, 2003).
36. EPA, *Mercury Study Report to Congress* (Office of Air Quality Planning and Standards and Office of Research and Development, U.S. Environmental Protection Agency, U.S. Government Printing Office, Washington, D.C., December 1997).
37. EPA, *Study of Hazardous Air Pollutant Emissions from Electric Utility Steam Generating Units—Final Report to Congress* (Office of Air Quality Planning and Standards, U.S. Environmental Protection Agency, U.S. Government Printing Office, Washington, D.C., February 1998).
38. EPA, *EPA ICR No. 1858: Information Collection Request for Electric Utility Steam Generating Unit Mercury Emissions Information Collection Effort* (U.S. Environmental Protection Agency, Washington, D.C., 1999).
39. EPA, Regulatory Finding on the Emissions of Hazardous Air Pollutants From Electric Utility Steam Generating Units, *Federal Register*, Vol. 65, No. 245, December 20, 2000, pp. 79825–79831.
40. EPA, *Clear Skies* (U.S. Environmental Protection Agency, Washington, D.C.), www.epa.gov/clearskies/ (accessed July 17, 2003).
41. EPA, *Clear Skies Basic Information* (U.S. Environmental Protection Agency, Washington, D.C.), www.epa.gov/clearskies/basic.html (accessed July 10, 2003).
42. Anon., Bush Initiative Faces Skepticism in Congress, *Chemical and Engineering News*, Vol. 81, No. 29, July 21, 2003, p. 5.
43. Tatsutani, M., Multi-Pollutant Proposals in the 108th Congress, paper presented at the OTC Annual Meeting, Philadelphia, PA, June 22, 2003.
44. EPA, Interstate Air Quality Rule (U.S. Environmental Protection Agency, Washington, D.C.), www.epa.gov/interstateairquality/ (accessed December 4, 2003).

45. Anon., Congress Considers Climate Change, *Chemical Engineering Progress*, Vol. 99, No. 6, June 2003, p. 23.
46. Anon., Global Initiative on CO_2 Storage, *Chemical and Engineering News*, Vol. 81, No. 26, June 30, 2003, p. 19.
47. Anon., Climate-Change Plan Released, *Chemical and Engineering News*, Vol. 81, No. 30, July 28, 2003, p. 39.
48. Anon., White House Seeks More Data on Global Climate Change, *Centre Daily Times*, July 24, 2003, p. A10.
49. Soud, H. N., *Developments in FGC* (IEA Coal Research, London, March 2000).
50. Wu, Z., *NO_X Control for Pulverized Coal Fired Power Stations* (IEA Coal Research, London, December 2002).
51. McConville, A., *An Overview of Air Pollution Emission Standards for Coal-Fired Plants Worldwide* (Coal and Slurry Technology Association, Washington, D.C., 1997), pp. 1–12.
52. Soud, H. N., *Developments in Particulate Control for Coal Combustion* (IEA Coal Research, London, April 1995).
53. Clarke, L. E. and L. L. Sloss, *Trace Elements: Emissions from Coal Combustion and Gasification* (IEA Coal Research, London, 1992).
54. Lutter, R. and E. Irwin, Mercury in the Environment: A Volatile Problem, *Environment*, Vol. 44, No. 9, November 2002, pp. 24–40.
55. United Nations, United Nations Framework Convention on Climate Change, 1992.
56. EPA, *States Guidance Document Policy Planning to Reduce Greenhouse Gas Emissions*, Second ed. (Office of Policy, Planning, and Evaluation, U.S. Environmental Protection Agency, Washington, D.C., May 1998).
57. Anon., *Global Warming in Brief* (Global Warming, Washington, D.C.), www.globalwarming.org/brochure.html (accessed November 2000).
58. Jackson, T. (editor), *Mitigating Climate Change: Flexibility Mechanisms* (Elsevier Science, Oxford, 2001), p. 17.
59. United Nations, Kyoto Protocol to the United Nations Framework Convention on Climate Change, 1997.
60. EIA, *International Energy Outlook 2002* (Energy Information Administration, U.S. Department of Energy, Washington, D.C., March 2002).
61. EPA, *Latest Findings on National Air Quality 2001 Status and Trends* (Office of Air Quality Planning and Standards, U.S. Environmental Protection Agency, U.S. Government Printing Office, Washington, D.C., September 2002).
62. EPA, *Latest Findings on National Air Quality 2002 Status and Trends* (Office of Air Quality Planning and Standards, U.S. Environmental Protection Agency, U.S. Government Printing Office, Washington, D.C., August 2003).
63. U.S. Department of Energy, Quicksilver, Don't Play With It, *FETC Focus*, Issue 2, March 1999, p 25.
64. Anon., Coal-Fired Plants Emit Most Mercury, *Chemical and Engineering News*, Vol. 81, No. 6, February 10, 2003, p. 20.
65. EIA, *Annual Energy Outlook 2003* (Energy Information Administration, U.S. Department of Energy, Washington, D.C., January 2003).

CHAPTER 5

Technologies for Coal Utilization

Historically, coal use in the United States, from the 1800s to the mid-1900s, was primarily for iron and steel production, locomotives for transportation, and household heat. In addition, many chemicals, including medicines, dyes, flavorings, ammonia, and explosives, were produced from coal. With electrification of the United States beginning around 1950 and the feedstocks for chemical production shifting from coal to oil, the primary applications for coal use from 1950 to the present have been for electricity generation and the production of iron and steel. Coal is used in the industrial sector for producing steam and to a lesser extent electricity, and some chemicals are produced from coal. The technologies used for generating power, heat, coke, and chemicals will be discussed in this chapter and include combustion, carbonization, gasification, and liquefaction, which have been referred to as the four "grand processes" of coal utilization [1]. The emphasis of this chapter is on coal combustion, as this technology is the single largest user of coal.

Coal Combustion

Burning coal to generate heat is the most straightforward way of using coal. The heat that is generated from burning coal is used for warmth, cooking, and industrial processes. The use of coal for warmth, cooking, and metal works has been around for thousands of years. While the Chinese are credited with using coal as early as 1000 B.C. [2] and the first documented use of coal in Western civilization was by the Greek philosophers Pliny, Aristotle, and Theophrastus in the fourth century [3], coal was probably used by prehistoric man, as coal can be found at outcrops and often is easy to ignite. As discussed in Chapter 2 (Past, Present, and Future Role of Coal), the use of coal increased substantially during the Industrial Revolution but decreased with the discovery of oil and its use as a home heating and transportation fuel. Today, the use of coal for direct residential heating and industrial processes represents a small percentage of total coal consumption. The primary use of coal is for burning in boilers to generate electricity. The electrification of U.S. households in the 1950s along with the electrification of U.S. industry (*i.e.*, electrometallurgical processes such as electric furnaces for steel and aluminum manufacturing, electric motors, computerized control of processes, and the widespread conversion from shaft power to electrical power) have resulted in a large and ever-increasing demand for electricity. A brief history of key technological advances of boilers and combustion systems is presented followed by discussions of steam fundamentals and how they apply to boiler development, chemistry of coal combustion, the types of combustion systems, and the influence of coal properties on utility boiler design, with an emphasis on coal ash properties.

Brief History of Boilers and Coal Combustion Systems

Harnessing steam power has been credited as being probably the most important technological advance to have contributed to the rise of industrial nations. Steam power was key to the Industrial Revolution and even today, after more than 100 years of development, boilers continue to dominate as a power source [4]. The first recorded use of a steam boiler dates back to 200 B.C. when Hero of Alexandria, a Greek mathematician and scientist, is credited with inventing a steam machine [5]. Hero's steam machine, shown in an artist's rendition in Figure 5-1, is simply a cauldron with a lid and a pipe for passing steam from the cauldron to a ball on a pivot. As steam exits the ball from outlet pipes, the ball spins [5]. This concept for using steam power is not practical, as it has an overall efficiency (*i.e.*, conversion of heat in the fuel to power output taking into account heat losses, friction, and steam leaking from joints) as low as 1% as demonstrated from working reconstructions [5]; however, it does demonstrate early recognition of steam as a power source. Unfortunately, Hero considered his invention a toy, and the failure of the Greeks and Romans to harness steam as a power source was without doubt one of the factors that prevented industrialization of their societies.

There is no record of practical steam application until the seventeenth century in England [6]. At this time, England required considerable fuel for space heating and cooking, industrial and military growth demanded greater amounts of fuel, and the forests were being rapidly depleted. It became necessary to find another source of energy, and coal mining expanded greatly. As large-scale coal mining developed, mines became deeper and often were flooded with water. England was desperately in need of a means to remove the water from the mines and, consequently, the first large-scale use of steam was made by mining engineers for steam-driven pumps. The first commercially successful steam engine was invented by Thomas Savery in 1698 and was developed for direct displacement of water [6]. This engine, however, was only moderately successful, as the height of water (*i.e.*, head pressure) that could be pumped was limited by the pressure that the boiler and vessels could withstand.

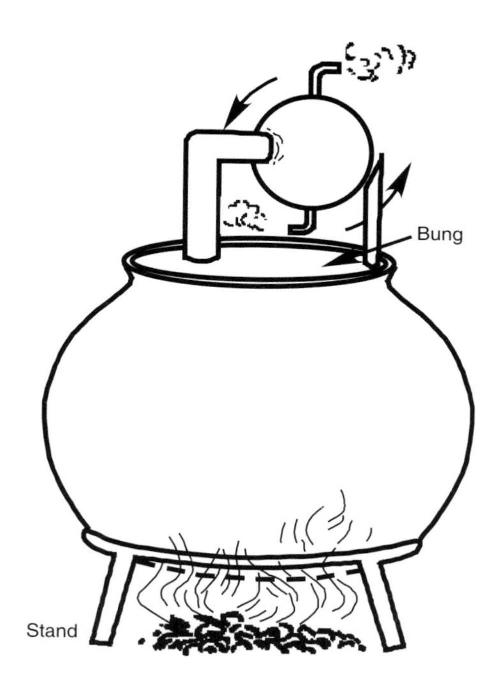

FIGURE 5-1. Hero's steam machine. (From Landels, J. G., *Engineering in the Ancient World*, University of California Press, Berkeley, 1978. With permission.)

In 1711, Thomas Newcomen overcame Savery's limitations by inventing a steam-driven pump using a piston and cylinder system and incorporating Denis Papin's 1690 invention of the safety valve. Papin invented a steam digester for culinary purposes using a boiler under pressure and invented the valve to avoid an explosion. Boiler-related accidents began to take a terrible toll on lives and equipment as boiler capacity was not keeping up with the demand for power. Papin is also credited with inventing a boiler with an internal firebox, the earliest record of such construction.

In the 1700s, developers of the steam engine noted that nearly half of the heat from the fire was lost because of short contact time between the hot gases and the boiler-heating surface. This concern for fuel efficiency led to the development of a boiler separate from the steam engine. In an effort to overcome the poor efficiency, John Allen developed, in 1730, an internal furnace with a flue winding through the water like a coil in a still and introduced the concept of forced draft by using bellows to force the gases through the flue [6]. During the latter half of the eighteenth century, James Watt made many significant improvements to the early steam engine and, along with Matthew Boulton, introduced the first boiler. In 1785, Watt took out a number of patents for variations in furnace construction. Richard Trevithick realized that the major problem of early steam systems was the manufacture of the boiler. Around 1800, Trevithick built a 650-psig (pounds per square

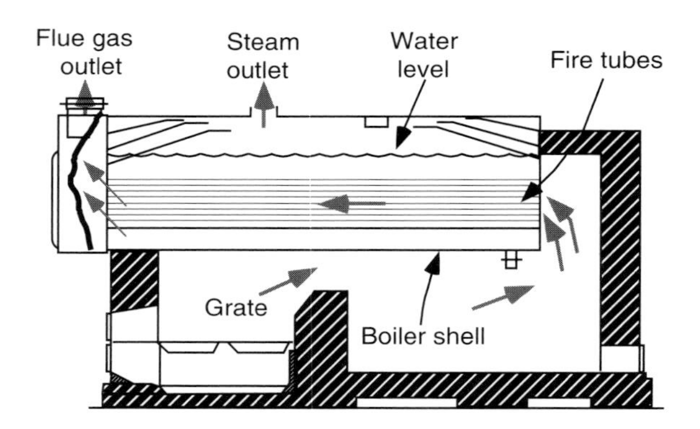

FIGURE 5-2. Typical firetube boiler design. (*Source:* Anon., Boilers and Auxiliary Equipment, *Power*, Special Edition, June 1988.)

inch gauge) engine with a high-pressure boiler [6]. The boiler was constructed of cast iron instead of copper, which was the material used up to this time, and the use of cast iron allowed for operation of the high-pressure steam engine.

Early boiler designs consisted of a simple shell for the boiler, with a feed pipe and steam outlet mounted on a brick setting. Fuel was burned on a grate within the setting and the heat released was directed over the lower surface before most of it went out the flue. Heating a single large vessel of water is inefficient, so it is necessary to bring more water into close contact with the heat. One way to do this is to direct the hot combustion products through tubes within the boiler shell, which increases the heating surface and helps distribute steam formation more uniformly through the water. This approach, with multiple flue pipes submersed in the water (*i.e.*, firetube boiler), was in widespread use up to about 1870. The firetube boiler is shown schematically in Figure 5-2. Firetube boilers, however, are limited in capacity and pressure and could not fulfill the requirements, which developed later, for higher pressures and larger unit sizes. The development of watertube boiler designs for steam generation, shown schematically in Figure 5-3, allowed for larger units and higher pressures. Watertube designs feature one or more relatively small drums with multiple tubes in which a steam/water mixture circulates. Heat flows from outside the tubes to the mixture. This subdivision of pressure parts makes large capacities and high pressures possible.

The development of the watertube boiler began in 1766 when a patent was granted to William Blakely that included a form of watertube design for the steam generator [6]. The first successful user of the watertube boiler was James Rumsey, who patented several forms of watertube designs in 1788. This was followed by a watertube boiler invention by John Stevens (1805),

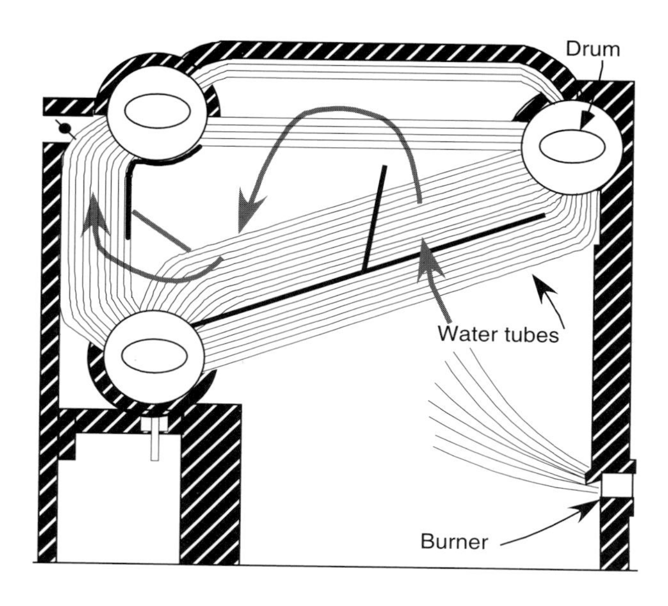

FIGURE 5-3. Typical watertube boiler design. (*Source:* Anon., Boilers and Auxiliary Equipment, *Power*, Special Edition, June 1988.)

another form of watertube design by John Cox Stevens (1805), and a watertube boiler design by Jacob Perkins (1822) that is the predecessor of the once-through steam generator. A significant breakthrough in watertube boilers occurred in 1856 when Stephen Wilcox proposed a design that allowed better water circulation and heat transfer. In 1866, George Babcock became associated with Stephen Wilcox and the first Babcock and Wilcox (B&W) boiler was patented in 1877. B&W boilers were considered to be the best engineered during the late 1800s and powered two of the earliest electric power generating stations [4]. In 1881, the Brush Electric Light Company in Philadelphia, Pennsylvania, began operations, and the Pearl Street Station opened in New York City in 1882.

Once electricity was recognized as a safe and reliable power source, the demand for boilers increased, and many generating stations were built to satisfy the needs of industrial and residential customers. With increasing demand for power came increasing demand on boiler manufacturers to improve output. In the period from 1900 to 1920, major strides were made as utilities and other boiler customers tried larger boilers and multiple boiler arrangements to achieve greater output [4]. Steam pressures and temperatures nearly doubled as manufacturers began to optimize the boiler for electric power generation. Key developments included using pulverized coal firing in place of stoker-fired boilers (*i.e.*, small coal particle size versus large coal particle size) in order to take advantage of the higher volumetric heat release rates of pulverized coal; increasing system efficiencies by using superheaters (heat-exchange surface to increase the steam temperature—see below), economizers (heat exchange surface to preheat the boiler feedwater), and combustion air preheaters (heat-exchange surface to preheat the combustion air); and improving materials of construction, allowing for steam generators to achieve steam pressures in excess of 1200 psig [4]. The relationship between system energy flows, superheated steam, and higher steam temperatures and pressures is illustrated later in this chapter.

The incorporation of superheaters into boiler designs resulted in great strides in the quest to improve steam pressures and temperatures in boiler systems in the United States. Much credit for incorporating superheaters is due to Earnest Foster, who founded the Power Specialty Company along with Pell Foster, and convinced boiler owners to install superheaters after he visited Europe and found that the United States was several years behind Europe in adopting superheated steam for power generation. The development of superheaters, reheaters, economizers, and air preheaters played a significant role in improving overall system efficiency because they utilized as much of the heat generated from burning the coal as possible. The separation of the steam from the water and the use of superheaters and reheaters allowed for higher boiler pressures and larger capacities. The implementation of pulverized coal firing over stoker firing, which became widespread by the mid-1920s, produced increased boiler capacity and improved combustion and boiler efficiencies over stoker-fired boilers, which were commonplace up until that time.

Two well-known companies also got their start in the early 1900s: Combustion Engineering, now known as Alstom Power, and Riley Stoker, Inc., now known as DB Riley, Inc. [4]. These companies led the technical development of fuel handling and the use of pulverized coal in the United States. Combustion Engineering was initially known for its stoker designs, including one for burning anthracite screenings and one for bituminous coal, but later expanded into a complete line of stokers. Similarly, Riley Stoker, known for producing large, multiple-retort underfeed stokers, expanded into a complete line of stokers for boilers of all sizes. Because of limitations placed on boiler capacity by the size restriction of stokers, Combustion Engineering developed coal pulverizing. This was a major technological improvement in steam generation. Traveling grate stokers had met their technical limits of steam generation at ~200,000 lb/hr, but by 1929 Combustion Engineering had erected the first steam generator unit to produce steam at one million lb/hr (using pulverized coal-firing) at New York Edison's East River Station [4].

In 1946, B&W introduced the cyclone furnace for use with slagging coals (*i.e.*, coals that contain inorganic constituents that will form a liquid ash at temperatures of ~2600°F or lower), which was the most significant advance in coal firing since the introduction of pulverized coal firing [6]. Cyclone furnaces provide the benefits obtained with pulverized coal firing but have the advantages of utilizing slagging coals, reducing costs due to less fuel preparation (*i.e.*, the coal can be coarser and does not need to be pulverized), and reducing the furnace size.

Fluidized-bed boilers for utilizing coal were originally developed in the 1960s and 1970s and offer several inherent advantages over conventional combustion systems, including the ability to burn coal cleanly by reducing sulfur dioxide emissions during combustion (*i.e.*, *in situ* sulfur capture) and generating lower emissions of nitrogen oxides. In addition, fluidized-bed boilers provide fuel flexibility, as a range of low-grade fuels can be burned efficiently. Today, all the major boiler manufacturers offer fluidized-bed boilers; however, options are limited for small, industrial-sized, fluidized-bed boilers, and they are not being produced at the very large steam capacities that pulverized coal-fired units are.

Advances in materials of construction, system designs, and fuel firing have led to increasing capacity and higher steam operating temperatures and pressures. In the United States, utilities typically choose between two basic pulverized coal-fired watertube steam generators: subcritical drum-type boilers with nominal operating pressures of either 1900 or 2600 psig or once-through supercritical units operating at 3800 psig [7]. These units typically range in capacity from 300 to 800 MW (*i.e.*, producing steam in the range of 2 to 7 million pounds per hour); however, ultra-supercritical units entered into service in 1988 and operate at steam pressures of 4500 psig and steam temperatures of 1050°F with capacities as high as 1300 MW.

Comparison of Industrial and Utility Boilers

Most coal that is consumed in the United States is used for generating electricity; however, a significant number of small non-electricity generating boilers in the United States burn coal. A brief discussion of industrial boilers is provided here and comparison to utility boilers is made. Utility boilers and industrial boilers are quite different. The major differences between a utility boiler and an industrial boiler include: (1) size of the boiler; (2) application of steam the boiler generates; (3) design of the boiler; (4) diversity of fuels including the use of by-product fuels; and (5) global competition for products produced. Comparatively, the typical utility boiler is much larger than the average industrial boiler. As a result, industrial boilers do not enjoy the economies of scale that utility boilers do and, in the case of emissions reduction, must pay more to remove a given amount of emissions.

Size and Number of Units The average new industrial boiler is considerably smaller than a utility boiler. A typical utility boiler produces about 3.5 million pounds of steam per hour (approximately 400 MW), while a typical industrial boiler produces about 100,000 pounds of steam per hour. Many industrial boilers are designed for less than 250,000 pounds of steam per hour but can be designed for greater than one million [6,8]. There are considerably more small industrial boilers than large utility boilers, and the industrial

TABLE 5-1
Distribution of Utility Boilers by Conventional Coal Combustion Technologies

Combustion Technology	*Number of Boilers*	*Capacity (Megawatts equivalent)*
Pulverized coal boilers	1068	294,035
Stokers	94	1077
Cyclones	89	25,727
Total	1251	320,839

boilers are tailored to meet the needs and constraints of widely varying industrial processes. The Council of Industrial Boiler Owners (CIBO) reports that the industrial boiler and process heater population (total and not just coal-fired units) consists of 42,000 and 15,000 units, respectively, ranging in size from 10,000 to 1,400,000 pounds of steam per hour with an average unit size of 100,000 pounds of steam per hour [9]. In comparison, there are about 4000 utility units, of which about 1250 boilers utilize conventional coal combustion technology (see Table 5-1), with an additional 67 utility-scale fluidized-bed boilers [10]. This is further illustrated in Figure 5-4, which shows the distribution of coal-fired boilers by capacity for conventional utility, conventional non-utility, fluidized-bed combustion (FBC) utility, and FBC non-utility boilers [10]. The Environmental Protection Agency (EPA) defines a non-utility boiler as a boiler whose primary product is not electricity but steam. Some of the non-utility boilers (both conventional and FBC) are cogeneration units in that they produce both steam and electricity.

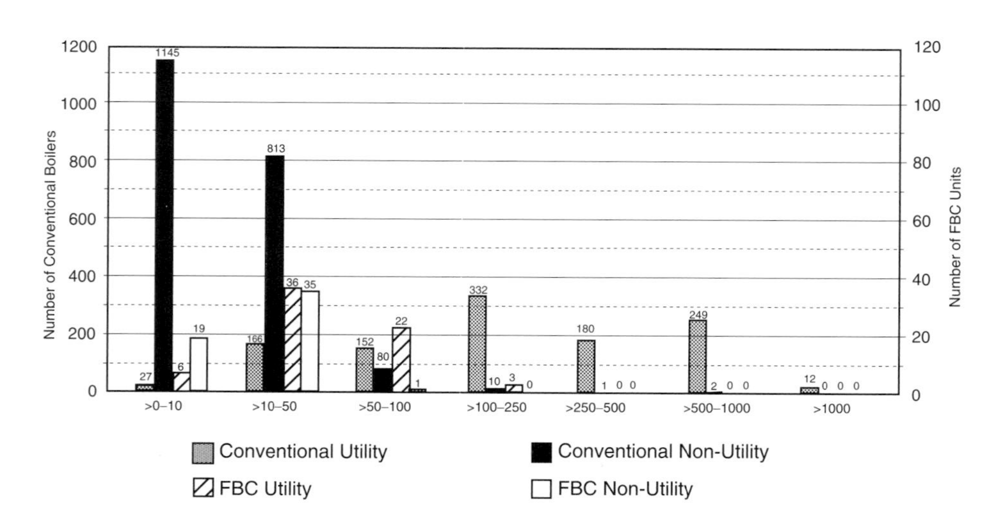

FIGURE 5-4. Distribution of coal-fired boilers by capacity (in MW). (Adapted from EPA [10].)

Application of Steam Industrial boilers are utilized in many different industries for a wide variety of purposes and the main product is process steam. Industrial boiler operation can vary significantly between seasons, daily, and even hourly depending on the steam demand. A utility boiler, however, generates steam for the sole purpose of powering turbines to produce electricity. A typical utility boiler (*i.e.*, a base-loaded unit) operates at a steady rate close to maximum capacity because of a constant demand for steam. Load swings from utility boilers that operate to meet utility load swings during the day or for seasonal peak demands (*i.e.*, peaking units) are more controlled than industrial boilers because they can balance their load over the complete electric production and distribution grid; consequently, utility boilers tend to have lower operating costs than industrial boilers that are similarly equipped. Utility units generally have a variety of backup alternatives for unscheduled outages. Industry, however, rarely has a backup system for steam generation because of the need to keep costs for steam production as low as possible; hence, industrial boilers routinely operate with reliability factors of around 98%.

Boiler Design Utility boilers are primarily field-erected units designed for high pressure and high temperature steam. Boiler designs, capacities, steam pressures, and temperatures, among other parameters, vary with the fuel and service conditions. An expanded discussion of the effect of coal type and characteristics on utility boiler design is provided later in this chapter. As previously mentioned, there are primarily two basic watertube boiler designs that are selected by utilities in the United States [7]: (1) subcritical drum-type boilers designed for a nominal operating pressure of 1900 or 2600 psig steam, or (2) once-through supercritical units designed for 3800 psig steam. There are many design criteria for steam generators and their auxiliary components, but the important issues are efficiency, reliability/availability, and cost. While some stoker and cyclone boilers are in operation, newer designs are primarily pulverized coal and fluidized-bed units. Industrial boilers and their incorporated combustion systems exhibit much variability in their designs and construction including low- and high-pressure steam production, variability in sizes, shop-assembled packaged boilers or field-erected units, and the capability of burning a wide variety of fuels. Industrial boilers consist of packaged and field-erected units of various boiler types: watertube, firetube, stokers, fluidized-bed, pulverized-coal, and cyclone units. Packaged units are available in capacities up to about 600,000 pounds of steam per hour, but boilers larger than 250,000 pounds of steam per hour typically cannot be shipped by rail, although they can be shipped by barge or ocean vessels [6]. The industrial boiler industry is influenced by several factors [7]: (1) the user's desire for fuel flexibility over the life of the unit; (2) the demands for ever-increasing emissions restrictions; (3) significant interest in burning low-quality fuels; (4) wider application of cogeneration; (5) the desire to optimize existing equipment in terms of efficiency, performance, and service life; and (6) the recognition that turndown (*i.e.*, ability to operate efficiently at reduced steam output/fuel firing rate) is as important over the boiler's lifetime as is maximum continuous rating.

Fuel Diversity and Global Product Competition Fuel diversity and global product competition are primarily of interest to the industrial sector; therefore, they are not discussed in detail here. While electricity is being sold throughout the United States as a result of deregulation in many states, electricity is not a global product (excluding any power sales to Mexico or Canada), whereas many industrial products must compete with international markets. Fuel diversity does affect utilities but not to the extent it does industrial boilers. Coal is the lowest price energy feedstock available and is used extensively in the power generation industry. Some fuels are cofired with coal, such as petroleum coke, tires, and biomass materials; however, utility boilers firing coal tend to utilize only coal. This may change in the future if legislation is passed requiring power generators to produce a percentage of their electricity from renewable energy. Industrial boiler users, on the other hand, are interested in using a wider variety of fuels as they experience more volatility in fuel availability and prices. Examples of industrial boiler fuels (not inclusive) include waste coals such as bituminous gob and anthracite culm, wood refuse, bagasse, digester (black) liquor, blast-furnace gas, petroleum coke, refining gas, carbon monoxide waste gas, peanut shells, palm fronds, rice husks, animal fats and proteins, and animal manure and litter [6,9,11–13]. Further discussion of utilizing biomass in utility boilers is provided in Chapter 8 (Coal's Role in Providing U.S. Energy Security) as part of a discussion on the role of coal in providing national security by protecting the U.S. food supply.

Basic Steam Fundamentals and Their Application to Boiler Development

Figures 5-5 through 5-7 are used to explain the concept of superheated steam and how it results in achieving higher temperatures and pressures. Figure 5-5 shows the general arrangement of a present-day watertube boiler, including the location of the superheaters, reheater, economizer, and air preheater. The steam generator energy flow depicted in Figure 5-6 shows how the various heat exchange surfaces are integrated with each other and the steam turbine. Figure 5-7 illustrates the relationships among temperature, pressure, and enthalpy (*i.e.*, heat content) of saturated and superheated steam. Heating water at any given pressure eventually will cause it to boil, and steam will be released. The heat required to bring the water from 32°F to the boiling point is the enthalpy, or heat content, of the liquid (measured in Btu/lb). When water boils, both it and the steam are at the same temperature, which is called the saturation temperature. For each boiling pressure, there is only

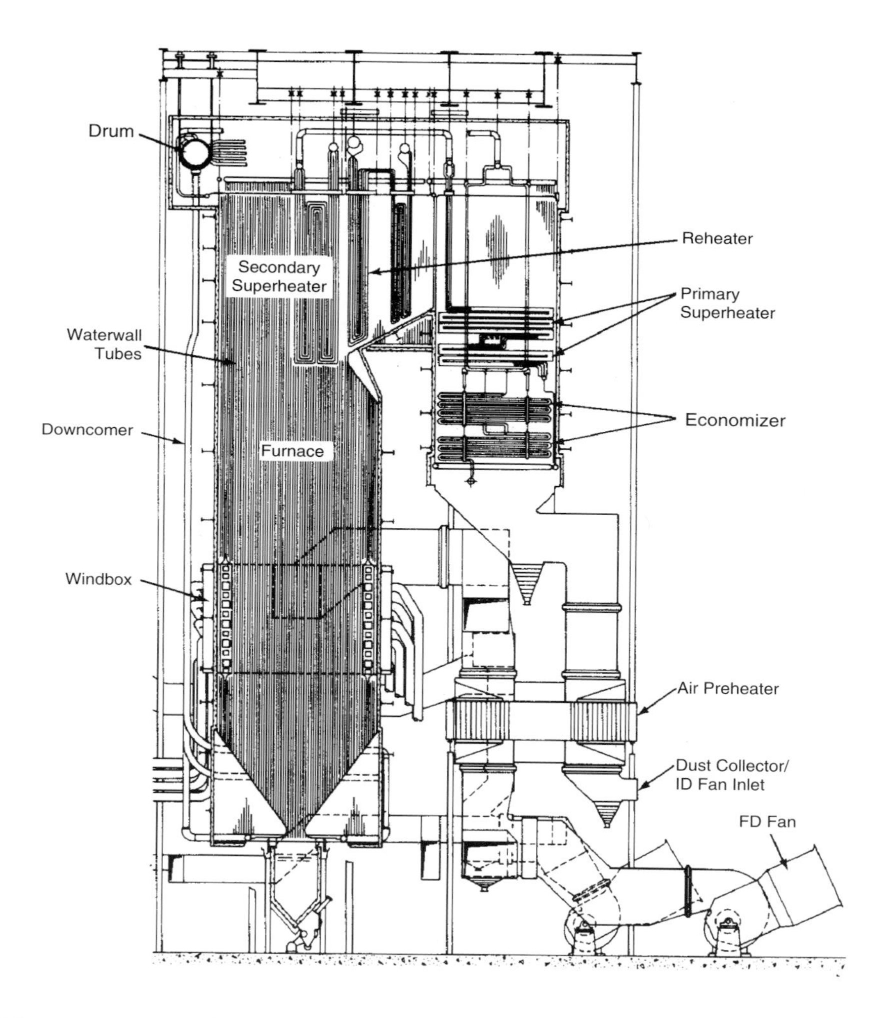

FIGURE 5-5. General arrangement of a watertube steam generator. (From Elliot, T. C., Ed., *Standard Handbook of Powerplant Engineering*, McGraw-Hill, New York, 1989. With permission.)

one saturation temperature and vice versa. During the boiling process, temperature remains constant as more heat is added, which is being used to change the water from the liquid to the vapor state. This heat is the enthalpy of evaporation and when added to the enthalpy of the saturated liquid gives the enthalpy of the saturated steam, which is the total amount of heat added to bring 32°F water up to 100% steam [7]. The temperature of the steam and water will remain the same as long as the two are in contact. To raise the temperature of the steam, it must be heated out of contact with water

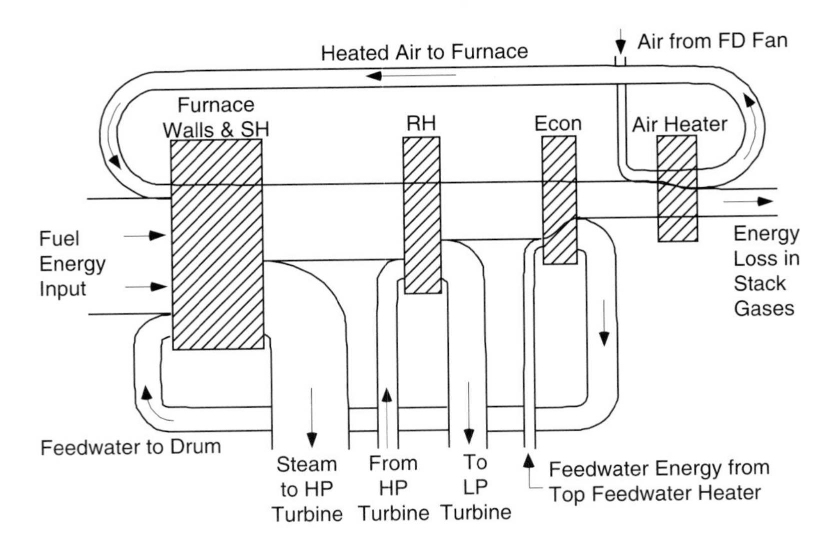

FIGURE 5-6. Steam generator energy flow. (From Elliot, T. C., Ed., *Standard Handbook of Powerplant Engineering*, McGraw-Hill, New York, 1989. With permission.)

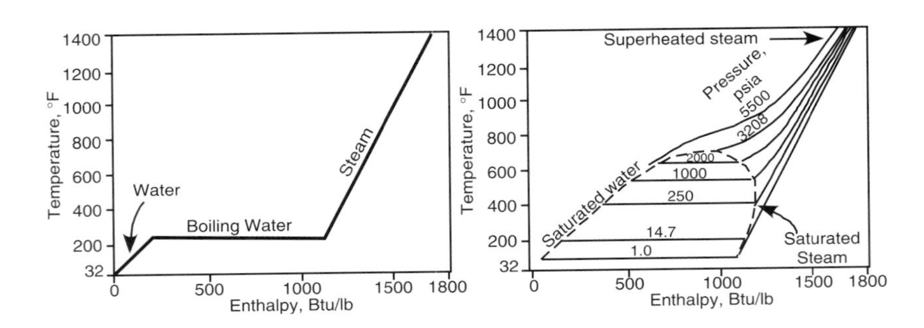

FIGURE 5-7. Water/steam enthalpy diagram. (*Source:* Anon., Boilers and Auxiliary Equipment, *Power*, Special Edition, June 1988.)

(*i.e.*, it must be superheated). The enthalpy of the steam will increase by the amount of heat added, and the temperature will rise. The temperature–pressure relationship is shown in Figure 5-7. The left diagram in Figure 5-7 shows the energy (in Btu/lb of water) required to heat water from 32°F to its boiling point of 212°F at atmospheric pressure (or 14.7 pounds per square inch absolute (psia)), the energy input to continue boiling the water until all the water is converted to steam, and the temperature rise of the steam (superheating) as more energy is put into the system. The right diagram illustrates that, as the pressure increases, the amount of heat required to raise the temperature of the water to its boiling point increases while the amount of heat necessary to vaporize it decreases.

Chemistry of Coal Combustion

Coal is burned in three ways: (1) as large pieces in a fixed bed or on a grate, (2) as smaller or crushed pieces in a fluidized bed, or (3) as very fine particles in suspension. Theoretically, any particle size can be burned by any of the three methods; however, engineering limitations establish preferred particle sizes for the three methods. Particle size has also been found to be the most important parameter with respect to the dominant reaction mechanism and other thermal behavior (*i.e.*, rate of heating, which can control volatiles yield and composition) [2]. The main characteristics of the three techniques are summarized in Table 5-2 [1,2].

The combustion process consists of several steps. As the coal particles are heated, moisture is driven off the coal particles. Next, the coal particles undergo devolatilization and release volatile organic constituents. The volatile matter is combusted in the gas phase (homogenous reaction). This can occur prior to and simultaneously with combustion of the char particles, which is the last step. Combustion of the char is a surface (heterogeneous) reaction. These reactions are for the most part sequential, and the slowest of these will determine the rate of the overall process.

TABLE 5-2
Comparison of Characteristics of Combustion Methods

Variables	*Combustion Method*		
	Fixed Bed (Stoker)	*Fluidized-Bed*	*Suspension*
Particle size			
Approximate top size	<2 in.	<0.2 in.	180 μm
Average size	0.25 in.	0.04 in.	45 μm
System/bed temperature	<1500°F	1500–1800°F	>2200°F
Particle heating rate	~1°/sec	10^3–10^4°/sec	10^3–10^6°/sec
Reaction time			
Volatiles	~100 sec	10–50 sec	<0.1 sec
Char	~1000 sec	100–500 sec	<1 sec
Reactive element description[a]	Diffusion-controlled combustion	Diffusion-controlled combustion	Chemically controlled combustion

[a]Described in text and illustrated in Figure 5-8.
Source: Adapted from van Krevelen [1] and Elliot [2].

Devolatilization of Pulverized Coal and Volatiles Combustion

The design of coal burners and furnaces is very dependent on the volatile matter released by the coal as it heats [15]. In flames, pulverized coal heats primarily by convective heat transfer with hot gases which are entrained and recirculated, with heating of only the coarsest particles being dominated by radiation from the hot regions of the flame. For large flames, in which coal remains for several hundred milliseconds, the extent of devlolatilization is strongly influenced by temperature rather than by limitations due to heating times or devolatilization kinetics. Studies have indicated that changes in heating rate (in the range of 1 to 50×10^3 °C/sec) have little effect on volatile yield and that the yield is more strongly influenced by the final temperature, with an increase in final temperature producing an increased yield [2]. Volatiles yield is also found to depend upon particle size, with smaller particles tending to yield more volatiles. Also, volatile yield can vary significantly within a given rank for coals that are similar in composition and mined adjacent to each other in the same coal basin (*e.g.*, subbituminous coals from neighboring Powder River Basin coal mines) [16]. The combustion of the volatiles is generally assumed to be a homogenous reaction, although the possibility of volatile matter burning heterogeneously has been suggested by Howard and Essenhigh [17]. The burning of the volatiles is a very fast process that is measured in milliseconds [18].

Char Combustion

Char combustion is a much slower process than devolatilization and therefore determines the time for complete combustion in a furnace, which is on the order of several seconds for pulverized coal at furnace temperatures. Studies have shown that the combustion of the char begins with chemisorption of oxygen at active sites on char surfaces and that the decomposition of the resultant surface oxides mainly generates carbon monoxide (CO) [2,19] (Some researchers think that an amount of CO_2 may also be released during this step.) The CO is then oxidized to CO_2 in a gaseous boundary zone around the char particle. Fresh reaction sites are continuously exposed as the surface oxides are decomposed. CO_2 then either moves off into the gas stream or is reduced to CO if it impinges on the char. The overall reaction mechanism is complex [2,20], but the combustion of char involves at least four carbon–oxygen reactions [19]:

$$C + \tfrac{1}{2}O_2 \longrightarrow CO \qquad (5\text{-}1)$$

$$CO + \tfrac{1}{2}O_2 \longrightarrow CO_2 \qquad (5\text{-}2)$$

$$CO_2 + C \longrightarrow 2CO \qquad (5\text{-}3)$$

$$C + O_2 + \longrightarrow CO_2 \qquad (5\text{-}4)$$

as well as the oxidation of non-carbon atoms, mainly:

$$S + O_2 \longrightarrow SO_2 \tag{5-5}$$

$$H_2 + \tfrac{1}{2}O_2 \longrightarrow H_2O \tag{5-6}$$

which may be followed by

$$H_2O + C \longrightarrow CO + H_2 \tag{5-7}$$

$$CO + H_2O \longrightarrow CO_2 + H_2 \tag{5-8}$$

Some species of the mineral matter can be volatilized during combustion, while others are left behind as ash. In both cases, the mineral matter is usually altered in composition and mineralogy.

The rate of char combustion is a complicated process, as it is influenced by mass transfer by diffusion through the pores and the surface reactions. The diffusion coefficients are strongly dependent on pore diameters and pressure, and the surface reaction is influenced by the formation of activated adsorption complexes and their decomposition [1].

The rate of char combustion is controlled by two processes: the chemical reaction rate of carbon and oxygen on the char surface and the rate of mass transfer of oxygen from the bulk gas stream through the boundary layer surrounding the particle to the particle surface. This is illustrated in Figure 5-8 [20], where the general relationship between temperature and reaction rate for a heterogeneous gas/solid system is shown. At low temperature (Region I), the chemical reaction rate is slow compared to the rate of diffusion through the pores; therefore, oxygen completely penetrates the char matrix. Combustion then takes place within the porous char, and the density of the char rather than the diameter changes. In this case, the oxygen concentration at the particle surface would be the same as that in the bulk gas stream, and the overall reaction rate would be limited by the inherent rate of the chemical reaction. In Region I, the rate of surface reaction is rate determining, and oxygen molecules diffuse fast enough to reach the whole internal surface. The reaction rate is given by:

$$q = -\frac{d}{6}\frac{\rho_p}{dt} \tag{5-9}$$

so that $q \propto d$, where q is the char combustion reaction rate (kg/m^2/sec), d is the diameter of the particle (m), ρ_p is particle density (kg/m^3), and t is burning time (sec).

The rate of chemical reaction may be expressed by a generalized expression of the type:

$$q = k_c = A\exp(-E/RT_p)P_s^m \tag{5-10}$$

where A is the true pre-exponential constant (kg C/m^2/sec [atm O_2]$^{-m}$), T is the particle temperature (K), R is the universal gas constant, and, because

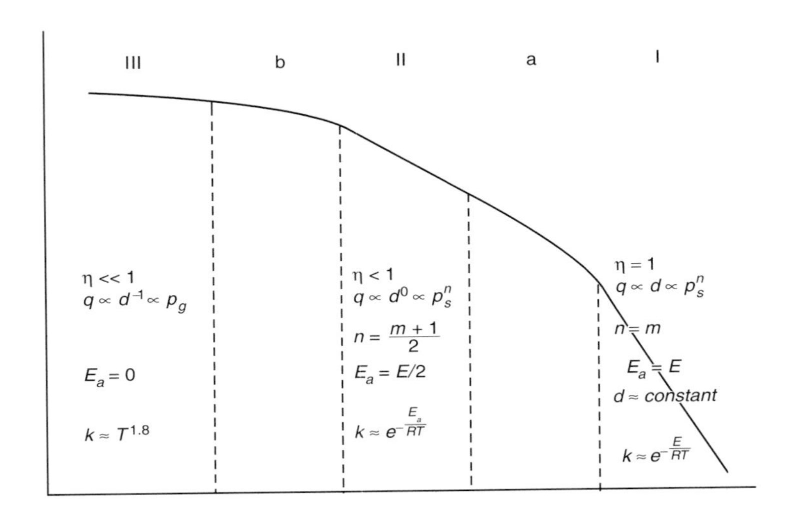

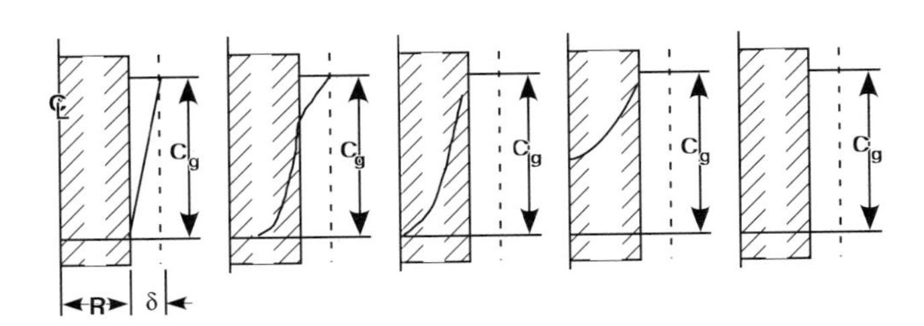

FIGURE 5-8. Relationship between temperature and reaction rate. C_g, concentration of oxygen in the bulk gas; δ, boundary layer thickness; η, effectiveness factor; E_a, apparent activation energy; E, true activation energy. (*Source:* Walker *et al.*, 1967.)

a chemical reaction controls the rate, P is the partial pressure of oxygen at the surface (atm), E is the true activation energy (J/mol), and m is the true reaction order.

As the temperature is increased, the chemical reaction becomes sufficiently rapid for the diffusion of oxygen through the pores to exert a notable rate-limiting effect. Under these conditions (Regime II), the diameter and the density of the particle will both change. The apparent activation energy and apparent order of reaction (n) are approximated by

$$E_a \approx 2, \quad n \approx (m+1)/2 \tag{5-11}$$

so that the apparent reaction rate does not change as rapidly with temperature.

A further increase in temperature eventually causes the chemical reactions to become so rapid that the oxygen is consumed as it reaches the outer surface of the particle. In this case (Regime III), the reaction is entirely controlled by the diffusion from the free stream to the particle, and only the diameter of the particle changes.

Field *et al.* [18] give an expression for the overall reaction rate coefficient k as:

$$k = \frac{1}{1/k_d + 1/k_c} \tag{5-12}$$

where k_d is the diffusional rate coefficient, and k_c is the chemical rate coefficient defined in Equation (5-2); the diffusional rate coefficient can be defined as:

$$k_d = \frac{24\phi D}{dRT_m} \tag{5-13}$$

where ϕ is a mechanism factor that takes the value of 1 for reaction to CO_2 and 2 for reaction to CO; D is the diffusion coefficient (cm^2/sec) of oxygen through the boundary layer at temperature T_m given by:

$$D = 3.49x\left(\frac{T_m}{1600}\right)^{1.75} \tag{5-14}$$

where x is the particle diameter (cm); R is the universal gas constant; and T_m is the mean temperature (K) for the boundary layer taken as the average of the surface temperature of the particle and the bulk gas temperature.

For the char sizes, porosities, internal surface areas, and temperatures typical to pulverized coal-fired furnaces, the char combustion rate is influenced by the chemical reactivity of the char, the external diffusion rate of oxygen from the bulk stream, and the internal diffusion of oxygen into the porous char matrix. Char ignition is likely to occur in Regime I or II when a large proportion of the internal surface is available for reaction. The final burn out is likely to occur in Regimes II and III, when external diffusion may have a significant influence on the combustion rate of large particles. The time for the char to burn out is proportional to the square of the initial size of the char particles from the coarse end of the grind [15].

The processes controlling the rate of char combustion in fluidized-bed systems differ slightly from those for pulverized coal-fired systems. In fluidized-bed combustion, particle sizes are larger, the processes by which oxygen is brought to the coal surface differ because of the presence of the surrounding bed particles, and the heat-transfer processes also differ from those in a pulverized coal-fired furnace [15].

All three regimes illustrated in Figure 5-8 are important in fluidized-bed combustion: Regime I, during ignition and for the smaller particles burning in the bed and freeboard; Regime II, for medium-sized particles; and Regime III, for large particles in the bed. The surface reaction rate, q (mass of carbon oxidized per unit area of particle outer surface per second), for the region separating Regimes II and III, which are of special interest in coal combustion, is defined as:

$$q = \frac{P_g}{(1/k_d + 1/k_c)} = kP_g \quad (5\text{-}15)$$

where P_g is the oxygen partial pressure in the gas outside of the boundary layer (kN/m^2). The mass transfer coefficient for the oxygen diffusion to the particles is:

$$h_m = \frac{ShD}{d} \quad (5\text{-}16)$$

where h_m is the mass flux of oxygen per unit area of surface per unit of concentration difference between that at the surface and that in the gas outside the boundary layer (m/sec), Sh is the Sherwood number (dimensionless), and D is the diffusion coefficient of oxygen through the gas mixture surrounding the particle (m^2/sec) [15].

Equation (5-15) can be used to derive the combustion rate q, in units of mass of carbon oxidized per unit time per unit area of particle outer surface:

$$q = k_d\left(P_g - P_s\right) + \frac{12\phi ShD}{dRT_m}\left(P_g - P_s\right) \quad (5\text{-}17)$$

where D is evaluated at a mean temperature T_m (K) in the diffusion layer, P_s is the oxygen partial pressure at the particle surface (kN/m^2), and the universal gas constant R has the units 8.31 J/mol K.

The chemical kinetic rate coefficient, k_c, is expressed by an Arrhenius-type equation:

$$k_c = A_a \exp\left(-E_a/RT_p\right) \quad (5\text{-}18)$$

where A_a is the apparent rate constant based on particle outer surface area (kg/m^{2}s per kN/m^2 of partial pressure of oxygen), and E_a is the apparent activation energy for Regime II combustion (kJ kg/mol).

Coal Combustion Systems

The manner in which coal is burned and the devices in which it is burned are primarily determined by the desired unit size or capacity (*i.e.*, required hourly steam production or electricity generation) and coal type and quality. The combustion methods, fixed-bed (*i.e.*, stokers), fluidized-bed, and suspension firing are discussed.

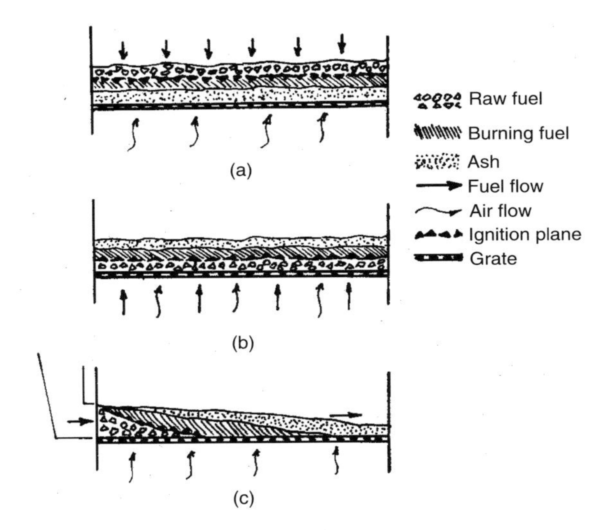

FIGURE 5-9. Patterns of feeding coal and combustion air to stokers: (a) overfeed; (b) underfeed; (c) crossfeed. (From Elliot, M. A., Ed., *Chemistry of Coal Utilization*, Secondary Suppl. Vol., John Wiley & Sons, New York, 1981. With permission.)

Fixed-Bed Combustion

Fixed-bed combustion covers a wide variety of applications, including domestic space heaters, underground gasification, and industrial stokers. It is the latter that are of interest for steam and power generation and are discussed here. Stokers were used to burn coal as early as the 1700s [7,14]. Stokers have evolved over the years from simple design to quite sophisticated devices to burn a variety of fuels including coal. Stokers are generally divided into three general groups, depending on how the fuel reaches the grate (*i.e.*, surface that contains the coal and allows combustion air to be introduced into the fuel bed) of the stoker for burning: underfeed, overfeed, and spreader designs. Three patterns of feeding coal and combustion air have been developed and are used singly or in combination in commercial equipment [2]. These patterns, illustrated in Figure 5-9, are referred to as:

- Overfeed—The fuel is fed onto the top of the bed and flows down as it is consumed while combustion air flows up through successive layers of ash, incandescent coke, and fresh coal;
- Underfeed—The flows of coal and combustion are parallel and usually upward;
- Crossfeed—The fuel moves horizontally and the combustion air moves upward at right angles to the fuel.

TABLE 5-3
Characteristics of Various Types of Stokers

Ability of the Unit to:	*Spreader*	*Overfeed*	*Underfeed*
Increase load rapidly	Excellent	Fair	Fair
Minimize carbon loss	Fair	Fair	Fair
Overcome coal segregation	Fair	Poor	Poor
Accept a wide variety of coals	Excellent, traveling grate; Fair, vibrating grate	Poor	Poor
Burn extremely fine coal	Poor	Poor	Poor
Permit smokeless combustion at all loads	Poor	Good	Good
Minimize fly ash discharge to stack	Poor	Good	Good
Maintain steam load under poor operating conditions	Good	Poor	Poor
Minimize maintenance	Good	Good	Fair
Minimize power consumption (stoker and boiler auxiliaries)	Good	Good	Good
Handle ash and cinders easily	Excellent	Good	Fair

Source: Power from Coal: Special Report, *Power*, February 1974. With permission.

Unfortunately, because there are many examples of commercial equipment with more than one pattern of combustion air and fuel feed, disagreement is sometimes found in the literature over which type predominates in a given stoker application, which also leads to differences in system classification. For our discussion here, underfeed stokers include single- and multiple-retort stokers; overfeed stokers include chain grate, traveling grate, and water-cooled vibrating grate stokers; and spreader stokers are classified into several groups, depending on the type of grate selected: stationary, dumping, reciprocating, vibrating, traveling, or water-cooled vibrating grate. Some characteristics of the stokers and their fuel requirements and capacities are given in Tables 5-3 through 5-5. Figure 5-10 illustrates the working principles of the three groups of stokers [19].

Underfeed Stokers Underfeed stokers are used primarily for burning coal in small boilers serving relatively constant steam loads of less than 30,000 lb/hr [7]. Coal is introduced beneath the active fuel bed and is moved from a storage hopper by means of a screw or ram into a retort. As coal is fed into the retort, the force of the incoming fuel causes the coal to rise in the retort and spill over onto the fuel bed or grate surface on either side of the retort. No air is supplied in the retort proper; it comes through openings, called tuyeres, in the grate section adjoining the trough. Underfeed stokers

TABLE 5-4
Stoker Fuel Requirements

Stoker	*Coal Types*	*Coal Sizes*
Underfeed	Bituminous coals or anthracite	Nut (2 × 3/4 in.) or prepared stoker (large—2 × 1/4 in.); smaller sizes acceptable if <50% – 1/4 in.
Overfeed	All coal ranks	Nut (2 × 3/4 in.) or prepared stoker (large or intermediate—2 × 1/4 or 1 × 1/8 in., respectively); <20% – 1/4 in.
Spreader	All coal ranks	Prepared stoker (small—3/4 × 1/16 in.); <30% – 1/4 in.

Source: Adapted from Berkowitz [19].

TABLE 5-5
Capacities of Coal-Fired Stokers

Stoker Type	*Recommended Grate Heat-Release Rates (1000 × Btu/hr-ft² of grate)*	*Steam Generation (1000 lb/hr)*
Underfeed stokers		
Single retort (ram feed)	250–475	2–50
Multiple retort	450–500	40–300
Spreader stokers	—	<10–400
Stationary and dumping grates	400–450	<50
Reciprocating and vibrating grates	600–650	Reciprocating gate, up to 75; vibrating gate, up to 100
Traveling grate	650–750	Up to 400
Overfeed stokers	400–425	10–300

Source: Adapted from Elliot [14].

are either of the single- or multiple-retort design, and water-cooled furnaces are preferred with underfeed stokers.

A relatively wide range of bituminous coals as well as anthracite can be burned on single- or multiple-retort stokers but typical specifications call for coal that is 3/4 × 1-1/4 in. with less than 50% of the fines passing through a 1/4-inch screen [14]. The free swelling index of the coal should be limited to 5 with single-retort stokers equipped with stationary tuyeres, and up to 7 on single-retort stokers with moving tuyeres as well as on multiple-retort stokers. It is normally recommended that the iron content in the ash be less than 20% as Fe_2O_3 with an ash fusion temperature above 2400°F and below 15% for coals having a lower ash fusion temperature.

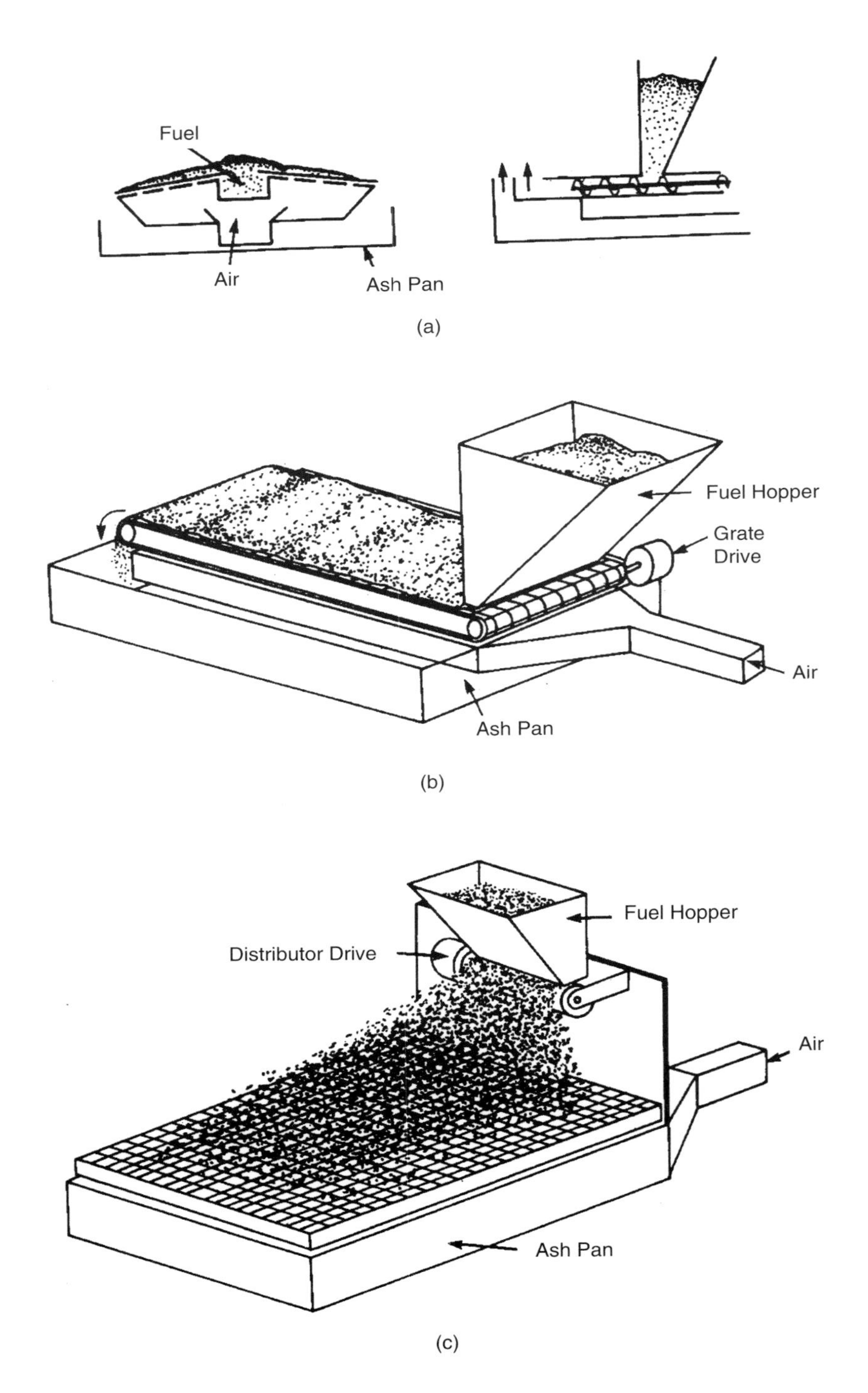

FIGURE 5-10. Working principles of mechanical stokers: (a) underfeed stoker; (b) overfeed stoker (traveling grate stoker); (c) spreader stoker. (From Berkowitz, N., *An Introduction to Coal Technology*, Academic Press, New York, 1979. With permission.)

Overfeed Stokers Overfeed, or mass-burning stokers, convey coal from the fuel hopper located at the front of the stoker. The depth of the fuel bed conveyed into the furnace is regulated by a vertical, adjustable feed gate across the width of the unit [7,14,21]. The fuel is conveyed into and through the furnace and passes over several combustion air zones. The ash is continuously discharged into a storage hopper at the rear end of the grate. Overfeed mass-burning stokers consist of three designs: chain grate, traveling grate, and water-cooled vibrating grate. Water-cooled furnaces are preferred with all moving-grate stokers to prevent slag formation on the furnace walls.

Chain grates consist of a wide chain with grate bars forming the links. The links are staggered and connected by rods extending across the stoker width. This chain assembly is continuously pulled or pushed through the furnace by an electric or hydraulic drive. The traveling grate has a chain drive (powered electrically or hydraulically) at the side of the grate with crossbars at intervals. Fingers, keys, or clips that form the grate surface are attached to these crossbars in an overlapping fashion to prevent ash from sifting through. The water-cooled, vibrating grate stoker consists of a grate surface mounted on, and in contact with, a grid of watertubes. These tubes are connected to the boiler circulatory system to ensure positive cooling. The structure is supported by a number of flexing plates, allowing the water-cooled grid and grate surface to move freely in a vibratory mode as the fuel bed moves through the furnace.

Chain grate stokers originally were developed for bituminous coal and traveling grate stokers for small sizes of anthracite [21]; however, almost any type of solid fossil fuel can be burned on the three stoker designs, including peat, lignite, subbituminous coal, non-caking bituminous coal, anthracite, and coke breeze [7]. Strongly caking bituminous coals may have a tendency to coke and prevent proper passage of combustion air through the fuel bed in the chain grate and traveling grate designs. In these designs, tempering (*i.e.*, the addition of water or steam) of the fuel bed is done in the fuel hopper to make the bed more porous, although the coal's heating value is decreased. The vibrating action of the grate in the water-cooled vibrating grate design, however, keeps the fuel bed uniform and porous without the addition of water or steam. Coal size ranges for overfeed mass-burning stokers are listed in Table 5-4. These stokers are quite sensitive to segregation of coal sizes or distribution of the coal. If the fuel size is not uniform across the width of the stoker, the fuel bed will not burn uniformly, resulting in unburned carbon being discharged into the ash hopper.

Spreader Stokers Spreader stokers are the most popular of the three types. One reason for this is that they are capable of burning all ranks of coal as well as many waste fuels [21]. In addition, they can accommodate a wide range of boiler sizes. Spreader stokers take fuel from feeders located across the front of the furnace and distribute it uniformly over the grate surface. The objective is to release an equal amount of energy from each square foot of active grate surface [7]. As the coal is spread over the grate, fines in the incoming coal stream burn in suspension while the large pieces fall to the grate, forming a fuel bed; hence, to a limited extent, spreader firing has characteristics similar to pulverized coal combustion. Primary air for combustion is admitted evenly throughout the active grate area, with an overfire air system providing secondary air and turbulence above the grate.

The fuel bed is normally thin, and there is rarely more than a few minutes' worth of coal on the grate. This, coupled with 25 to 50% of the coal being burned in suspension, allows the spreader stoker to respond quickly to load swings. This makes the spreader stoker well suited for industrial applications where process loads fluctuate rapidly. The most common types of grates used today for spreader-stoker firing are the vibrating (or oscillating), traveling, and water-cooled vibrating grates. The water-cooled vibrating grate stoker is designed primarily for refuse burning (although conceivably could be used for coal) and is not discussed here. Stationary, dumping, and reciprocating grates see limited service. Not all of these grates are suited for coal firing.

The intermittent cleaning types of grates are stationary and dumping [14]. The stationary grate is seldom used because of hazards to the operator when removing ash through an open fire door. The dumping grate is seldom used for coal because the cleaning process results in high opacity in the stack. When it is used, it is for capacities of under 50,000 pounds of steam per hour and a heat release rate from the grate of no more than 450,000 Btu/hr-ft^2.

The reciprocating grate discharges ash by a slow back-and-forth motion of moving grates alternating with stationary grates, which causes the fuel bed to move forward, dumping the ash into a pit at the front of the boiler. The grate can be used on boilers from 5000 to 75,000 pounds of steam per hour and can accommodate a wide range of bituminous coals or lignite without preparation other than sizing. Because of the stepped nature of the reciprocating grate, it is used only for fuels with sufficient ash quantity to provide an adequate ash depth for insulation on the top of the grates.

The vibrating or oscillating grate is suspended on flexing plates, and an eccentric drive or weights are used to impart a vibrating action to the grate surface, which conveys the ash to the front of the stoker and discharges them into an ash pit [14]. This grate type is well suited for coal.

The traveling grate spreader stoker is the most popular type. The endless grate moves at speeds between 4 and 20 feet per hour, depending on the steam demand, toward the front of the boiler, discharging ash continuously into an ash pit. The return grate then passes underneath in the air chamber. Traveling grate spreader stokers are designed to handle a wide range of coals.

Fluidized-Bed Combustion

Fluidized-bed combustion (FBC) is an emerging technology for the combustion of fossil and other fuels and is attractive because of several inherent advantages it has over conventional combustion systems. These advantages include fuel flexibility, low NO_x emissions, and *in situ* control of SO_2 emissions. The fluidized-bed concept was first used around 1940 in the chemical industry to promote catalytic reactions. In the 1950s, the pioneering work on coal-fired fluidized-bed combustion was begun in Great Britain, particularly by the National Coal Board and the Central Electricity Generating Board [2,22]. The U.S. Department of Interior's Office of Coal Research, one of the predecessors of the current Department of Energy (DOE), began studying the fluidized-bed combustion concept in the early 1960s (and still continues sponsoring research into advanced fluidized-bed combustion systems) because it recognized that the fluidized-bed boiler represented a potentially lower cost, more effective, and cleaner way to burn coal [23]. Around 1990, atmospheric fluidized-bed combustion crossed the commercial threshold and every major U.S. boiler manufacturer currently offers fluidized-bed boilers as a standard package. Fluidized-bed coal combustors have been called the "commercial success story of the last decade in the power generation business" and are perhaps the most significant advance in coal-fired boiler technology in a half century.

Fluidized-bed combustion technology is used in both the utility and non-utility sectors and comprises approximately 1% of fossil fuel-fired capacity. Approximately half of the facilities using FBC technologies are utilities or independent power producers. Facilities in the food products and pulp and paper industries, along with educational institutions, make up most of the non-utility FBC facilities [10]. FBC technology accounts for a small proportion of capacity but the technology has increased dramatically over the last 20 years [10]. In 1978, four plants in the United States had four FBC boilers; however, as of December 1996, 84 facilities had 123 FBC boilers, representing 4951 MW of equivalent electrical generating capacity. Because of the fuel flexibility, efficiency, and emissions characteristics of the FBC boilers, this technology is predicted to increase in the future, and additional units are being installed, both commercial units and advanced concepts through cofunded DOE programs (which are discussed in more detail in Chapter 7, Future Power Generation). Figure 5-11 shows the geographic distribution of the FBC facilities [10]. These facilities are distributed throughout the United States; however, Pennsylvania and California account for the largest numbers of plants. Pennsylvania accounts for more than 20% of capacity and California more than 10%. Pennsylvania is a leader in utilizing coal wastes (anthracite culm and bituminous coal gob) in FBC boilers.

In a typical FBC, solid, liquid, or gaseous fuel (or fuels), an inert material such as sand or ash (referred to as bed material), and limestone are kept suspended through the action of combustion air distributed below the

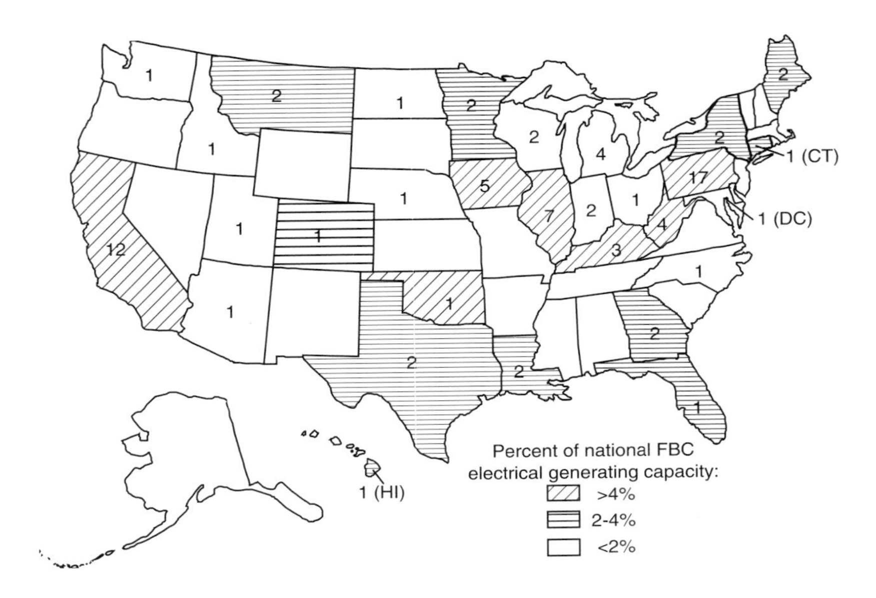

FIGURE 5-11. Number of fluidized-bed combustion facilities by state. (From EPA, *Report to Congress: Wastes from Combustion of Fossil Fuels*, Vol. 2, *Methods, Findings, and Recommendations,* U.S. Environmental Protection Agency, U.S. Government Printing Office, Washington, D.C., March 1999, chap. 3.)

combustor floor [14]. The primary functions of the inert material are to disperse the incoming fuel particles throughout the bed, heat the fuel particles quickly to the ignition temperature, act as a flywheel for the combustion process by storing a large amount of thermal energy, and provide sufficient residence time for complete combustion. The FBC concept is attractive because it increases turbulence and permits lower combustion temperatures. Turbulence is promoted by fluidization making the entire mass of solids behave much like a liquid. Improved mixing (and hence enhanced heat transfer to the bed material) permits the generation of heat at a substantially lower and more uniformly distributed temperature than occurs in conventional systems such as stoker-fired units or pulverized coal-fired boilers. The bed temperature in an FBC boiler is typically 1450 to 1650°F. This operating temperature range is well below that at which significant thermally-induced NO_x production occurs. Staged combustion can be applied to minimize fuel-bound NO_x formation as well. With regard to SO_2 emissions, the operating temperature range is where the reactions of SO_2 with a suitable sorbent, commonly limestone, are thermodynamically and kinetically balanced [24]. The percent capture for a given sorbent addition rate drops significantly outside the 1450 to 1650°F range. An additional reason why the bed temperature must be kept above 1400°F is that carbon utilization decreases with decreasing temperature, thereby reducing combustion efficiency.

Role of Sorbents in an FBC Process In an FBC system, the sorbent, usually limestone but sometimes dolomite (a double carbonate of calcium and magnesium), undergoes a thermal decomposition commonly known as calcination. When using limestone, the calcination reaction is:

$$CaCO_3 + \text{heat} \longrightarrow CaO + CO_2 \quad (5\text{-}19)$$

$$(\text{limestone} + \text{heat} \longrightarrow \text{lime} + \text{carbon dioxide})$$

Calcination of limestone is an endothermic reaction that occurs when limestone is heated above 1400°F. Calcination is thought to be necessary before the limestone can absorb and react with gaseous sulfur dioxide.

Capture of the gaseous sulfur dioxide is accomplished via the following equation to produce a solid product, calcium sulfate:

$$CaO + SO_2 + \tfrac{1}{2}O_2 \longrightarrow CaSO_4 \quad (5\text{-}20)$$

$$\text{lime} + \text{sulfur dioxide} + \text{oxygen} \longrightarrow \text{calcium sulfate}$$

The limestone is continuously reacted; therefore, it is necessary to continuously feed limestone with the fuel. The sulfation reaction requires an excess amount of limestone to always be present. The amount of excess limestone that is required is dependent upon a number of factors, such as the amount of sulfur in the fuel, the temperature of the bed, and the physical and chemical characteristics of the limestone.

The primary role of the sorbent in an FBC process is to maintain air quality compliance; however, the sorbent is also important in bed inventory maintenance, which affects the heat-transfer characteristics and affects the quality and handling characteristics of the ash. Depending on the sulfur content of the fuel, the limestone can comprise up to 50% of the bed inventory, with the remaining portion being fuel ash or other inert material. This is especially true of FBC systems firing refuse from bituminous coal cleaning plants that contain high levels of sulfur. When the bed is comprised of a large quantity of calcium (oxide or carbonate), there is the potential for ash disposal concerns as the pH of the ash can become very high.

Comparison of Bubbling and Circulating Fluidized-Bed Combustion Boilers The principle of FBC systems can be explained by examining Figure 5-12. The fundamental distinguishing feature of all FBC units is the velocity of the air through the unit as illustrated in Figure 5-12 [7,14]. Bubbling beds have lower fluidization velocities, and the concept is to prevent solids from elutriating (*i.e.*, carrying over) from the bed into the convective passes. Circulating fluidized-bed units apply higher velocities to promote solids elutriation. Bubbling fluidized-bed units characteristically operate with a mean particle size between 1000 and 1200 μm and fluidizing velocities between the minimum fluidizing velocity and the entraining velocity of the fluidized solid particles (*i.e.*, 4 to 12 ft/sec). Under these conditions,

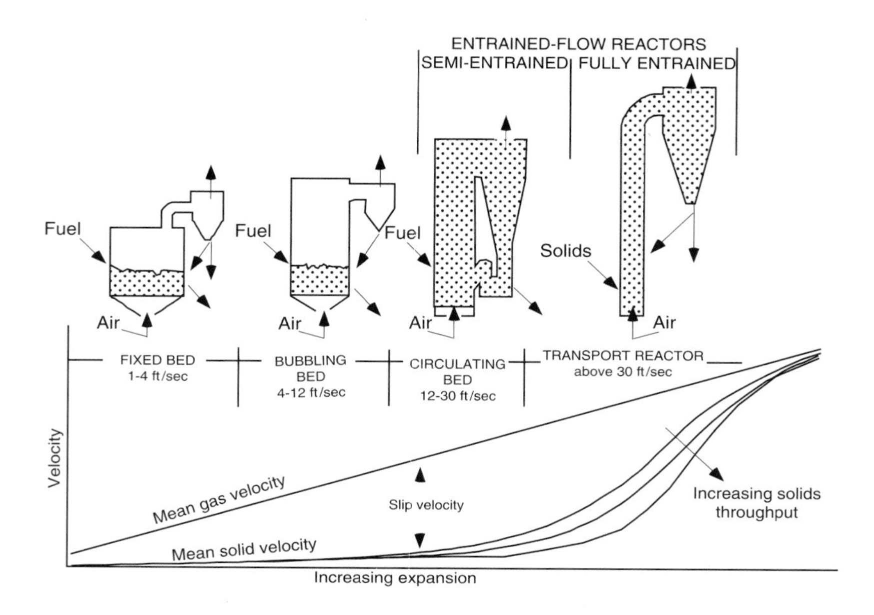

FIGURE 5-12. Fluidizing velocity of air for various bed systems. (Adapted from *Power* [7] and Elliot [14].)

a defined bed surface separates the high solids-loaded bed and the low solids-loaded freeboard regions. Most bubbling-bed units, however, utilize reinjection of the solids escaping the bed to obtain satisfactory performance. Some bubbling-bed units have the fuel and air distribution configured so that a high degree of internal circulation occurs within the bed [25]. A generalized schematic of a bubbling fluidized-bed boiler is shown in Figure 5-13 [26].

Circulating fluidized-bed (CFB) units operate with a mean particle size between 100 and 300 μm and fluidizing velocities up to about 30 ft/sec. A generalized schematic of a CFB boiler is given in Figure 5-14 [26]. Because CFBs promote elutriation and the solids are entrained at a high rate by the gas, bed inventory can be maintained only by recirculation of solids separated by the off-gas by a high efficiency process cyclone. Notwithstanding the high gas velocity, the mean solids velocity in the combustor is lowered due to the aggregate behavior of the solids. Clusters of solids are continuously formed, flow downward against the gas stream, are dispersed, are reentrained, and form clusters again. The solids thus flow upward in the combustor at a much lower mean velocity than the gas. The slip velocity between gas and solids is very high with corresponding high heat and mass transfer. This is further illustrated in Figure 5-15, which shows that CFBs achieve higher rates of heat transfer from the solids to the boiler tubes than do bubbling fluidized-bed units [14].

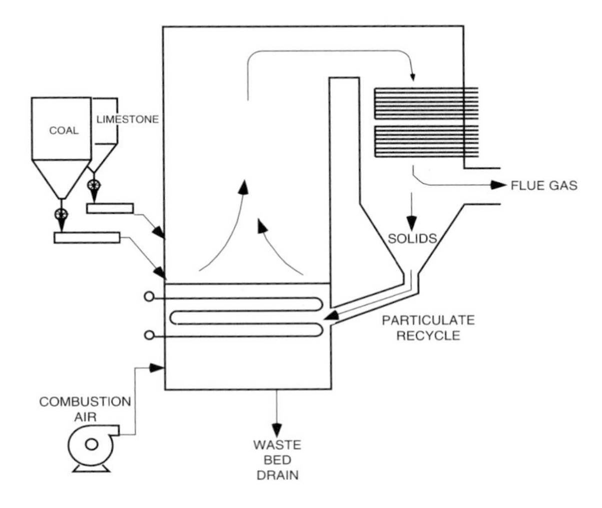

FIGURE 5-13. Generalized schematic diagram of a bubbling fluidized-bed boiler. (From Gaglia, B. N. and A. Hall, Comparison of Bubbling and Circulating Fluidized-Bed Industrial Steam Generation, in *Proc. of the International Conference on Fluidized-Bed Combustion*, May 3–7, 1987.)

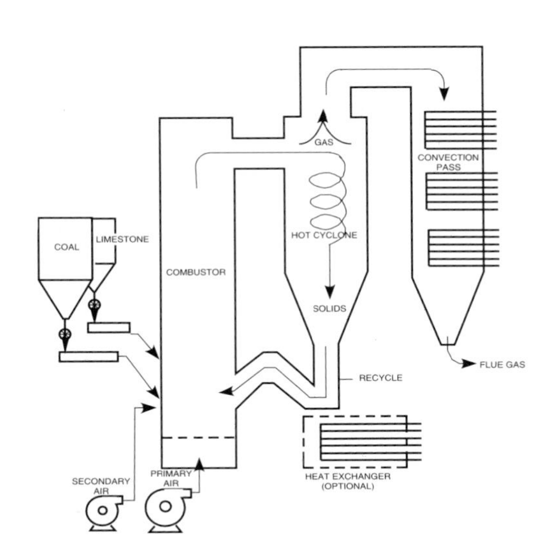

FIGURE 5-14. Generalized schematic diagram of a circulating fluidized-bed boiler. (From Gaglia, B. N. and A. Hall, Comparison of Bubbling and Circulating Fluidized-Bed Industrial Steam Generation, in *Proc. of the International Conference on Fluidized-Bed Combustion*, May 3–7, 1987.)

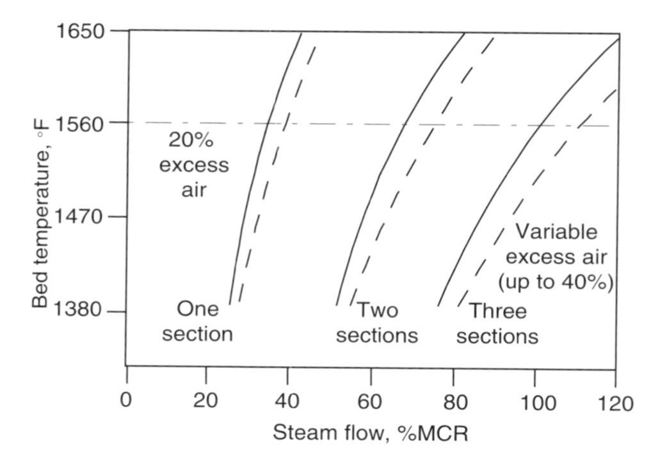

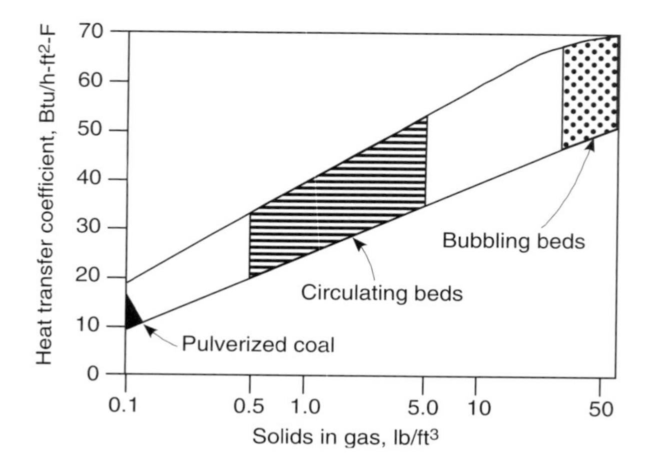

FIGURE 5-15. Relationship between heat transfer and solids loading/bed density. (From Elliot, T. C., Ed., *Standard Handbook of Powerplant Engineering*, McGraw-Hill, New York, 1989. With permission.)

Circulating fluidized-beds include these major components: a refractory-lined combustor bottom section with fluidizing nozzles on the floor above the windbox; an upper combustor section, usually with water-walls; a transition pipe, including a hot-solids separator and reentry down-comer; convective boiler section; and, in some designs, an external heat exchanger [14]. An external heat exchanger is a refractory-lined box containing an air distribution grid and an immersed tube bundle designed to cool material from the hot-solids separator and that is used to compensate for variations in the heat absorption rate caused by changes in fuel properties and load conditions. The solids separators are refractory-lined cyclones that are used to keep the solids circulating. The solids reinjection device, called an L-valve, J-valve, loop seal, Fluoseal, or a sealpot, depending on the manufacturer and configuration, provides a simple, nonmechanical hydraulic barometric seal against the combustor shell.

Suspension Firing

Pulverized coal-firing is the method of choice for large industrial boilers (*e.g.*, >250,000 pounds of steam per hour) and coal-fired electric utility generators because pulverized coal-fired units can be constructed to very large sizes (*i.e.*, up to ~1300 MW or ~9.5 million pounds of steam per hour), and, unlike stoker units where some designs have coal restrictions, they can accommodate virtually any coal with proper design provisions. The coal size distribution for pulverized coal-fired units is typically <2% by weight greater than 50 mesh (300 μm) with 65 to 70% less than 200 mesh (74 μm) for lignites and subbituminous coals and 80 to 85% less than 200 mesh for bituminous coals [19]. After the coal is pulverized, it is pneumatically transported to the burners using a portion of the combustion air, typically 10% of the total combustion air (the remaining combustion air is introduced at or near the burner), in a manner that permits stable ignition, effective control of flame shape and travel, and thorough and complete mixing of fuel and air.

Pulverized coal-fired units are typically classified into two types, depending on the furnace design for ash removal. In dry-bottom furnaces, the ash is removed from the system in dry form; in wet-bottom or slag-tap furnaces, the ash is removed in molten form. Dry-bottom furnaces are the more common of the two types and are now almost the only type sold in the United States. Dry-bottom furnaces are simpler to operate, more flexible with respect to fuel properties, and more reliable than slag-tap furnaces [2]. Dry-bottom furnaces are larger (hence, more costly) than wet-bottom furnaces since they must be sized to accommodate the ash where most of it (>80%) remains entrained in the flue gas and must be removed by particulate control devices at the back end of the system. Slag-tap units were developed to reduce the amount of fine fly ash that had to be handled by producing a heavier, granular ash and retaining most of the ash (up to 80%) in the furnace.

Dry-Bottom Firing The most frequently used dry-bottom furnace and burner configurations are shown in Figure 5-16 [2]. These arrangements cover firing systems suitable for all ranks of coal and coal qualities, including high ash or moisture content, low heating value, low ash fusion temperature, and high potential for ash deposition. Dry-bottom furnaces are designed to remove the ash as a solid; therefore, the rate of heat transfer and temperature in the furnace must be controlled. The dry-bottom furnaces are designed such that the heat release rates are much lower than wet-bottom and cyclone furnaces, and this, coupled with maintaining the furnace exit gas temperature below the ash fusion temperature, results in larger furnace designs.

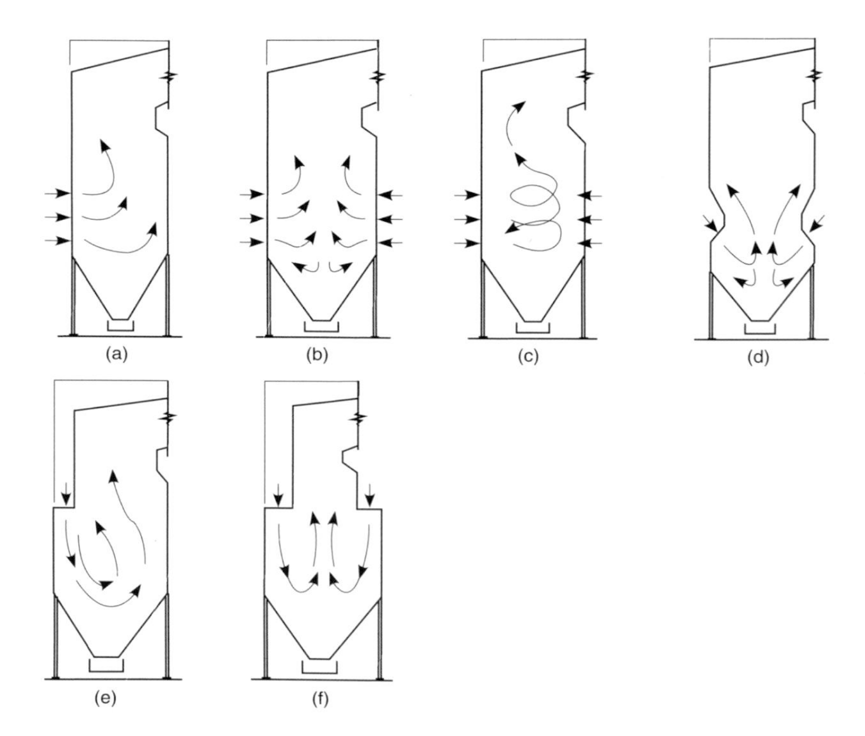

FIGURE 5-16. Dry-bottom furnace and burner configurations: (a) horizontal (front or rear); (b) opposed horizontal; (c) tangential (or corner firing); (d) opposed inclined; (e) single U-flame; (f) double U-flame. (From Elliot, M. A., Ed., *Chemistry of Coal Utilization*, Secondary Suppl. Vol., John Wiley & Sons, New York, 1981. With permission.)

Flame temperatures in the pulverized coal-fired units are typically around 2750°F. Heat is lost primarily by radiation in the furnace to the waterwalls and superheater/reheater tubes suspended in the furnace, and the temperature of the flue gas exiting the furnace is typically 1850°F.

Horizontal and opposed horizontal furnaces (Figure 5-16a,b) are usually fired by circular burners spaced uniformly across the width of the furnace on the front or rear wall or on both front and rear walls. Each burner has its own flame envelope, and the firing system can be designed so that an individual burner may be placed in service, adjusted, or removed from service independently of the other burners. In front or rear wall firing, the burners are arranged in such a way as to promote turbulence. In opposed firing, the burners in opposite walls of the furnace impinge their flames against each other to increase turbulence [7].

In tangential and, to a lesser extent, opposed inclined furnaces (Figure 5-16c,d), the burner turbulence is replaced by the overall furnace turbulence. In these furnaces, a single flame envelope promotes combustion stability and avoids the high flame temperatures that tend to favor NO_X

formation. In addition, the burners in the tangential furnace, where the fuel and air are admitted at all four corners and at different levels of the furnace and the burners, can be tilted upward or downward by 20° from the horizontal, thereby changing the temperature of the flue gas by as much as 150°F [7]. This allows for changing the combustion volume of the furnace to control superheat and reheat temperatures.

Single and double U-flame furnaces (Figure 5-16e,f) are used for firing difficult-to-ignite and slowly burning fuels such as anthracite and coke. In these designs, the fuel is fired downward, and radiation from the rising portion of the flames and from the burners in the opposite arch (in the double U-flame units) assists in maintaining a stable flame over a wide load range.

Wet-Bottom Firing Early wet-bottom furnaces (*i.e.*, in the 1920s) were simply open, single-stage furnaces with burners located close together and near the furnace floor to achieve the high temperatures necessary for melting the ash [2]. The furnace type used was usually one of those shown in Figure 5-16a–d, which was modified to accommodate the molten ash, and it satisfactorily used favorable coals where limited turndown was required. For coal ash that is difficult to melt and when a larger turndown range is required, two-stage designs have been developed. Examples of two-stage, slag-tap firing are shown in Figure 5-17 [2]. The primary advantage of wet-bottom furnaces

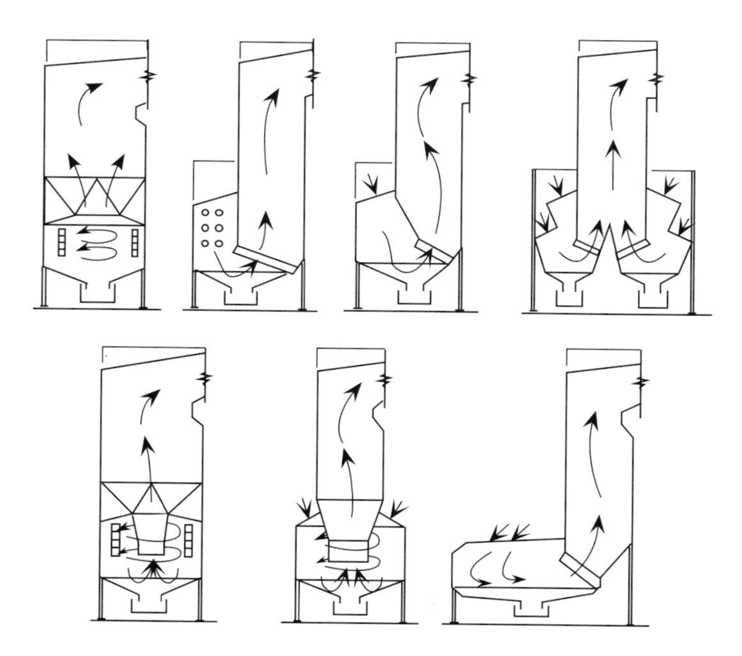

FIGURE 5-17. Furnace and burner configurations for two-stage slag-tap firing. (From Elliot, M. A., Ed., *Chemistry of Coal Utilization*, Secondary Suppl. Vol., John Wiley & Sons, New York, 1981. With permission.)

is easier ash handling and disposal; however, the disadvantages of using wet-bottom furnaces have led to its decline in the United States. These disadvantages include lower boiler efficiency through sensible heat loss of the slag, less fuel flexibility, higher incidences of ash fouling and external corrosion of pressure parts, decreased average steam generator availability, and higher levels of NO_X emissions [2].

Cyclone Furnaces Cyclone-furnace firing, shown in Figure 5-18, is a form of two-stage, wet-bottom design although some do not classify it as suspension firing because a large portion of the fuel is burned on the surface of a moving slag layer [2]. In cyclone firing, one or more combustors are mounted on the wall of the main furnace. Most cyclone furnaces in the United States are fired with coal crushed to about 1/4-inch top size while foreign practice uses partially pulverized coal (*e.g.*, 25% finer than 200 mesh [74 μm]). In the screened-furnace type, the gases exiting the cyclone pass through a small chamber and slag screen before entering the main furnace. This design has been largely replaced by the open-furnace arrangements as larger units have been developed. The development of the horizontal cyclone furnace occurred rapidly in the United States in the mid-1940s, and B&W was the leader in this technology development. The interest in the cyclone furnace is due to the several good features that it has, including a very high rate of heat production

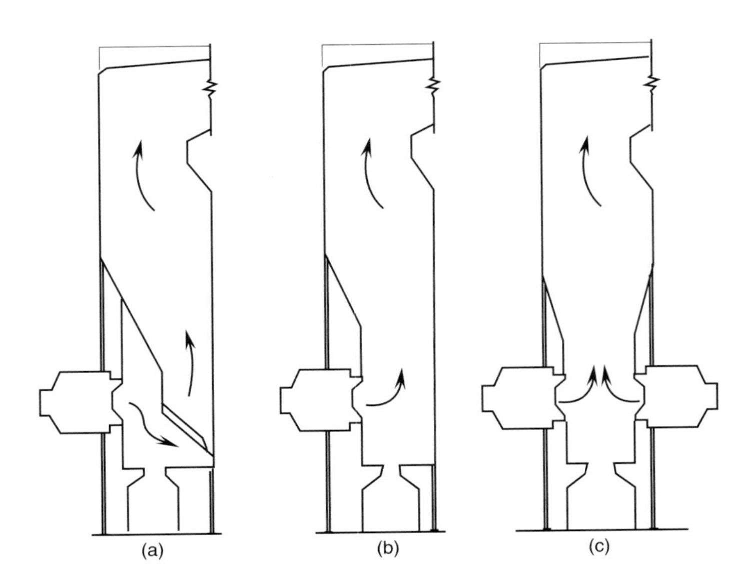

FIGURE 5-18. Horizontal cyclone furnace arrangements: (a) screened furnace; (b) one-wall open furnace; (c) opposed open furnace. (From Elliot, M. A., Ed., *Chemistry of Coal Utilization*, Secondary Suppl. Vol., John Wiley & Sons, New York, 1981. With permission.)

(*i.e.*, up to 500,000 Btu/hr-ft^2 compared to 150,000 and 400,000 Btu/hr-ft^2 in dry-bottom and slag-tap furnaces, respectively); high flame temperatures (~3000°F) to melt the ash sufficiently; ability to utilize coarser particles than pulverized coal-fired units, which results in lower system costs because pulverizers are not required; and the ability to be designed to use almost any coal type as well as opportunity fuels such as tires, petroleum coke, and others. The fuel characteristics of greatest interest in cyclone firing are the ash fusibility and viscosity of the ash lining the walls of the cyclone. The composition of the ash must be such that the ash will melt, coat the walls of the cyclone, and tap (*i.e.*, be fluid and exit steadily) from the cyclone. In addition, the moisture content of the fuel, such as in lignites, is important because high moisture fuels will consume heat while the moisture is being evaporated, which can affect temperature of the cyclone and hence the fluid behavior of the slag. As previously mentioned, the elevated temperatures produced in wet-bottom furnaces result in the generation of high levels of NO_X. Because of this, the use of cyclone furnaces in future installations is unlikely, and more attractive alternatives are pulverized-coal and fluidized-bed systems.

Influence of Coal Properties on Utility Boiler Design

The design of a utility steam-generating plant requires a technical and economic evaluation. Parameters that must be considered to arrive at a final design include the heat release rate, fuel properties (*e.g.*, ash fusion temperatures, volatile matter, ash content), percentage of excess air, production of emissions (*e.g.*, NO_X), boiler efficiency, and steam temperature [14], with the most important item to consider being the fuel burned [8]. This section discusses the influence of coal properties on boiler design, specifically as they relate to suspension firing, as this is the primary combustion technique used by the electric-generating industry today.

The coal properties that influence the design of the overall boiler system include but are not limited to coal and ash and handling, coal pulverizing, boiler size and configuration, burner details, amount of heat recovery surface and its placement, types and sizing of pollution control devices, and auxiliary components such as forced and induced-draft fan sizes, water treatment, and preheaters. The discussion in this section focuses on the influence of coal properties on furnace design consideration.

Furnace Design

Furnaces for burning coal are more liberally sized than those for gas or fuel oil firing, as illustrated in Figure 5-19 [8]. This is necessary to complete combustion within the furnace and to prevent the formation of fouling or slagging deposits. A furnace is designed to take advantage of the high radiant heat flux near the burners [14]. Because the flue gas temperature at the exit of the boiler (*i.e.*, entrance to the convective section) must be at least 100°F

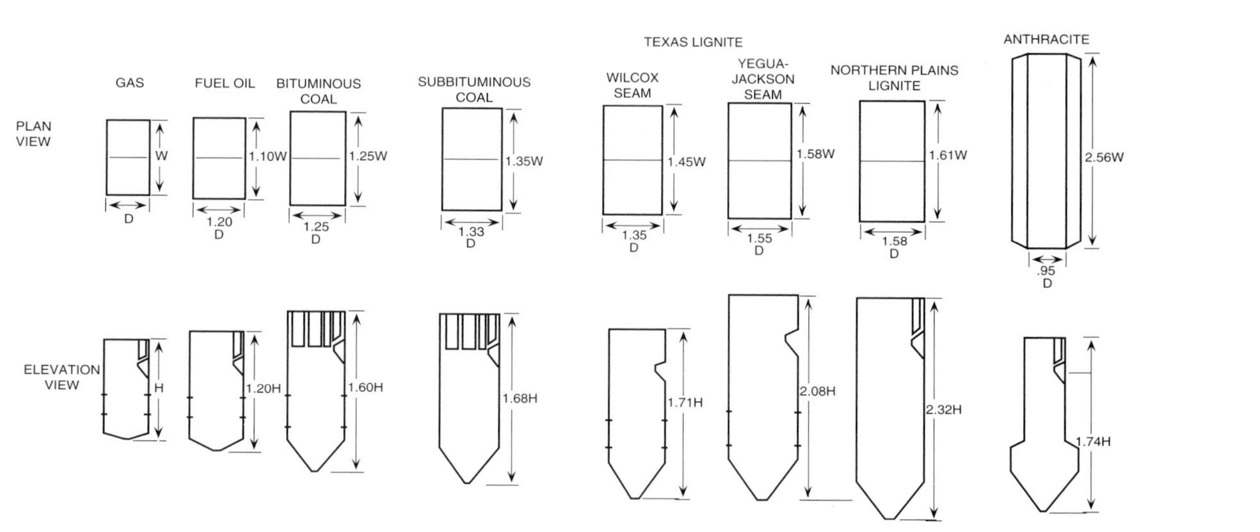

FIGURE 5-19. Effect of fuel type on furnace sizing assuming a constant heat input. H, distance between centerlines of lowest hopper headers and furnace roof tubes; W, width; D, depth. (Adapted from Singer [8]. With permission of ALSTOM Power, Inc.)

below the ash-softening temperature (which varies from ~2000 to 2500°F), the radiant heat-transfer surface in a coal-fired boiler must be increased by 15% or more in order to achieve a steep reduction in temperature from the flame temperature of ~2750°F.

The convective section of the boiler is designed to extract the maximum amount of heat from the partially cooled flue gas exiting the boiler. The flue gas velocity should not exceed about 60 ft/sec in the convective section, when firing coal, to minimize erosion of the tubes from the fly ash [14]. Sootblowers are required in coal-fired power plants to keep the heat-transfer surface clean.

The furnace size and shape must allow for adequate coal residence time within the furnace to achieve complete combustion. Sufficient heat must be contained in the flue gas exiting the boiler to enable efficient designs of superheater, reheater, economizer, and air heater heat-transfer surfaces. The flue gas exiting the stack must be low to minimize heat losses from the system but must be above the dew point of the acid gases so metal corrosion is not experienced.

In general, ignition stability in a pulverized coal-fired furnace varies directly with the ratio of volatile matter to fixed carbon [14]; hence, coals such as anthracite are typically fired in U-flame furnaces. Coals with higher volatile matter can be more easily burned in suspension, which allows for a lower furnace temperature but requires a larger furnace heat-release area. Also, coals with higher volatile matter tend to have lower ash fusion temperatures (*i.e.*, higher fouling and slagging tendencies), thereby requiring lower furnace temperatures [14].

Figure 5-19 illustrates the relationship between furnace exit gas temperature requirements and heat release rates for typical coals as compared to natural gas and fuel oil. To avoid slagging problems, a lower heat release rate is required for a coal having a relatively low ash fusion temperature than for a coal having a higher ash fusion temperature. The furnace exit gas temperature is primarily a function of heat release rate, which is the available heat divided by the equivalent water-cooled furnace surface [14].

Ash Characteristics

Coal is a very heterogeneous substance, and the mineral matter distributed throughout the coal exists in various forms, compositions, and associations and cannot be simply represented by composition (*i.e.*, elemental oxides) and a single set of melting temperatures (*i.e.*, ash fusion temperatures). Empirical indices have been developed using coal ash chemistry; however, they are successfully used only part of the time and are not applicable to ranks of coal beyond that for which they were developed. Some of these historical indices that are important when evaluating coal ash behavior, as they affect furnace slagging and fouling on both the furnace walls and convective surfaces (which vary among boiler manufacturers), include but are not limited to ash fusibility temperatures, dolomite concentrations, total ash concentration and composition, and base-to-acid, iron-to-calcium, and silica-to-aluminum ratios [8,27,28]. These parameters indicate the slagging and fouling potential of an ash. In addition, slag viscosity is an important parameter for cyclone furnace operation.

Slag Viscosity The viscosity of the slag formed from the coal ash is an important parameter for cyclone-fired furnaces. Slag will just flow on a horizontal surface at a viscosity of 250 poises [6]. The temperature at which this viscosity occurs (T_{250}) is used as the criterion to determine the suitability of a coal for a cyclone furnace. The T_{250} can be either calculated from a chemical analysis of the coal ash or, more preferably, determined experimentally using a high-temperature viscometer, and a value of 2600°F is considered maximum. Coals with a slag viscosity of 250 poises at 2600°F or lower are considered candidates for cyclone furnaces provided the ash analysis does not indicate excessive formation of iron or iron pyrites [6]. The T_{250} index is used for all ranks of coal.

In dry-bottom furnaces, the formation of slag must be avoided so as not to adversely affect the unit's operation. A relationship between furnace slagging and T_{250} has been developed [27]:

Slagging Rating	T_{250}
Low	>2325°F
Medium	2550–2100°F
High	2275–2050°F
Severe	<2200°F

Slagging and Fouling Potential The potential for slagging (*i.e.*, fused slag deposits that form on furnace walls and other surfaces exposed to predominately radiant heat) is temperature and ash composition related. Slagging potential affects furnace sizing, arrangement of radiant and convective heating surfaces, and the number of sootblowers required [28]. Fouling deposits form primarily in lower temperature regions of the furnace and convective section and affect the design and maintenance of superheaters, reheaters, furnace waterwalls, air heaters, and the number of sootblowers required. The potential for fouling is linked to the alkaline content of the ash, specifically the active alkalis.

Ash fusibility has long been recognized as a tool for measuring the performance of coals related to slagging and deposit buildup. American Society for Testing Materials (ASTM) Standard D1857 specifies the experimental procedure to use to determine the ash fusion temperatures. The test is based on the gradual thermal deformation of a pyramid-shaped ash sample in either an oxidizing or reducing atmosphere. Four temperatures are obtained during the test: (1) initial deformation temperature, which is the temperature at which the tip of the pyramid begins to show evidence of deformation; (2) softening or fusion temperature, which is the temperature at which the ash sample has fused and the height equals the width; (3) hemispherical temperature, which is the temperature at which the sample has fused into a hemispherical shape for which the height is equal to half of the width at the base; and (4) fluid temperature, which is the temperature at which the sample has fused down into a nearly flat layer. The ash-softening temperature is related to the type and ease of deposit removal from heat-transfer surfaces. If ash particles arrive at heat-absorbing surfaces at temperatures below their softening temperature, they will not form a bonded structure, and ash removal is relatively easy. If the ash particles arrive at these surfaces after they have been subjected to temperatures above their softening temperature and have become plastic or liquid, the resulting deposit will be tightly bonded and more difficult to remove [8]. Also, the temperature difference between the initial deformation and fluid temperatures provides information on the type of deposit to expect on furnace tube surfaces [8]. A small temperature difference indicates a thin, running, tenacious, difficult-to-remove slag, whereas wider temperature differences indicate less-adhesive deposits.

Another measure of ash viscosity that is used to predict furnace slagging is the temperature of critical viscosity (T_{cv}) [29]. This is the temperature at which the viscosity properties of the molten slag change on cooling from those of a Newtonian fluid to those of a Bingham plastic and is believed to be the temperature at which solid phases start to crystallize from the melt.

Several slagging and fouling indices have been developed based on the ash composition. Many of these indices have been developed for eastern U.S.-type coals. These are characterized as high iron and low alkali and alkaline-earth content coals. The ash composition of such coals reflects a Fe_2O_3 level exceeding the combined CaO, MgO, Na_2O, and K_2O percentage. In addition, sulfur percentages are higher than western United States coals. The acidic ash constituents (reported as weight percent on an oxide basis) SiO_2, Al_2O_3, and TiO_2 are generally considered to produce high-melting-temperature ashes. Temperatures will be lowered by the relative amounts of basic oxides, Fe_2O_3, CaO, MgO, Na_2O, and K_2O, available in the ash. The base (B)/acid (A) ratio defined as:

$$\frac{B}{A} = \frac{Fe_2O_3 + CaO + MgO + Na_2O + K_2O}{SiO_2 + Al_2O_3 + TiO_2} \tag{5-21}$$

has been developed as an indictor to predict the relative performance of coal ash in the furnace. This index can be used for all ranks of coal. A base/acid ratio in the range of 0.4 to 0.7 reflects low-fusibility temperatures and a higher slagging potential [8].

The base/acid ratio has also been used to define a slagging index expressed as $R_S = B/A \times S$, where S is the weight percent sulfur in the dry coal.

The slagging index has been used with success to identify four types of slagging coals [30]:

Slagging Type	*Slagging Index (R_S)*
Low	<0.6
Medium	0.6–2.0
High	2.0–2.6
Severe	>2.6

The influence of alkalis, notably sodium (Na_2O) and potassium (K_2O), on fusibility and slagging are proportional to the quantity in the coal ash. For sodium-containing coals, the rate of buildup is a function of the sodium concentration. The alkalis can be present in the coal in various forms. Alkalis that vaporize during combustion are classified as active or mobile alkalis and are free to react or condense in the boiler and consist primarily of simple inorganic salts and organically bound alkalis. More stable forms of alkalis exist in impurities such as clays and shales and remain inert during combustion. The mode of occurrence of the alkalis in coal are determined through a coal leaching process via sequential washings using water, acetic acid, and hydrochloric acid. The most active alkalis are those soluble in water and acetic acid. The fouling potential of coals is directly related to the soluble concentration of sodium and has been shown to vary from ~0.001 lb soluble sodium per lb ash per million Btu fired for a low-fouling coal to ~0.044 lb soluble sodium per lb ash per million Btu fired for coals with high or severe fouling potential [8].

Two fouling indices related mainly to sodium have been proposed to predict the extent of fouling of convective heat-transfer surfaces. Both apply to eastern U.S. coals, rather than to the lignites in which the $CaO + MgO$ content of the ash may be greater than the Fe_2O_3 content [29,30]. The indices are:

- R_F = Base/acid $\times$ Na_2O (ASTM ash);
- R'_F = Base/acid $\times$ water-soluble Na_2O (LTA, low-temperature ash).

Sodium is determined conventionally on the ASTM ash and on the water-soluble portion of the LTA. The fouling characteristics of coals are divided into four categories [29,30]:

Fouling Tendency	*R_F*	*R'_F*
Low	<0.2	<0.1
Medium	0.2–0.5	0.1–0.25
High	0.5–1.0	0.25–0.7
Severe	>1.0	>0.7

Numerous fouling studies, especially those using low-rank coals, have shown a relationship of the sodium in the ash and the fouling rate. This is particularly true of coals with lignitic-type ash (low-iron, high-alkali, and high-alkaline earth metals) as found primarily in the United States west of the Mississippi river, although sodium does contribute to deposition in higher rank coals as well. A fouling index based on the percent of Na_2O in coal ash is [27]:

Lignitic Ash		*Bituminous Ash*	
Percent Na_2O	*Fouling Potential*	*Percent Na_2O*	*Fouling Potential*
<2.0	Low	<0.5	Low
2–6	Medium	0.5–1.0	Medium
6–8	High	1.0–2.5	High
8	Severe	>2.5	Severe

Chlorine content of the coal is also used by some to predict fouling [29]. The validity of this parameter is in doubt, but the following values are recognized as representative of those used by industry:

Total Chlorine in Coal (%)	*Fouling Rating*
<0.2	Low
0.2–0.3	Medium
0.3–0.5	High
>0.5	Severe

Additional fouling and slagging parameters have been developed; however, many of these serve as guides to be used in conjunction with other parameters and the operating experience of the various coals and boiler units. Some of these factors include:

1. The silica/alumina ratio (SiO_2/Al_2O_3) can provide additional information relating to ash fusibility [8]. The general range of values is between 0.8 and 4.0; for two coals having similar base/acid ratios, the one with a higher silica/alumina ratio should have lower fusibility temperatures.

2. The iron/calcium ratio (Fe_2O_3/CaO) indicates the fluxing (*i.e.*, lowering of the ash fusion temperature) potential of the iron and calcium in the ash. Iron/calcium ratios between 1.0 and 0.2 have a marked effect on lowering the fusibility temperatures of coal ash, and extreme effects are evident between ratios of 3.0 and 0.3 [8].

3. A dolomite percentage (DP) is defined as:

$$DP = \left(\frac{CaO + MgO}{Fe_2O_3 + CaO + MgO + Na_2O + K_2O} \right) \times 100 \qquad (5\text{-}22)$$

which is used primarily for coal ashes with a basic oxide content over 40% (*i.e.*, western U.S. coals). It has been empirically related to the viscosity of coal ash slags and, at a given basic concentration, a higher *DP* usually results in higher fusion temperatures and higher slag viscosities [8].

4. A silica percentage (*SP*) is defined as:

$$SP = \left(\frac{SiO_2}{SiO_2 + \text{Equiv.}\,Fe_2O_3 + CaO + MgO} \right) \times 100 \qquad (5\text{-}23)$$

which has been empirically correlated with the viscosity of coal ash slags. As *SP* increases, the slag viscosity increases [8].

The development of a reliable coal-screening tool has long been a goal of the utility industry. The indices based on ASTM coal and ash analysis have provided useful information to boiler designers and operators but are not refined enough to be applied to all coals and all boilers. Also, because many of these indices were developed for bituminous coals, they are poor indicators of performance when applied to low-rank coals; consequently, different analytical techniques are being applied and new indices are continually being developed. Many of these new indices rely heavily on computer-controlled scanning electron microscopy coupled with coal and coal ash analyses. The technique of automated or computer-controlled scanning electron microscopy has enabled the identification and sizing of coal mineral matter *in situ* and is also used to determine the inherent or extraneous nature of the mineral matter in coal. The direct analysis of mineral matter in coal, size determinations, and observation of the association of mineral matter with coal particles are essential to determining the behavior of mineral matter during combustion of pulverized coal. Examples of indices include wall slagging, convective pass fouling based on either sulfates or silicates being the primary bonding component, cyclone slagging, and deposit strength [31]. Many of the indices developed by industry are proprietary.

As emphasized in this section, a major factor affecting furnace performance in large coal-fired utility boilers is the inorganic matter of the fuel. Most problems with the ash are associated with its effect on heat transfer by thermally insulating furnace wall tubes and convective pass tube banks. Accumulation of ash deposits can decrease the heat transfer rate to the tube surface, resulting in high flue gas temperatures.

Boiler manufacturers have developed their own empirical relationships to predict furnace performance and the effect of ash slagging and fouling. Also, testing coals in pilot-scale slagging and fouling combustors or boilers is routinely performed by boiler manufacturers, universities, and other test facilities to assess deposition performance. Each boiler manufacturer has its own criteria to allow for the effects of the ash characteristics; these are based on sound engineering judgment and years of experience with boilers of similar size.

Carbonization

Carbonization is the process by which coal is heated and volatile products—gaseous and liquid—are driven off, leaving a solid reside called char or coke. Coal carbonization processes are classified into high-temperature operations if they are performed at temperatures greater than 1650°F or low-temperature operations if they are conducted below 1350°F. These temperatures are somewhat arbitrary, as they reflect the pronounced physical changes that coal undergoes at temperatures between 1110 and 1470°F [19]. Carbonization processes reaching into the 1350 to 1650°F range are termed medium-temperature processes.

Coals of a very definite range of rank soften on heating, swell on decomposition, and resolidify on continued degasification [1]. Devolatilization is a continuous process but a distinction can be made between the primary carbonization stage, in which mainly tar is evolved, and the secondary carbonization stage, in which only gas is split off. The characteristic temperatures and stages of the carbonization process are illustrated in Figure 5-20 [1].

The main purpose of high-temperature carbonization is the production of metallurgical coke for use in blast furnaces and foundries. Some coke is used for the manufacture of calcium carbide and electrode carbons, as reductant in certain ferrous and nonferrous open-hearth operations, and in foundries to produce cast iron; however, more than 90% of the coke produced is used in blast furnaces to smelt iron ore and produce pig iron, and modern coke-making practices are virtually dictated by the coke quality in this market. Low-temperature carbonization has been mainly used to provide town gas for residential and street lighting, tars for use in chemical production, and smokeless fuels for domestic and industrial heating.

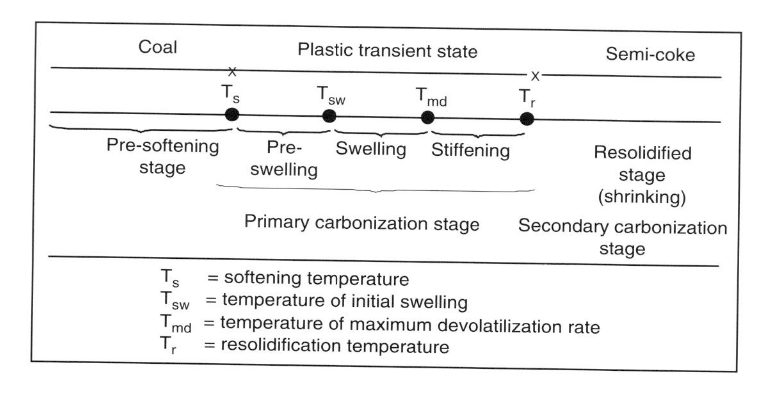

FIGURE 5-20. Characteristic carbonization temperatures and stages. (From Van Krevelen, D. W., *Coal: Typology–Physics–Chemistry–Constitution*, Third ed., Elsevier Science, Amsterdam, 1993. With permission.)

To put the use of coal used for carbonization into perspective with the electrical power generation industry (*i.e.*, using coal in combustion technologies), of the ~1065.8 million short tons of coal consumed in the United States in 2002, ~22.5 million short tons (or ~2%) was used in coke plants, as compared to ~975.9 million short tons (or ~92%) consumed by the electric power sector [32]. This section provides a brief history of high-temperature carbonization, specifically as it was developed for the iron and steel industry; reviews coking processes; discusses coal properties of interest for producing coking coals and the uses of coke; and concludes with a review of low-temperature carbonization.

Brief History of Carbonization (High-Temperature)

The carbonization of coal has its historical roots in the iron and steel industry. The ironmaking processes developed around the Mediterranean Sea and spread northward through Europe [33]. The Phoenicians, Celts, and Romans all helped spread ironmaking technology, and the Romans brought one of the ironmaking techniques as far north as Great Britain. Originally, charcoal produced from wood was the fuel used to melt the iron ore, and a tremendous amount of wood was needed for this industry. For example, one type of furnace (the Stuckofen) used in fourteenth-century Germany could produce 4000 lb of iron per day with a fuel rate of 250 lb of charcoal per 100 lb of iron produced [33]. This was an early version of the charcoal blast furnace, and these furnaces that developed in Continental Europe soon spread to Great Britain. By 1615, there were 800 furnaces, forges, or iron mills in Great Britain, 300 of them blast furnaces. The rate of growth in the number of these furnaces was so great that, during the 1600s, parliament passed laws to protect the remaining forests; consequently, many blast furnaces were shut down, alternative fuels were sought, and England encouraged the production of iron in its North American colonies, which had abundant supplies of wood and iron ore. The first successful charcoal blast furnace in the New World was constructed at Saugus, Massachusetts, outside of Boston, in 1645.

Due to the depletion of virgin forests in Great Britain to sustain the charcoal iron, the iron masters were forced to look at alternative fuel sources. The alternative fuels included bituminous coal, anthracite, coke, and even peat [33]. The development of coke and anthracite ironmaking paralleled each other and coexisted with charcoal production during the 1700s and 1800s, while bituminous coal and peat never became major ironmaking fuels. The widespread use of coke in place of charcoal came about in the early 1700s when Abraham Darby and his son showed that coke burned more cleanly and with a hotter flame than coal [19]. Up until 1750, the only ironworks using coke on a regular basis were two furnaces operated by the Darby family [33]; however, from 1750 to 1771 the use of coke spread, and 27 coke furnaces were in production. The use of coke increased iron production because it was stronger than charcoal and could support the weight of more raw materials, thus furnace size was increased.

The use of coke then spread to Continental Europe: Creussot, France, in 1785; Gewitz, Silesia, in 1796; Seraing, Belgium, in 1826; Mulhiem, Germany, in 1849; Donete, Russia, in 1871; and Bilbao, Spain, in 1880 [33]. In North America, the first attempt to use coke as the exclusive fuel was in the Mary Ann furnace in Huntington, Pennsylvania, although coke was mixed with other fuels as early as 1797 in U.S. blast furnaces.

The efficient use of coke and anthracite in producing iron was accelerated by the use of steam-driven equipment, the invention of equipment to preheat air from entering the blast furnace, and the design of the tuyeres and tuyere composition [33]. The evolution of both coke and anthracite ironmaking progressed in the United States during the 1800s and, by 1856, 121 anthracite furnaces were in operation in the United States. With coke being the strongest and most available fuel, the evolution of 100% coke furnaces continued, with major steps being made in the Pittsburgh area between 1872 and 1913. The Carnegie Steel Company and its predecessor firms developed technological process improvements at its Monongahela Valley ironmaking furnaces that ultimately made it possible for the United States to take over worldwide leadership in iron production. This is not true today, however, as much of the steel production has shifted overseas, beginning in the 1960s and early 1970s.

Coking Processes

Early processes for the production of coke were similar to those employed for the production of wood charcoal. Bituminous coal was built up into piles and ignited in such a way that only the outside layers actually burned while the central portion was carbonized [2]. Piles, also called kilns, appeared for the first time in England in 1657 and spread from there to other European bituminous coal producing regions. Around 1850, half-open brick kilns (*i.e.*, the Schaumburg kiln) were constructed from which circular mounds of coal emitted tar-containing volatilized gases directly into the atmosphere [2]. The next development was the closed beehive oven, which in its original form discharged the distillation and flue gases through a chimney at a greater height. Beehive ovens were used in England until the end of the 1800s; today, some of these ovens are still in operation in South Africa, South America, Australia, and the United States. The beehive oven is a simple domed brick structure into which coal can be charged through an opening at the top and then leveled through a side door to form on a bed ~2 feet thick [19]. Heat is supplied by burning the volatile matter released from the coal, and carbonization progresses from the top down through the charge. Approximately 5 to 6 short tons of coal can be charged, and a period of 48 to 76 hours is required for carbonization. Some beehive ovens are still in operation because of system improvements and the addition of waste heat boilers to recover heat from the combustion products. Similarly, the heat required for coking in the pile and the Schaumburg kiln is produced by partial combustion of the coal, which results in a substantial loss of material by combustion, with a coke yield in these ovens (including the beehive oven) being at most 55% of the coal. Flame ovens, in which the coal was coked in chambers heated from the outside, were developed in 1850 in Belgium and the Saar District in Germany [2]. The high-heating-value, volatilized gases were burned in flues in the walls of the ovens to produce coke yields of ~75%, with coking times of 48 hours.

The first coke ovens that produced satisfactory blast furnace or foundry coke as the main product, and tar, ammonia, and later benzene as by-products, were built around 1856 and were known as by-product recovery ovens [2]. Modifications to the design has continued but the basic design of these ovens, essentially the modern coke oven, was completed by the 1940s [1]. The horizontal slot-type coke (by-product recovery) oven, in which higher temperatures can be attained and better control over coke quality can be exercised, has superceded other designs and is used for coking bituminous coal [19]. Modern slot-type coke ovens are comprised of chambers 50 to 55 feet long, 20 to 22 feet high, and ~18 inches wide. A number of these chambers (from 20 to 100) alternating with similar cells that accommodate heating flues serve as a battery. Coal, crushed to 80% minus 1/8 in. with a top size of 1 in., is loaded along the top of the ovens using a charging car on rails and is leveled by a retractable bar. Coking takes place in completely sealed ovens, and when carbonization is completed (after 15–20 hours) the oven doors are opened and a ram on one side pushes the red-hot coke into a quenching car or onto a quenching platform. Coke yield is about 75%. By-product gas and tar vapors are removed from the oven to collector mains for further processing or for use in the battery.

A block flow diagram of the recovery of by-products from a coke oven is shown in Figure 5-21 [22]. From a ton of coal, a modern by-product coke oven yields about 1500 lb of coke, 11,000 ft^3 of gas, 8 to 10 gallons of light oil, and 25 lb of chemicals, mostly ammonium compounds. The by-product gas and tar vapors leaving the coke oven undergo a separation process to remove the tars from the gas. The gas then is treated to recover ammonia, as ammonium sulfate or phosphate, while the tars are fractionated by distillation into three oil cuts, which are designated as light, middle (or tar acid), or heavy oil. The gas, mainly a mixture of hydrogen and methane, has about one-half the heating value of natural gas and is used on-site as fuel in the flue chambers in the coke ovens or in the furnaces used for heat-treating finished steel [22].

The light-oil cut (boiling point, <430°F) from the distillation of the coal tar consists primarily of benzene (45–72%), toluene (11–19%), xylenes (3–8%), styrene (1–1.5%), and indene (1–1.5%); it is either processed into gasoline and aviation fuels or fractionated to provide solvents and feedstocks for chemical industries [19]. In either case, sulfur compounds, nitrogen bases, and undesirable unsaturates are removed. Middle oils are usually cut to

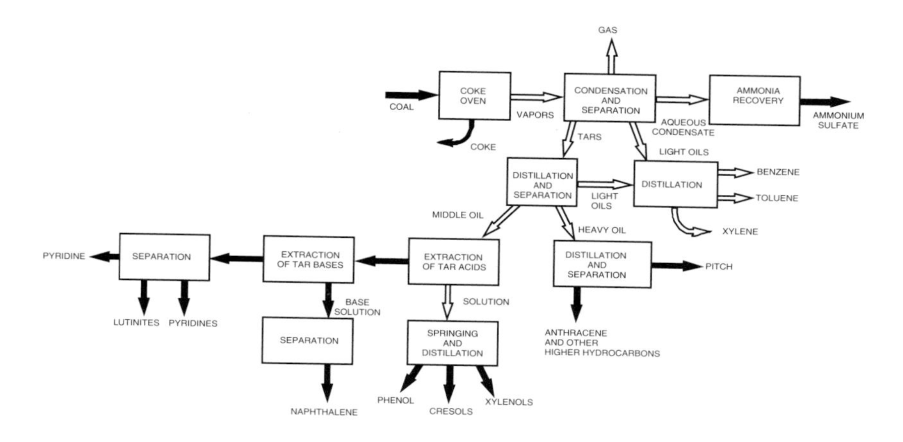

FIGURE 5-21. Simplified block diagram illustrating the by-products recovered and processing performed to produce useful chemicals. (Adapted from Schobert [22].)

boil between 430 and 710°F, and, after sequential extraction of tar acids, tar bases, and naphthalene, they are processed to meet specifications for diesel fuels, kerosene, or creosote. The tar acids are mostly comprised of phenols, while the compounds produced from the tar bases include pyridine, picolines, lutidines, anilines, quinoline, isoquinoline, and methylquinolines [19]. The temperature at which distillation of the heavy oils is performed depends on what type of pitch residue is desired but usually is between 840 and 1040°F [19]. The distillate is a rich source of hydrocarbons, mainly anthracene, phenanthrene, carbazole, acenaphthene, fluorine, and chrysene. The remaining heavy oils are marketed as fuel oils or blended with pitches to meet specifications for various grades of road tar. The residual coal tar pitches are complex mixtures of over 5000 compounds. They have economic importance because of their resistance to water and weathering and are used as briquetting binders and as binders in the preparation of carbon electrodes and other carbon artifacts.

Coal Properties for Coke Production

The properties of the coke are influenced by the coal or coal blend from which the coke is made as well as by the carbonizing conditions used. The rank of the coal or the average rank of the coal blend used affects the properties of the coke product. The choice of a suitable coal is crucial to quality control of the coke, and coke producers blend coals to manipulate coke characteristics. The practice of blending coals was originally developed to eliminate difficulties that were encountered when carbonizing highly fluid coals but was later used to stretch supplies of scarce or costly ideal-caking coals and quickly became the principal method for manipulating coke characteristics.

TABLE 5-6
Coal Properties for Coke Production

Coal Parameter	*Typical Values*
Carbon content (dry, ash-free) (wt.%)	~85
Hydrogen content (dry, ash-free) (wt.%)	~5.25
Volatile matter content (dry, ash-free) (wt.%)	24–28
Ash content (wt.%)	<10
Sulfur content (wt.%)	~0.5
H/C ratio	0.725
O/C ratio	0.04
Heating value (moist, mineral-matter-free) (Btu/lb)	~15,500
Vitrinite reflectance (%)	~1.25
Vitrinite/inertinite/exinite (dry basis) (%)	~55/35/10
Free swelling index	6.5–8
Maximum fluidity (dial divisions per minute)	~1000
Roga index	~45
Gray–King assay	≥G4

It is common practice to blend high-volatile coals with low-volatile and/or medium-volatile coals to improve the strength of the coke.

The ability of a coal to melt upon heating and to form a coherent residue on cooling is termed caking; an essential prerequisite for a coking coal is that it should cake or fuse when heated. Coals that are low in rank, such as lignites, or high in rank, such as anthracites, do not cake and therefore are not capable of forming coke. Several properties of coals are measured to identify appropriate coking coals, including swelling, fluidity, composition, maceral analyses, and vitrinite reflectance. These coal characteristics are listed in Table 5-6 along with typical values [1,19,34]. The mineral matter, or ash, content of the coal is of interest in coke production because the ash dilutes the coal and affects its caking properties. The composition of the ash is also important, as the quality of iron and steel is affected by the sulfur and phosphorus content. It is necessary for the formation of coke that some of the coal's organic constituents or macerals melt when the coal is heated. A caking coal behaves as if it were a pseudo-liquid when this occurs, and the viscosity of this material at various temperatures plays an important role in coking operations. The relative proportions of reactive and inert materials in a coking coal affect the strength of the final carbonized product.

Coking Conditions

In addition to the coal properties listed in the previous section, the carbonization conditions influence the properties of the resultant coke. Carbonization conditions of interest include the particle size of the coal charged to the coke oven, charge density, rate of heating, oven design, and special conditions such as preheating and partial briquetting. It is important to pulverize the coals (measured as the quantity of coal passing a 1/8-in. screen, which is typically 80% but can vary from 50 to 100%) to reduce the inert particles of the coal as well as to reduce the size of coal particles that exhibit low fluidities [2]. Higher levels of pulverization tend to make a more homogenous mixture of the reactive and inert components of the coals blended. The bulk density of the coal charged to the oven is also adjusted in order to produce denser or more homogenous cokes. In addition to varying the coal particle size, this adjustment is accomplished by adding small amounts of water or oil to the blend.

There has been much effort to increase the productivity of coke ovens through improved oven design and operating practices. The rate of heating the coal charge has been shown to be important in coke yield and properties [2]. As coking rate is increased, the coke size, shatter index, and stability decrease while the hardness factor increases. Coke size becomes more uniform with increased heating rate. Faster coking rates and the resultant *in situ* crushing are advantages to the iron makers who crush coke to produce a uniformly sized burden for the furnace [2]. Preheating high-oxygen coals or marginally coking coals can substantially improve the quality of coke from these coals and reduce the required coking times [2]. Coking times for these coals can be reduced by 30 to 45%, while preheating strengthens the resultant cokes.

In an effort to reduce the cost of blast furnace coke and to extend the range of ranks and types of coals that can be incorporated into blast furnace coke blends, formed coke processes have been developed. Formed coke processes involve carbonizing coal or a blend of coals (that may contain low-rank or highly volatile coals) that have been compressed into shaped briquettes.

Low-Temperature Carbonization

Low-temperature carbonization was originally developed to provide town gas for residential and street lighting and to manufacture a smokeless fuel for domestic and industrial heating. The by-product tars were economically important and were often essential feedstocks for the chemical industry or were refined to gasoline, heating oils, and lubricants [19]. Low-temperature carbonization evolved and was used extensively in industrialized European countries but was eventually abandoned after 1945 as oil and natural gas became widely available. These early processes used fixed- and moving-bed technology, operated in batch or continuous mode, and consisted of vertical or horizontal retorts with direct or indirect heating [1]. In the 1970s, interest in low-temperature carbonization was resurrected after the oil crisis. Most of the techniques utilized now, however, are different and mainly consist of using fluidized-bed or entrained-flow pyrolyzers. Some of the low-temperature carbonization technologies are the FMC Coke, COED, U.S. Steel Coke, Occidental Pyrolysis, Lurgi-Ruhrgas, Coalite,

Phurnacite, and Home Fire processes [2,35]. In addition, technologies such as the liquids-from-coal (LFC) process and the advanced coal conversion process (ACCP), are examples of commercial/near-commercial technologies to upgrade low-rank coals [36–38].

The preferred coals for low-temperature carbonization are typically lignites, subbituminous coal, or highly volatile bituminous coal, which, when pyrolyzed at temperatures between 1100 and 1300°F, yield a porous char with reactivities that are typically not much lower than those of their parent coals [19]. These reactive chars are easily ignited and are used as smokeless fuels or as feedstocks to gasification processes, are blended with coals to make coke-oven feed, or are used as a power plant fuel [19,22,37,38]. The tars that are produced during low-temperature carbonization are much different than those from high-temperature carbonization. High-temperature carbonization tends to produce mainly aromatic compounds, whereas those produced during low-temperature carbonization are predominately aliphatic compounds, hence the different end-use applications of the tar by-products. Gas yield and composition are also different during low-temperature carbonization, with gas yields being ~25% of those produced during high-temperature carbonization, but the gas contains more methane and less hydrogen, giving it a higher heating value [22].

Smokeless Fuel Commercial Processes

The primary application of low-temperature carbonization is to make smokeless fuels for use in homes and small industrial boilers in areas that have high population density and rely on coal as a fuel, particularly coal that has a high volatile-matter content. This is especially true in Great Britain, which has regulated smoke-controlled areas, and several commercial plants are producing smokeless fuels for open fires, room heaters, multifuel stoves, cookers, and independent boilers [39]. These fuels are marketed under names such as Coalite, Sunbrite, Phurnacite, Taybrite, and Home Fire.

Examples of two processes used in Great Britain are the Coalite and the Home Fire processes. The Coalite process uses moving-bed, vertical-retort technology and is a continuous process using indirect heating [1]. The Coalite works are located at Bolsover in Derbyshire and started operation in 1937. In the Coalite process, deep-mined British bituminous coal is carbonized in several batteries, each battery consisting of 40 metal retorts, assembled in two rows of 20, at a temperature of ~1200°F [40]. The coal charge to each retort, which is ~660 lb, remains in the retort for 4 hours, after which time a ram pushes the Coalite into a cooler. Typically, 1 metric ton of coal blend will produce 1100 to 1870 lb of smokeless coal or semi-coke, 5300 to 6350 standard cubic feet (scf) of gas, 18 to 20 gallons of coal oil, 3 to 5 gallons of light oil, and 45 to 48 gallons of aqueous liquor. The gas is recycled for onsite use to heat the batteries, generate steam, and general heating. Oils are distilled to yield pitch (for use as a boiler fuel), heavy oil (to produce creosotes and disinfectants), and middle and light oils (to produce phenols, cresols, and xylenols). The liquor contains dissolved chemicals—principally, ammonia, monohydric and dihydric phenols—and through extraction and fractionation a wide range of chemicals is produced, including catechol, resorcinol, and methyl resorcinol. Coalite is the leading manufacturer of smokeless fuel in the United Kingdom. The Home Fire process uses a blend of bituminous coals and fluidized-bed technology. At the Home Fire plant, located near Coventry, the coal is crushed to 1/4-in. particles, dried, and devolatilized for 20 minutes at 800°F in a fluidized-bed reactor [22,35]. The hot char is extruded into hexagonal briquettes, cooled, and quenched.

Low-Rank Coal Upgrading

The U.S. Department of Energy (DOE) cofunded two programs through their Clean Coal Technology (CCT) program (which is discussed in detail in Chapter 7, Future Power Generation) in which low-rank coals were upgraded using low-temperature pyrolysis/thermal upgrading technologies as part of the process. Western U.S. low-rank coals, primarily subbituminous coals and lignite, are generally low in sulfur, making them (specifically, the subbituminous coals) attractive as power plant fuels in place of high-sulfur eastern U.S. coals; however, disadvantages of the low-rank coals are high moisture content and low heating value. Consequently, two new processes—ENCOAL's liquids-from-coal process and Rosebud SynCoal Partnership's advanced coal conversion process—have been developed and successfully demonstrated, and are ready for commercialization.

The LFC technology uses a mild pyrolysis or mild gasification process to produce a low-sulfur, high-heating value fuel and a coal-derived liquid [36–38,41]. In the process, coal that has undergone some coal drying is fed into a pyrolyzer that is operated near 1000°F to remove any remaining moisture and release volatile gases. The solid fuel—process-derived fuel (PDF)—is used as a boiler fuel. The pyrolysis gas stream is sent through a cyclone to remove entrained particles and then cooled to condense the desired hydrocarbons—coal-derived liquids (CDLs)—and stop any secondary reactions. The CDLs were utilized at seven industrial fuel users and one steel mill blast furnace during the 5-year demonstration; however, studies were performed on upgrading the CDLs to produce cresylic acids, petroleum refinery feedstock for producing transportation fuels, oxygenated liquids, and pitch. The process was demonstrated for approximately 5 years in a plant feeding 1000 short tons coal per day located near the Buckskin Mine (Triton Coal Company) in Gillette, Wyoming, and produced over 83,000 short tons of solid fuel product and 4.9 million gallons of liquid product. The process is considered commercial and is actively being marketed in the United States (in the Powder River Basin, Alaska's Beluga field, and the lignite fields of North Dakota and Texas) and abroad (in China, Indonesia, and Russia) [37,41,42]; five detailed commercial feasibility studies have been completed [42].

A large-scale commercial plant has been designed, with participation from Mitsubishi Heavy Industries, to utilize 15,000 metric tons coal feed. The plant, located near Gillette, Wyoming, has received the Industrial Siting Permit and an Air Quality Construction Permit, but is on hold due to lack of funding [42].

The ACCP process is an advanced thermal conversion process coupled with physical cleaning techniques to upgrade high-moisture, low-rank coals to produce a high-quality, low-sulfur fuel. In this process, coal is fed to a vibratory fluidized-bed reactor to remove surface moisture. It then flows to a second vibratory reactor, where coal is heated to 600°F to remove chemically bound water, carboxyl groups, volatile sulfur compounds, and a small amount of tar. In this process, the volatiles are not collected but are used in a process heater. The technology was demonstrated from 1993 to 2001 (longer than the planned 5-year period) in a 45-short ton/hour facility located adjacent to a unit train load-out facility at Western Energy Company's Rosebud coal mine near Colstrip, Montana, in which 2.8 million short tons of raw coal were processed. Nearly 1.9 million short tons of SynCoal were produced and shipped to various customers, including cement and lime kilns and utility boilers, and the product was used as a betonite additive in the foundry industry. Three different feedstocks were tested at the ACCP facility—two North Dakota lignites and a subbituminous coal—and the products were fired in a utility boiler located in North Dakota and three utility boilers located in Montana [43]. The technology is being marketed and promoted worldwide, and a project was actively pursued for Minnkota's Milton R. Young Power Station, which had test fired SynCoal produced from one of the North Dakota lignites; however, the project was suspended due to a lack of equity investors.

Gasification

Gasification is a process to upgrade a solid feedstock, which is difficult to handle, by removing undesirable impurities and converting it into a gaseous form that can be purified and used directly as a fuel or further reacted to produce other gaseous or liquid fuels, or chemicals. There are many reasons for the interest in gasification as a process for utilizing coal. Liquid and gaseous fuels are easier to handle and use than coal, whether the fuel is used for heating, cooking, transportation, or power production. Shipping coal can be difficult and labor intensive and can have negative environmental impacts. Impurities in coal can be more readily removed through gasification than when utilized directly. Synthetic fuels burn more cleanly than coal, and fewer sulfur and nitrogen oxides are formed during combustion. Carbon capture and sequestration is easier in a gasification system than in a combustion system, which is important with carbon dioxide being considered for regulation as a pollutant. Gasification of coal is especially attractive to nations that have coal reserves and lack reserves of oil and gas or are depleting them. And, finally, gasification of coal reduces the concerns of volatile swings in availability and cost of gaseous or liquid fuels that are experienced with petroleum or natural gas. Although the use of coal gasification is currently rather limited, this technology is poised to be the technology of the future for the production of electricity, steam, chemicals, and fuels such as hydrogen. This is further expanded upon in Chapter 7 (Future Power Generation).

This section summarizes gasification as a technology for utilizing coal, beginning with a brief introduction of the history of gasification. This is followed by a discussion of gasification principles and a review of gasifier types, specifically those that are commercially available. The influence of coal properties on gasification is presented and the regional distribution of gasifiers in the world is briefly discussed, followed by brief descriptions of the main gasification technologies currently being used.

Brief History of Coal Gasification

The discovery of gases from coal, as verified by written records, dates back to the early 1600s. In 1609, a Belgian chemist, Jan van Helmont, observed that gas was evolved from coal when it was heated [22]. At the end of the 1600s, a clergyman from Yorkshire, England, John Clayton, experimented with collecting gas from coal. In 1792, William Murdock, a Scottish engineer, pioneered the commercial gasification of coal using the technique of heating coal in a retort in the absence of air to convert coal to gas and coke. The gas produced in this manner (*i.e.*, carbonization) has several names: coal gas, town gas, city gas, or illuminating gas. By 1798, gas-lighting systems were being installed in factories and mills in England; these applications were followed by their installation as street-lighting systems beginning in 1807. By 1816, most of London was lit by gas.

In 1816, the Baltimore Gas Company, the first coal gasification company in the United States, was established. Gas lighting spread throughout the east coast: Boston in 1821, New York in 1823, and Philadelphia in 1841 [22]. By the mid-1920s, about 20% of the gas supply in the United States was being produced from coal [22]. Prior to World War II, at least 20,000 gasifiers were operating in the United States. In the 1940s, the increasing availability of low-cost natural gas led to its substitution for gases derived from coal and the demise of the gasifiers. In the 1950s and 1960s, petroleum dominated the market, and no new gasification processes emerged; few coal gasification plants were installed in the United States [2]. It was not until the late 1960s and early 1970s, when the United States began to experience natural gas shortages and the oil embargo of the 1970s occurred, that the significance of coal reserves was recognized. This recognition led to a tremendous surge in interest in coal utilization, primarily in the areas of gasification and liquefaction (*i.e.*, production of coal to liquid products).

Principles of Coal Gasification

Carbonization of coal to produce coal gas is a relatively simple process to perform and is done in a retort in the absence of air. The composition of the gas being produced varies depending on the coal being used but is typically comprised of hydrogen (40–50%) and methane (30–40%), with minor amounts (2–10%) of nitrogen, carbon monoxide, ethylene, and carbon dioxide. The gas yield is approximately 10,000 scf per short ton of coal carbonized with a heating value of 550 to 700 Btu/scf. When carbonizing a bituminous coal, about 20% of the weight of the coal is converted to gas [22]. This gas is used as a fuel at coking operations.

Although carbonization of coal is a simple process, only a small fraction of the coal is converted to gas; consequently, processes to convert all of the carbon in the coal to gas were developed. In one of these processes, air is slowly passed through a hot bed of coal, converting most of the carbon to carbon monoxide, with some carbon dioxide being formed. Some of the carbon dioxide is then converted to carbon monoxide by reacting with hot fuel carbon. The reactions that occur are:

$$C + O_2 \longrightarrow CO_2 \tag{5-4}$$

(Combustion of carbon: $\Delta H = +170.0 \times 10^3$ Btu/lb mole of carbon gasified)

and

$$C + CO_2 \longrightarrow 2CO \tag{5-3}$$

(Boudouard reaction: $\Delta H = -72.19 \times 10^3$ Btu/lb mole of carbon gasified)

resulting in:

$$2C + O_2 \longrightarrow 2CO \tag{5-1}$$

($\Delta H = +97.81 \times 10^3$ Btu/lb mole of carbon gasified)

Reactions (5-3) and (5-4) are collectively sufficiently exothermic to sustain reaction with the reactants fed at ambient conditions.

The gas produced by this method is called producer gas, and when a bituminous coal is used the gas composition is typically 20 to 25% carbon monoxide, 55 to 60% nitrogen, 2 to 8% carbon dioxide, and 3 to 5% hydrocarbons [22]. Unfortunately, the producer gas is diluted with nitrogen, and the heating value of the gas is only about 100 to 150 Btu/scf. The yield of producer gas is 150,000 to 170,000 scf per short ton of coal. Producer gas was used in a variety of industrial applications such as open-hearth furnaces in steel mills, glass-making furnaces, and pottery kilns; however, the demand for producer gas has been reduced with the demise of open-hearth furnaces in the steel industry and the development of natural gas and electric furnaces.

The temperatures developed in the fuel bed during Reactions (5-3) and (5-4) can be very high, and, when the ash in the bed is fusible, the endothermic carbon–steam reaction must be imposed by adding steam to the air:

$$C + H_2O \longrightarrow CO + H_2 \tag{5-7}$$

(Carbon–steam reaction: $\Delta H = -58.35 \times 10^3$ Btu/lb mole of carbon gasified)

This reaction moderates the temperature and yields hydrogen in the product gas. This mixture of carbon monoxide and hydrogen is also called water gas. The water gas process is cyclic and involves a gas-making period during which the fuel bed is blown with steam to produce carbon monoxide and hydrogen, followed by an air-blowing period during which the heat is generated in the fuel bed. A typical water gas contains 50% hydrogen, 40% carbon monoxide, and small amounts of carbon dioxide and nitrogen and has a heating value of 300 Btu/scf. When a water-gas generator is being blown with air to reheat the bed, producer gas is made from the reaction of the hot carbon with oxygen, yielding about 35,000 scf of water gas and 80,000 scf of producer gas from one short ton of coal [22]. Water gas is a useful starting material for synthesizing chemicals or liquid fuels and is a good source of hydrogen. Treating the water gas with steam oxidizes the carbon monoxide to carbon dioxide and increases the amount of hydrogen by the equation:

$$CO + H_2O \longrightarrow CO_2 + H_2 \tag{5-8}$$

(Water-gas shift reaction: $\Delta H = +3.83 \times 10^3$ Btu/lb mole of carbon gasified)

The carbon dioxide can be removed from the product stream, leaving reasonably pure hydrogen.

Hydrogen can also be reacted with carbon at elevated pressures by the carbon hydrogenation or hydrogasification reaction:

$$C + 2H_2 \longrightarrow CH_4 \tag{5-24}$$

($\Delta H = +39.38 \times 10^3$ Btu/lb mole of carbon gasified)

and whenever the carbon source generates volatile matter, further quantities of methane will form by thermal cracking.

Gasifier Types

Gasification processes are classified on the basis of the method used to bring the coal into contact with the gasifying medium (air or oxygen). The three principal commercial modes are fixed-bed, fluidized-bed, and entrained-flow systems. Their principal features are illustrated in Figure 5-22 [2] and their important characteristics are listed in Table 5-7 [44].

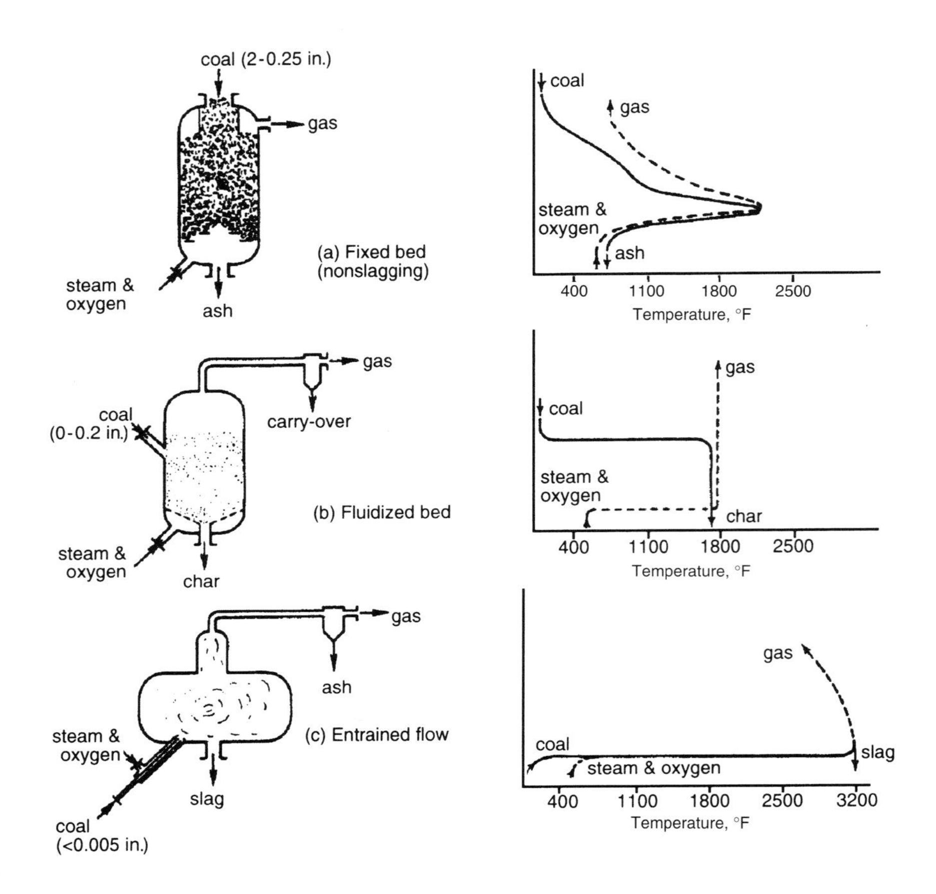

FIGURE 5-22. Classification and characteristics of the commercial gasification systems. (From Elliot, M. A., Ed., *Chemistry of Coal Utilization*, Secondary Suppl. Vol., John Wiley & Sons, New York, 1981. With permission.)

Fixed-Bed Gasifier

In a fixed-bed gasifier, 1/4- to 2-in. coal is supplied countercurrent to the gasifying medium. Coal moves slowly down (sometimes this type of gasifier is called a moving-bed gasifier), ideally in plug flow against an ascending stream of gasifying medium. Reaction zones, shown in Figure 5-23, typically consist of drying and devolatilization, reduction, combustion, and ash zones. In the drying and devolatilization zone, located at the top of the gasifier, the entering coal is heated and dried and devolatilization occurs. In the reduction/ gasification zone, the devolatilized coal is gasified by reactions with steam and carbon dioxide. In the combustion zone, oxygen reacts with the remaining char, and this zone is characteristic of high temperatures. The ash is

TABLE 5-7
Characteristics of Generic Gasifier Types

	Gasifier Type				
	Fixed Bed		*Fluidized-Bed*		*Entrained Flow*
	Ash Conditions				
	Dry Ash	*Slagging*	*Dry Ash*	*Agglomerating*	*Slagging*
Fuel characteristics					
Fuel size limits	1/4–2 in.	1/4–2 in.	<1/4 in.	$\ll$1/4 in.	<0.005 in.
Caking coal acceptable?	Yes, with modifications	Yes	Possibly	No, noncaking only	Yes
Preferred feedstock	Lignite, reactive bituminous coal, anthracite, wastes	Bituminous coal, anthracite, petcoke, wastes	Lignite, reactive bituminous coal, anthracite, wastes	Lignite, bituminous coal, anthracite, cokes, biomass, wastes	Lignite, reactive bituminous coal, anthracite, petcokes
Ash content limits	No limitation	<25% preferred	No limitation	No limitation	<25% preferred
Preferred ash melting temperature (°F)	>2200	<2370	>2000	>2000	<2372
Operating characteristics					
Exit gas temperature (°F)	Low[a] (800–1200)	Low (800–1200)	Moderate (1700–1900)	Moderate (1700–1900)	High (>2300)
Gasification pressure (psig)	435+	435+	15	15–435	<725
Oxidant requirement	Low	Low	Moderate	Moderate	High
Steam requirement	High	Low	Moderate	Moderate	Low
Unit capacities (MW_{th} equiv.)	10–350	10–350	100–700	20–150	Up to 700
Key distinguishing characteristics	Hydrocarbon liquids in raw gas	Large char recycle			Large amount of sensible heat energy in the hot raw gas
Key technical issue	Utilization of fines and hydrocarbon liquids	Carbon conversion			Raw gas cooling

[a]Fixed-bed gasifiers operating on low-rank coals have exit temperatures lower than 800°F.

Source: Ratafia-Brown, J. *et al.*, *Major Environmental Aspects of Gasification-Based Power Generation Technologies*, Office of Fossil Energy, U.S. Department of Energy, Washington, D.C., December 2002.

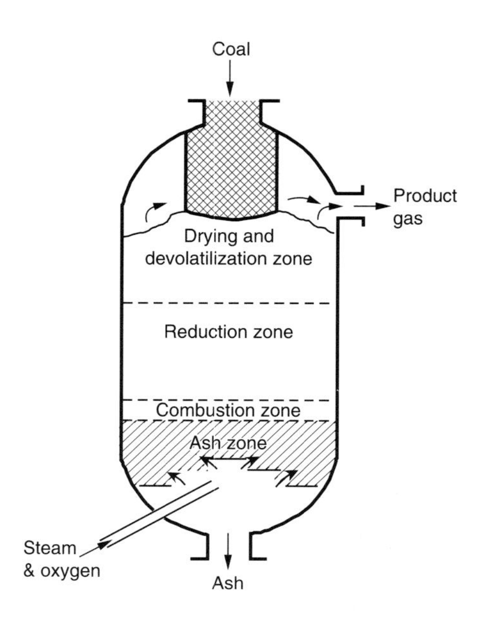

FIGURE 5-23. Reaction zones in a fixed-bed gasifier. (From Elliot, M. A., Editor, *Chemistry of Coal Utilization,* Secondary Suppl. Vol., John Wiley & Sons, New York, 1981. With permission.)

removed from the bottom of the gasifier either in dry form if the temperature in the gasifier is controlled with excess steam to maintain the temperature below the ash fusion point or as liquid slag. Both the ash and the product gas leave at modest temperature as a result of heat exchange with the entering gasifying medium and fuel, respectively. Fixed-bed gasifiers have the following characteristics [44]:

- Low oxidant requirements;
- Design modifications required for handling caking coal;
- High cold-gas thermal efficiency when the heating value of the hydrocarbon liquids is included;
- Limited ability to handle fines.

Fluidized-Bed Gasifier

In a fluidized-bed gasifier, coal crushed to less than 1/8 to 1/4 in. in size enters the side of the reactor and is kept suspended by the gasifying medium. Similar to a fluidized-bed combustor, mixing and heat transfer are rapid, resulting in uniform composition and temperature throughout the bed. The temperature is sustained below the ash fusion temperature, which avoids clinker formation and possible slumping (*i.e.,* de-fluidization of the bed). Some char particles are entrained in the product gas as it leaves the gasifier, but they are recovered and recycled back into the gasifier via a cyclone. The ash is discharged with the char, and the product gas and char temperatures are high, with some heat transfer occurring with the incoming steam and recycled gas. Fluidized-bed gasifiers have the following characteristics [44]:

- Acceptance of a wide range of solid feedstock (including solid waste, wood, and high ash content coals);
- Uniform temperature;
- Moderate oxygen and steam requirements;
- Extensive char recycling.

Entrained-Flow Gasifier

In the entrained-flow gasifier, pulverized coal (<0.005 in.) is entrained with the gasifying medium to react in co-current flow in a high-temperature flame. Residence time in this type of gasifier is very short. Entrained-flow gasifiers generally use oxygen as the oxidant and operate at high temperatures, well above ash-slagging conditions, to ensure high carbon conversion. The ash exits the system as a slag. The product gas and slag exit close to the reaction temperature. Entrained-flow gasifiers have the following characteristics [44]:

- Ability to gasify all coals regardless of coal rank, caking characteristics, or amount of coal fines, although feedstocks with lower ash contents are favored;
- Uniform temperatures;
- Very short fuel residence times in the gasifier;
- Very finely sized and homogenous solid fuel required;
- Relatively large oxidant requirements;
- Large amount of sensible heat in the raw gas;
- High-temperature slagging operation;
- Entrainment of some molten slag in the raw gas.

Influence of Coal Properties on Gasification

Coal properties have a major influence on the process and gasifier design and include moisture; ash, volatile matter, and fixed carbon content; caking tendencies; reactivity; ash fusion characteristics; and particle size distribution. Some of these properties are listed in Table 5-7, and each is briefly discussed in the following sections.

Moisture

Fixed-bed gasifiers can accommodate moisture contents of up to 35%, provided the ash content is not in excess of about 10% [2]. Predrying may be performed if the moisture and ash contents are above these amounts. Entrained-flow or fluidized-bed gasifiers require the moisture content to be reduced to less than ~5% by drying to improve coal handling. In the entrained-flow system, the residual moisture contributes to the gasification steam but requires heat to evaporate it.

Ash

Ash should be kept at a minimum because provisions must be made for introducing it to and withdrawing it from the system, provisions that add to the complexity and cost of the overall system. Ash can be used as a heat transfer medium, either by its flow countercurrent to the products of gasification and gasifying agents in fixed-bed systems or by provisions in entrained-flow systems [2]. In fixed-bed systems, ash accumulates at the base of the fuel bed and is withdrawn by a mechanical grate if unfused or through a taphole if it is a liquid slag. In the entrained-flow system, it is removed as a liquid slag. In fluidized-bed systems, the ash is mixed with the char and the ash is separated either by sintering and agglomeration of the ash or circulation from the bed through a fully-entrained combustor to melt and separate the ash as a liquid slag. Fluidized-bed and entrained-flow gasifiers tend to have higher losses of carbon in the ash than the fixed-bed systems. Ash constituents are important in the selection of materials of construction, particularly in slagging combustors. In addition, proper ash composition, or its chemical manipulation through the addition of fluxing agents, is necessary for desirable slagging operations.

Volatile Matter

The volatile matter from the coal can add to the products of gasification without incurring steam decomposition or oxygen consumption. The volatile matter, which can vary from less than 5% (on a moist, ash-free basis) for anthracite to over 50% for subbituminous coal or lignites, can consist of carbon oxides, hydrogen, and traces of nitrogen compounds [2]. The volatile matter composition, the type of coal, and the conditions under which the volatile matter is driven off affect the nature of the residual fixed carbon or char that remains.

Fixed Carbon

The nature of the fixed carbon, which is the major component of the char after the moisture and volatile matter are driven off, is important to the performance of the gasifier and can vary physically and chemically. Properties such as density, structure, friability/strength, and reactivity depend primarily on the original coal but they are influenced by the pressure, rate at which the coal is heated, and its final temperature [2].

Caking Tendencies

The caking tendencies of the coal—strongly caking and swelling, weakly caking, and noncaking—must be considered in the gasifier and process design [2].

Some gasifiers can be designed to handle caking and swelling coals, but others will require the coal to be pretreated. Table 5-7 lists the acceptability of caking coal to the generic types of gasifiers [44].

Ash Fusion

The ash fusion temperature is a measure of when the ash will melt and transform from a solid to a liquid state. This temperature is an important parameter for the design and operation of gasification systems—for those that operate below the ash fusion temperature so as not to incur fusion, sintering, or clinkering of the ash, as well as those gasification systems that operate above the ash fusion temperature to promote slag production. Typical preferred ash melting temperatures for the generic types of gasifiers are listed in Table 5-7. Another important ash characteristic is the relationship between temperature and ash viscosity as the flow characteristics of the slag are critical. The importance of slag viscosity was discussed earlier in the chapter and is as relevant to slagging gasifiers as it is for slagging combustors.

Reactivity

Coals vary in their reactivity to steam and to a lesser extent to hydrogen [2]. Reactive coals, or their chars, decompose steam more rapidly and sustain that decomposition down to a lower temperature than do less reactive coals. Reactivity has three important influences: (1) it favorably influences methane formation; (2) it reduces oxygen consumption by allowing steam decomposition down to a lower temperature; and (3) it allows less steam to be used per volume of oxygen or air than with less reactive coals without incurring ash clinkering.

Coal Size Distribution

The coal size limits are important gasifier system design considerations. In fixed-bed gasification systems, provisions have to be made for the fines that are generated from mining, transportation, and processing. This may include steam and power generation or briquetting, extrusion, or injection to allow them to be supplied to the fixed-bed gasifier [2]. The size distribution is less critical with fluidized-bed and entrained-flow gasification systems.

Regional Distribution of Gasification Systems

In 1999, the first World Gasification Survey was conducted by SFA Pacific, Inc., with support from the DOE and in cooperation with member countries of the Gasification Technologies Council [45,46]. The survey identified 160 commercial gasification plants in operation, under construction, or in planning and design stages in 28 countries in North and South America, Europe, Asia, Africa, and Australia. The total capacity of these facilities,

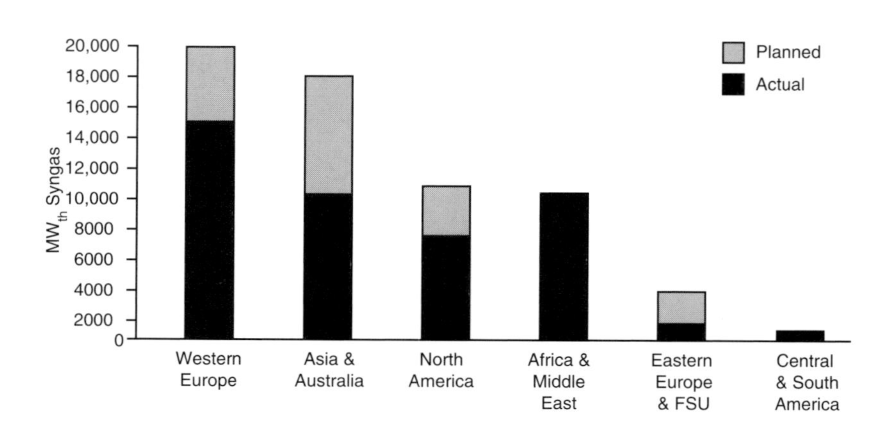

FIGURE 5-24. Distribution of gasification projects by geographic region. (From SFA Pacific, Inc., and U.S. DOE, *Gasification: Worldwide Use and Acceptance*, Office of Fossil Energy, U.S. Department of Energy, Washington, D.C., January 2000.)

when in operation, will be more than 15,000 million scf (or 4500 billion Btu) of syngas per day. Figure 5-24 shows the distribution of gasifier projects by geographic region. A summary of the survey that includes the number of gasifiers, applications (chemicals, gaseous fuels, and power), and production by country is provided in Appendix D [46].

Historically, syngas from gasification has been used primarily as feedstock for the production of chemicals. In 1989, chemical production accounted for ~50% of syngas use worldwide; however, this figure is changing as more power generation projects are being constructed and planned [46]. For new capacity added between 1990 and 1999, the power-to-chemical syngas volume ratio was ~1.4:1. The post-2000 ratio is 3:1 in favor of power generation, and by 2005 power production is projected to be nearly that of chemical production—approximately 5300 million scf per day. Fuels and gases are projected to be approximately 2800 and 1400 million scf per day, respectively. The distribution of gasification applications is illustrated in Figure 5-25 on a megawatt thermal (MW$_{th}$) syngas basis.

Coal and petroleum-based materials provide the majority of feedstocks for world gasification capacity, which is projected to rise to about 90% of the total capacity. In 1999, coal and petroleum feedstocks accounted for 4900 and 3200 scf per day synfuel production, respectively, which is projected to increase to approximately 6700 scf per day synfuel production for each feedstock type by 2005 [46]. This is illustrated in Figure 5-26 on an MW$_{th}$ syngas basis [45].

Commercial Gasification Systems

Gasification of abundant U.S. coal provides an alternative to coal-fired combustion systems as it is more efficient and environmentally friendly.

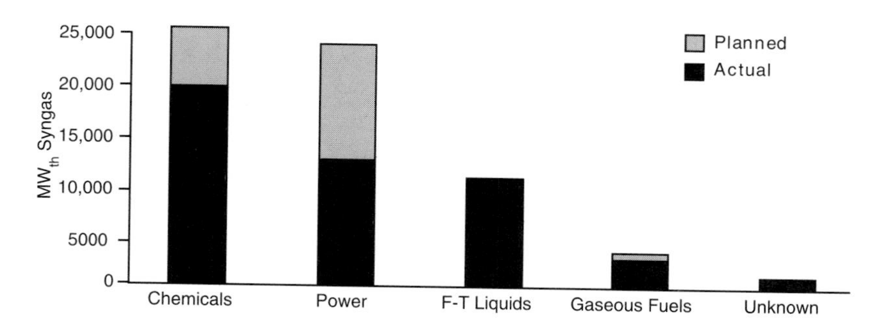

FIGURE 5-25. Distribution of gasification applications. (From SFA Pacific, Inc., and U.S. DOE, *Gasification: Worldwide Use and Acceptance*, Office of Fossil Energy, U.S. Department of Energy, Washington, D.C., January 2000.)

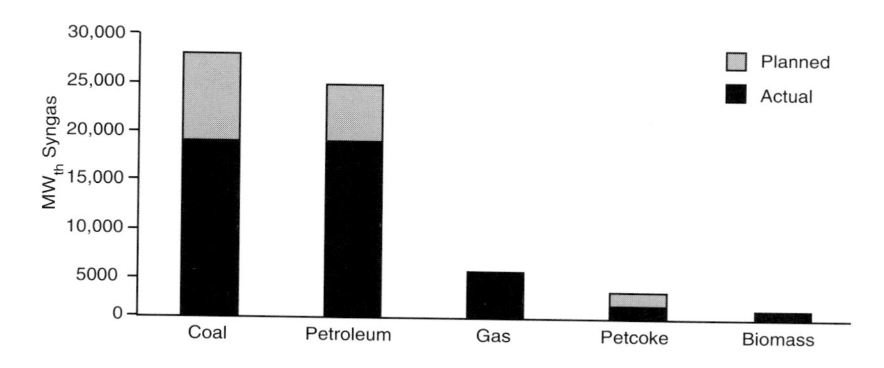

FIGURE 5-26. Gasification by primary feedstocks. (From SFA Pacific, Inc., and U.S. DOE, *Gasification: Worldwide Use and Acceptance*, Office of Fossil Energy, U.S. Department of Energy, Washington, D.C., January 2000.)

Coal gasification is a well-proven technology that began with the production of coal gas for towns and progressed to the production of fuels, such as oil and synthetic natural gas (SNG), chemicals, and more recently, to large-scale integrated gasification combined cycle (IGCC) power generation. Currently, the number of operating IGCC power plants is small; however, as listed in Appendix D, several power projects are under construction or in the planning and design stages, and gasification technology is the center of future U.S. energy complexes that are under development and discussed in detail in Chapter 7 (Future Power Generation). This section briefly discusses the gasification technologies used in commercial applications. A description of an IGCC power plant and details of the IGCC power plants in operation and proposed for the future are given in Chapter 7.

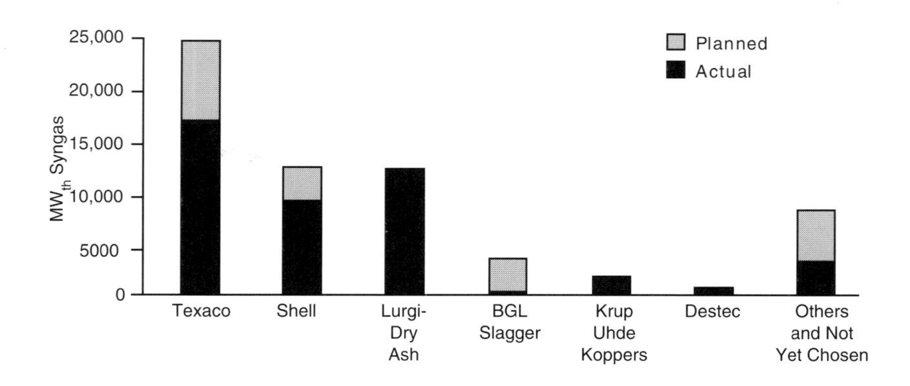

FIGURE 5-27. Gasification by technology. (From SFA Pacific, Inc., and U.S. DOE, *Gasification: Worldwide Use and Acceptance*, Office of Fossil Energy, U.S. Department of Energy, Washington, D.C., January 2000.)

The commercially well-proven ChevronTexaco, Shell, and Lurgi dry-ash gasification technologies represent a major portion of the worldwide gasification capacity, as illustrated in Figure 5-27 [45]. In addition, these three technologies represent 27 of the 30 largest commercial projects in the world (see Table 5-8) [45]. ChevronTexaco is the leading licensor of gasification technology based on total capacity, followed by Lurgi dry-ash and Shell technologies. There appears to be little interest in new Lurgi dry-ash gasifiers, which is probably due to limitations regarding the feedstock, such as the use of petroleum coke, and the extensive waste liquid clean-up requirements. These limitations do not exist for the ChevronTexaco and Shell technologies.

Fixed-Bed Gasifiers

Fixed-bed gasifiers differ in exit gas conditions and in special design configurations. There are two main commercial fixed-bed gasifier technologies. The Lurgi dry-ash gasifier was originally developed in the 1930s and has been used extensively for town gas production and in South Africa by Sasol (South African Coal Oil and Gas Corporation) for chemicals from coal. In fact, of the chemicals produced worldwide from syngas, Sasol produces nearly a third of them and uses Lurgi gasifiers to do so. Also, Lurgi gasifiers are used in Dakota Gasification's Great Plains Synfuels Plant located near Beulah, North Dakota, which is the only facility in the United States that manufactures a high-Btu synthetic natural gas from lignite; 17,000 short tons of lignite are processed daily to produce 148 million scf per day of SNG along with by-product chemicals. In the dry-ash Lurgi gasifier, the temperature at the bottom of the bed is kept below the ash fusion point so that the ash is removed as a solid. In the 1970s, Lurgi and the British Gas Corporation

TABLE 5-8
Largest 30 Commercial Gasification Projects in the World

Gasification Plant Owner	*Location*	*Gasification Technology*	*Output (MW_t equiv.)*	*Startup Year*	*Feed/Product*
Sasol-II	South Africa	Lurgi dry ash	4130	1977	Subbituminous coal/F-T liquids
Sasol-III	South Africa	Lurgi dry ash	4130	1982	Subbituminous coal/F-T liquids
Respol/Iberdrola	Spain	ChevronTexaco	1654	2004	Vacuum residue/electricity
Dakota Gasification Company	United States	Lurgi dry ash	1545	1984	Lignite and refinery residue/SNG
SARLUX srl	Italy	ChevronTexaco	1067	2000	Visbreaker residue/electricity and H_2
Shell MDA Sdn. Bhd.	Malaysia	Shell	1032	1993	Natural gas/mid-distillates
Linde AG	Germany	Shell	984	1997	Visbreaker residue/H_2 and methanol
ISAB Energy	Italy	ChevronTexaco	982	1999	ROSE asphalt/electricity and H_2
Sasol-I	South Africa	Lurgi dry ash	911	1955	Subbituminous coal/F-T liquids
Total France/EdF/ Texaco	France	ChevronTexaco	895	2003	Fuel oil/electricity and H_2
Unspecified owner	United States	ChevronTexaco	656	1979	Natural gas/methanol and CO
Shell Nederland Raffinaderij BV	The Netherlands	Shell	637	1997	Visbreaker residue/H_2 and electricity
SUV/EGT	Czech Republic	Lurgi dry ash	636	1996	Coal/electricity and steam
Chinese Petroleum Corporation	Taiwan	ChevronTexaco	621	1984	Bitumen/H_2 and CO
Hydro Agri Brunsbüttel	Germany	Shell	615	1978	Heavy vacuum residue/ammonia
Public Service of Indiana	United States	E-Gas (Destec)	591	1995	Bituminous coal/electricity
VEBA Chemie AG	Germany	Shell	588	1973	Vacuum residue/ammonia and methanol
Elcogas SA	Spain	Prenflow	588	1997	Coal and petcoke/electricity
Motiva Enterprises LLC	United States	ChevronTexaco	558	1999	Fluid petcoke/electricity and steam
API Raffineria di Anocona S.p.A.	Italy	ChevronTexaco	496	1999	Visbreaker residue/electricity
Chempoetrol a.s.	Czech Republic	Shell	492	1971	Vacuum residue/ammonia and methanol
Demkolec BV	Netherlands	Shell	466	1994	Bituminous coal/electricity
Tampa Electric Company	United States	ChevronTexaco	455	1996	Coal/electricity
Ultrafertil S.A.	Brazil	Shell	451	1979	Asphalt residue/ammonia
Shanghai Pacific Chemical Corp.	China	ChevronTexaco	439	1995	Anthracite/methanol and town gas
Exxon USA, Inc.	United States	ChevronTexaco	436	2000	Petcoke/electricity and syngas
Shanghai Pacific Chemical Corp.	China	IGT U-GAS	410	1994	Bituminous coal/fuel gas and town gas
Gujarat National Fertilizer Co.	India	ChevronTexaco	405	1982	Refinery residue/ammonia and methanol
Esso Singapore Pty. Ltd.	Singapore	ChevronTexaco	364	2000	Residual oil/electricity and H_2
Quimigal Adubos	Portugal	Shell	328	1984	Vacuum residue/ammonia

Note: F-T, Fischer–Tropsch synthesis; SNG, synthetic natural gas.

Source: SFA Pacific, Inc., and U.S. DOE, *Gasification: Worldwide Use and Acceptance*, Office of Fossil Energy, U.S. Department of Energy, Washington, D.C., January 2000.

developed a slagging version of the gasifier, referred to as the BGL gasifier, in which the temperature at the bottom of the gasifier is sufficient for the ash to melt.

Lurgi Gasifier The most successful fixed-bed gasifier is the Lurgi gasifier, which was developed in Germany during the 1930s as a means to produce town gas. The first commercial plant was built in 1936. It was initially used for lignites, but process developments in the 1950s allowed for the use of bituminous coals as well. The Lurgi gasification process has been used extensively worldwide.

The Lurgi dry-ash gasifier, shown schematically in Figure 5-28, is a pressurized gasifier typically operating at 30 to 35 atm. Sized coal enters the top of the gasifier through a lock hopper and moves down through the bed. Steam and oxygen enter at the bottom and react with the coal as the gases move up the bed. Ash is removed at the bottom of the gasifier by a rotating grate and lock hopper and is kept in a dry state through the injection of steam

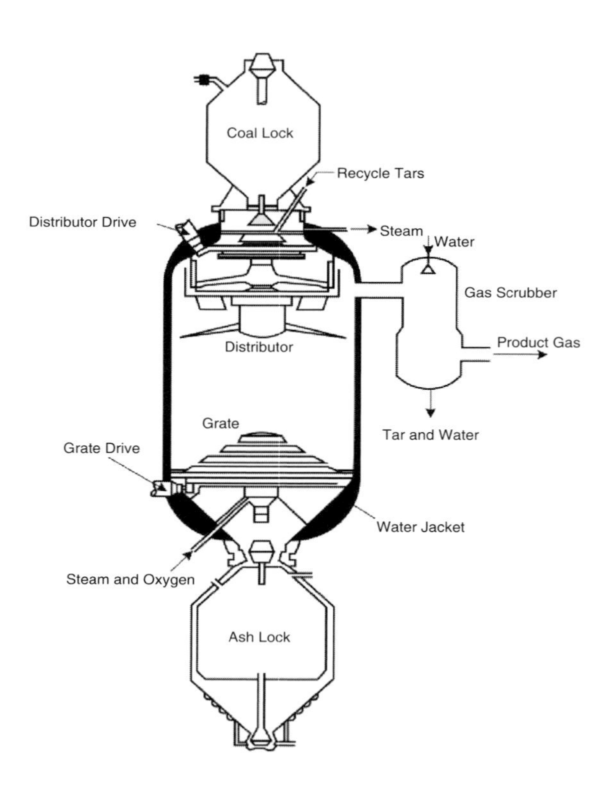

FIGURE 5-28. Schematic diagram of a modern Lurgi dry-ash gasifier. (From Berkowitz, N., *An Introduction to Coal Technology*, Academic Press, New York, 1979. With permission.)

to cool the bed below the ash fusion point. As the coal moves down the gasifier it goes through sequential stages of drying and devolatilization with the resultant char undergoing gasification and combustion. The countercurrent operation results in a temperature drop in the gasifier. Gas temperatures are approximately 500 to 1000°F in the drying and devolatilization zone, 1800°F in the gasification zone, and 2000°F in the combustion zone [22,44]. The raw syngas, a mixture of carbon monoxide and hydrogen, which also contains tar, exits the gasifier at 570 to 930°F.

BGL Gasifier The BGL fixed-bed gasifier was developed in the 1970s to provide a syngas with a high methane content in order to provide an efficient means of manufacturing SNG from coal. The BGL fixed-bed gasifier, shown schematically in Figure 5-29, is a dry-feed, pressurized, slagging gasifier. The operational concept is similar to the Lurgi dry-ash gasifier with two notable differences. The BGL fixed-bed gasifier is more fuel flexible in that it can use run-of-mine coal (rather than sized coal), and the gasifier is operated at temperatures above the ash fusion point to form a liquid slag.

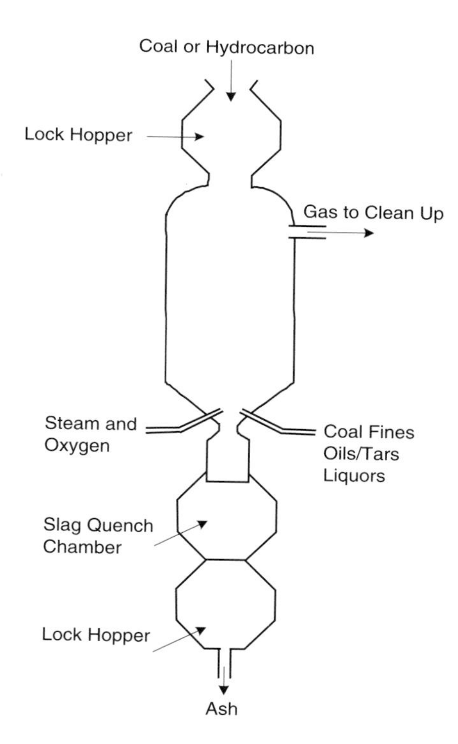

FIGURE 5-29. Schematic diagram of the BGL fixed-bed gasifier. (Adapted from Ratafia-Brown, J. *et al.* [44].)

Slag is withdrawn from the slag pool through an opening in the grate. The slag flows into a quench chamber and lock hopper in series. Syngas exits the gasifier at ~1040°F and passes into a water quench vessel and a boiler feedwater preheater designed to lower the gas temperature to approximately 300°F. Soluble hydrocarbons, such as tars, oils, and naphtha, are recovered from the aqueous liquor in a gas–liquor separation unit and recycled to the gasifier [44].

Fluidized-Bed Gasifiers

Fluidized-bed gasifiers may differ in ash conditions, dry or agglomerating, and in design configurations for improving char use. Commercial versions of this type of gasifier include the high-temperature Winkler (HTW) and Kellogg–Rust–Westinghouse (KRW) designs.

HTW Gasifier The high-temperature Winkler (HTW) gasifier, shown schematically in Figure 5-30, is a dry-feed, pressurized, fluidized-bed, dry-ash gasifier. The HTW process was developed by Rheinbraun in Germany during the 1920s to utilize coal with a small particle size and too friable for use in existing fixed-bed gasifiers. The HTW technology is capable of gasifying a variety of feedstocks, including reactive low-rank coals with a higher

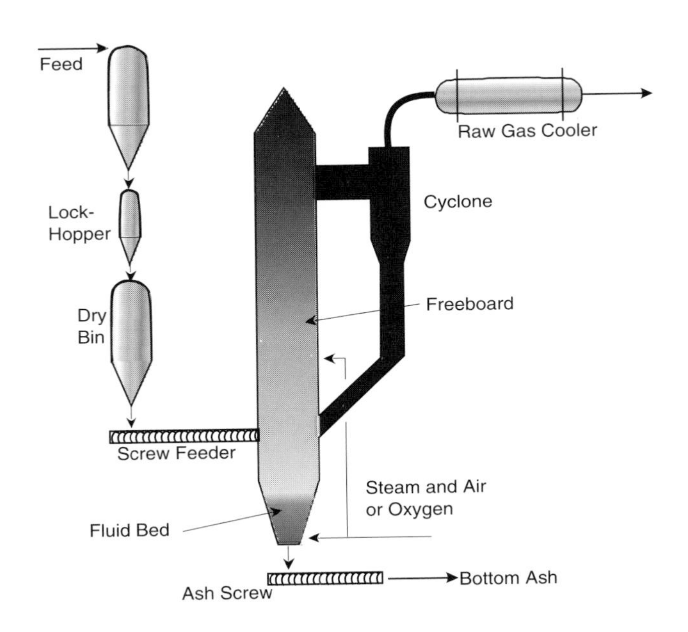

FIGURE 5-30. Schematic diagram of an HTW fluidized-bed gasifier. (From Ratafia-Brown, J. *et al.*, *Major Environmental Aspects of Gasification-Based Power Generation Technologies*, Office of Fossil Energy, U.S. Department of Energy, Washington, D.C., December 2002.)

ash-softening temperature and reactive caking and noncaking bituminous coals. Coal with a particle size less 1/8 in. is fed into the gasifier using a screw feeder. The upward flow of the gasifying medium, air or oxygen, keeps the particles of coal, ash, and semi-coke/char in a fluidized state. Gas and elutriated solids flow up the gasifier, and additional air or oxygen is added in this region to complete the gasification reactions. Fine ash particles and char that are entrained in the gas are removed in a cyclone and recycled to the gasifier. Ash is removed from the base of the gasifier by means of an ash screw. The syngas exiting the gasifier is at a high temperature so it does not contain any high-molecular-weight hydrocarbons such as tars, phenols, and other substituted aromatic compounds [44]. The gasifier fluid bed is operated at about 1470 to 1650°F, and the temperature is controlled to ensure that it does not exceed the ash-softening point. The temperature in the freeboard can be significantly higher, up to 2000°F. The operating pressure can vary between 145 psig for syngas manufacture and 360 to 435 psig for an IGCC application [44].

KRW Gasifier The Kellogg–Rust–Westinghouse gasifier, shown schematically in Figure 5-31, is a pressurized, dry feed, fluidized-bed, slagging gasifier

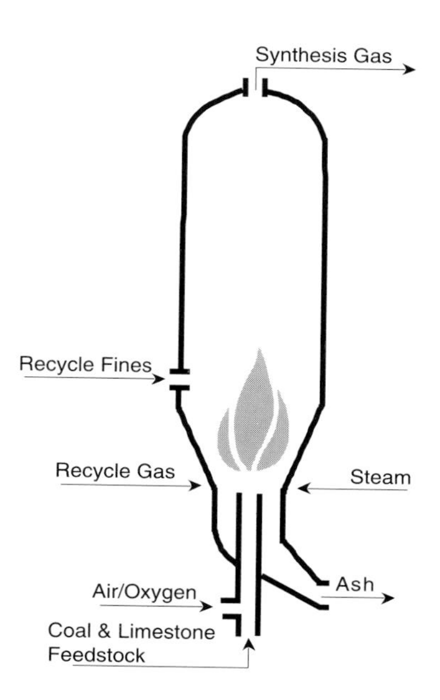

FIGURE 5-31. Schematic diagram of the KRW gasifier. (From Ratafia-Brown, J. *et al., Major Environmental Aspects of Gasification-Based Power Generation Technologies*, Office of Fossil Energy, U.S. Department of Energy, Washington, D.C., December 2002.)

developed by the M.W. Kellogg Company. The KRW gasifier is capable of gasifying all types of coals, including high-sulfur, high-ash, low-rank, and high-swelling coals. Coal and limestone, crushed to less than 1/4 in., are fed into the bottom of the gasifier, and air or oxygen enters through concentric, high-velocity jets [44]. This process ensures thorough mixing of the fuel and air or oxygen. The coal immediately releases its volatile matter upon entering the gasifier, and it oxidizes rapidly to produce the heat for the gasification reactions. An internal recirculation zone is established, with the coal/char moving down the sides of the gasifier and back into the central jet. Steam that is introduced with the air or oxygen and through jets in the side of the gasifier reacts with the char to form the syngas. Fine ash particles that are carried out of the bed are captured in a high-efficiency cyclone and reinjected into the gasifier. The internal recycling of the larger char particles results in the char becoming enriched in ash, and the low-melting components of the ash cause the ash particles to agglomerate. As the ash particles become larger, they begin to migrate toward the bottom of the gasifier where they are removed along with spent sorbent (*i.e.*, limestone that has reacted with sulfur to form calcium sulfide (CaS)) and some unreacted char. The ash, char, and spent sorbent flow into a fluidized-bed sulfator, where the char and calcium sulfide are oxidized. The calcium sulfide forms calcium sulfate, $CaSO_4$, which is chemically stable and can be disposed of in a landfill.

Entrained-Flow Gasifiers

Differences among entrained-flow gasifiers include the coal feed systems (coal–water slurry or dry coal), internal designs to handle the very hot reaction mixture, and heat-recovery configurations. Entrained-flow gasifiers have been selected for nearly all the coal-based IGCCs currently in operation or under construction [44]. The major commercial entrained-flow gasifiers include the ChevronTexaco, Shell, Prenflo, and E-Gas gasifiers. Of these, the ChevronTexaco gasifier and the Shell gasifier technologies are in use in over 100 units worldwide [44].

ChevronTexaco Gasifier The ChevronTexaco gasifier, shown schematically in Figure 5-32, is a single-stage, down-fired, entrained-flow gasifier [44]. A fuel–water slurry (*e.g.*, 60–70% coal) and 95% pure oxygen are fed to the pressurized gasifier. The coal and oxygen react exothermally at a temperature ranging from 2200 to 2700°F and a pressure greater than 20 atm to produce syngas and molten ash. Operation at the high pressures eliminates the production of hydrocarbon gases and liquids in the syngas. The hot gases are cooled using either a radiant syngas cooler located inside the gasifier to produce high pressure steam or an exit gas quench. Slag drops into the water pool at the bottom of the gasifier, is quenched and separated from the blackwater, and is removed through a lockhopper. The ChevronTexaco technology has operated commercially for over 40 years with feedstocks such as natural gas,

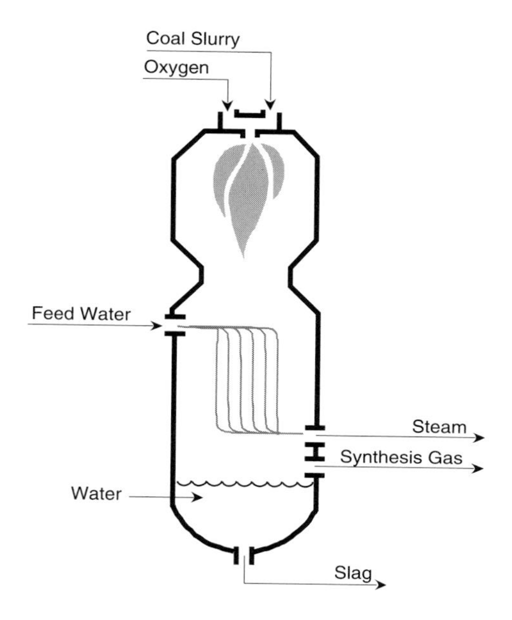

FIGURE 5-32. Schematic diagram of the ChevronTexaco gasifier. (From Ratafia-Brown, J. *et al.*, *Major Environmental Aspects of Gasification-Based Power Generation Technologies*, Office of Fossil Energy, U.S. Department of Energy, Washington, D.C., December 2002.)

heavy oil, coal, and petroleum coke. Currently, 60 commercial plants are in operation, with 12 using coke and coal, 28 using oil, and 20 using a gas feedstock [44].

Shell Gasifier The Shell gasifier, the successor to the Koppers–Totzek process and then the Shell–Koppers process, is shown schematically in Figure 5-33 and is a dry-feed, pressurized, entrained-flow, slagging gasifier that can operate on a wide variety of feedstocks. Pulverized coal is pressurized in lock hoppers and fed to the gasifier by dense-phase conveying with transport gas, which can be nitrogen or syngas [44]. Preheated oxygen is mixed with steam and used as a temperature moderator prior to feeding to the fuel injector. Temperatures and pressures in the gasifier are 2700 to 2900°F and 350 to 650 psig, respectively. A syngas is produced that contains mainly hydrogen and carbon monoxide with little carbon dioxide. Elevated temperatures eliminate the production of hydrocarbon gases and liquids in the product gas. The high temperature converts the ash into molten slag, which runs down the refractory walls into a water bath, where it is quenched and the ash/water slurry is removed through a lock hopper. The raw gas leaving the gasifier at 2500 to 3000°F contains a small quantity of char and about half of the molten ash. The hot gas is partially cooled to temperatures below

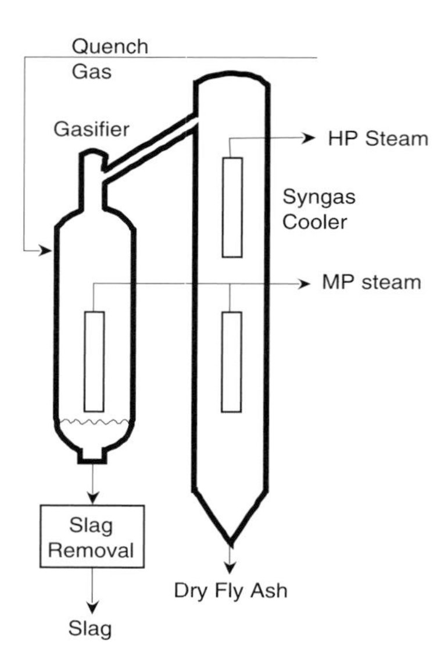

FIGURE 5-33. Schematic diagram of the Shell gasifier. (From Ratafia-Brown, J. *et al.*, *Major Environmental Aspects of Gasification-Based Power Generation Technologies*, Office of Fossil Energy, U.S. Department of Energy, Washington, D.C., December 2002.)

the ash fusion point by quenching it after it leaves the gasifier. The syngas undergoes further cooling before the particles are removed in a wet scrubber. The first Shell gasification process units were commissioned in the 1950s. Shell started development work with coal in 1972.

Prenflo Gasifier The Prenflo gasifier, developed by Uhde (formerly Krupp Uhde) and shown schematically in Figure 5-34, is a pressurized, dry-feed, entrained-flow, slagging gasifier [44]. Coal, ground to ~100 μm, is pneumatically conveyed by nitrogen to the gasifier. The coal is fed through injectors located in the lower part of the gasifier with oxygen and steam. Syngas, produced at temperatures of ~2900°F, is quenched with recycled cleaned syngas to reduce its temperature to ~1500°F in an internal syngas cooler. The syngas is further cooled to ~700°F through evaporator stages before exiting the gasifier. The molten slag flows down the walls into a water bath, where it is quenched and granulated before removal through a lock hopper system.

E-Gas Gasifier The E-Gas gasifier, shown schematically in Figure 5-35, is a slurry-feed, pressurized, entrained-flow gasifier. It is an upward flow gasifier with two-stage operation. The coal is slurried via wet crushing, with coal concentrations ranging from 50 to 70 wt.%, and about 75% of the total slurry feed is fed to the first stage of the gasifier, which operates at 2600°F and

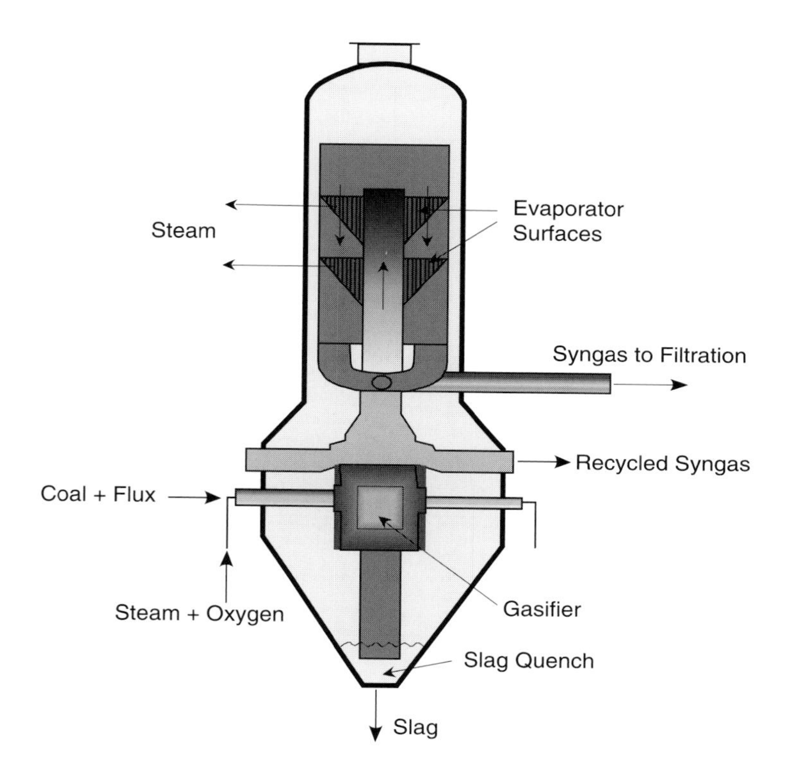

FIGURE 5-34. Schematic diagram of the Prenflo gasifier. (From Ratafia-Brown, J. *et al.*, *Major Environmental Aspects of Gasification-Based Power Generation Technologies*, Office of Fossil Energy, U.S. Department of Energy, Washington, D.C., December 2002.)

400 psig. Oxygen is combined with the slurry and injected into the first stage of the gasifier, where the highly exothermic gasification/oxidation reactions occur. Operation at the elevated temperatures eliminates the production of hydrocarbon gases and liquids in the product gas. The molten ash flows into a water bath, where it is quenched and removed. The raw gas from the first stage enters the second stage, where the remaining 25% coal slurry is injected. The endothermic gasification/devolatilization reactions occur in this stage at a temperature of ~1900°F, and some hydrocarbons are added to the product gas. Char is produced in the second stage and is recycled to the first stage, where it is gasified. The syngas exits the gasifier and undergoes further cooling and cleaning.

Liquefaction

Liquefaction is the conversion of coals into liquid products. The three methods by which liquids can be derived from coals are pyrolysis, indirect

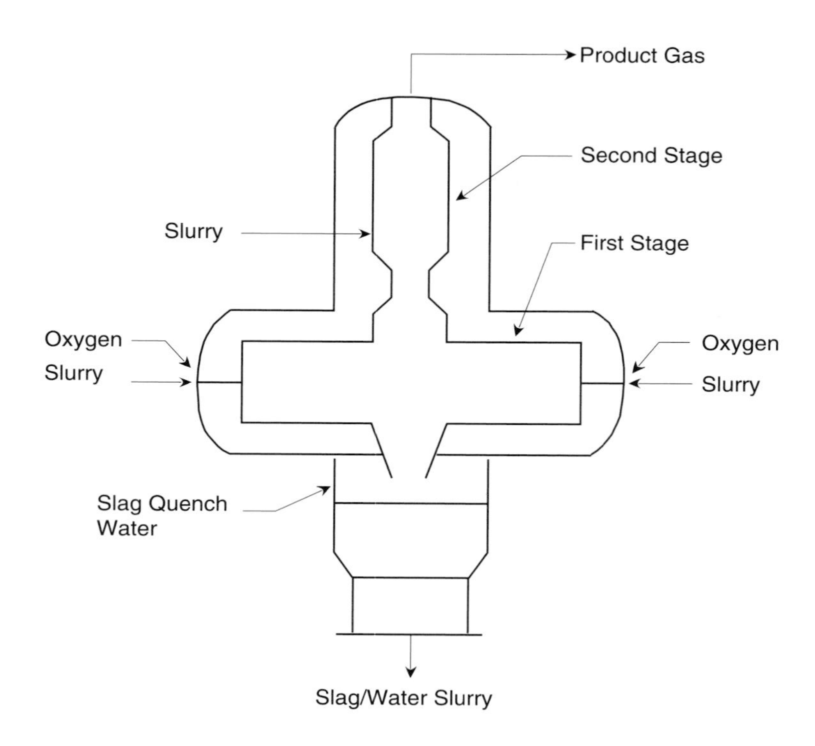

FIGURE 5-35. Schematic diagram of the E-Gas gasifier. (From Ratafia-Brown, J. *et al.*, *Major Environmental Aspects of Gasification-Based Power Generation Technologies*, Office of Fossil Energy, U.S. Department of Energy, Washington, D.C., December 2002.)

liquefaction, and direct liquefaction. In pyrolysis processes, the liquids are a by-product of coke production. The term liquefaction refers to the conversion of the coal to a product that is primarily a liquid. In indirect liquefaction, the coal is gasified into a mixture of carbon monoxide and hydrogen (*i.e.*, syngas), which was discussed earlier. The syngas is then processed into liquid products using Fischer–Tropsch synthesis. In direct liquefaction, also referred to as coal hydrogenation, coal is mixed with a hydrogen-donor solvent and reacted with hydrogen or syngas under elevated pressures and temperatures to produce a liquid fuel. Indirect liquefaction is used quite extensively throughout the world, as illustrated in Table 5-8, while direct liquefaction has not been able to compete with other liquid or gaseous fuels. Liquefaction processes are technically feasible; Germany and South Africa—countries with abundant coal supplies but little or no petroleum resources—have demonstrated that a country can meet much of its liquid fuel needs through liquefaction. Germany did so during World War II, while South Africa became self-reliant during its years of apartheid which made it susceptible to oil embargoes. This section provides a brief history of the development of liquefaction, followed by a discussion of liquefaction, mainly indirect liquefaction. As previously discussed in this chapter and in Chapter 7 (Future Power Generation), gasification integrated with chemicals and fuels production is the technology anticipated to be utilized for future power complexes. Direct liquefaction is a much-researched technology with proven processes operated at the pilot and demonstration scale; however, it is not a widely accepted commercial technology, although facilities are under construction in China. Thus, a brief review of key direct liquefaction processes will be presented here but direct liquefaction will not be discussed in great detail.

The Beginning of the Synthetic Fuel Industry

Coal was hydrogenated in the laboratory by Berthelot as early as 1869 [47]. The reaction was carried out with hydriodic acid at 520°F for 24 hours, and a 67% yield of oil containing aromatics and naphthenes was obtained. Major advances were made by the Germans in the early 1900s. Germany, a country with abundant coal reserves but virtually no petroleum deposits, was becoming increasingly dependent on gasoline and diesel engines. To ensure that Germany would never lack a plentiful supply of liquid fuels, German scientists and engineers invented and developed two processes that enabled them to produce synthetic petroleum from their coal supplies and to establish the world's first technologically successful synthetic liquid fuel industry [48]. In 1911, Friedrich Bergius obtained oil by hydrogenating coal without a catalyst under hydrogen pressure at 570 to 660°F. In 1913, he applied for the first patent on coal hydrogenation, and in 1931 he was awarded the Nobel Prize in Chemistry [19,47]. Bergius also observed that coal paste could be injected readily into a vessel under pressure. The role of catalysts in the hydrogenation of coal was not realized until later.

At the end of 1925, I.G. Farben, a chemical company, hydrogenated coal using a molybdenum oxide catalyst. The presence of the catalyst allowed the hydrogenation of coal in the presence of excess hydrogen at low pressure and at temperatures of 750 to 840°F. In the following year, I.G. Farben conducted the liquefaction process in two steps because high-molecular-weight materials in the intermediate hydrogenation product fouled the catalyst. Coal was mixed with the catalyst and hydrogenated in the liquid phase to middle oil, which was further hydrogenated to gasoline in vapor phase over a fixed bed of catalyst [47].

About 10 years after Bergius began his work, Franz Fischer and Hans Tropsch at the Kaiser–Wilhelm Institute for Coal Research in Mülheim, invented a second process for the synthesis of liquid fuel from coal [48]. By the mid-1930s, I.G. Farben and other chemical companies such as Ruhrchemie had started to industrialize synthetic liquid fuel production, resulting in the construction of 12 coal hydrogenation and 9 Fischer–Tropsch plants by the time World War II ended. The processes were complementary in that coal hydrogenation produced high-quality aviation and motor gasoline, while

Fischer–Tropsch syntheses produced diesel and lubricating oil, waxes, and some lower quality motor gasoline [48].

Germany's successful synthetic fuel industry continued to grow through the 1930s, and from 1939 to 1945 it produced 18 million metric tons of liquids from coal and tar, and another three million metric tons of liquids from Fischer–Tropsch synthesis [48]. Neither coal-to-oil process could produce a synthetic liquid fuel at a cost competitive with natural petroleum; however, they persevered because they provided the only path Germany could follow in its search for petroleum independence.

Similarly, South Africa, fearing boycott as a result of their racial policies, decided to proceed with a synthetic fuels plant in 1951, although this process would produce liquid fuels that were more expensive than petroleum refined products. The South Africans selected Fischer–Tropsch technology because the Germans had successfully used it and because direct liquefaction had not been used on a scale that was as large as the South Africans were planning [22]. This led to the creation of Sasol and government subsidies that exist even today [49]. However, now that Sasol has the equipment in place, crude oil is readily available, and apartheid has been abolished, political and economic considerations have forced the South African government to phase out the subsidy for transportation fuels. Sasol anticipated losing its subsidies and has gradually shifted its emphasis from producing only transportation fuels to deriving a significant fraction of its profits from the sale of chemicals and petrochemical feedstocks [49]. Plants were constructed in South Africa in three stages: 1955, 1982, and 1992 (see Table 5-8) and are discussed later in this chapter. All three of these plants are currently operating.

In the United States, interest in converting coal to liquid products has been cyclic and affected by the cost and availability of petroleum. In the beginning of the industrial revolution, coal was the major source of energy in the United States and continued to dominate the U.S. energy supply for the next 100 years. Petroleum, however, quickly became the preferred energy source after its discovery in Pennsylvania in 1859 and rapid commercial production in the early 1900s. By the early 1920s, worries that oil supplies were becoming depleted along with an expanding automobile industry resulted in a short-lived interest in coal liquefaction; when oil was discovered in Texas in the mid-1920s, further interest in coal liquefaction ceased. After World War II, consumption of petroleum and natural gas in the United States exceeded that of coal, the United States experienced petroleum shortages, and coal liquefaction was again considered as an alternative. A sizeable research effort resulted, but discovery of massive petroleum reserves in the Middle East in the mid-1940s once again made coal liquefaction an uneconomical alternative. In 1972, the U.S. petroleum production began to decline and unrest developed in the Middle East. The limited availability of domestic supplies of natural gas and crude oil and the desire to reduce the country's dependence on foreign sources of energy led to significant liquefaction research by the federal government, private industry, and universities. The major processes are discussed below.

Indirect Liquefaction: Fischer–Tropsch Synthesis

Fischer–Tropsch synthesis is the conversion of carbon monoxide and hydrogen to liquid hydrocarbons and related oxygenated compounds over variously promoted group-VII catalysts and is defined by the reaction:

$$CO + 2H_2 \rightarrow (-CH_2-) + H_2O \tag{5-25}$$

Fischer–Tropsch synthesis can, in principle, provide almost all hydrocarbons conventionally obtained from petroleum. The actual product mix depends on the temperature, pressure, CO/H_2 ratio, and catalysts used, and there are several variants of the process. Table 5-9 illustrates how the product composition can be influenced [19], and Table 5-10 lists some technical information on three industrial processes [1].

Medium-pressure synthesis, conducted at 430 to 640°F and 5 to 50 atm over iron catalysts, yields mainly gasoline, diesel oils, and heavier paraffins [19]. The proportion of gasoline in the product mix increases, and gasoline quality improves as the hydrogen-to-carbon monoxide ratio in the feed gas is increased, and the overall product composition is influenced by the type of reactor: fixed catalyst beds or a fluid-bed reactor. High-pressure synthesis, conducted at 210 to 300°F and 50 to 1000 atm over ruthenium catalysts, furnishes mainly straight-chain paraffin waxes with molecular weights up to 105,000 and melting ranges up to 270 to 273°F; however, the formation of lower-molecular-weight hydrocarbons can be increased by increasing the proportion of hydrogen in the feed gas, raising the reaction temperature, and decreasing the pressure.

TABLE 5-9
Influence of Operating Parameters on the Composition of the Product of Medium-Pressure Fischer–Tropsch Synthesis

Increase in:	*Mean Molecular Weight of Products*	*Yield[a]*	
		Oxygenated Products	*Olefins*
Pressure	+	+	–
Temperature	–	–	+
CO conversion	+	–	–
Flow rate	–	+	+
Gas recycle ratio	–	+	+
H_2:CO ratio	–	–	–

[a] +, Increasing; –, decreasing.

Source: Berkowitz, N., *An Introduction to Coal Technology*, Academic Press, New York, 1979. With permission.

TABLE 5-10
Technical Information on Three Industrial Fischer–Tropsch Processes

Process	*Cobalt F-T (Normal/Medium Pressure)*	*ARGE*[a] *(Sasol I)*	*Synthol*[b] *(Sasol I and II)*
Reactor	Fixed bed	Fixed bed	Entrained
Stages	2–3	1	1
Pressure (atm.[c])	1–12	23–25	24–28
Temperature (°F)	380	430–480	610–640
Catalyst	Co	Fe (Cu)[d]	Fused Fe
Promotor	MgO, ThO_2	K_2O	K_2O
Flow rate (h^{-1})	100	500–700	Not available
Recycle ratio	0	2.5	2.0–2.4
CO conversion	90–95	73	77–85
Yield (wt.%)			
C_1–C_2	Not available	7	20
C_3–C_4	7	14 (of which 45% are olefins)	23 (of which 87% are olefins)
C_5–C_{12}	30	24.8 (of which 50% are olefins)	39 (of which 70% are olefins)
C_{13}–C_{20}	10	14.7 (of which 40% are olefins)	5 (of which 60% are olefins)
C_{20+}	11	36.2 (of which 15% are olefins)	5
Oxygen composition	Not available	4	8
Alcohols/ketones	Not available	2.3	7.8
Acids	Not available	Not available	1.0

[a] ARGE: Arbeitsgermeinschaft Ruhrchemie-Lurgi.
[b] Circulating entrained-flow reactor.
[c] 1 standard atmosphere = 14.7 psia (pounds per square inch absolute).
[d] (Cu) = about 5% copper.

Source: Van Krevelen, D. W., *Coal: Typology–Physics–Chemistry–Constitution*, Third edition, Elsevier Science, Amsterdam, 1993. With permission.

Iso-synthesis is a reaction that converts carbon monoxide and hydrogen to branched-chain hydrocarbons and is usually conducted at 750 to 930°F and 100 to 1000 atm over thoria or K_2CO_3-promoted thoria–alumina catalysts and yields predominately low-molecular-weight (C_4 and C_5) isoparaffins [19]. Iso-synthesis at temperatures much above 750°F promotes coproduction of aromatics, while oxygenated compounds are formed at temperatures below 750°F. The synthol process is used for the production of oxygenated compounds. In this process, the feed gas is reacted at 750 to 840°F and ~140 atm over an alkalized iron catalyst [19]. Synol synthesis involves the interaction of carbon monoxide and hydrogen at 355 to 390°F and 5 to 50 atm in the presence of highly reduced ammonia catalysts to produce a product with 40 to 50% oxygenated straight-chain compounds. Oxyl synthesis is carried out at 355 to 390°F and 20 to 50 atm over a precipitated iron catalyst to produce a product with over 30% oxygenated straight-chain compounds. The oxo-synthesis process is used for producing aldehydes and occurs with the reaction between an olefin and syngas at 210 to 390°F and 100 to 500 atm over cobalt carbonyl catalysts. The oxo-synthesis process has become the most common industrial method for producing C_3 to C_{16} aldehydes.

Direct Liquefaction

Direct liquefaction was developed in the 1930s by Bergius and collaborators and involved reacting pulverized coal or coal/oil slurries with gaseous hydrogen at high pressures and temperatures. It was used extensively in Germany during World War II and in the former Soviet Union and Czechoslovakia for several years after the war. Plants were also constructed in Japan and Great Britain but on a much smaller scale. These plants were either destroyed during the war or shut down shortly after because they were not economical to operate. Direct liquefaction research and development continued after the war and increased significantly after the oil crisis in the 1970s. Direct liquefaction research funding essentially ended in the mid- to latter 1980s. The funding for the research came from a host of agencies, companies, and states, and examples of funding sources include (but the list is obviously not inclusive): the DOE and its predecessors, the Office of Coal Research (OCR) and Energy Research and Development Administration (ERDA); Electric Power Research Institute (EPRI); Edison Electric Institute (EEI); oil companies, such as Amoco, Ashland, Conoco, Mobil, Exxon, Phillips, and Gulf; major utilities, such as Southern Services, Inc.; other companies such as the Great Northern Railroad; and major coal-producing states such as Kentucky. With the exception of the coal-to-liquids plants that Hydrocarbon Technologies, Inc. (HTI) is currently constructing in China, development of direct liquefaction technologies has achieved only demonstration-scale status, primarily due to the availability of lower priced petroleum-refined fuels. This section will introduce some of the main processes that have been developed. From a chemical standpoint, coal has a lower hydrogen-to-carbon (H/C)

TABLE 5-11
Information on Select Liquefaction Processes

Process	*I.G. Farben/ Bergius*	*SRC-I*	*SRC-II*	*H-Coal*	*EDS*	*Costeam*	*HTI DCL*
Temperature (°F)							
Stage 1	900	840	860	840	700–900	750–840	800
Stage 2	750	NA[a]	NA[a]	NA[a]	500–840	—	?[b]
Pressure (atm)							
Stage 1	350–700	140	140	210	20–170	270	170
Stage 2	300	NA[a]	NA[a]	NA[a]	80–210[c]	NA[a]	?
Carbon content of coal (wt.%)	70–83	70–85	~75	~78	75–80	60–70	?
H_2 consumption (wt.% on coal)							
Stage 1	6	2.4	4.7	5.5	—	6.5+	?
Stage 2	4	NA[a]	NA[a]	NA[a]	6[c]	NA[a]	?
Scale of operation (short tons per day)	150–1800	50	50	600	250	5–10 (lb/hr)[d]	4300[e]
Sulfur coal	<3.0	3	3	1–3	1–3	1	?
Product (wt.%)	0.1	0.7	0.3–0.7	0.3–0.7	0.3	Low	Low

[a]Not applicable.
[b]Unknown.
[c]Recycle oil hydrogenation.
[d]Small scale.
[e]Per train; three trains planned.
Source: Adapted from van Krevelen [1]; Miller [50]; HTI [51]; Elliot [2].

ratio than petroleum: ~0.7 to >1.2. Direct liquefaction transforms coal into liquid hydrocarbons by directly adding hydrogen to the coal. Examples of some of the operating parameters of the primary processes are summarized in Table 5-11 [1,50].

Bergius/I.G. Farben Process

The Bergius process was put into commercial practice by I.G. Farben in Leuna, Germany, in 1927, and additional plants were erected in the 1930s. The process operated in two stages. Liquid-phase hydrogenation first transformed the coal into middle oils, with boiling points between 300 and 615°F, and subsequent vapor-phase hydrogenation then converted these oils to gasoline, diesel fuel, and other relatively light hydrocarbons [19]. The Bergius process converts 1 short ton of coal to 40 to 45 gallons of gasoline, 50 gallons of diesel fuel, and 35 gallons of fuel oil [22]. The gasoline fraction contains 75 to 80% paraffins and olefins and 20 to 25% aromatic compounds. Liquid-phase hydrogenation of low-rank coals was usually accomplished at 890 to 905°F under pressures of 250 to 300 atm using an iron-oxide hydrogenation catalyst. For liquid-phase hydrogenation of bituminous coals, temperatures were similar, but pressures were in the range of 350 to 700 atm.

Solvent Refining Processes

The solvent-refined coal (SRC) process is considered the least complex of the various process schemes. Hydrogenation of the coal in the SRC process occurs at elevated temperatures and pressures in the presence of hydrogen. Catalysts other than the minerals contained in the coal are not used [2].

SRC-I Process In the SRC-I process, ground and dried coal is fed to the reactor as a slurry. The transport liquid, or process solvent, is a distillate fraction recovered from the coal hydrogenation product and serves as the hydrogen donor. The coal slurry and hydrogen are introduced into a dissolver at 840°F and 1500 psig. The process produces a low-ash (~0.1%), low-sulfur (~0.3%), solid fuel with a heating value of ~16,000 Btu/lb. The SRC-I fuel was envisioned as a boiler fuel to replace natural gas and fuel oil. Testing of this process started with Gulf Oil in the 1960s. Much of the developmental work on this process was performed at Wilsonville, Alabama, in a 6 short ton/day plant with funding from the EEI, Southern Services, the DOE, and the EPRI. A 50 short ton/day plant was built at Fort Lewis, Washington, and testing was performed by Gulf Oil.

SRC-II Process The SRC-I process was modified to eliminate a number of processing steps and to produce an all-distillate, low-sulfur fuel oil from coal rather than a solid fuel. In the SRC-II process, pulverized and dried coal is mixed with recycled slurry solvent from the process. The slurry mixture is pumped to reaction pressure (~140 atm), preheated to about 700 to 750°F, and fed into the dissolver, which is operated at 820 to 870°F [2]. The dissolver effluent is separated into vapor- and liquid-phase fractions. The overhead vapor stream undergoes several stages of cooling and separations, and the condensed liquid is distilled to produce naphtha and a middle distillate oil, which are converted to gasoline and diesel fuel, respectively. The gaseous products are purified to remove hydrogen sulfide and carbon dioxide, and the hydrogen-rich gas is then recycled to the reactor with make-up hydrogen. The liquid-phase product acts as the solvent for the SRC-II process. SRC-II testing was performed at the Tacoma, Washington, pilot-plant by Gulf Oil. Funding for this work was provided by the DOE.

Costeam Process The costeam process, investigated at the Pittsburgh Energy Research Center (DOE) and the University of North Dakota/Grand Forks Energy Research Center was intended to produce low-sulfur liquid products from lignites and subbituminous coals. This process uses crude syngas containing about 50 to 60% carbon monoxide and 30 to 50% hydrogen, rather than pure hydrogen. The high moisture content of the low-rank coals provides the water that is converted to steam. Ground coal is slurried with recycled product oil from the process. The slurried coal is pumped to unit pressure (4000 psig), mixed with syngas, heated to 750 to 840°F, and maintained at conditions for periods of 1 to 2 hours to liquefy the coal. Severe operating conditions are required for the dissolution of the lignite. These operating conditions were a drawback for the process, as long residence times and high pressures are costly and present major engineering problems. This process, although a potential candidate for liquefying low-rank coals, was only tested in small-scale equipment. It is included in this discussion because it was considered a potential process for utilizing low-rank coals.

Catalytic Processes

The most important process of this group is the H-Coal process, developed by Hydrocarbon Research, Inc. (HRI) as an outgrowth of previous work on the hydrogenation of petroleum fractions. The development of this process was sponsored by the ERDA and a large group of oil companies [1].

H-Coal Process In the H-coal process, pulverized coal is slurried with recycled oil and, along with hydrogen, fed to an ebullated-bed reactor, a feature that distinguishes this process from others. Reactor conditions are normally in the range of 825 to 875°F and 2500 to 3500 psig. The reactor contains a bed of catalytic particles, cobalt molybdate on alumina oxide. The products from the reactor include hydrocarbon gases, light and heavy distillate oils, and bottoms slurry. Variations of the processing scheme can produce fuel oil, naphtha, synthetic crude, ammonia, and fuel gas. Further processing can produce gasoline and jet fuel. HRI operated several small-scale reactors in their Trenton, New Jersey, test facility, and a pilot plant producing 800 short tons coal per day was constructed in Cattletsburg, Kentucky.

In 1995, when it became an employee-owned company, HRI changed its name to Hydrocarbon Technologies, Inc. (HTI). HTI has modified its process, now known as HTI's direct coal liquefaction (DCL) process and shown schematically in Figure 5-36, and it has gone from a single- to double-stage reactor system [51]. In the HTI DCL process, pulverized coal is dissolved in recycled, coal-derived, heavy-process liquid at about 2500 psig and 800°F while hydrogen is added. Most of the coal structure is broken down in the first-stage reactor. Liquefaction is completed in the second-stage reactor, at a slightly higher temperature and lower pressure. A proprietary GelCat catalyst is dispersed in the slurry for both stages. The process produces diesel fuel and gasoline. HTI has entered into an agreement with Shenhua Group Corporation, Ltd., for a direct coal liquefaction plant to be constructed in the People's Republic of China [52]. The plant will be located approximately 80 miles south of Baotou, at Majata, Inner Mongolia in China. The plant will have an ultimate capacity of 50,000 barrels per day of ultra-clean, low-sulfur, diesel fuel and gasoline produced from Chinese coal. The plant is

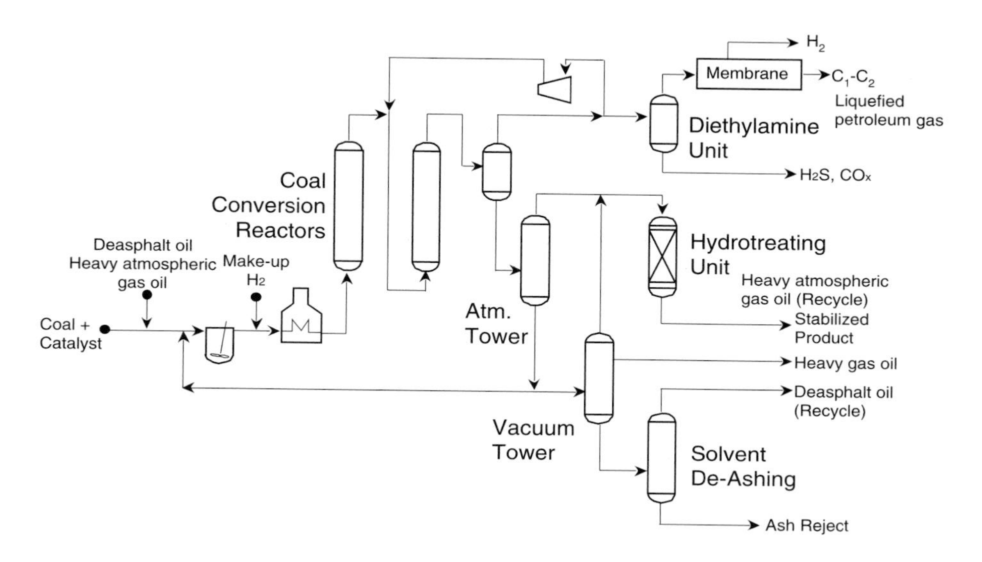

FIGURE 5-36. HTI's direct coal liquefaction process scheme. (From HTI, Coal Liquefaction Technology (Hydrocarbon Technologies, Inc., Lawrenceville, NJ), www.htinj.com/products/liquefaction.html (accessed April 2004). With permission.)

expected to be operational in 2005. Shenhua Group intends to construct three additional plants in the Shengdong Coalfield of China, which spans Shaanxi Province and Inner Mongolia. These would be the only commercial, large-scale direct coal liquefaction plants operational in the world since the World War II-vintage plants constructed by the Germans.

Donor Solvent Processes

The main representative of this process is the Exxon donor solvent (EDS) process. In a donor solvent process, the solubilization of the coal is done by the hydrogen donor liquid, and hydrogen molecules come from the donor liquid and not gaseous hydrogen. The donor liquid is then recycled and hydrotreated to add hydrogen back into the donor liquid.

Exxon Donor Solvent Process In the EDS process, finely ground coal is mixed with the donor solvent, and the coal is liquefied in a noncatalytic tubular plug-flow reactor in the presence of molecular hydrogen and a hydrogen-rich donor solvent [2]. The liquefaction reactor operates at 800 to 880°F and 1700 to 2300 psig. The products from the liquefaction reactor are separated by distillation into light hydrocarbon gases, ranging from methane to propane and methylpropane, a naphtha fraction, a heavy distillate, and a bottoms fraction [22]. The naphtha and heavy distillate fractions are treated by conventional petroleum-refining technology. About 85% of the naphtha is recovered as gasoline, and about 50% of the heavy distillate is recovered as a mixture of benzene, toluene, and xylenes. Further processing of the heavy distillate produces fractions comparable to jet fuel and heating fuel [22]. A portion of the heavy distillate is hydrotreated and recycled to slurry the coal. In addition, the bottoms are either coked in a fluid coking plant or recycled to the liquefaction reactor. Recycling to the liquefaction unit results in a dramatic increase in the conversion of the coal to liquid products. In addition to small-scale test reactors, Exxon, the DOE, the EPRI, and an international group of industrial sponsors installed a pilot plant in Baytown, Texas, that produces 250 short tons coal per day. This was the minimum size needed for confident scale-up of the critical process and mechanical features of the EDS process.

Concluding Statements

Several direct liquefaction processes were introduced in this chapter. Although these have been some of the most important and, in most cases, the most successful processes, it must be noted that many processes have been conceived and researched. Due to economics, however, none has been able to compete with available liquid and gaseous fuels, especially in industrialized countries. The work of the Germans in the 1930s and 1940s and the facilities planned for China illustrate that large-scale, direct coal liquefaction commercial plants are technically feasible. They can be constructed and operated when necessary if dictated by political mandates, whether during a period of isolation in wartime or due to a decision to be self-sufficient and use indigenous resources.

References

1. Van Krevelen, D. W., *Coal: Typology–Physics–Chemistry–Constitution*, 3rd ed. (Elsevier Science, Amsterdam, 1993).
2. Elliot, M. A. (editor), *Chemistry of Coal Utilization*, Secondary Suppl. Vol. (John Wiley & Sons, New York, 1981).
3. Moore, E. S., *Coal: Its Properties, Analysis, Classification, Geology, Extraction, Uses, and Distribution* (John Wiley & Sons, New York, 1922), p. 124.
4. Kuehn, S. E., Power for the Industrial Age: A Brief History of Boilers, *Power Engineering*, Vol. 100, No. 2, February 1996, pp. 5–19.
5. Landels, J. G., *Engineering in the Ancient World* (University of California Press, Berkley, 1978).
6. B&W, *Coal: Its Generation and Use*, 39th ed. (The Babcock & Wilcox Company, New York, 1978).
7. Anon., Boilers and Auxiliary Equipment, *Power*, Special Edition, June 1988.

8. Singer, J. G. (editor), *Combustion: Fossil Power Systems* (Combustion Engineering, Inc., Windsor, CT, 1981).
9. CIBO, *Industrial Combustion Boiler and Process Heater MACT Summary Sheet* (Council of Industrial Boiler Owners, Burke, VA, July 3, 2002).
10. EPA, *Report to Congress: Wastes from Combustion of Fossil Fuels*, Vol. 2, *Methods, Findings, and Recommendations* (U.S. Environmental Protection Agency, U.S. Government Printing Office, Washington, D.C., March 1999), chap. 3.
11. Miller, B. G., S. Falcone Miller, and A. W. Scaroni, Utilizing Agricultural By-Products in Industrial Boilers: Penn State's Experience and Coal's Role in Providing Security for our Nation's Food Supply, paper presented at the Nineteenth Annual International Pittsburgh Coal Conference (University of Pittsburgh, September 23–27, 2002).
12. Miller, B. G., S. Falcone Miller, R. E. Cooper, N. Raskin, and J. J. Battista, A Feasibility Study for Cofiring Agricultural and Other Wastes with Coal at Penn State University, paper presented at the Nineteenth Annual International Pittsburgh Coal Conference (University of Pittsburgh, September 23–27, 2002).
13. Miller, B. G. and S. Falcone Miller, Utilizing Biomass in Industrial Boilers: The Role of Biomass and Industrial Boilers in Providing Energy/National Security, in *Proc. of the First CIBO Industrial Renewable Energy and Biomass Conference* (Washington, D.C., April 7–9, 2003).
14. Elliot, T. C. (editor), *Standard Handbook of Powerplant Engineering* (McGraw-Hill, New York, 1989).
15. Lawn, C. J. (editor), *Principles of Combustion Engineering for Boilers* (Academic Press, London, England, 1987).
16. Tillman, D. A., B. G. Miller, D. K. Johnson, and D. J. Clifford, Structure, Reactivity, and Nitrogen Evolution Characteristics of a Suite of Coals and Solid Fuels, in *Proc. of the 29th International Technical Conference on Coal Utilization and Fuel Systems* (Coal Technology Association, Gaithersburg, MD, 2004).
17. Howard, J. B., and R. H. Essenhigh, Simultaneous Gas Phase Volatiles Combustion, in *Proc. of the 11th Int. Symp. on Combustion* (The Combustion Institute, Pittsburgh, PA, 1967), pp. 399–408.
18. Field, M. A., D. W. Gill, B. B. Morgan, and P. G. W. Hawksley, *Combustion of Pulverized Coal*, The British Coal Utilisation Research Association, Cheney & Sons Ltd, Leatherhead, England, 1967.
19. Berkowitz, N., *An Introduction to Coal Technology* (Academic Press, New York, 1979).
20. Walker, P. L., F. Rusinko, and L. G. Austin, *Gas Reactions of Carbon, Advances in Catalysis and Related Subjects*, Vol. XI (Academic Press, New York), 1959.
21. Anon., Power from Coal: Special Report, *Power*, February 1974.
22. Schobert, H. H., *Coal: The Energy Source of the Past and Future* (American Chemical Society, Washington, D.C., 1987).

23. DOE, Fluidized-Bed Combustion: An R&D Success Story (U.S. Department of Energy, Washington, D.C.), www.fossil.energy.gov/programs/powersystems/combustion/fluidizedbed_success.shtml (accessed September 17, 2003).
24. Tang, J. T. and F. Engstrom, Technical Assessment on the Ahlstrom Pyroflow Circulating and Conventional Bubbling Fluidized-Bed Combustion Systems, in *Proc. of the International Conference on Fluidized-Bed Combustion*, p. 37 (May 3–7, 1987).
25. Virr, M. J., The Development of a Modular System To Burn Farm Animal Waste To Generate Heat and Power, in *Proc. of the Seventeenth Annual International Pittsburgh Coal Conference* (University of Pittsburgh, September 11–14, 2000).
26. Gaglia, B. N. and A. Hall, Comparison of Bubbling and Circulating Fluidized-Bed Industrial Steam Generation, in *Proc. of the International Conference on Fluidized-Bed Combustion*, p. 18 (May 3–7, 1987).
27. Bryers, R. W., Fireside Behavior of Minerals in Coal, paper presented at the Symposium on Slagging and Fouling in Steam Generators (January 31, 1987), 63 pp.
28. Kumar, K. S., R. E. Sommerland, and P. L. Feldman, Know the Impacts from Switching Coals for CAAA Compliance, *Power*, Vol. 135, No. 5, May 1991, pp. 31–38.
29. Winegartner, E. C., *Coal Fouling and Slagging Parameters, Research Department* (American Society of Mechanical Engineers, New York, 1974), 34 pp.
30. Attig, R. C. and A. F. Duzy, Coal Ash Deposition Studies and Application to Boiler Design, in *Proc. of the American Power Conference*, Vol. 31 (1969), pp. 290–300.
31. Bonson, S., MI Introduces New Indices to Assist Prediction of Ash Deposition, *Microbeam*, Vol. 14, Winter 2003, pp. 1–3.
32. EIA, *Quarterly Coal Report January–March 2003* (Energy Information Administration, U.S. Department of Energy, Washington, D.C., June 2003).
33. Wakelin, D. H. (editor), *The Making, Shaping and Treating of Steel*, 11th ed., Ironmaking Vol. (The AISE Steel Foundation, Pittsburgh, PA, 1999).
34. Ward, C. R. (editor), *Coal Geology and Coal Technology* (Blackwell Scientific, Melbourne, 1984).
35. U.K. Coal, *Section 270: Coking and Carbonisation Plant* (United Kingdom Coal), http://open.voa.gov.uk/Instructions/chapters/Rating/vol15/sect270/s270.htm (accessed September 26, 2003).
36. DOE, *Clean Coal Technology: Upgrading Low-Rank Coals*, Technical Report No. 10 (Office of Fossil Energy, U.S. Department of Energy, Washington, D.C., August 1997).
37. ENCOAL, *ENCOAL Mild Coal Gasification Project: ENCOAL Project Final Report*, prepared for U.S. Department of Energy, No. DE-FC21-90MC27339 (ENCOAL Corp., Gillette, WY, September 1997).
38. DOE, *Clean Coal Technology Demonstration Program: Program Update 2001 as of September 2001* (Office of Fossil Energy, U.S. Department of Energy, Washington, D.C., July 2002).

39. Coal Merchants, Listing of Economy Fuels, www.coalmerchantsfederation.co.uk/Authorised%20Fuels.htm, accessed September 26, 2003.
40. Coalite Smokeless Fuels, *The Process* (Coalite, Chesterfield, U.K.), www.coalite.co.uk/page4.html (accessed September 26, 2003).
41. Frederick, J. P. and B. A. Knottnerus, Role of the Liquids From Coal Process in the World Energy Picture, paper presented at the Fifth Annual Clean Coal Technology Conference, Tampa, Florida (January 8, 1997).
42. DOE, *ENCOAL Mild Coal Gasification Project: Project Fact Sheet* (U.S. Department of Energy, Washington, D.C.), www.lanl.gov/projects/cctc/ factsheets/encol/encoaldemo.html (last modified January 13, 2003).
43. DOE, *Advanced Coal Conversion Process Demonstration: Project Fact Sheet* (U.S. Department of Energy, Washington, D.C.), www.lanl.gov/projects/cctc/factsheets/rsbud/adcconvdemo.html (last modified December 2, 2002).
44. Ratafia-Brown, J., L. Manfredo, J. Hoffman, and M. Ramezan, *Major Environmental Aspects of Gasification-Based Power Generation Technologies* (Office of Fossil Energy, Washington, D.C., December 2002).
45. SFA Pacific, Inc., and DOE, *Gasification: Worldwide Use and Acceptance* (Office of Fossil Energy, U.S. Department of Energy, Washington, D.C., January 2000).
46. Gasification Technologies Council, *Gasification: A Growing, Worldwide Industry* (Gasification Technologies Council), Arlington, Virginia, www.gasification.org/ (accessed October 2, 2003).
47. Wu, W. R. K. and H. H. Storch, Hydrogenation of Coal and Tar, *U.S. Bureau of Mines Bulletin*, No. 633, 1968, pp. 1–10.
48. Stranges, A. N., Germany's Synthetic Fuel Industry 1927–1945, *Energia*, Vol. 12, No. 5, 2001, p. 2.
49. Davis, B., Fischer-Tropsch Synthesis: The CAER Perspective, *Energia*, Vol. 8, No. 3, 1997, p. 1.
50. Miller B. G., Autoclave Studies of Lignite Liquefaction (M.S. Thesis, University of North Dakota, 1982).
51. HTI, Coal Liquefaction Technology (Hydrocarbon Technologies, Inc., Lawrenceville, NJ), www.htinj.com/products/liquefaction.html (accessed April 2004).
52. HTI, *Headwater's HTI Subsidiary Signs License Agreement with Shenhua Group—China's Largest Coal Company* (Hydrocarbon Technologies, Inc., Lawrenceville, NJ), www.htinj.com/news/061802shenhualicagmnt.html (accessed June 2002).

CHAPTER 6

Emissions Control Strategies for Power Plants

For more than the last quarter century, power plant operators in the United States have been installing new pollution-control technologies to meet ever-tightening regulatory standards for clean air. The Clean Air Act of 1970 (details of which can be found in Chapter 4, where the history of legislative action in the United States is discussed) established national standards to limit levels of air pollutants such as sulfur dioxide, nitrogen oxides, carbon monoxide, ozone, lead, and particulate matter. The act and its amendments in 1977 resulted in the development and installation of particulate matter and sulfur dioxide control technologies for coal-fired boilers. Particulate control devices, specifically electrostatic precipitators (ESPs) and fabric filter baghouses, began to be installed on power plants, and efforts to develop new control technology, including flue gas desulfurization units, commonly called scrubbers, to remove sulfur from flue gas led to the installation of such units on many power generation facilities. In addition, technologies to reduce nitrogen oxides began to be developed.

The 1990 Clean Air Act Amendments contained major revisions to the Clean Air Act and required further reductions in power plant emissions, especially sulfur- and nitrogen-containing pollutants that contribute to acid rain. Some of the resulting sulfur dioxide control strategies that were implemented include switching to low-sulfur fuels, installation of flue gas desulfurization units, and the development of a new market-based cap-and-trade system that requires power plants to either reduce their emissions or acquire allowances from other companies to achieve compliance. To meet the more stringent nitrogen oxide standards from the 1990 Clean Air Act Amendments, several technologies, specifically low-NO_X burners, selective catalytic and noncatalytic reduction, cofiring, and reburning have been developed with varying levels of their implementation occurring. With impending legislation (*e.g.*, Clear Skies Act, Multipollutant Control, fine particulate control) additional NO_X control is anticipated and is being planned for by the power generating industry.

This chapter begins by summarizing the progress that has been made over the last ~30 years in reducing emissions from coal-fired power plants. Commercial control strategies for pollutants that are currently regulated, such as sulfur dioxide, nitrogen oxides, and particulate matter, are discussed. Control technologies that are under development for reducing mercury emissions, where regulations will be promulgated in 2004 and full compliance expected by 2007, are also discussed. A summary of control options under development for carbon dioxide removal is provided. The chapter concludes with a discussion of multi-pollutant control technologies.

Currently Regulated Emissions

The pollutants of primary interest and currently regulated in the power generation industry include sulfur dioxide, nitrogen oxides, and particulate matter. Although pollutants such as carbon monoxide and volatile organic compounds (which lead to the production of ozone when reacted with nitrogen oxides) are important and are tracked nationally, no specific control technologies exist for these pollutants as they are formed from incomplete combustion in the boiler. Power plants are minor contributors to nationwide carbon monoxide emissions. Similarly, power plants are insignificant contributors to nationwide hydrocarbon emissions. Carbon monoxide and hydrocarbon emissions are discussed in Chapter 4 (Coal-Fired Emissions and Legislative Action in the United States).

Emissions of sulfur dioxide, nitrogen oxides, and particulate matter in the United States have been dramatically reduced over the last ~30 years due to legislative mandate and technological advances achieved through federal, state, and industrial efforts. This decline is illustrated in Figures 6-1 and 6-2, which show the decreases in overall as well as individual emissions despite increases in the use of coal for power generation. The steady decrease in emission rates is clearly shown in Figure 6-1, where the emission rates (given on a per billion kWh basis) are reported in 5-year increments [1]. The decrease in the emission rates of these three pollutants is also shown in Figure 6-2, along with projected near-term emission rates [2]. The emission rates in Figure 6-2 are compared with the coal use for power generation for the same time periods and illustrate how pollutant emissions per unit of coal burned have decreased significantly while at the same time coal use has increased. The technologies that are being used to achieve these reductions are discussed in the following sections.

Sulfur Dioxide (SO_2)

Sulfur dioxide is one of the most abundant air pollutants emitted in the United States, totaling about 16 million short tons in 2002. Of the total

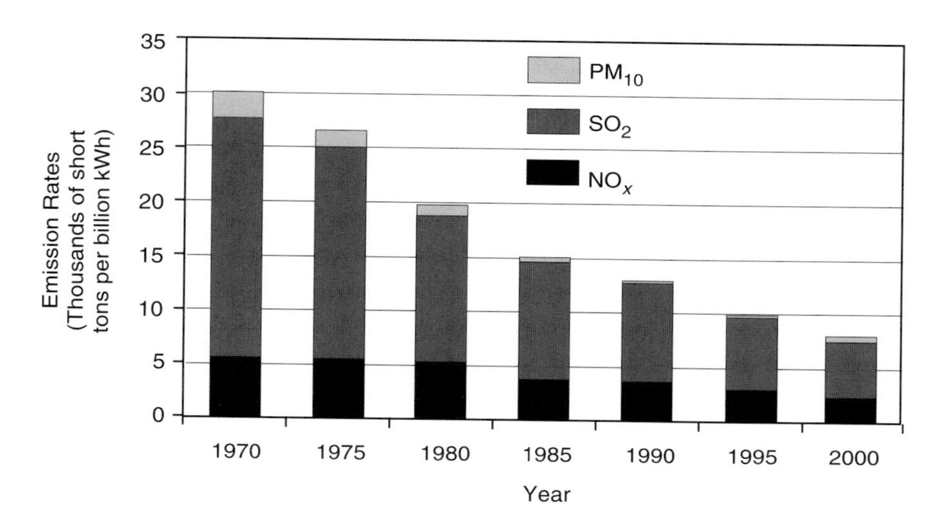

FIGURE 6-1. Emission rates of sulfur dioxide, nitrogen oxides, and particulate matter from coal-fired power plants for the period 1970 to the present. (From DOE, *National Energy Technology Laboratory Accomplishments FY 2002*, Office of Fossil Energy, U.S. Department of Energy, Washington, D.C., August 2003.)

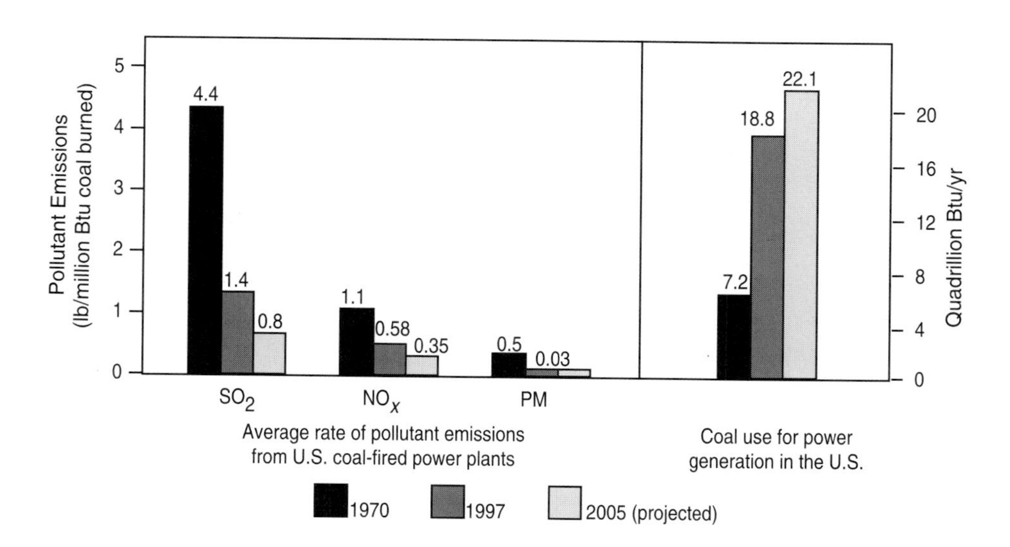

FIGURE 6-2. Past, present, and future emission rates of sulfur dioxide, nitrogen oxides, and particulate matter from coal-fired power plants. (From DOE, *National Energy Technology Laboratory Accomplishments FY 2002*, Office of Fossil Energy, U.S. Department of Energy, Washington, D.C., August 2003.)

anthropogenic emissions, fossil fuel combustion accounted for approximately 63% (*i.e.*, 10 million short tons). Sulfur dioxide emissions have decreased 33 and 31% for the periods from 1983 to 2002 and from 1993 to 2002, respectively [1]. Reductions in SO_2 emissions and concentrations since 1990 are primarily due to controls implemented under the U.S. Environmental Protection Agency's (EPA's) Acid Rain Program beginning in 1995. As of 2000, 192 coal-fired electric generators were equipped with scrubbers and provided a total of nearly 90,000 MW generating capacity [3]. It must be noted, however, that there is variability in reported generating capacity under SO_2 control among the various agencies, so the quantity ranges from ~90,000 to 102,000 MW [3–5]. The chemistry of sulfur dioxide formation is reviewed in this section, followed by technologies used to control SO_2 emissions. Control technologies will focus on commercially available and commercially used systems. Industry deployment of the SO_2 removal process worldwide is discussed as are the economics of flue gas desulfurization.

Chemistry of Sulfur Oxide (SO_2/SO_3) Formation

Sulfur in coal occurs in three forms: as pyrite, organically bound to the coal, or as sulfates. The sulfates represent a very small fraction of the total sulfur while pyritic and organically-bound sulfur comprise the majority. The distribution between pyritic and organic sulfur is variable with up to approximately 40% of the sulfur being pyritic. During combustion, the pyritic and organically bound sulfur are oxidized to sulfur dioxide with a small amount of sulfur trioxide (SO_3) being formed. The SO_2/SO_3 ratio is typically 40:1 to 80:1 [6].

The overall reaction for the formation of sulfur dioxide is:

$$S + O_2 \longrightarrow SO_2 \quad \Delta H_f = 128{,}560 \text{ Btu/lb mole} \tag{6-1}$$

and the overall reaction for the formation of sulfur trioxide is:

$$SO_2 + \tfrac{1}{2}O_2 \longleftrightarrow SO_3 \quad \Delta H_f = 170{,}440 \text{ Btu/lb mole} \tag{6-2}$$

It is proposed that sulfur monoxide, SO, is formed early in the reaction zone from sulfur-containing molecules and is an important intermediate product [6]. The major SO_2 formation reactions are believed to be:

$$SO + O_2 \longrightarrow SO_2 + O \tag{6-3}$$

and

$$SO + OH \longrightarrow SO_2 + H \tag{6-4}$$

with the highly reactive O and H atoms possibly entering the reaction scheme later.

The reactions involving SO_3 are reversible. The major formation reaction for SO_3 is the three-body process:

$$SO_2 + O + M \longrightarrow SO_3 + M \tag{6-5}$$

where M is a third body that is an energy absorber [6]. The major steps for removal of SO_3 are thought to be the following:

$$SO_3 + O \longrightarrow SO_2 + O_2 \tag{6-6}$$

$$SO_3 + H \longrightarrow SO_2 + OH \tag{6-7}$$

$$SO_3 + M \longrightarrow SO_2 + O + M \tag{6-8}$$

Sulfur Dioxide Control

Methods to control sulfur dioxide emissions from coal-fired power plants include switching to a lower sulfur fuel, cleaning the coal to remove the sulfur-bearing components such as pyrite, or installing flue gas desulfurization systems. In the past, building tall stacks to disperse the pollutants was a control method; however, this practice is no longer an alternative, as tall stacks do not remove the pollutants; they only dilute the concentrations to reduce the ground-level emissions to acceptable levels.

When fuel switching or coal cleaning is not an option, flue gas desulfurization (FGD) is selected to control sulfur dioxide emissions from coal-fired power plants (except for fluidized-bed combustion systems, which are discussed later in this chapter). FGD has been in commercial use since the early 1970s and has become the most widely used technique to control sulfur dioxide emissions next to the firing of low-sulfur coal. Many FGD systems are currently in use and others are under development. This section summarizes the worldwide application of FGD systems with an emphasis on the United States. FGD processes are generally classified as wet scrubbers or dry scrubbers but can also be categorized as follows [4]:

- Wet scrubbers;
- Spray dryers;
- Dry (sorbent) injection processes;
- Regenerable processes;
- Circulating fluid-bed and moving-bed scrubbers;
- Combined SO_2/NO_x removal systems.

Based on the nature of the waste/by-product generated, a commercially available throwaway FGD technology may be categorized as wet or dry. A wet FGD process produces a slurry waste or a salable slurry by-product. A dry FGD process application results in a solid waste, the transport and disposal of which is easier compared to the waste/by-product from wet FGD applications. Regenerable FGD processes produce a concentrated SO_2

by-product, usually sulfuric acid or elemental sulfur. The recent focus on mercury removal in FGD systems is discussed later in this chapter.

Worldwide Deployment of FGD Systems Post-combustion control of sulfur dioxide emissions from pulverized coal combustion began in the early 1970s in the United States and Japan. Western Europe followed in the 1980s. In the 1990s, the application of FGD became more widespread, and countries in Central and Eastern Europe and Asia, for example, have installed FGD systems. Table 6-1 lists various control technologies and the amount of electricity generation that is being controlled in countries throughout the world [4]. According to Soud [4], as of 1999, 680 FGD systems were installed in 27 countries, and 140 systems are currently under construction or planned in nine countries.

Worldwide, approximately 30,000 MW of generating capacity were controlled in 1980 compared to no controlled capacity in 1970. Controlled generating capacity subsequently increased to ~130,000 MW in 1990 and ~230,000 MW in 2000. In the United States, controlled capacity rose from zero in 1970 to 25,000 MW in 1980, to ~75,000 MW in 1990, and to ~100,000 MW in 2000. FGD systems were installed (as of 1999) to control sulfur dioxide emissions from over 229,000 MW of generating capacity worldwide. Of these systems, ~87% consisted of wet FGD technology and 11% dry FGD technology, with the balance consisting of regenerable technology [5]. Of the worldwide capacity controlled with FGD technology, ~44% is in the United States alone, as shown in Table 6-2. In the United States, ~100,000 MW of capacity were equipped with FGD technology in 1999. Of these FGD systems, approximately 83, 14, and 3% consisted of wet FGD, dry FGD, and regenerable technology, respectively. Worldwide, out of the 668 units equipped with FGD in 1999, 522 were equipped with wet FGD, 124 with dry FGD, and 22 with regenerable FGD.

Of the U.S. wet FGD technology population today, 69% are limestone processes [5]. Abroad, limestone processes comprise as much as 93% of the total wet FGD technology installed. Of the worldwide capacity equipped with dry FGD technology, 74% use spray drying processes. This compares with 80% for spray drying processes in the United States. A summary of the FGD systems in the United States, by process, is given in Table 6-3 for 1989 (actual capacity) and 2010 (projected capacity). The three primary processes are throwaway-product systems, including the two wet scrubbing systems using limestone and lime where a synthetic gypsum ($CaSO_4$) is produced. A lack of commercial markets for the gypsum results in this material being disposed of rather than utilized. Characteristics of these processes are provided in the next section. A variety of FGD processes exist, and the selection of a system is dependent upon site-specific consideration, economics, and other criteria. Elliot [8] provides a ranking of various FGD processes used in the United States in Tables 6-4 and 6-5 where cost, performance,

TABLE 6-1
Existing and Future FGD Systems

Country	*Wet Lime/ Limestone/ Gypsum (MW)*	*Wet Lime/ Limestone/ Other (MW)*	*Spray Dry Scrubbers (MW)*	*Sorbent Injection (MW)*	*CFB and Moving-Bed Scrubbers (MW)*	*Regenerable Systems (MW)*	*Combined SO_2/NO_x Removal (MW)*
Existing							
Austria	825	—	835	—	260	—	—
Canada	1495	—	—	600	—	—	—
China	2485	—	510	300	—	—	325
Czech Republic	2230	—	—	—	—	—	—
Denmark	2260	—	1200	—	—	—	305
Finland	1875	—	530	250	—	—	—
France	1800	—	—	600	—	—	—
Germany	43,670	445	2245	25	295	885	400
Greece	300	—	—	—	—	—	—
Italy	5420	—	—	75	—	—	—
India	—	500	—	—	—	—	—
Japan	21,725	365	—	350	185	—	—
Korea, Republic	7500	—	—	—	—	—	—
Netherlands	3915	—	—	—	—	—	—
Norway	—	30	—	—	—	—	—
Poland	7390	50	315	1720	—	—	—
Russian Federation	510	—	—	—	—	—	—
Slovakia	330	—	—	—	—	—	—
Slovenia	275	—	—	—	—	—	—
Spain	1930	—	385	—	—	—	—

(continued)

TABLE 6-1
(continued)

Country	*Wet Lime/ Limestone/ Gypsum (MW)*	*Wet Lime/ Limestone/ Other (MW)*	*Spray Dry Scrubbers (MW)*	*Sorbent Injection (MW)*	*CFB and Moving-Bed Scrubbers (MW)*	*Regenerable Systems (MW)*	*Combined SO_2/NO_x Removal (MW)*
Sweden	—	—	360	130	—	—	—
Taiwan	7100	—	—	—	—	—	—
Thailand	2100	300	—	—	—	—	—
Turkey	3355	—	—	—	—	—	—
Ukraine	150	—	—	—	—	—	—
United Kingdom	5960	—	—	—	—	—	—
United States	21,680	62,475	12,000	970	80	2870	1845
Future							
China	720	—	—	200	—	—	—
Denmark	480	—	—	—	—	—	—
Germany	5035	—	—	—	—	—	—
Israel	1100	—	—	—	—	—	—
Japan	8300	—	—	—	—	—	—
Korea, Republic	2000	—	—	—	—	—	—
Sri Lanka	—	300	—	—	—	—	—
Turkey	470	—	—	—	—	—	—
United States	801	1780	300	520	—	—	—

Source: From Soud, H. N., *Developments in FGD*, IEA Coal Research, London, 2000. With permission.

TABLE 6-2
Worldwide Electrical Generating Capacity with Flue Gas Desulfurization Technology

Technology	*United States (MW)*	*Abroad (MW)*	*Total (MW)*
Wet	82,859	116,374	199,233
Dry	14,386	11,008	25,394
Regenerable	2798	2059	4857
Total	100,043	129,441	229,484

Source: [5].

and flexibility of application are assessed. Wet limestone systems have been installed across the United States at plants of all sizes firing all ranks of coal with sulfur contents varying from low to high. Wet lime systems have been installed at power plants of all sizes firing both low- and high-sulfur coals with plants predominantly in the Ohio River valley [8]. Some plants in the West use wet lime systems, where the cost of lime delivered to the plant is less than limestone. Some plants firing high-sulfur coal in the Midwest have selected wet sodium-based dual-alkali systems. Dry scrubbing systems have typically been selected at power plants firing low-sulfur coals. Generally, dry scrubbing systems are considered more economical for power plants firing low-sulfur coal, while wet-based systems are selected for high-sulfur coal applications.

Techniques to Reduce Sulfur Dioxide Emissions

The primary methods used to control sulfur dioxide emissions from coal-fired power plants are to switch to a lower-sulfur fuel or install flue gas desulfurization systems and, to a lesser extent, clean coal to remove the sulfur-bearing components. These techniques are discussed in this section with an emphasis on flue gas desulfurization technologies.

Using Low-Sulfur Fuels One option for reducing sulfur dioxide emissions is to switch to fuels containing less sulfur. Fuel switching includes using natural gas, liquefied natural gas, low-sulfur fuel oils, or low-sulfur coals in place of high-sulfur coals. In coal-fired boilers, switching from a high-sulfur coal to lower sulfur non-coal fuels may make sense from both an economic and technological standpoint for smaller sized industrial and utility boilers; however, the practice of switching power generation units from coal to natural gas is a questionable one. While this option may make good business sense (at least at the time), it is neither good energy policy nor advisable energy security to use a premium fuel for power generation; this topic will be discussed in more detail in Chapter 8 (Coal's Role in Providing U.S. Energy Security).

TABLE 6-3
Summary of Flue Gas Desulfurization Processes in the United States

Throwaway Product	Active Material (for Scrubbing)	By-Product	% of MW (1989)	% of MW (2010)
Wet scrubbing				
Dual alkali	Na_2SO_3 solution regenerated by CaO or $CaCO_3$	$CaSO_3/CaSO_4$	3.4	2.3
Lime	$Ca(OH)_2$ slurry	$CaSO_3/CaSO_4$	16.3	13.5
Lime/alkaline fly ash	$Ca(OH)_2$/fly ash slurry	$CaSO_3/CaSO_4$	7.0	4.9
Limestone	$CaCO_3$ slurry	$CaSO_3/CaSO_4$	48.2	43.9
Limestone/alkaline fly ash	$CaCO_3$/fly ash slurry	$CaSO_3/CaSO_4$	2.4	1.6
Sodium carbonate	Na_2CO_3/Na_2SO_4 slurry	Na_2SO_4	4.0	3.3
Spray drying				
Lime	Slaked $Ca(OH)_2$ slurry	$CaSO_3/CaSO_4$	8.8	7.9
Sodium carbonate	Na_2CO_3	Na_2SO_4	4.0	3.3
Reagent type not selected	Undecided	—	0.7	2.1
Dry injection				
Lime	$Ca(OH)_2$ (dry)	$CaSO_3/CaSO_4$	0.2	0.1
Sodium carbonate	Na_2CO_3	Na_2SO_4	0	0.2
Reagent type not selected	Undecided	—	0	2.2
Process not selected	—	—	0	2.2
Salable Product	***Active Material (for Scrubbing)***	***By-Product***	***% of MW (1989)***	***% of MW (2010)***
Wet scrubbing				
Lime	$Ca(OH)_2$ slurry	$CaSO_4$	<0.1	<0.1
Limestone	$CaCO_3$ slurry	$CaSO_4$	4.1	4.6
Magnesium oxide	Mg(OH) slurry	Sulfuric acid	1.4	1.0
Wellman Lord	Na_2SO_3 solution	Sulfuric acid	3.1	2.1
Spray drying				
Lime	Slaked $Ca(OH)_2$ slurry	Dry scrubber waste	0	0.3
Process undecided	—	—	0	7.8

Source: Adapted from Wark *et al.* [6] and Davis [7].

Fuel switching to lower sulfur coals is chosen by many power generators to achieve emissions compliance. In the United States, the replacement of high-sulfur Eastern or Midwestern bituminous coals with lower sulfur Appalachian region bituminous coals or Powder River Basin coals is a control option that is widely exercised, as illustrated in Chapter 2 (Past, Present,

TABLE 6-4
Assessment Relative to Cost and Performance

FGD Process	*Criterion*				
	Operating Cost	*Capital Cost*	*SO_2 Removal*	*Reliability*	*Commercial Use*
Limestone					
Natural oxidation	M	M	M	M	H
Forced oxidation	M	M	M	M	H
MgO-lime	M	M	H	H	H
High-calcium lime	M	M	M	M	M
Dual-alkali					
Lime	M	M	H	H	M
Limestone	L	M	H	—	—
Dry scrubbing	H	M	L	H	H
Dry injection	M	L	L	—	—
Wellman–Lord	H	H	H	M	M
Regenerable MgO	M	H	H	M	L

Note: H, high; M, medium; L, low.
Source: Elliot, T. C., Ed., *Standard Handbook of Powerplant Engineering*, McGraw-Hill, New York, 1989. With permission.

TABLE 6-5
Assessment with Respect to Flexibility of Application

FGD Process	*Criterion*				
	High Sulfur	*Low Sulfur*	*Retrofit Ease*	*Waste Management*	*SO_2/NO_x Removal*
Limestone					
Natural oxidation	H	H	M	L	L
Forced oxidation	H	M	M	L	L
MgO-lime	H	L	M	L	M
High-calcium lime	H	M	M	L	L
Dual-alkali					
Lime	H	L	M	L	M
Limestone	H	L	M	L	L
Dry scrubbing	M	H	M	M	M
Dry injection	L	H	H	L	L
Wellman–Lord	H	L	L	M	M
Regenerable MgO	H	M	M	H	M

Note: H, high; M, medium; L, low.
Source: Elliot, T. C., Ed., *Standard Handbook of Powerplant Engineering*, McGraw-Hill, New York, 1989. With permission.

TABLE 6-6
Estimated Recoverable Coal Reserves in the United States by Sulfur Range and Major Coal-Producing Region (as of January 1, 1997)

Coal-Producing Region	Sulfur Content Categories (Pounds of Sulfur per Million Btu)							
	Low Sulfur (≤0.60)		Medium Sulfur (0.61–1.67)		High Sulfur (≥1.68)		Total	
	Million Short Tons	Percent of Total	Million Short Tons	Percent of Total	Million Short Tons	Percent of Total	Million Short Tons	Percent of Total
Appalachia	11,675	11.6	20,337	24.0	23,283	25.9	55,295	20.1
Interior	769	0.8	10,041	11.8	57,966	64.4	68,776	25.0
Western	87,775	87.6	54,529	64.2	8768	9.7	151,072	54.9
Total	100,219	—	84,907	—	90,017	—	275,143	—

Source: U.S. Energy Information Administration, *U.S. Coal Reserves: 1997 Update,* U.S. Department of Energy, Office of Coal, Nuclear, Electric and Alternate Fuels, U.S. Government Printing Office, Washington, D.C., February 1999.

and Future Role of Coal), where coal production by region is discussed. The option of using lower sulfur coal has resulted in a large increase in western coal production and use. Table 6-6 illustrates the distribution of sulfur in U.S. coals by region [9].

The relationship between sulfur content in the coal and pounds of sulfur per million Btu is provided in Table 6-7 for comparison. This listing, developed by the U.S. Department of Energy's (DOE's) Energy Information Administration (EIA), is used for approximate correlations with New Source Performance Standards (NSPS) and 1990 Clean Air Act Amendments criteria. With the exception of the low-sulfur coal, which meets NSPS requirements, the medium- and high-sulfur coals require control strategies. This includes emission reduction technologies or offsets through sulfur dioxide allowances.

TABLE 6-7
Comparison of Sulfur Content in Coal with Pounds of Sulfur per Million Btu

Qualitative Rating	Pounds of Sulfur per Million Btu	Approximate Range of Coal Sulfur Content (%)	
		High-Grade Bituminous Coal	High-Grade Lignite
Low sulfur	≤0.4 to 0.6	≤0.5 to 0.8	≤0.3 to 0.5
Medium sulfur	0.61 to 1.67	0.8 to 2.2	0.5 to 1.3
High sulfur	1.68 to >2.50	2.2 to >3.3	1.3 to >1.9

Source: U.S. Energy Information Administration, *U.S. Coal Reserves: 1997 Update,* U.S. Department of Energy, Office of Coal, Nuclear, Electric and Alternate Fuels, U.S. Government Printing Office, Washington, D.C., February 1999.

Low-sulfur coals are also imported to the United States, specifically to coastal areas such as Florida or the Eastern seaboard. Similar to the replacement of high-sulfur coals with non-coal fuels in power generation units, this practice of importing lower sulfur coals to the United States for sulfur dioxide compliance also needs to be questioned from the standpoint of energy security.

Coal Cleaning Coal preparation, or beneficiation, is a series of operations that remove mineral matter (*i.e.*, ash) from coal. Preparation relies on different mechanical operations (not discussed in detail here) to perform the separation, such as size reduction, size classification, cleaning, dewatering and drying, waste disposal, and pollution control. Coal preparation processes, which are physical processes, are designed mainly to provide ash removal, energy enhancement, and product standardization [8]. Sulfur reduction is achieved because the ash material removed contains pyritic sulfur. Coal cleaning is used for moderate sulfur dioxide emissions control, as physical coal cleaning is not effective in removing organically bound sulfur. Chemical coal cleaning processes are being developed to remove the organic sulfur; however, these are not used on a commercial scale. An added benefit of coal cleaning is that several trace elements, including antimony, arsenic, cobalt, mercury, and selenium, are generally associated with pyritic sulfur in raw coal and they, too, are reduced through the cleaning process. As the inert material is removed, the volatile matter content, fixed carbon content, and heating value increase, thereby producing a higher quality coal. The moisture content, a result of residual water from the cleaning process, can also increase, which lowers the heating value, but this reduction is usually minimal and has little impact on coal quality. Coal cleaning does add additional cost to the coal price; however, among the several benefits of reducing the ash content are lower sulfur content; less ash to be disposed of; lower transportation costs, as more carbon and less ash is transported (coal cleaning is usually done at the mine and not the power plant); and increases in power plant peaking capacity, rated capacity, and availability [10]. Developing circumstances are making coal cleaning more economical and a potential sulfur control technology and include [8]:

- Higher coal prices and transportation costs;
- Diminishing coal quality because of less selective mining techniques;
- The need to increase availability and capacity factors at existing boilers;
- More stringent air quality standards;
- Lower costs for improving fuel quality versus investing in extra pollution control equipment.

Wet Flue Gas Desulfurization Wet scrubbers are the most common FGD method currently in use (or under development); they include a variety of processes and involve the use of many sorbents and are manufactured by a large number of companies. The sorbents used by wet scrubbers include calcium-, magnesium-, potassium-, or sodium-based sorbents, ammonia, or seawater. Currently, no commercial potassium-based scrubbers are in use, and only a limited number of ammonia or seawater systems are in use or being demonstrated. The calcium-based scrubbers are by far the most popular, and this technology is discussed in this section along with the use of sodium- and magnesium-based sorbents.

Limestone- and Lime-Based Scrubbers Wet scrubbing with limestone or lime is the most popular commercial FGD system. The inherent simplicity, the availability of an inexpensive sorbent (limestone), production of a usable by-product (gypsum), reliability, availability, and the high removal efficiencies obtained (which can be as high as 99%) are the main reasons for the popularity of this system. Capital costs are typically higher than other technologies, such as sorbent injection systems; however, the technology is known for its low operating costs as the sorbent is widely available and the system is cost effective.

In a limestone/lime wet scrubber, the flue gas is scrubbed with a 5 to 15% (by weight) slurry of calcium sulfite/sulfate salts along with calcium hydroxide ($Ca(OH_2)$) or limestone ($CaCO_3$). Calcium hydroxide is formed by slaking lime (CaO) in water according to the reaction:

$$CaO(s) + H_2O(l) \longrightarrow Ca(OH)_2(s) + \text{heat} \qquad (6\text{-}9)$$

In the limestone and lime wet scrubbers, the slurry containing the sulfite/sulfate salts and the newly added limestone or calcium hydroxide is pumped to a spray tower absorber and sprayed into it. The sulfur dioxide is absorbed into the droplets of slurry and a series of reactions occur in the slurry. The reactions between the calcium and the absorbed sulfur dioxide create the compounds calcium sulfite hemihydrate ($CaCO_3 \cdot \frac{1}{2}H_2O$) and calcium sulfate dihydrate ($CaSO_4 \cdot 2H_2O$). Both of these compounds have low solubility in water and precipitate from the solution. This enhances the absorption of sulfur dioxide and further dissolution of the limestone or hydrated lime.

The reactions occurring in the scrubbers are complex. The simplified overall reaction for a limestone scrubber is:

$$SO_2(g) + CaCO_3(s) + \tfrac{1}{2}H_2O(l) \longrightarrow CaSO_3 \cdot \tfrac{1}{2}H_2O(s) + CO_2(g) \qquad (6\text{-}10)$$

and the reaction for a lime scrubber is:

$$SO_2(g) + Ca(OH)_2(s) + H_2O(l) \longrightarrow CaSO_3 \cdot \tfrac{1}{2}H_2O(s) + \tfrac{3}{2}H_2O(l) \qquad (6\text{-}11)$$

The calcium sulfite hemihydrate can be converted to the calcium sulfate dihydrate with the addition of oxygen by the reaction:

$$CaSO_3 \cdot \tfrac{1}{2}H_2O(s) + \tfrac{3}{2}H_2O(l) + \tfrac{1}{2}O_2(g) \longleftrightarrow CaSO_4 \cdot 2H_2O(s) \qquad (6\text{-}12)$$

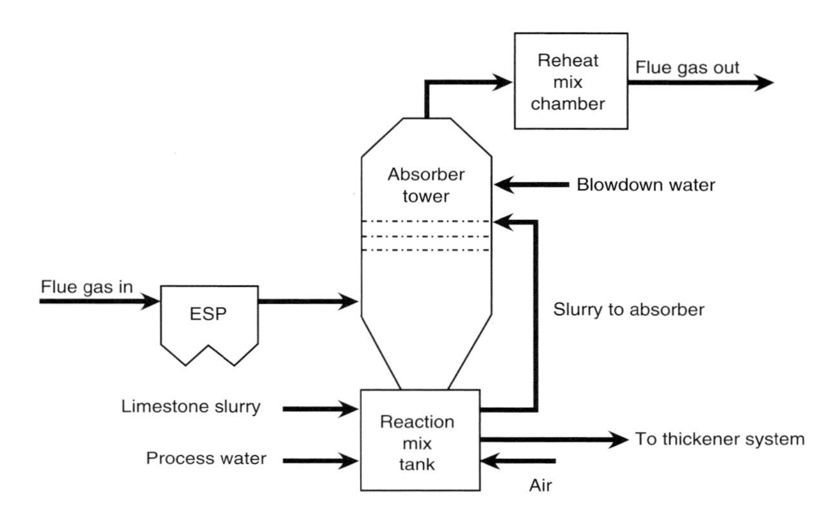

FIGURE 6-3. Limestone scrubber system with forced oxidation.

The actual reactions that occur, however, are much more complex and include a combination of gas–liquid, solid–liquid, and liquid–liquid ionic reactions. In the limestone scrubber, the following reactions describe the process [11]. In the gas–liquid contact zone of the absorber (see Figure 6-3 for a typical schematic diagram of a limestone scrubber system), sulfur dioxide dissolves into the aqueous state:

$$SO_2(g) \longleftrightarrow SO_2(l) \tag{6-13}$$

and is hydrolyzed to form ions of hydrogen and bisulfite:

$$SO_2(l) + H_2O(l) \longleftrightarrow HSO_3^- + H^+ \tag{6-14}$$

The limestone dissolves in the absorber liquid forming ions of calcium and bicarbonate:

$$CaCO_3(s) + H^+ \longleftrightarrow Ca^{++} + HCO_3^- \tag{6-15}$$

which is followed by acid–base neutralization:

$$HCO_3^- + H^+ \longleftrightarrow CO_2(l) + H_2O(l) \tag{6-16}$$

stripping of the CO_2 from the slurry:

$$CO_2(l) \longleftrightarrow CO_2(g) \tag{6-17}$$

and dissolution of the calcium sulfite hemihydrate:

$$CaSO_3 \cdot \tfrac{1}{2}H_2O(s) \longleftrightarrow Ca^{++} + HSO_3^- + \tfrac{1}{2}H_2O(l) \tag{6-18}$$

In the reaction tank of a scrubber system, the solid limestone is dissolved into the aqueous state (Reaction (6-15)), acid-base neutralization occurs (Reaction (6-16)), the CO_2 is stripped out (Reaction (6-17)), and the calcium sulfite hemihydrate is precipitated by the reaction:

$$Ca^{++} + HSO_3^- + \tfrac{1}{2}H_2O(l) \longleftrightarrow CaSO_3 \cdot \tfrac{1}{2}H_2O(s) + H^+ \qquad (6\text{-}19)$$

The dissolution of the calcium sulfite in the gas–liquid contact zone in the absorber is necessary in order to minimize scaling of the calcium sulfite hemihydrate in the absorber [6]. The equilibrium pH for calcium sulfite is $\sim$6.3 at a CO_2 partial pressure of 0.12 atm, which is the typical concentration of CO_2 in flue gas. Typically, the pH is maintained below this level to keep the calcium sulfite hemihydrate from dissolving (*i.e.*, keep Reaction (6-18) from proceeding to the right). The slurry returning from the absorber to the reaction tank can have a pH as low as 3.5, which is increased to 5.2 to 6.2 by the addition of freshly prepared limestone slurry to the tank [6]. The pH in the reaction tank must be maintained at a pH that is less than the equilibrium pH of calcium carbonate in water, which is 7.8 at 77°F.

The reaction equations for the lime scrubber are similar to those for the limestone scrubber, with the exception that the following reactions are substituted for Reactions (6-15) and (6-16), respectively [11]:

$$Ca(OH)_2(s) + H^+ \longleftrightarrow CaOH^+ + H_2O(l) \qquad (6\text{-}20)$$

$$CaOH^+ + H^+ \longleftrightarrow Ca^{++} + H_2O(l) \qquad (6\text{-}21)$$

Limestone with Forced Oxidation (LSFO) Limestone scrubbing with forced oxidation (LSFO) is one of the most popular systems in the commercial market. A limestone slurry is used in an open spray tower with *in situ* oxidation to remove SO_2 and form a gypsum sludge. The major advantages of this process, relative to a conventional limestone FGD system (where the product is calcium sulfite rather than calcium sulfate (gypsum)), are easier dewatering of the sludge, more economical disposal of the scrubber product solids, and decreased scaling on the tower walls. LSFO is capable of greater than 90% SO_2 removal [12].

In the LSFO system, the hot flue gas exits the particulate control device, usually an ESP, and enters a spray tower where it comes into contact with a sprayed dilute limestone slurry. The SO_2 in the flue gas reacts with the limestone in the slurry via the reactions listed earlier to form the calcium sulfite hemihydrate. Compressed air is bubbled through the slurry, which causes this sulfite to be naturally oxidized and hydrated to form calcium sulfate dihydrate. The calcium sulfate can be first dewatered using a thickener or hydrocyclones then further dewatered using a rotary drum filter. The gypsum is then transported to a landfill for disposal. The formation of the calcium sulfate crystals in a recirculation tank slurry also helps to reduce the chance of scaling.

The absorbing reagent, limestone, is normally fed to the open spray tower in an aqueous slurry at a molar feed rate of 1.1 mol of $CaCO_3$ per mol of SO_2 removed. This process is capable of removing more than 90% of the SO_2 present in the inlet flue gas. The advantages of LSFO systems include [12]:

- The scaling potential is lower on lower internal surfaces due to the presence of gypsum seed crystals and reduced calcium sulfate saturation levels; this in turn provides greater reliability of the system;
- The gypsum product is filtered more easily than the calcium sulfite ($CaSO_3$) produced with conventional limestone systems;
- The chemical oxygen demand is lower in the final disposed product;
- The final product can be safely and easily disposed of in a landfill;
- The forced oxidation allows greater limestone utilization than in conventional systems;
- Costs of the raw material (limestone) used as an absorbent are lower;
- LSFO is an easier retrofit than natural oxidation systems because the process uses smaller dewatering equipment.

A disadvantage of this system is the high energy demand due to the relatively higher liquid-to-gas ratio necessary to achieve the required SO_2 removal efficiencies.

Limestone with Forced Oxidation Producing a Wallboard Gypsum By-Product In the limestone/wallboard (LS/WB) gypsum FGD process, a limestone slurry is used in an open spray tower to remove SO_2 from the flue gas. The flue gas enters the spray tower where the SO_2 reacts with the $CaCO_3$ in the slurry to form calcium sulfite. The calcium sulfite is then oxidized to calcium sulfate in the absorber recirculation tank. The calcium sulfate produced with this process is of a higher quality so that it may be used in wallboard manufacture.

There are a few differences with this process in order to achieve a higher quality gypsum. The LS/WB system uses horizontal belt filters to produce a drier product and provides sufficient cake washing to remove residual chlorides. Because the by-product is a higher quality, the use of the product handling system is replaced with by-product conveying and temporary storage equipment. Sulfuric acid addition is used in systems with an external oxidation tank. The acid is used to control the pH of the slurry and neutralizes unreacted $CaCO_3$.

The limestone feed rate in this process is 1.05 mol $CaCO_3$ per mol of SO_2 removed, which is slightly lower than the feed rate for the LSFO system [12]. Other advantages of this process are that the disposal area is kept to a minimum because most of the by-product is reusable. The gypsum can be sold to cement plants and agricultural users. Also, SO_2 removal is slightly enhanced because of the high sulfite-to-sulfate conversion.

There are some disadvantages to this process. Few full-scale operating systems actually produce quality gypsum in the United States. To produce quality gypsum, specific process control and tight operator attention are constantly needed to ensure that chemical impurities do not lead to off-specification gypsum. Another disadvantage is the inability to use cooling tower blowdown as system make-up water due to chloride limits in the gypsum by-product.

Limestone with Inhibited Oxidation In the limestone with inhibited oxidation process, the hot flue gas exits the particulate control device and enters an open spray tower, where it comes into contact with a dilute $CaCO_3$ slurry. This slurry contains thiosulfate ($Na_2S_2O_3$), which inhibits natural oxidation of the calcium sulfite. The calcium sulfite is formed from the reaction with SO_2 in the flue gas and the $CaCO_3$ slurry. The slurry absorbs the SO_2, then drains down to a recirculation tank below the tower. By inhibiting natural oxidation of the sulfite, gypsum scaling on process equipment is reduced along with gypsum relative saturation, which is reduced below 1.0. Thiosulfate is either added directly as $Na_2S_2O_3$ to the feed tank or is generated *in situ* by the addition of emulsified sulfur. In some cases, thiosulfate has the ability to increase the dissolution of the calcium carbonate and enlarge the size of the sulfite crystals to improve solids dewatering [12]. This process is capable of removing more than 90% of the SO_2 in the flue gas. The calcium sulfite slurry product is thickened, stabilized with fly ash and lime, and then sent to a landfill. The calcium carbonate feed rate is 1.10 mol Ca per mol of SO_2 removed. The effectiveness of thiosulfate is site specific because the amount of thiosulfate required to inhibit oxidation strongly depends on the chemistry and operating conditions of each FGD system. Variables such as saturation temperature, dissolved magnesium, chlorides, flue gas inlet SO_2 and O_2 concentrations, and slurry pH affect the thiosulfate effectiveness [12].

Thiosulfate has been shown to increase limestone utilization when added to the system. This occurs because the thiosulfate reduces the gypsum relative saturation level, which in turn reduces the level of calcium dissolved in the liquor. The dissolution rate is increased by lowering the calcium concentration in the slurry. Thiosulfate also improves the dewatering characteristics of the sulfite product. By preventing the high concentrations of sulfate, the thiosulfate allows the calcium sulfite to form larger, single crystals. This increases the settling velocity of the crystals and improves the filtering characteristics, which results in a higher solid content of dewatered product.

There are a few disadvantages of the process. The thiosulfate/sulfur reagent requires additional process equipment and storage facilities. Also, the reagent can cause corrosion of many stainless steels under scrubber conditions. Another disadvantage is that the thiosulfate is fairly temperature dependent, thus requiring the system to operate within a particular temperature range.

Magnesium-Enhanced Lime In the magnesium-enhanced lime (MagLime) process, the hot flue gas exits the particulate control device and enters a spray tower, where it comes into contact with a magnesium sulfite/lime slurry. Magnesium lime, such as thiosorbic lime (which contains 4–8% MgO), is fed to the open spray tower in an aqueous slurry at a molar feed rate of 1.1 mol CaO per mol of SO_2 removed. The SO_2 is absorbed by the reaction with magnesium sulfite, forming magnesium bisulfite. This occurs through the following reactions [12]:

$$SO_2(g) + H_2O(l) \longrightarrow H_2SO_3(aq) \longrightarrow H^+ + HSO_3^- \qquad (6\text{-}22)$$

$$H^+ + MgSO_3(s) \longrightarrow HSO_3^- + Mg^{++} \qquad (6\text{-}23)$$

The magnesium sulfite absorbs the H^+ ion and increases the HSO_3^- concentration in Reaction (6-23). This allows the scrubber liquor to absorb more of the SO_2. The absorbed SO_2 reacts with hydrated lime to form solid-phase calcium sulfite. The magnesium sulfite is reformed by the following reactions:

$$Ca(OH)_2(s) + 2HSO_3^- + Mg^{++} \longrightarrow Ca^{++}SO_3^{--} + 2H_2O(l) + MgSO_3(s) \qquad (6\text{-}24)$$

$$Ca^{++}SO_3^{--} + \tfrac{1}{2}H_2O \longrightarrow CaSO_3 \cdot \tfrac{1}{2}H_2O(s) \qquad (6\text{-}25)$$

Inside the absorber, some magnesium sulfite present in the solution is oxidized to sulfate. This sulfite reacts with the lime to form calcium sulfate solids. Calcium sulfite and sulfate solids are the main products of the MagLime process. The calcium sulfite sludge is dewatered using thickener and vacuum filter systems then fixated using fly ash and lime prior to disposal in a lined landfill. The magnesium remains dissolved in the liquid phase.

Some advantages of the MagLime process compared to the LSFO process include [12]:

- High SO_2 removal efficiency at low liquid-to-gas ratios;
- Lower gas-side pressure drop due to lower liquid-to-gas ratios;
- Reduced potential for scaling, which improves reliability of the system;
- Lower power consumption due to a lower slurry recycle rate;
- Lower capital investment due to smaller reagent handling equipment and no oxidation air compressor;
- Reduction in freshwater use because the process water may be recycled for the mist eliminator wash.

The three major disadvantages of the process are the expense of the lime reagent compared to the limestone, the use of fresh water for lime slaking, and the difficult dewatering characteristics of the calcium sulfite/sulfate sludge. The sulfite can be oxidized to produce gypsum, but this requires extensive equipment and process control.

Limestone with Dibasic Acid The dibasic acid enhanced limestone process is very similar to the LSFO process. The hot flue gas exits the particulate control device and enters a spray tower, where it comes into contact with a diluted limestone slurry. The SO_2 in the flue gas reacts with the limestone and water to form hydrated calcium sulfite:

$$SO_2(g) + CaCO_3(s) + \tfrac{1}{2}H_2O(l) \longrightarrow CaSO_3 \cdot \tfrac{1}{2}H_2O(s) + CO_2(g) \qquad (6\text{-}10)$$

This equation is rate limited by the absorption of SO_2 into the scrubbing liquor:

$$SO_2(l) + H_2O(l) \longleftrightarrow HSO_3^- + H^+ \qquad (6\text{-}14)$$

The dissolved SO_2 ions then react with the calcium ions to form calcium sulfite. The hydrogen ions in solution are partly responsible for reforming the SO_2.

After absorbing the SO_2, the slurry drains from the tower to a recirculation tank. Here, the calcium sulfite is oxidized to calcium sulfate dihydrate using oxygen:

$$CaSO_3 \cdot \tfrac{1}{2}H_2O(s) + \tfrac{3}{2}H_2O(l) + \tfrac{1}{2}O_2(g) \longleftrightarrow CaSO_4 \cdot 2H_2O(s) \qquad (6\text{-}12)$$

Dibasic acid acts as a buffer by absorbing free hydrogen ions formed by Reaction (6-14), which shifts the reaction to the right to form more sulfite ions, thus removing more SO_2. Alkaline limestone is added to replace the buffering capabilities of the acid; therefore, there is no net consumption of the dibasic acid during SO_2 absorption. The limestone dissolution rate is increased by increasing the SO_2 removal efficiency at a low slurry pH. This results in a lower reagent consumption due to an increase in calcium carbonate availability in the recirculation tank.

The dibasic acid process offers some advantages compared to the LSFO process [12]:

- Increased SO_2 removal efficiency;
- Reduced liquid-to-gas ratio and the potential to decrease the reagent feed rate, which lowers capital and operating costs for the limestone grinding equipment, slurry handling, and landfill requirements;
- Reduced scaling because of the low pH and reduced gypsum relative saturation levels;
- Increased system reliability by reducing the maintenance requirements and increasing the flexibility of the system.

Disadvantages of the process include:

- More process capital is needed for the dibasic acid feed equipment;
- There is the potential for corrosion and erosion due to the low system pH;
- Odorous by-products are produced by the dibasic acid degradation; although the SO_2 absorption reactions do not consume the dibasic acid, the acid does degrade by carboxylic oxidation into many short chain molecules, one of which is valeric acid, which has a musty odor;
- Control problems may be caused by foaming in the recirculation and oxidation tanks due to the presence of the dibasic acid.

Sodium-Based Scrubbers Wet sodium-based systems have been in commercial operation since the 1970s. These systems can achieve high SO_2 removal efficiencies while burning coals with medium to high sulfur content. A disadvantage of these systems, however, is the production of a waste sludge that requires disposal.

Lime Dual Alkali In the lime dual alkali process, the hot flue gas exits the particulate control device and enters an open spray tower where the gas comes into contact with a sodium sulfite (Na_2SO_3) solution that is sprayed into the tower [12]. An initial charge of sodium carbonate (Na_2CO_3) reacts directly with the SO_2 to form sodium sulfite and CO_2. The sulfite then reacts with more SO_2 and water to form sodium bisulfite ($NaHSO_3$). Some of the sodium sulfite is oxidized by excess oxygen in the flue gas to form sodium sulfate (Na_2SO_4). This does not react with SO_2 and cannot be reformed by the addition of lime to form calcium sulfate. The above process is described by the following reactions:

$$Na_2CO_3(s) + SO_2(g) \longrightarrow Na_2SO_3(s) + CO_2(g) \tag{6-26}$$

$$Na_2SO_3(s) + SO_2(g) + H_2O(l) \longrightarrow 2NaHSO_3(s) \tag{6-27}$$

$$Na_2SO_3(s) + \tfrac{1}{2}O_2(g) \longrightarrow Na_2SO_4(s) \tag{6-28}$$

along with the minor reaction:

$$2NaOH(aq) + SO_2(g) \longrightarrow Na_2SO_3(s) + H_2O(l) \tag{6-29}$$

The calcium sulfites and sulfates are reformed in a separate regeneration tank and are formed by mixing the soluble sodium salts (bisulfate and sulfate) with slaked lime. The calcium sulfites and sulfates precipitate from the solution in the regeneration tank. The scrubber liquor then has a pH of 6 to 7 and consists of sodium sulfite, sodium bisulfite, sodium sulfate, sodium hydroxide, sodium carbonate, and sodium bicarbonate [12].

The lime dual alkali process has several advantages over the LSFO process, including the following [12]:

- The system has a higher availability because there is less potential for scaling and plugging of the soluble absorption reagents and reaction products;
- Corrosion and erosion are prevented with the use of a relatively high pH solution;
- Maintenance labor and material requirements are lower because of the high reliability of the system;
- The main recirculation pumps are smaller because the absorber liquid/gas feed rate is less;
- Power consumption is lower due to the smaller pump requirements;
- There is no process blowdown water discharge stream;
- The highly reactive alkaline compounds in the absorbing solution allow for better turndown and load following capabilities.

There are two main disadvantages of the process compared to the LSFO system: The sodium carbonate reagent is more expensive than limestone, and the sludge must be disposed of in a lined landfill because of sodium contamination of the calcium sulfite/sulfate sludge.

Regenerative Processes Regenerative FGD processes regenerate the alkaline reagent and convert the SO_2 to a usable chemical by-product. Two commercially accepted processes are discussed in this section. The Wellman–Lord process is the most highly demonstrated regenerative technology in the world, while the regenerative magnesia scrubbing process is in commercial service in the United States. Other processes have undergone demonstrations, are used on a limited basis, or are currently under development and include ammonia-based scrubbing, an aqueous carbonate process, and the citrate process.

Wellman–Lord Process The Wellman–Lord process uses sodium sulfite to absorb SO_2, which is then regenerated to release a concentrated stream of SO_2. Most of the sodium sulfite is converted to sodium bisulfite by reaction with SO_2 as in the dual alkali process. Some of the sodium sulfite is oxidized to sodium sulfate. Prescrubbing of the flue gases is necessary to saturate and cool the flue gas to about 130°F. This removes chlorides and remaining fly ash and prevents excessive evaporation in the absorber. A schematic of the system is shown in Figure 6-4 [8]. The basic absorption reaction for the Wellman–Lord process is:

$$SO_2(g) + Na_2SO_3(aq) + H_2O(l) \longrightarrow 2NaHSO_3(aq) \qquad (6\text{-}30)$$

The sodium sulfite is regenerated in an evaporator–crystallizer through the application of heat. A concentrated SO_2 stream (*i.e.*, 90%) is produced at the

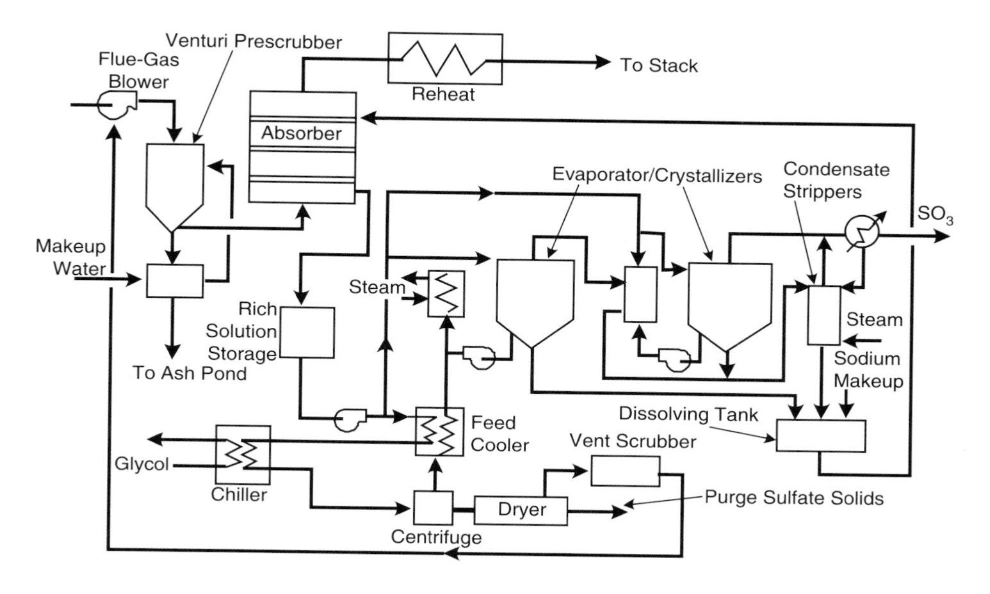

FIGURE 6-4. Wellman–Lord process. (From Elliot, T. C., Ed., *Standard Handbook of Powerplant Engineering*, McGraw-Hill, New York, 1989. With permission.)

same time. The overall regeneration reaction is:

$$2NaHSO_3(aq) + \text{heat} \longrightarrow Na_2SO_3(s) + H_2O(l) + SO_2(conc.) \qquad (6\text{-}31)$$

The concentrated SO_2 stream that is produced may be compressed, liquefied, and oxidized to produce sulfuric acid or reduced to elemental sulfur. A small portion of collected SO_2 oxidizes to the sulfate form and is converted in a crystallizer to sodium sulfate solids that are marketed as salt cakes [8].

The advantages of this process include minimal solid waste production, low alkaline reagent consumption, and the use of a slurry rather than a solution, which prevents scaling and allows the production of a marketable by-product. Disadvantages of the process are the high energy consumption and high maintenance due to the complexity of the process, as well as the large area required for the system. Another disadvantage is that a purge stream of about 15% of the scrubbing solution is required to prevent build-up of the sodium sulfate. Thiosulfate must be purged from the regenerated sodium sulfite.

Regenerative Magnesia Scrubbing In the magnesium oxide process, MgO in the slurry is used in a manner similar to the use of limestone or lime in the lime scrubbing process. The primary difference between the processes is that the magnesium oxide process is regenerative, whereas lime scrubbing is generally a throwaway process.

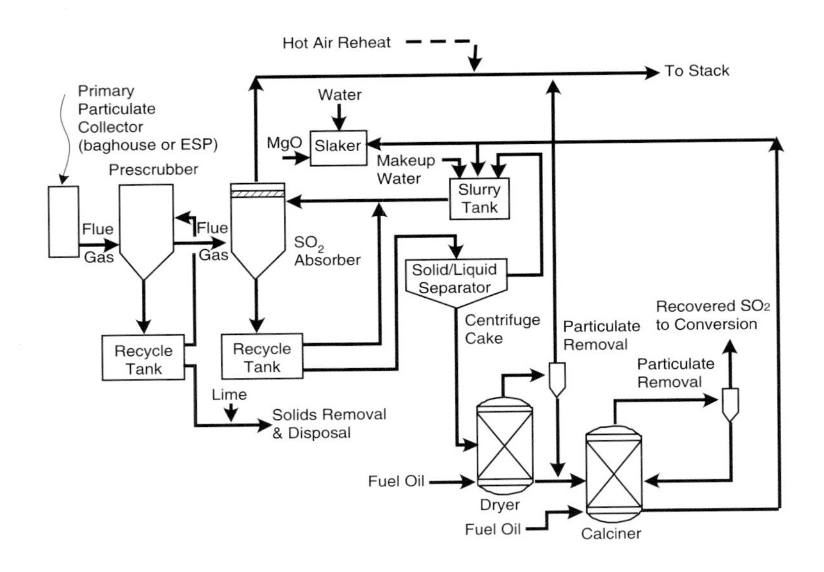

FIGURE 6-5. Regenerative magnesia scrubbing process. (From Elliot, T. C., Ed., *Standard Handbook of Powerplant Engineering*, McGraw-Hill, New York, 1989. With permission.)

The magnesium oxide process, shown in Figure 6-5, uses a slurry of slaked magnesium oxide ($Mg(OH)_2$) to remove SO_2 from the flue gas to form magnesium sulfite and sulfate via the basic reactions:

$$Mg(OH)_2(s) + SO_2(g) \longrightarrow MgSO_3(s) + H_2O(l) \tag{6-32}$$

$$MgSO_3(s) + \tfrac{1}{2}O_2(g) \longrightarrow MgSO_4(s) \tag{6-33}$$

A bleed stream of scrubber slurry is centrifuged to form a wet cake containing 75 to 90% solids, which is then dried to form a dry, free-flowing mixture of magnesium sulfite and sulfate. This mixture is heated to decompose most of the magnesium sulfite/sulfate to SO_2 and MgO. A stream of 10 to 15% SO_2 is produced. Coke, or some other reducing agent, is added in the calcination step to reduce any sulfate present. The regenerated MgO is slaked and used in the absorber. The regeneration reactions are:

$$MgSO_3(s) + \text{heat} \longrightarrow MgO(s) + SO_2(g) \tag{6-34}$$

$$MgSO_4(s) + \tfrac{1}{2}C(s) + \text{heat} \longrightarrow MgO(s) + SO_2(g) + \tfrac{1}{2}CO_2(g) \tag{6-35}$$

The SO_2 product gas is generally washed and quenched and fed to a contact sulfuric acid plant to produce concentrated sulfuric acid by-product. Sulfur production is possible but would be expensive because the SO_2 stream is dilute.

Advantages of this process include high SO_2 removal efficiencies (up to 99%), minimum impact of fluctuations in inlet SO_2 levels on removal efficiency, low chemical scaling potential, the capability to regenerate the sulfate (which simplifies waste management), and more favorable economics compared to other available regenerative processes [8]. The main disadvantages of the process are its complexity and the need for a contact sulfuric acid plant to produce a salable by-product.

Dry Flue Gas Desulfurization Technology Dry FGD technology includes lime or limestone spray drying; dry sorbent injection, including furnace, economizer, duct, and hybrid methods; and circulating fluidized-bed scrubbers. These processes are characterized with dry waste products that are generally easier to dispose of than waste products from wet scrubbers. All dry FGD processes are throwaway types.

Spray Dry Scrubbers Spray dry scrubbers are the second most widely used method for controlling SO_2 emissions in utility coal-fired power plants. Prior to 1980, the removal of SO_2 by absorption was usually performed using wet scrubbers. Wet scrubbing requires considerable equipment, so alternatives to wet scrubbing have been developed, including spray dry scrubbers. Lime (CaO) is usually the sorbent used in the spray drying process, but hydrated lime ($Ca(OH)_2$) is also used. This technology is also known as semi-dry flue gas desulfurization and is generally used for sources that burn low- to medium-sulfur coal. In the United States, this process has been used in both retrofit applications and new installations on units burning low-sulfur coal [5,6].

In this process, the hot flue gas exits the boiler air heater and enters a reactor vessel. A slurry consisting of lime and recycled solids is atomized/sprayed into the absorber. The slurry is formed by the reaction:

$$CaO(s) + H_2O(l) \longrightarrow Ca(OH)_2(s) + \text{heat} \tag{6-9}$$

The SO_2 in the flue gas is absorbed into the slurry and reacts with the lime and fly ash alkali to form calcium salts:

$$Ca(OH)_2(s) + SO_2(g) \longrightarrow CaSO_3 \cdot \tfrac{1}{2}H_2O(s) + \tfrac{1}{2}H_2O(v) \tag{6-36}$$

$$Ca(OH)_2(s) + SO_3(g) + H_2O(v) \longrightarrow CaSO_4 \cdot 2H_2O(s) \tag{6-37}$$

Hydrogen chloride (HCl) present in the flue gas is also absorbed into the slurry and reacts with the slaked lime. The water that enters with the slurry is evaporated, which lowers the temperature and raises the moisture content of the scrubbed gas. The scrubbed gas then passes through a particulate control device downstream of the spray drier. Some of the collected reaction product, which contains some unreacted lime, and fly ash is recycled to the slurry feed system while the rest is sent to a landfill for disposal. Factors affecting the absorption chemistry include the flue gas temperature, SO_2 concentration in the flue gas, and the size of the atomized slurry droplets. The residence time in the reactor vessel is typically about 10 to 12 seconds.

The lime spray dryer process offers a few advantages over the LSFO process [12]. Only a small alkaline stream of scrubbing slurry must be pumped into the spray dryer. This stream contacts the gas entering the dryer instead of the walls of the system. This prevents corrosion of the walls and pipes in the absorber system. The pH of the slurry and dry solids is high, allowing for the use of mild steel materials rather than expensive alloys. The product from the spray dryer is a dry solid that is handled by conventional dry fly ash particulate removal and handling systems, which eliminates the need for dewatering solids handling equipment and reduces associated maintenance and operating requirements. Overall power requirements are decreased because less pumping power is required. The gas exiting the absorber is not saturated and does not require reheating, thereby reducing capital costs and steam consumption. Chloride concentration increases the SO_2 removal efficiencies (whereas, in wet scrubbers, increasing chloride concentration decreases efficiency), which allows the use of cooling tower blowdown for slurry dilution after completing the slaking of the lime reagent. The absorption system is less complex, so operating, laboratory, and maintenance manpower requirements are lower than those required for a wet scrubbing system.

There are some disadvantages of the lime spray dryer compared to the LSFO system, and these, along with the advantages, must be evaluated for specific applications [12]. A major product of the lime spray dryer process is calcium sulfite, as only 25% or less oxidizes to calcium sulfate. The solids handling equipment for the particulate removal device has to have a greater capacity than conventional fly ash removal applications. Fresh water is required in the lime slaking process, which can represent approximately half of the system's water requirement. This differs from wet scrubbers, where cooling tower water can be used for limestone grinding circuits and most other makeup water applications. The lime spray dryer process requires a higher reagent feed ratio than the conventional systems to achieve high removal efficiencies. Approximately 1.5 mol CaO per mol of SO_2 removed are needed for 90% removal efficiency. Lime is also more expensive than limestone; therefore, the operating costs are increased. These costs can be reduced if higher coal chloride levels and/or calcium chloride spiking are used because chlorides improve removal efficiency and reduce reagent consumption. A higher inlet flue gas temperature is needed when a higher sulfur coal is used, which in turn reduces the overall boiler efficiency.

Combining spray dry scrubbing with other FGD systems such as furnace or duct sorbent injection and particulate control technology such as a pulse-jet baghouse allows the use of limestone as the sorbent instead of the more costly lime [4]. Sulfur dioxide removal efficiencies can exceed 99% with such a combination.

Sorbent Injection Processes A number of dry injection processes have been developed to provide moderate SO_2 removal that are easily retrofitted to existing facilities and feature low capital costs. Of the five basic processes, two are associated with the furnace—furnace sorbent injection and convective pass (economizer) injection—and three are associated with injection into the ductwork downstream of the air heater—in-duct injection, in-duct spray drying, and hybrid systems. Combinations of these processes are also available. Sorbents include calcium- and sodium-based compounds; however, the use of calcium-based sorbents is more prevalent. Furnace injection has been used in some small plants using low-sulfur coals. Hybrid systems may combine furnace and duct sorbent injection or introduce a humidification step to improve removal efficiency. These systems can achieve as high as 70% removal and are commercially available [4]. Process schematics for dry-injection SO_2 control technologies are illustrated in Figure 6-6.

Figure 6-7 provides a representation of the level of SO_2 removal that the dry calcium-based sorbent injection processes achieve and the temperature regimes in which they operate [13]. The peak at approximately 2200°F represents furnace sorbent injection, the peak at about 1000°F represents convective pass/economizer injection, and the peak at the low temperature represents all of the processes downstream of the air heater.

Another dry limestone injection technique, limestone injection into a multistage burner (LIMB), was developed from the 1960s to the 1980s but has not been adopted on a commercial scale for utility applications and is not discussed in detail here. In this process, which offers low capital costs and which is used in some industrial-scale applications where low SO_2 removal is required, limestone is added to the coal stream and fed with the coal directly to the burner. This process gives poor SO_2 removal (typically ~15% but in rare cases as much as 50%), experiences dead-burning (*i.e.*, sintering or melting of the sorbent which reduces surface area and lowers sulfur capture), is difficult to introduce in a uniform manner, and can cause operational problems such as tube fouling and impairment of ESP performance because of excessive sorbent addition [8]. Sorbents under development are also not discussed in this section; rather, this section focuses mainly on commercial applications. Calcium organic salts (*e.g.*, calcium acetate, calcium magnesium acetate, and calcium benzoate), pyrolysis liquor, and other sorbents are under development for use in injection processes.

Furnace Sorbent Injection (FSI) With the exception of LIMB, furnace sorbent injection (FSI) is the simplest dry sorbent process. In this process, illustrated in Figure 6-6a, pulverized sorbents, most often calcium hydroxide and sometimes limestone, are injected into the upper part of the furnace to react with the SO_2 in the flue gas. The sorbents are distributed over the entire cross section of the upper furnace, where the temperature ranges from 1400 to 2400°F and the residence time for the reactions is 1 to 2 seconds. The sorbents decompose and become porous solids with high

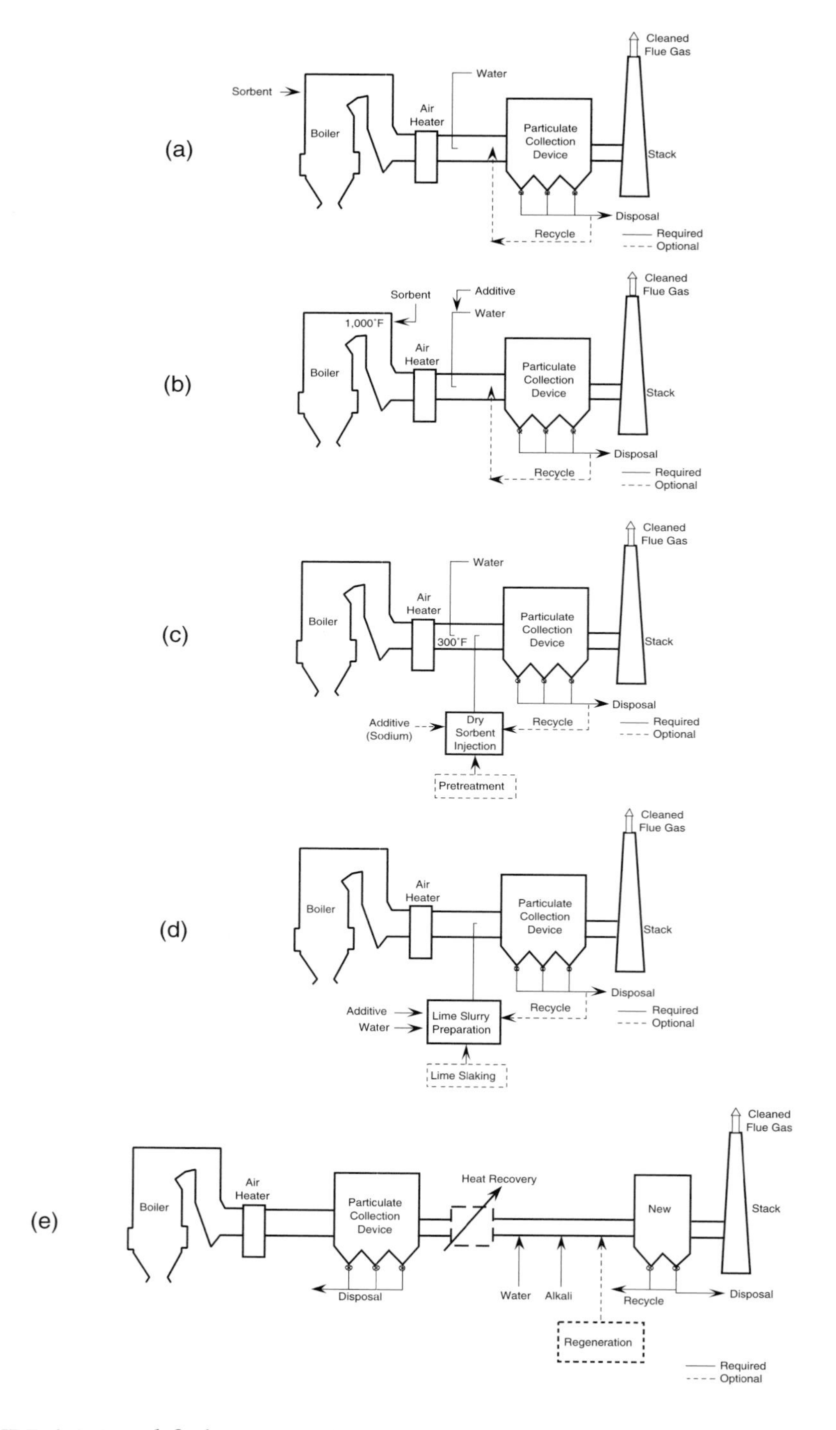

FIGURE 6-6. Simplified process schematics for dry-injection SO_2 control technologies. (Adapted from Rhudy *et al.* [13].)

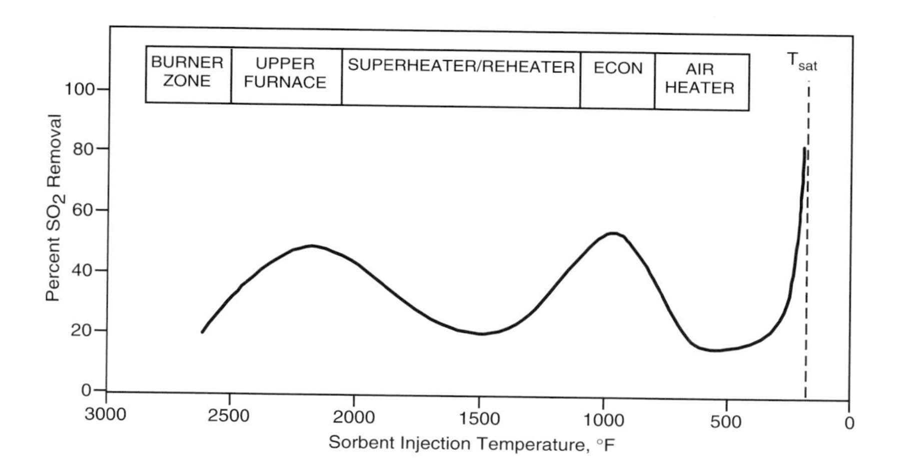

FIGURE 6-7. SO_2 capture regimes for hydrated calcitic lime at a Ca/S molar ratio of 2.0. (From Rhudy, R. *et al.*, Status of Calcium-Based Dry Sorbent Injection SO_2 Control, in *Proc. of the Tenth Symposium on Flue Gas Desulfurization*, November 17–21, 1986, pp. 9-69–9-84.)

surface area. At temperatures higher than ~2300°F, dead-burning or sintering is experienced.

When limestone is used as the sorbent, it is rapidly calcined to quicklime when it enters the furnace:

$$CaCO_3(s) + \text{heat} \longrightarrow CaO(s) + CO_2(g) \tag{6-38}$$

Sulfur dioxide diffuses to the particle surface and heterogeneously reacts with the CaO to form calcium sulfate:

$$CaO(s) + SO_2(g) + \tfrac{1}{2}O_2(g) \longrightarrow CaSO_4(s) \tag{6-39}$$

Sulfur trioxide, although present at a significantly lower concentration than SO_2, is also captured using calcium-based sorbents:

$$CaO(s) + SO_3(g) \longrightarrow CaSO_4(s) \tag{6-40}$$

Approximately 15 to 40% SO_2 removal can be achieved using Ca/S in the flue gas molar ratio of 2.0. The optimum temperature for injecting limestone is ~1900 to 2100°F.

The calcium sulfate that is formed travels through the rest of the boiler flue gas system and is ultimately collected in the existing particulate control device with the fly ash and unreacted sorbent. Some concerns exist regarding increased tube deposits as a result of injecting solids into the boiler, and the extent of calcium deposition is influenced by overall ash chemistry, ash loading, and boiler system design.

The following overall reactions occur when using hydrated lime as the sorbent:

$$Ca(OH)_2(s) + \text{heat} \longrightarrow CaO(s) + H_2O(v) \quad (6\text{-}41)$$

$$CaO(s) + SO_2(g) + \tfrac{1}{2}O_2(g) \longrightarrow CaSO_4(s) \quad (6\text{-}39)$$

$$CaO(s) + SO_3(g) \longrightarrow CaSO_4(s) \quad (6\text{-}40)$$

Approximately 50 to 80% SO_2 removal can be achieved using hydrated lime at a Ca/S molar ratio of 2.0. The hydrate is injected at very nearly the same temperature window as limestone, and the optimum range is 2100 to 2300°F.

The FSI process can be applied to boilers burning low- to high-sulfur coals. The factors that affect the efficiency of the FSI system are flue gas humidification (to condition the flue gas to counter degradation that may occur in ESP performance from the addition of significant quantities of fine, high-resistivity sorbent particles), type of sorbent, efficiency of ESP, and temperature and location of the sorbent injection. The process is better suited for large furnaces with lower heat release rates [12]. Systems that use hydrated calcium salts sometimes have problems with scaling; however, this can be prevented by keeping the approach to adiabatic saturation temperature above a minimum threshold.

The FSI system has several advantages [12]. One advantage is simplicity of the process; the dry reagent is injected directly into the flow path of the flue gas in the furnace, and a separate absorption vessel is not required. The injection of lime in a dry form allows for a less complex reagent handling system, which lowers operating labor and maintenance costs and eliminates the problems of plugging, scaling, and corrosion found in slurry handling. Power requirements are lower because less equipment is needed. Steam is not required for reheat, whereas most LSFO systems require some form of reheat to prevent corrosion of downstream equipment. The sludge dewatering system is eliminated because the FSI process produces a dry solid, which can be removed by conventional fly ash removal systems.

The FSI process has a few disadvantages when compared to the LSFO process [12]. One major disadvantage is that the process only removes up to 40 and 80% SO_2 when using limestone and hydrated lime, respectively, at a Ca/S molar ratio of 2.0, whereas the LSFO process can remove more than 90% SO_2 using 1.05 to 1.1 mol CaO per mol SO_2 removed. This is further illustrated in Figure 6-8, which shows calcium utilization (defined as the percent SO_2 removed divided by the Ca/S ratio) of hydrated lime and limestone at various injection temperatures [13]; hence, more sorbent is needed in the FSI process, and lime, which works better than limestone, is more expensive than limestone. There is a potential for solids deposition and boiler convective pass fouling, which occurs during the humidification step due to the impact of solid droplets on surfaces. Also, there is a potential for corrosion at the point of humidification and in the ESP, downstream ductwork, and

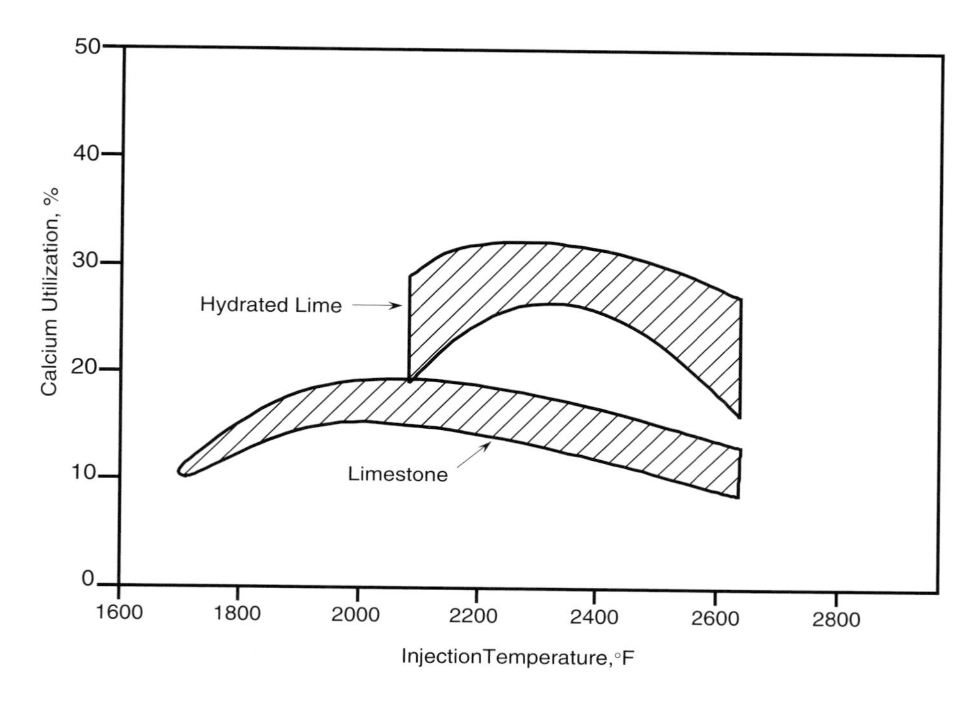

FIGURE 6-8. Calcium utilization as a function of sorbent injection temperature for furnace sorbent injection. (Adapted from Rhudy *et al.* [13].)

stack. The corrosion at the point of humidification is caused by operating below the acid dewpoint, whereas downstream corrosion is caused by the humidified gas temperature being close to the water saturation temperature. Plugging can also occur, thereby affecting system pressures. The efficiency of an ESP can be reduced by increased particulate loading and changes in the ash resistivity. This can, in turn, lead to the installation of additional particulate collection devices. Sintering of the sorbent is a concern if it is injected at too high of a temperature (*e.g.*, >2300°F for hydrated lime). Multiple injection ports in the furnace wall may be needed to ensure proper mixing and follow boiler load swings and hence shifting temperature zones. Hydration of the free lime in the product may be required. Lime is very reactive when exposed to water and can pose a safety hazard for disposal areas.

Economizer Injection In an economizer injection process (shown in Figure 6-6b), hydrated lime is injected into the flue gas stream near the economizer inlet where the temperature is between 950 and 1050°F. This process is not commercially used at this time but was extensively studied because it was found that the reaction rate and extent of sulfur capture (see Figure 6-7) are comparable to FSI. However, the economizer temperatures are too low for dehydration of the hydrated lime (only about 10% of the hydrated lime forms quicklime), and the hydrate reacts directly with the SO_2 to form calcium sulfite:

$$Ca(OH)_2(s) + SO_2(g) \longrightarrow CaSO_3(s) + H_2O(v) \qquad (6\text{-}42)$$

This process is best suited for older units in need of a retrofit process and can be used for low- to high-sulfur coals. The advantages and disadvantages of this system are similar to the FSI process (but will not be discussed in detail here as this process is not currently being used in the power industry) with the notable exception that no reactive CaO is contained in the waste.

Duct Sorbent Injection: Duct Spray Drying Spray dry scrubbers are the second most widely used method for controlling SO_2 emissions in utility coal-fired power plants. Lime is usually the sorbent used in this technology, but sodium carbonate is also used, specifically in the western United States. Spray dryer FGD systems have been installed on over 12,000 MW of total FGD capacity, as shown in Table 6-1, as well as numerous industrial boilers.

The first commercial dry scrubbing system on a coal-fired boiler in the United States was installed in mid-1981 at the Coyote station (jointly owned by Montana–Dakota Utilities, Northern Municipal Power Agency, Northwestern Public Service Company, and Ottertail Power Company) near Beulah, North Dakota. The 425 MW unit burns lignite from a mine-mouth plant and initially used soda ash (Na_2CO_3) as the sulfur removal reagent. The spray dryer was modified about 10 years later, and the unit currently uses lime as the reagent. The second dry scrubbing system on a coal-fired utility boiler was installed on two 440 MW units; it became operational in 1982 and 1983 at the Basin Electric Power Cooperative's Antelope Valley station, also located near Beulah, North Dakota. These units fire minemouth lignite and use a slaked lime slurry to remove SO_2 in the spray dryer.

A slaked-lime slurry is sprayed directly into the ductwork to remove SO_2 (see Figure 6-6c). The reaction products and fly ash are captured downstream in the particulate removal device. A portion of these solids is recycled and reinjected with the fresh sorbent. Dry spray drying (DSD) is a relatively simple retrofit process capable of 50% SO_2 removal at a Ca/S ratio of 1.5. The concept is the same as conventional spray drying except that the existing ductwork provides the residence time for drying instead of a reaction vessel. The main difference is that the residence time in the duct is much shorter (*i.e.*, 1–2 sec, compared to 10–12 sec in a spray drying vessel).

The slaked lime is produced by hydrating raw lime to form calcium hydroxide. This slaked lime is atomized and absorbs the SO_2 in the flue gas. The SO_2 reacts with the slurry droplets as they dry to form equimolar amounts of calcium sulfite and calcium sulfate. The water in the lime slurry improves SO_2 absorption by humidifying the gas. The reaction products, unreacted sorbent, and fly ash are collected in the particulate control device located downstream. Some of the unreacted sorbent may react with a portion of the CO_2 in the flue gas to form calcium carbonate. Also, a little more SO_2

removal is achieved in the particulate control device. The reactions occurring in the process are:

$$CaO(s) + H_2O(l) \longrightarrow Ca(OH)_2(s) + \text{heat} \quad (6\text{-}9)$$

$$Ca(OH)_2(s) + SO_2(g) \longrightarrow CaSO_3 \cdot \tfrac{1}{2}H_2O(s) + \tfrac{1}{2}H_2O(v) \quad (6\text{-}36)$$

$$Ca(OH)_2(s) + SO_2(g) + \tfrac{1}{2}O_2(g) + H_2O(v) \longrightarrow CaSO_4 \cdot 2H_2O(s) \quad (6\text{-}43)$$

$$Ca(OH)_2(s) + CO_2(g) \longrightarrow CaCO_3(s) + H_2O(v) \quad (6\text{-}44)$$

There are two different methods for atomizing the slurry. One method is the use of rotary atomizers, with the ductwork providing the short gas residence time of 1 to 2 seconds. When using this atomizer, the ductwork must be sufficiently long to allow for drying of the slurry droplets. There must also be no obstructions in the duct. The second method for atomizing the slurry is the use of dual-fluid atomizers, where compressed air and water are used to atomize the slurry. This process is referred to as the confined zone dispersion (CZD) process. The dual-fluid atomizer has been shown to be more controllable due to the adjustable water flow rate. This atomizer is also relatively inexpensive and has a long and reliable operating life with little maintenance. The spray is confined in the duct, which allows better mixing with the flue gas rather than impinging on the walls.

The DSD process has several advantages compared to wet processes. The DSD process is less complex because the reagent is injected directly into the flow path of the flue gas, and a separate absorption vessel is not needed. Less equipment is needed so power requirements are lower. The waste from this process does not contain reactive lime, as the FSI process does, and therefore does not require special handling.

Some of the problems encountered by the DSD system are also common to other dry processes. A main disadvantage of the system includes limited SO_2 removal efficiency (*i.e.*, ~50%) and low calcium utilization compared to wet processes. Quicklime is more expensive than limestone. If an ESP is used, there is the potential for reduced efficiency due to changes in fly ash resistivity and the increased dust loading in the flue gas. Additional collection devices may be needed as well as humidification to improve ESP collection efficiency. There must be sufficient length (*i.e.*, residence time of the ductwork) to ensure complete droplet vaporization prior to the particulate collection device. This is necessary for good sulfur capture and to avoid plugging and deposition, which in turn results in an increased pressure drop that the induced draft fans must overcome.

Duct Sorbent Injection: Dry Sorbent Injection Dry sorbent injection (DSI), also referred to as in-duct dry injection, is illustrated in Figure 6-6d. Hydrated lime is the sorbent typically used in this process, especially for power generation facilities; however, sodium-based sorbents have been tested extensively, including full-scale utility demonstrations, and are used in industrial systems such as municipal and medical waste incinerators for acid gas control.

When hydrated lime is used in this process, it is injected either upstream or downstream of a flue gas humidification zone. In this zone, the flue gas is humidified to within 20°F of the adiabiatic saturation temperature by injecting water into the duct downstream of the air preheater [12]. The SO_2 in the flue gas reacts with the calcium hydroxide to form calcium sulfate and calcium sulfite:

$$Ca(OH)_2(s) + SO_2(g) + \tfrac{1}{2}O_2(g) + H_2O(v) \longrightarrow CaSO_4 \cdot 2H_2O(s) \qquad (6\text{-}43)$$

$$Ca(OH)_2(s) + SO_2(g) \longrightarrow CaSO_3 \cdot \tfrac{1}{2}H_2O(s) + \tfrac{1}{2}H_2O(v) \qquad (6\text{-}36)$$

The water droplets are vaporized before they strike the surface of the wall or enter the particulate control device. The unused sorbent, products, and fly ash are all collected in the particulate control device. About half of the collected material is shipped to a landfill, while the other half is recycled for injection with the fresh sorbent into the ducts [12].

The DSI system offers many of the same advantages and disadvantages that other dry systems offer [12]. The process is less complex (*i.e.*, no slurry recycle and handling, no dewatering system, fewer pumps, and no reactor vessel) than a wet system, specifically LSFO. The humidification water and hydrated lime are injected directly into the existing flue gas path. No separate SO_2 absorption vessel is necessary. The handling of the reagent is simpler than in wet systems. DSI systems have less equipment to install so operating and maintenance costs are reduced. The waste product is free of reactive lime so no special handling is required.

Some of the problems encountered by the DSI system and its disadvantages, as compared to the LSFO system, are common to other dry processes. Sulfur dioxide removal efficiencies are lower (as is calcium utilization) than wet systems and range from 30 to 70% for a Ca/S ratio of 2.0. Quicklime is more expensive than limestone. When an ESP is used for particulate control, there is the potential for reduced efficiency due to increased fly ash resistivity and dust loading in the flue gas. Additional collection devices may be required. A sufficient length of ductwork is necessary to ensure a residence time of 1 to 2 seconds in a straight, unrestricted path. Plugging of the duct can occur if the residence time is insufficient for droplet vaporization, leading to increased system pressure drop.

In the dry sodium desulfurization process, a variety of sodium-containing crystalline compounds may be injected directly into the flue gas. The main compounds of interest include [14]:

- Sodium carbonate (Na_2CO_3), a refined product of ~98% purity;
- Sodium bicarbonate ($NaHCO_3$), a refined product of ~98% purity;
- Nacholite ($NaHCO_3$), a natural material of ~76% purity containing high levels of insolubles;

- Sodium sesquicarbonate ($NaHCO_3 \cdot Na_2CO_3 \cdot 2H_2O$), a refined product of ~98% purity;
- Trona ($NaHCO_3 \cdot Na_2CO_3 \cdot 2H_2O$), a natural material of ~88% purity containing high levels of insolubles.

Sodium bicarbonate and sodium sesquicarbonate have been the most extensively tested in pilot-, demonstration-, and full-scale utility applications due to proven success and commercial availability. In addition, sodium bicarbonate is extensively used in industrial applications for acid gas control. Of the compounds listed above, sodium bicarbonate has demonstrated the best sulfur capture in coal-fired boiler applications, as shown in Figure 6-9, which illustrates SO_2 removal as a function of normalized stoichiometric ratio (NSR) for sodium bicarbonate injection into a coal-fired, pilot-scale test facility and industrial boiler equipped with fabric filter baghouses. Note that the NSR represents the molar ratio between the injected sodium compound and the initial SO_2 concentration in the flue gas, considering that it takes 2 mol of sodium to react with only 1 mol of SO_2.

The flue gas stream must be above 240°F for rapid decomposition of the sodium bicarbonate when it is injected or little SO_2 capture will occur. While SO_2 will react directly with the sodium bicarbonate, in the presence of nitric oxide (NO) this reaction is inhibited and does not result in significant sulfur capture; therefore, for acceptable SO_2 capture to progress rapidly and attain acceptable levels of utilization, the bicarbonate component must begin to decompose. As flue gas temperatures increase into the optimum range of 240 to 320°F, the carbonate is decomposed, and the subsequent sulfation reaction occurs [14]. When the bicarbonate component decomposes, carbon dioxide and water vapor are evolved from the particle interior, creating a network of void spaces. Sulfur dioxide and NO can diffuse to the fresh sorbent surfaces, where the heterogeneous reactions to capture SO_2 (and to a lesser extent NO) take place. The decomposition and sulfation reactions are:

$$2NaHCO_3(s) + \text{heat} \longrightarrow Na_2CO_3(s) + H_2O(v) + CO_2(g) \tag{6-45}$$

$$SO_2(g) + Na_2CO_3(s) + \tfrac{1}{2}O_2(g) \longrightarrow Na_2SO_4(s) + CO_2(g) \tag{6-46}$$

Lower NO_x emissions also result from injection of dry sodium compounds [4]. The mechanism is not well understood, but reductions up to 30% have been demonstrated which are a function of SO_2 concentration and NSR ratio. There is a side effect of this reduction, though. Nitric oxide is oxidized to NO_2 (a reddish-brown gas), and not all of the NO_2 is reacted with the sorbent. As the NO_2 concentration increases in the stack, an undesirable coloration in the plume can be created.

Hybrid Systems Hybrid sorbent injection processes are typically a combination of FSI and DSI systems with the goal of achieving greater SO_2 removal and sorbent utilization [4]. Various types of configurations have

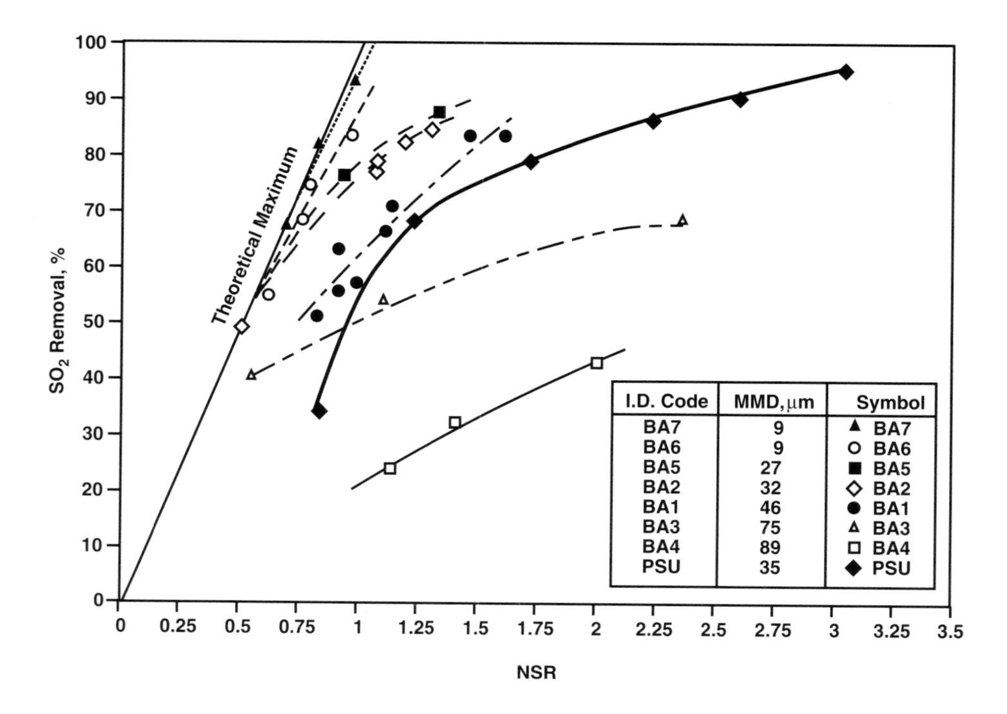

FIGURE 6-9. Comparison of SO_2 removal as a function of NSR and particle size when using sodium bicarbonate. Tests identified by the BA code were performed in a coal-fired, pilot-scale facility, and the tests coded PSU were performed in a coal-fired, industrial boiler system. Both facilities were equipped with fabric filter baghouses. The Penn State tests were performed with a flue gas temperature of ~380°F, and the pilot-scale tests were performed at ~300°F. All tests were performed with sorbent from the same vendor. (From Bland, V. V. and C. E. Martin, *Full-Scale Demonstration of Additives for NO_2 Reduction with Dry Sodium Desulfurization*, EPRI GS-6852, Electric Power Research Institute, Palo Alto, CA, June 1990; Miller, B. G. *et al.*, *The Development of Coal-Based Technologies for Department of Defense Facilities: Phase II Final Report*, DE-FC22-92PC92162, prepared for the U.S. Department of Energy, Federal Energy Technology Center, Pittsburgh, PA, July 31, 2000, 784 pages.)

been tested including injecting secondary sorbents, such as sodium compounds, into the ductwork or humidifying the flue gas in a specially designed vessel. Humidification reactivates the unreacted CaO and can increase the SO_2 removal efficiency. Advantages of hybrid processes include high SO_2 removal, low capital and operating costs, less required space (thus lending itself to easy retrofit), easy operation and maintenance, and no wastewater treatment [4].

In some hybrid systems, a new baghouse is installed downstream of an existing particulate removal device (generally an ESP). The existing

ESP continues to remove the ash, which can be either sold or disposed. Sulfur dioxide removal is accomplished in a manner similar to in-duct injection, with the sorbent injection occurring upstream of the new baghouse [13].

The potential advantages of this system include the potential for toxic substances control because a baghouse is the last control device (this is further discussed later in the section on mercury), easier waste disposal, the potential for sorbent regeneration, separate ash and product streams, and more efficient recycle without ash present [13]. The major issue is the high capital cost of adding a baghouse, although the concept of adding one with a high air-to-cloth ratio (3–5 actual cubic feet per minute (acfm)/ft^2) can minimize this cost. Hybrid systems are discussed in more detail later in this chapter.

Circulating Fluidized-Bed Scrubbers Circulating fluidized-bed (CFB) scrubbers are the least commonly used commercial option, and few or no systems are planned in the future (see Table 6-1). CFB scrubbers include dry and semi-dry systems [4]. Commercial application of the dry CFB system is the more widespread of the two processes, and it can achieve SO_2 removal efficiencies of 93 to 97% at a Ca/S molar ratio of 1.2 to 1.5. In this system, hydrated lime is injected directly into a CFB reactor along with water to obtain operation close to the adiabatic saturation temperature. The main features of this process include simplicity and reliability, with proven high availability; mild steel construction, which does not require a lining; no moving parts or slurry nozzles; handling of only dry solids; water injection independent of reagent feed; moderate space requirements; and greater flexibility to handle varying SO_2 and SO_3 concentrations [4]. The CFB scrubber uses hydrated lime rather than the less expensive limestone commonly used in wet FGD technology processes. Additionally, due to a higher particulate matter concentration downstream of the scrubber, improvements to the particulate removal device, specifically an ESP, may be needed to meet the required particulate emission levels.

Fluidized-Bed Combustion

Fluidized-bed combustion is not an SO_2 control technology *per se*; however, this combustion technology does offer the capability to control SO_2 emissions during the combustion process rather than after combustion where FGD systems need to be installed. Fluidized-bed combustion and the role of sorbents in controlling SO_2 are discussed in Chapter 5 (Technologies for Coal Utilization). The sorbents used in a fluidized-bed combustor are usually limestones, but sometimes dolomites (a double carbonate of calcium and magnesium) are used. The calcination and sulfation chemistries are also discussed in Chapter 5.

Economics of Flue Gas Desulfurization

The costs of an FGD system are site specific and include capital and operating costs. The capital costs of an FGD system depend on many factors, including [16]:

- Market conditions;
- Geographical location;
- Preparatory site work required;
- Volume of flue gas to be scrubbed;
- Concentration of SO_2 in the flue gas;
- Extent of SO_2 removal required;
- Quality of the products produced;
- Process and waste water treatment;
- The need for flue gas reheat;
- The degree of reliability and redundancy required;
- Life of the system.

The capital costs of wet FGD systems have been declining in the United States over the last 30 years. The prototype FGD systems of the 1970s cost $400/kW and experienced many problems [17]. With standardization, better chemistry, and improved materials, the cost of FGD systems in the 1980s dropped to $275/kW. Additional developments, such as reduced redundancy, fewer modules, increased competition from foreign vendors, and the use of well-engineered packages, has dropped capital costs to about $100/kW today.

Operating costs are divided into variable and fixed costs [16]. Variable costs include the costs of the sorbents/reagents, costs associated with disposal or utilization of the by-products, and steam, power, and water costs. Fixed costs include costs of operating labor, maintenance, and administration. The operating and maintenance costs of an FGD system can be significant. According to Soud [4], operating and maintenance costs for the various subsystems in a pulverized coal-fired power generating facility with state-of-the-art environmental protection are 78% for the boiler/turbine/generator, 10% for the FGD system, 6% for a selective catalytic reduction system, 4% for wastewater treatment, and 2% for an ESP.

In the following sections, the costs associated with different FGD systems are briefly discussed. It must be noted, however, that it is difficult to compare costs (whether capital or operating) between systems because costs have many site-specific factors, are dependent on the age of the system, are influenced by economies of scale, and are higher for retrofit applications than new installations.

Wet Processes In a compilation of costs for various processes, Wu [16] reported that the capital and operating costs (for new installations) for a limestone wet FGD process are approximately $100/kW and $100/kWh, respectively, while the capital and operating costs are approximately $50/kW

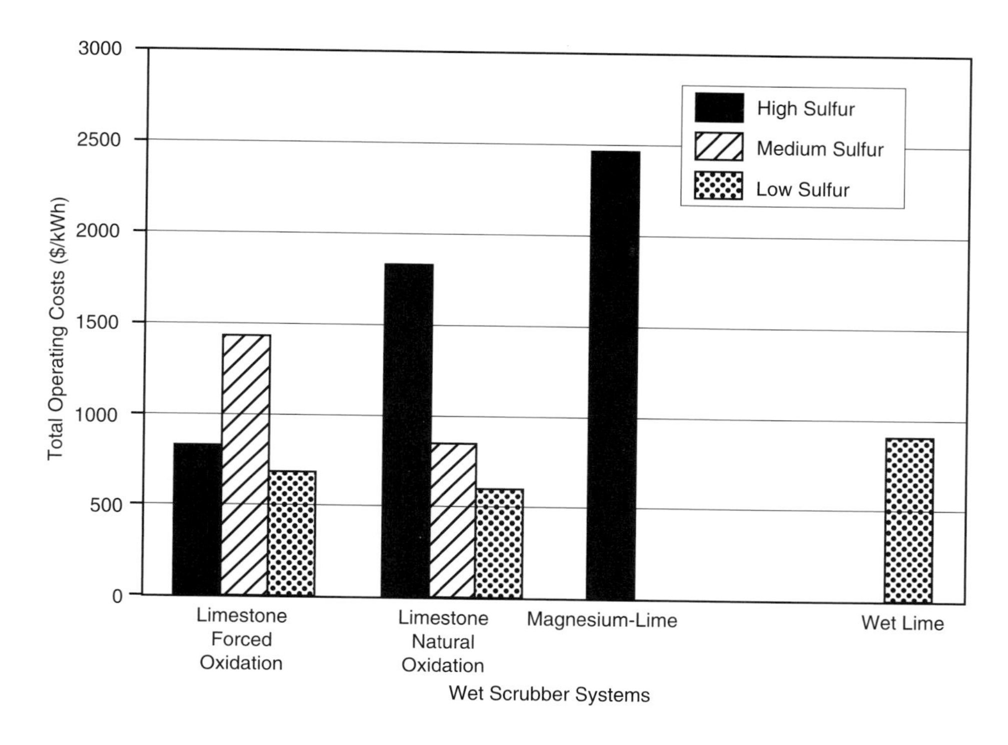

FIGURE 6-10. Operating costs for wet lime/limestone FGD systems with different sulfur content coals.

and $125/kWh, respectively, for magnesium-based systems and $50/kW and $200/kWh, respectively, for sodium-based scrubbers. The variability in operating costs, though, is evident in data reported by Blythe *et al.* [18] and summarized in Figure 6-10, which shows the total operating costs for various systems for low-, medium-, and high-sulfur coals. Similarly, capital costs for recent LSFO retrofits can vary from $180 to $348/kW [16].

Spray Dry Processes The spray dry process generally has lower capital costs but higher and more expensive sorbent use (typically lime) than wet processes. This process is used primarily for small- to medium-sized plants firing low- to medium-sulfur coals and is preferable for retrofits [16]. For example, Blythe *et al.* [18] reported total operating costs for lime spray dry processes of $836/kWh compared to $695/kWh for LSFO systems (see Figure 6-10) for low-sulfur coal applications.

Sorbent Injection Processes The sorbent injection process, with a moderate SO_2 removal efficiency, has a relatively low capital cost. The capital costs for furnace sorbent injection are approximately $70 to $120/kW [16].

Circulating Fluidized-Bed Processes The circulating fluidized-bed processes have relatively low capital costs, similar to those for the spray dry process. The process also has low to moderate fixed operating costs, but the variable operating costs are relatively high [16].

Regenerative Processes Regenerative processes generally have high capital costs and power consumption [16]. The net variable operating costs are moderate because the processes produce salable by-products; however, the fixed operating costs are substantially higher than other processes.

Nitrogen Oxides (NO_X)

Approximately 10 million short tons of NO_X were emitted from fuel combustion in 2002, with power plants contributing less than 4.5 million short tons [19,20]. All sources affected by the EPA's Acid Rain Program NO_X requirements reduced their combined NO_X emissions by 27% over the period 1990 to 2002 [20]. These reductions have been achieved while the amount of fuel burned to produce electricity, as measured by heat input, increased 28% since 1990. In 2002, more than 1000 units complied with emission rate limits. NO_X formation mechanisms are reviewed in this section, followed by technologies used to control NO_X emissions. Similar to the discussion on SO_2 control technologies, NO_X control technologies will focus on commercially available, commercially used systems, with the focus on pulverized coal-fired boiler systems.

NO_X Formation Mechanisms

NO_X formation from coal combustion was introduced in Chapter 3 (The Effect of Coal Use on Human Health and the Environment), and the discussion is expanded in this section. NO_X formation during pulverized coal combustion is also discussed, as this is the most widely used process for power generation. The majority of nitrogen oxides emitted from power plants are in the form of nitric oxide (NO), with only a small fraction as nitrogen dioxide (NO_2) and nitrous oxide (N_2O). Collectively, these oxides are referred to as NO_X. NO originates from the coal-bound nitrogen and nitrogen in the air used in the combustion process and is produced through three mechanisms: thermal NO, prompt NO, and fuel NO. Fuel-bound nitrogen accounts for 75 to 95% of the total NO generated, while thermal and prompt NO account for the balance, with prompt NO being no more than 5% of the total NO [21].

The factors that influence NO_X emissions in pulverized coal-fired boilers can be generally categorized as boiler design, boiler operation, and coal properties [21]; however, NO_X formation is complex, and many parameters influence its production [21,22]. Boiler design factors include boiler type, capacity, burner type, number and capacity of the burners, burner zone heat release, residence times, and presence of overfire air ports. Similarly, boiler operation factors include load, mills in operation, excess air level, burner tilt, and burner operation. Coal properties that influence NO_x production include the release of volatiles and nitrogen partitioning, ratio of combustibles-to-volatile matter, heating value, rank, and nitrogen content.

Thermal NO Thermal NO formation involves the high-temperature ($> 2370°F$) reaction of oxygen and nitrogen from the combustion air [21]. The principal reaction governing the formation of NO is the reaction of oxygen atoms formed from the dissociation of O_2 with nitrogen. These reactions, referred to as the Zeldovich mechanism, are:

$$N_2 + O\bullet \longleftrightarrow NO + N\bullet \tag{6-47}$$

$$N\bullet + O_2 \longleftrightarrow NO + O\bullet \tag{6-48}$$

These reactions are sensitive to temperature, local stoichiometry, and residence time. High temperature is required for the dissociation of oxygen and to overcome the high activation energy for breaking the triple bond of the nitrogen molecule. These reactions dominate in fuel-lean, high-temperature conditions. Under fuel-rich conditions, hydroxyl and hydrogen radical concentrations are increased which initiates oxidation of the nitrogen radicals, so at least one additional step should be included in this mechanism:

$$N\bullet + OH\bullet \longleftrightarrow NO + H\bullet \tag{6-49}$$

Reactions (6-47) through (6-49) are usually referred to as the extended Zeldovich mechanism. In addition, the following reactions can occur in fuel-rich conditions [21]:

$$H\bullet + N_2 \longleftrightarrow N_2H \tag{6-50}$$

$$N_2H + O\bullet \longleftrightarrow NO + NH\bullet \tag{6-51}$$

Thermal NO is of greater significance in the post-flame region than within the flame; consequently, several technologies have been developed for reducing thermal NO by lowering the peak temperature in the flame, minimizing the residence time in the region of the highest temperature, and controlling the excess air levels.

Prompt NO Prompt NO is the fixation of atmospheric (molecular) nitrogen by hydrocarbon fragments in the reducing atmosphere in the flame zone [21]. The proposed mechanism is:

$$CH\bullet + N_2 \longleftrightarrow HCN + N\bullet \tag{6-52}$$

$$HCN + O\bullet \longleftrightarrow NH\bullet + CO \tag{6-53}$$

$$NH\bullet + O\bullet \longleftrightarrow NO\bullet + H\bullet \tag{6-54}$$

The main reaction product of hydrocarbon radicals with N_2 is HCN, and the amount of NO formed is governed by the reactions of the nitrogen atoms with available radical species. In fuel-rich environments, therefore, the formation of N_2 is favored due to the reduced concentrations of hydroxyl and oxygen radical concentrations [21].

Fuel NO Nitrogen in the coal, which typically ranges from 0.5 to 2.0 wt.%, occurs mainly as organically-bound heteroatoms in aromatic rings or clusters [21]. Pyrrolic (5-membered ring) nitrogen is the most abundant form and contributes 50 to 60% of the total nitrogen. Pyridinic (6-membered ring) nitrogen comprises about 20 to 40% of the total nitrogen. The remaining 0 to 20% nitrogen is thought to be in amine or quaternary nitrogen form. Coal nitrogen is first released during volatilization in the coal flame as an element in aromatic compounds referred to as tar. The tar undergoes pyrolysis to convert most of the nitrogen to HCN as well as some NH_3 and NH. Some nitrogen is expelled from the char as HCN and occasionally NH_3; however, this occurs at a much slower rate than evolution from the volatiles. The partitioning of nitrogen between volatiles and char is important in NO_x formation.

Nitric oxide formation proceeds along two paths [21]. The nitrogen from the char reacts with oxygen to form NO. The NH_3 and NH released from the volatile matter and, to a lesser extent, the coal reacts with oxygen atoms to form NO. HCN is converted to NO via a pathway of hydrogen abstraction to form ammonia species and subsequently NO. Volatile nitrogen species can also be converted to nitrogen atoms through a series of fuel-rich pyrolysis reactions. Also, reactions between NO and volatile nitrogen species and carbon particles can result in the formation of nitrogen molecules:

$$C + NO \longleftrightarrow \tfrac{1}{2}N_2 + CO \qquad (6\text{-}55)$$

$$CH\bullet + NO \longleftrightarrow HCN + O\bullet \qquad (6\text{-}56)$$

In a fuel-rich environment, the main product of the reaction of NO with hydrocarbon radicals is HCN, which is then converted to N_2 in an oxygen-deficient environment. This is the basis for reburning, discussed later in this chapter, where a secondary hydrocarbon fuel is injected into combustion products containing NO.

Approximately 15 to 40% of the fuel nitrogen is converted to NO, and the formation of NO is influenced by stoichiometry, flame temperature, coal nitrogen content, and coal volatile matter content [21]. Approximately 25% of the char nitrogen is converted to NO. The reason for the fuel NO dominance (*i.e.*, 75–95% of total NO_x production) is because the N–H and N–C bonds, common in fuel-bound nitrogen, are weaker than the triple bond in molecular nitrogen, which must be dissociated to produce thermal NO.

Nitrogen Dioxide and Nitrous Oxide Small amounts of nitrogen dioxide (NO_2) and nitrous oxide (N_2O) are formed during coal combustion, but they comprise less than 5% of the total NO_x production. The oxygen levels are too low and the residence times are too short in high-temperature coal flames for much of the NO to be oxidized to NO_2. Nitrous oxide, however, can be formed in the early part of fuel-lean flames by gas-phase reactions [21]:

$$O\bullet + N_2 \longleftrightarrow N_2O \tag{6-57}$$

$$NH\bullet + NO \longleftrightarrow N_2O + H\bullet \tag{6-58}$$

$$NCO\bullet + NO \longleftrightarrow N_2O + CO \tag{6-59}$$

NO_x Control in Pulverized Coal Combustion

Technologies for control of NO_x emissions from pulverized coal-fired power plants can be divided into two groups: (1) combustion modifications where the NO_x production is reduced during the combustion process, and (2) flue gas treatment, which removes the NO_x from flue gas following its formation. Sometimes the practice of injecting reducing agents to reduce NO_x to molecular nitrogen (N_2) is classified separately; however, in this section it is included as a flue gas treatment. Table 6-8 lists various NO_x control technologies with a summary of their attributes [23]. The abatement or emission control principles for these various control methods include reducing peak flame temperatures, reducing the residence time at peak flame temperatures, chemically reducing NO_x, oxidizing NO_x with subsequent absorption, removing nitrogen, using a sorbent, or a combination of these methods.

Reducing combustion temperature is accomplished by operating at non-stoichiometric conditions to dilute the available heat with an excess of fuel, air, flue gas, or steam [23]. The combustion temperature is reduced by using fuel-rich mixtures to limit the availability of oxygen, using fuel-lean mixtures to dilute energy input, injecting cooled oxygen-depleted flue gas into the combustion air to dilute energy, injecting cooled flue gas with the fuel, or injecting water or steam.

Reducing residence times at high combustion temperatures is accomplished by restricting the flame to a short region to prevent the nitrogen from becoming ionized. Fuel, steam, more combustion air, or recirculating flue gas is then injected immediately after this region. Chemically reducing NO_x removes oxygen from the nitrogen oxides. This is accomplished by reducing the valence level of nitrogen to zero after the valence has become higher. Oxidizing NO_x intentionally raises the valence of the nitrogen ion to allow water to absorb to it. This is accomplished by using a catalyst, injecting hydrogen peroxide, creating ozone within the air flow, or injecting ozone into the air flow. Removing nitrogen from combustion is accomplished by removing nitrogen as a reactant either by using low nitrogen content fuels or using oxygen instead of air. The ability to vary coal nitrogen contents, however, is limited. Treatment of flue gas by injection sorbents such as ammonia, limestone, aluminum oxide, or carbon can remove NO_x and other pollutants. This type of treatment has been applied in the combustion chamber,

TABLE 6-8
NO_X Control Technologies

Technique	*Description*	*Advantages*	*Disadvantages*	*Impacts*	*Applicability*
Less excess air (LEA)	Reduces oxygen availability	Easy modification	Low NO_X reduction	High CO; flame length; flame stability	All fuels
Off stoichiometric	Staged combustion	Low cost; no capital cost for BOOS		Flame length; fan capacity; header pressure	All fuels; multiple burners required for BOOS
a. Burners out of service (BOOS) b. Overfire air (OFA)			a. Higher air flow for CO reduction b. High capital cost		
Low NO_X burner	Internal staged combustion	Low operating cost; compatible with FGR	Moderately high capital cost	Flame length; fan capacity; turndown capability	All fuels
Flue gas recirculation (FGR)	<30% Flue gas recirculated with air, decreasing temperature	High NO_X reduction potential for low nitrogen fuels	Moderately high capital and operating costs; affects heat transfer and system pressures	Fan capacity; furnace pressure; burner pressure drop; turndown stability	All fuels
Water/steam injection	Reduces flame temperature	Moderate capital cost; NO_X reduction similar to FGR	Efficiency penalty; fan power higher	Flame stability; efficiency penalty	All fuels
Reduced air preheat	Air not preheated, reduces flame temperature	High NO_X reduction potential	Significant efficiency loss (1%/40°F)	Fan capacity; efficiency penalty	All fuels

(continued)

TABLE 6-8
(continued)

Technique	Description	Advantages	Disadvantages	Impacts	Applicability
Selective catalytic reduction (SCR)	Catalyst located in air flow and promotes reaction between ammonia and NO_X	High NO_X removal	Very high capital cost; high operating cost; catalyst siting; increased pressure drop; possible water wash required	Space requirements; ammonia slip; hazardous materials; disposal	All fuels
Selective non-catalytic reduction (SNCR)	Injects reagent to react with NO_X				All fuels
a. Urea		a. Low capital cost; moderate NO_X removal; non-toxic chemical	a. Temperature dependent; NO_X reduction less at lower loads	a. Furnace geometry; temperature profile	
b. Ammonia		b. Low operating cost; moderate NO_X removal	b. Moderately high capital cost; ammonia storage, handling, injection system	b. Furnace geometry; temperature profile	
Fuel reburning	Injects fuel to react with NO_X	Moderate cost; moderate NO_X removal	Extends residence time	Furnace temperature profile	All fuels (pulverized solid)
Combustion optimization	Changes efficiency of primary combustion	Minimal cost	Extends residence time	Furnace temperature profile	All fuels
Inject oxidant	Chemical oxidant injected into flow	Moderate cost	Nitric acid removal	Add-on	All fuels
Oxygen instead of air	Uses oxygen as oxidizer	Moderate to high cost; intense combustion; eliminate thermal NO_X	Eliminate prompt NO_X; furnace alteration	Equipment to handle oxygen	All fuels

(continued)

TABLE 6-8
(continued)

Technique	*Description*	*Advantages*	*Disadvantages*	*Impacts*	*Applicability*
Ultra-low nitrogen fuel	Uses low-nitrogen fuel	Eliminates fuel NO_X; no capital cost	Possible rise in operating cost	Minimal change	All ultra-low nitrogen fuels
Sorbent injection (combustion; duct to baghouse; duct to ESP)	Uses a chemical to absorb NO_X or an adsorber to capture or reduce it	Can control other pollutants as well as NO_X; moderate operating cost	Cost of sorbent; space for the sorbent storage and handling	Add-on	All fuels
Air staging	Admits air in separated stages	Reduces peak combustion temperature	Extends combustion to a longer residence time at lower temperature	Add ducts and dampers to control air; furnace modification	All fuels
Fuel staging	Admits fuel in separated stages	Reduces peak combustion temperature	Extends combustion to a longer residence time at lower temperature	Adds fuel injectors to other locations; furnace modification	All fuels

Source: EPA, *Technical Bulletin, Nitrogen Oxides (NO_X): Why and How They Are Controlled*, Office of Air Quality Planning and Standards, U.S. Environmental Protection Agency, U.S. Government Printing Office, Washington, D.C., November 1999.

flue gas, and particulate control device. Many of these methods can be combined to achieve a lower NO_X concentration than can be achieved alone by any one method. In some cases, technologies that are used to control other pollutants, such as SO_2, can also reduce NO_X.

Combustion Modifications Primary NO_X control technologies involve modifying the combustion process. Several technologies have been developed and applied commercially and include:

- Low-NO_X burners;
- Furnace air staging;
- Flue gas recirculation;
- Fuel staging (*i.e.*, reburn);
- Process optimization.

Options to control NO_X during combustion and their effects are different for new and existing boilers. For new boilers, combustion modifications are easily made during construction, whereas for existing boilers viable alternatives are more limited. Modifications can be complicated, and unforeseen problems may arise. When combustion modifications are made, it is important to avoid adverse impacts on boiler operation and the formation of other pollutants such as N_2O or CO. Issues pertaining to low-NO_X operation include:

- Safe operation (*e.g.*, stable ignition over the desired load range);
- Reliable operation to prevent corrosion, erosion, deposition, and uniform heating of the tubes;
- Complete combustion to limit formation of other pollutants such as CO, polyorganic matter, or N_2O;
- Minimal adverse impact on the flue gas cleaning equipment;
- Low maintenance costs.

Combustion modification technologies redistribute the fuel and air to slow mixing, reduce the availability of oxygen in the critical NO_X formation zones, and decrease the amount of fuel burned at peak flame temperatures. In addition, reburning chemically destroys the NO_X formed by hydrocarbon radicals during the combustion process. The commercially applied technologies are discussed in detail in the following sections. One technology listed in Table 6-8, low excess air (LEA), is the simplest of the combustion control strategies but is not discussed in detail here because it has achieved only limited success in coal-fired applications (*i.e.*, 1–15%). In this technique, excess air levels are reduced until there are adverse impacts on CO formation and flame length and stability. Similarly, a technique known as burners out of service (BOOS) has met with limited success with coal and is not discussed in detail. In this technique, the fuel flow to the selected burner is stopped but airflow is maintained to create staged combustion in the furnace. The remaining burners operate fuel rich, which limits oxygen availability, lowers peak flame temperatures, and reduces NO_X formation. The unreacted products combine with air from the burners out of service to complete burnout before exiting the furnace.

Low-NO_X Burners Prior to concerns being raised regarding NO_X emissions in the early 1970s, coal burners were designed to provide highly turbulent mixing and combustion at peak flame temperatures to ensure high combustion efficiency, a condition that is ideal for NO_X formation [24]. In 1971, industry began developing low-NO_X burners for coal-fired boilers with the promulgation of New Source Performance Standards (NSPSs). By the mid-1970s, low-NO_X burners were being demonstrated, and commercial operation started in the late 1970s [25]. They have undergone considerable improvements in design spurred by the 1990 Clean Air Act Amendments Title IV, Phase II acid rain regulations and Title I ozone regulations [26,27]. The technology is well proven for NO_X control in both wall- and tangentially-fired boilers and is commercially available; a significant number of them are installed worldwide.

Low-NO_X burners work under the principle of staging the combustion air within the burner to reduce NO_X formation. Rapid devolatilization of the coal particles occurs near the burner in a fuel-rich, oxygen-starved environment to produce NO. NO_X formation is suppressed because oxygen molecules are not available to react with the nitrogen released from the coal and present in the air, and the flame temperature is reduced. Hydrocarbon radicals that are generated under the sub-stoichiometric conditions then reduce the NO that is formed to N_2. The air required to complete the burnout of the coal is added after the primary combustion zone where the temperature is sufficiently low so that additional NO_X formation is minimized.

Larger and more branched flames are produced by staging the air [21]. This flame structure limits coal and air mixing during the initial devolatilization stage while maximizing the release of volatiles from the coal. The more volatile nitrogen that is released with the volatiles and the longer the residence time in the fuel-rich zone, the lower the amount of fuel NO that is produced. An oxygen-rich layer is produced around the flame that aids in carbon burnout. An example of this concept is shown in Figure 6-11, which is a schematic of a low-NO_X burner (*i.e.*, Ahlstom Power's Radially Stratified Fuel Core burner) that illustrates a typical flowfield emanating from it [28]. A photograph of the burner, which is a 20 million Btu/hr prototype used for developmental work at Penn State prior to its commercialization and worldwide deployment, is shown in Figure 6-12, which depicts the various dampers and air scoops for channeling and controlling the quantity and degree of swirl of the various air streams [29].

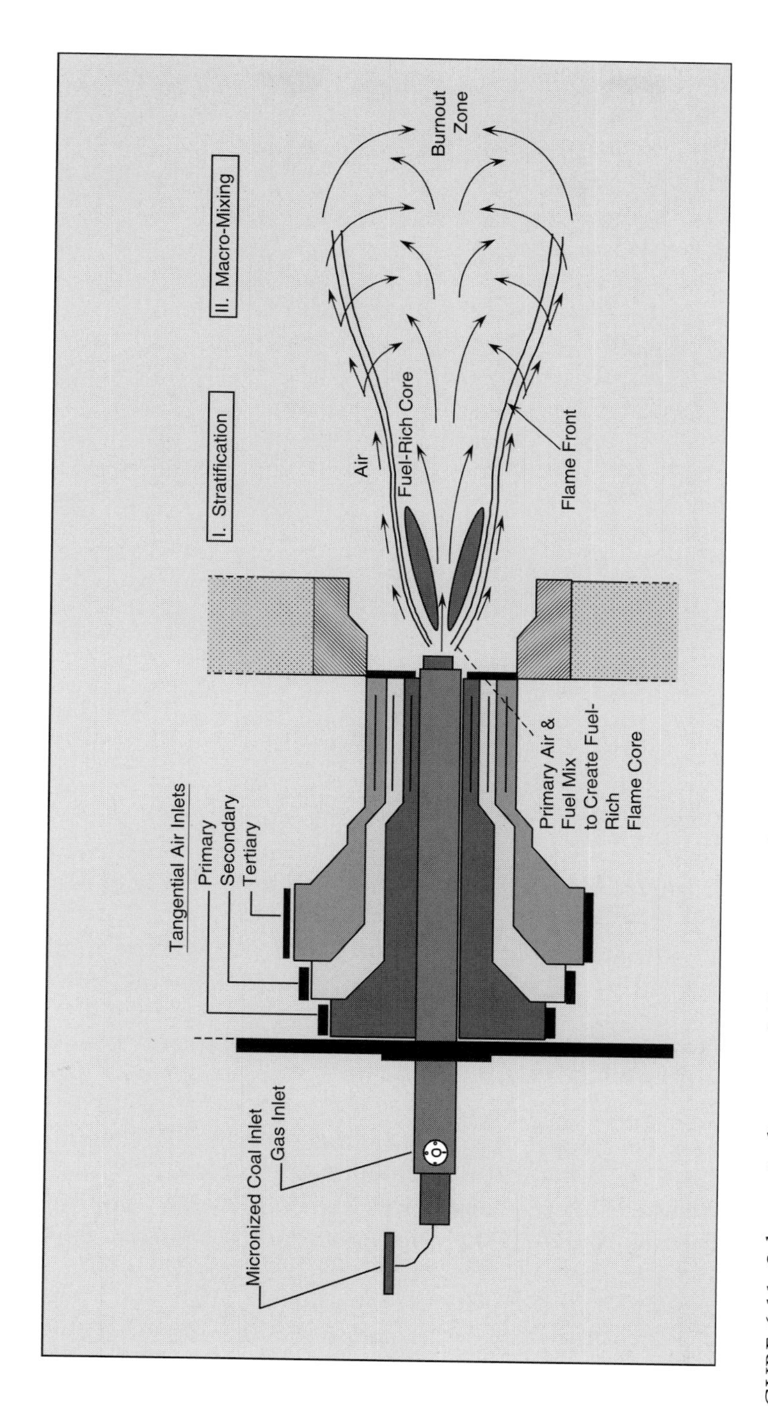

FIGURE 6-11. Schematic diagram of Alstom's RSFC burner depicting flow fields. (From Patel, R. L. *et al.*, Firing Micronized Coal with a Low NO_x RSFC Burner in an Industrial Boiler Designed for Oil and Gas, in *Proc. of the Thirteenth Annual International Pittsburgh Coal Conference*, 1996.)

FIGURE 6-12. Photograph of the RSFC burner showing internal components.

Low NO_X burners are designed to accomplish the following [30]:

- Maximize the rate of volatiles evolution and total volatile yield from the fuel with the fuel nitrogen evolving in the reducing part of the flame;
- Provide an oxygen-deficient zone where the fuel nitrogen is evolved to minimize its conversion to NO_X but sufficient oxygen is available to maintain a stable flame;
- Optimize the residence time and temperature in the reducing zone to minimize conversion of the fuel nitrogen to NO_X;
- Maximize the char residence time under fuel-rich conditions to reduce the potential for NO_X formation from the nitrogen remaining in the char after devolatilization;
- Add sufficient air to complete combustion.

All low-NO_X burners employ the air-staging principle, but the designs vary widely between manufacturers. All of the major boiler manufacturers have one or more versions of low-NO_X burners employed in boilers throughout the world. Mitchell [21] reported that over 370 units worldwide were fitted with low-NO_X burners at a total generating capacity of more than 125 GW prior to 1998. The number of units installing low-NO_X burners has increased significantly, as the DOE reports that low-NO_X burners are currently found on more than 75% of U.S. coal-fired power capacity [1]. This is significant, as the DOE reported that ~1030 coal-fired, steam–electric generators today have a nameplate capacity of 328 GW and produce more than 1515 billion kWh of electricity [3,31].

Low-NO_X burners, based on air-staging alone, are capable of achieving 30 to 60% NO_X reduction. In addition, they should perform in such a way that [30]:

- The overall combustion efficiency is not significantly reduced;
- Flame stability and turndown limits are not impaired;
- The flame has an oxidizing envelope to minimize the potential for high temperature corrosion at the furnace walls;
- Flame length is compatible with furnace dimensions;
- The performance should be acceptable for a wide range of coals.

The major concern with low-NO_X burners is the potential for reducing combustion efficiency and thereby increasing the unburned carbon level in the fly ash. An increase in the unburned carbon level will lower the fly ash resistivity, which can reduce the efficiency of an ESP. In addition, it may also affect the sale of the ash. Some operating parameters that can be adjusted to mitigate the impact of the unburned carbon include [30]:

- Fire coal with high reactivity and high volatile matter content;
- Reduce the size of the coal particles;
- Balance coal distribution to the burners;
- Use advanced combustion control systems.

Furnace Air Staging One technique to stage combustion is to install secondary and even tertiary overfire air (OFA) ports above the main combustion zone. This is a well-proven, commercially-available technology for NO_X reduction at coal-fired power plants and is applicable to both wall- and tangentially-fired boilers [30]. When OFA is employed, 70 to 90% of the combustion air is supplied to the burners with the coal (*i.e.*, primary air), and the balance is introduced to the furnace above the burners (*i.e.*, overfire air). The primary air and coal produce a relatively low-temperature, oxygen-deficient, fuel-rich environment near the burner which reduces the formation of fuel-NO_X. The overfire air is injected above the primary combustion zone to produce a relatively low-temperature secondary combustion zone that limits the formation of thermal NO_X.

Overfire air in combination with low NO_X burners can reduce NO_X emissions by 30 to 70%. Advanced OFA systems, such as separated overfire air (SOFA), where the overfire air is introduced some distance above the burners, and close-coupled overfire air (CCOFA), where the overfire air nozzles are immediately above the burners, can achieve higher NO_X reduction efficiency [21]. Mitchell [21] reports that furnace air staging is used in ~300 pulverized coal-fired units with a total generating capacity of over 100 GW. A number of advanced overfire air systems are commercially available and designs vary among suppliers [30]. Furnace air staging can increase unburned carbon levels in the ash by 35 to 50%, with the degree of increase being dependent on the reactivity of the coal used [30]. In addition, operational problems can be experienced, including waterwall corrosion, changes in slagging and fouling patterns, and a loss in steam temperature.

Flue Gas Recirculation Flue gas recirculation (FGR) involves recirculating part of the flue gas back into the furnace or the burners to modify conditions in the combustion zone by lowering the peak flame temperature and reducing the oxygen concentration, thereby reducing thermal NO_X formation. FGR has been used commercially for many years at coal-fired units; however, unlike gas- and oil-fired boilers, which can achieve high NO_X reduction, coal-fired boilers typically realize less than 20% NO_X reduction due to a relatively low contribution of thermal NO_X to total NO_X. In conventional FGR applications, 20 to 30% of the flue gas is extracted from the boiler outlet duct upstream of the air heater (at ~570 to 750°F) and is mixed with the combustion air. This process reduces thermal NO_X formation without any significant effect on fuel NO_X. A major consideration of FGR is the impact on boiler thermal performance [30]. The reduced flame temperature lowers heat transfer, potentially limiting the maximum heating capacity of the unit, which results in a reduction in steam-generating capacity.

Fuel Staging (Reburn) Reburn is a comparatively new technology that combines the principles of air and fuel staging. In this technology, a reburn fuel (*e.g.*, coal, oil, gas, orimulsion, biomass, coal–water mixtures) is used as a reducing agent to convert NO_X to N_2. The process does not require modifications to the existing main combustion system and can be used on wall-, tangential-, and cyclone-fired boilers. Reburn is a combustion hardware modification in which the NO_X produced in the main combustion zone is reduced downstream in a second combustion zone (*i.e.*, the reburn zone). This, in turn, is followed by a zone where overfire air is introduced to complete burnout. This is illustrated in Figure 6-13 [32].

In the primary combustion zone, the burners are operated at a reduced firing rate with low excess air (stoichiometry of 0.9 to 1.1) to produce lower fuel and thermal NO_X levels. The reburn fuel, which can be 10 to 30% of the total fuel input on a heat input basis, is injected above the main combustion zone to create a fuel-rich zone (stoichiometry of 0.85–0.95) [32]. In this zone, most of the NO_X reduction occurs, with hydrocarbon radicals formed

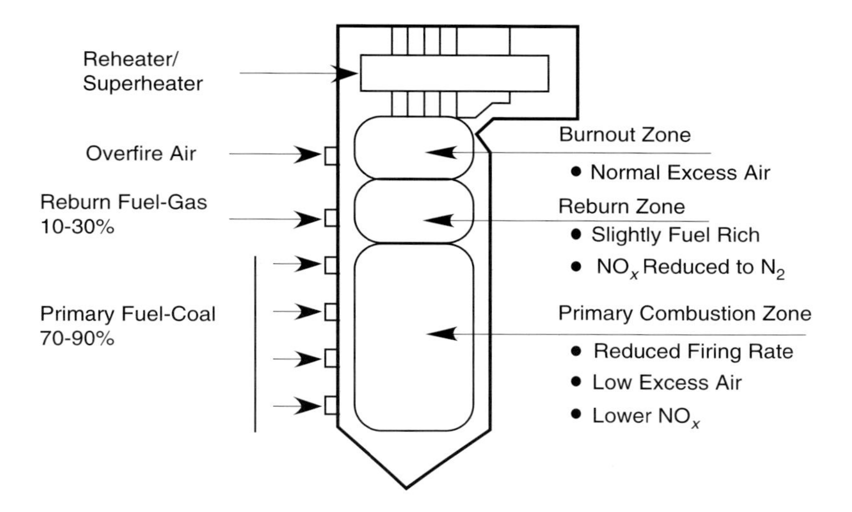

FIGURE 6-13. Schematic of the reburn process. (From EPA, *Control of NO_X Emissions by Reburning*, Office of Research and Development, U.S. Environmental Protection Agency, U.S. Government Printing Office, Washington, D.C., February 1996.)

in the reburn zone reacting with the NO_X to form N_2 and water vapor. The temperature in this zone must be greater than 1800°F. The remaining combustion air is injected above the reburn zone to produce a fuel-lean burnout zone.

Reburn technology is considered relatively new, but numerous pilot-scale tests and full-scale demonstrations have been conducted and the concept was proposed in the late 1960s [33]. The concept was based on the principle of Myerson *et al.* [34] that CH fragments can react with NO. The major chemical reactions for the reburn process are [32]:

$$\text{Hydrocarbon fuel} \xrightarrow[\text{Heat and } O_2 \text{ deficiency}]{} \bullet CH_2 \tag{6-60}$$

where hydrocarbon radicals are produced due to the pyrolysis of the fuel in the oxygen-deficient, high-temperature reburn zone. The hydrocarbon radicals then mix with the combustion gases from the primary combustion zone:

$$\bullet CH_3 + NO \longrightarrow HCN + H_2O \tag{6-61}$$

$$N_2 + \bullet CH_2 \longrightarrow NH_2 + HCN \tag{6-62}$$

$$\bullet H + HCN \longrightarrow \bullet CN + H_2 \tag{6-63}$$

The radicals then react with the NO to form molecular nitrogen:

$$NO + \bullet NH_2 \longrightarrow N_2 + H_2O \tag{6-64}$$

$$NO + \bullet CN \longrightarrow N_2 + CO \quad (6\text{-}65)$$

$$2NO + 2CO \longrightarrow N_2 + 2CO_2 \quad (6\text{-}66)$$

An oxygen-deficient atmosphere is critical for Reactions (6-61) through (6-63) to occur. If oxygen levels are high, the NO_x reduction reactions will not occur, and the following will predominate:

$$CN + O_2 \longrightarrow CO + NO \quad (6\text{-}67)$$

$$NH_2 + O_2 \longrightarrow H_2O + NO \quad (6\text{-}68)$$

To complete the combustion process, air is introduced above the reburn zone. Some NO_x is formed from conversion of HCN and ammonia compounds; however, the net effect is to significantly reduce the total quantity of NO_x emitted from the boiler. The reactions with HCN and ammonia are:

$$HCN + \tfrac{5}{4}O_2 \longrightarrow NO + CO + \tfrac{1}{2}H_2O \quad (6\text{-}69)$$

$$NH_3 + \tfrac{5}{4}O_2 \longrightarrow NO + \tfrac{3}{2}H_2O \quad (6\text{-}70)$$

$$HCN + \tfrac{3}{4}O_2 \longrightarrow \tfrac{1}{2}N_2 + CO + \tfrac{1}{2}H_2O \quad (6\text{-}71)$$

$$NH_3 + \tfrac{3}{4}O_2 \longrightarrow \tfrac{1}{2}N_2 + \tfrac{3}{2}H_2O \quad (6\text{-}72)$$

Reburn offers the advantages of being able to operate over a wide range of NO_x reduction values using a variety of reburn fuels. A reburn system can be varied from relatively low levels of reduction (25 to 30%) using an overfire air system without any reburn fuel to higher levels of reduction (~70%) when reburn fuel is added [21,30]. This allows for fine-tuning to meet emissions limits.

Concerns regarding the use of reburn technology are similar to those for other combustion modification processes. This includes concerns about incomplete combustion (*i.e.*, CO and hydrocarbon production and unburned carbon in the fly ash), changes in slagging and fouling characteristics, different ash characteristics and fly ash loadings, corrosion of boiler tubes in reducing atmospheres, higher fan power consumption, and pulverizer constraints (if pulverized coal is used as the reburn fuel).

Cofiring Cofiring is the practice of firing a supplementary fuel, such as coal-water slurry fuel (CWSF) or biomass, with a primary fuel (*i.e.*, coal) in the same burner or separately but into the main combustion zone. This technology was originally developed to utilize opportunity fuels; however, various levels of NO_x reduction were achieved and provide an option for NO_x reduction without investing in a post-combustion system when the emissions are near the regulatory requirements. This technique is not currently used as a commercial means for NO_x reduction; however, it is briefly discussed in this section because several demonstrations of this technology have been conducted, with a few still ongoing. In addition, it is considered a viable option for NO_x trimming especially if used in conjunction with legislation that mandates a percentage of electricity be generated from renewable/sustainable sources. Such legislation has been seriously discussed in the United States and has been included in congressional bills although they have not yet passed.

The CWSF technology was originally developed as a fuel oil replacement and underwent considerable research and development from the late 1970s to the late 1980s. During the late 1980s and early 1990s, coal suppliers and coal-fired utilities began to evaluate the production of CWSF using bituminous coal fines from coal cleaning circuits in an effort to reduce dewatering/drying costs and/or to recover and utilize low-cost impounded coal fines [35,36]. This marked a philosophical change in the driving force behind utilizing CWSF in the United States as well as the CWSF characteristics of these two fuel types, as cofire CWSFs are quite different from fuel oil-replacement CWSFs: Cofire CWSFs have a low solids content (50%) and no additive package to wet the coal, provide stability, and modify rheology, whereas fuel oil-replacement CWSFs have a high solids content (~70%) and an expensive additive package. Extensive testing performed by several companies and universities culminated in waste impoundment characterizations and several utility demonstrations in pulverized coal-fired boilers (both wall- and tangentially-fired units) and cyclone-fired boilers. Funding for these demonstrations was provided by industry, the DOE, the Electric Power Research Institute (EPRI), and state agencies. Penn State provided fuel support in all but one of these demonstrations, as summarized in a CWSF preparation and operation manual prepared by Morrison *et al.* [37], where the CWSFs were being developed to provide coal preparation plants a means for utilizing difficult-to-dewater fines, cleaning up waste coal impoundments to reduce coal mine liability, and supplying utilities with a low-cost fuel that also serves as a low-cost NO_X reduction technology. NO_X reductions were achieved that varied from ~11% in cyclone-fired boilers [38] to ~30% in wall-fired boilers [39,40] to ~35% in tangentially-fired boilers [41]. Several mechanisms were responsible for the NO_X reduction, including lower flame temperature from the addition of the water, staged combustion from cofiring in low-NO_X burners, and the CWSF acting as a reburn fuel when injected in upper level burners.

Biomass cofiring has been demonstrated and deployed at a number of power plants in the United States and Western Europe using a variety of materials, including sawdust, urban wood waste, switchgrass, straw, and other similar materials [42]. Biomass fuels have been cofired with all ranks of coal: bituminous and subbituminous coals and lignites. The benefits of biomass cofiring include reduced NO_X, fossil CO_2, SO_2, and mercury emissions.

Cofiring biomass, particularly sawdust and urban wood waste but also switchgrass to a lesser extent, in large-scale pulverized coal-fired and cyclone-fired units has been demonstrated at several utilities with seven commercial installations in the United States [43,44]. Many of the demonstrations were conducted to achieve NO_X reductions, which can vary significantly but can be as high as ~35%. Tillman [42] noted that the dominant mechanism for NO_X reduction is to support deeper staging of combustion when staging has not been particularly extensive. When biomass can introduce or accentuate staging by early release of volatile matter, then NO_X reduction can be significant [42,45]. A secondary mechanism for NO_X reduction is the influence of cofiring on furnace exit gas temperature (FEGT). Data indicate that cofiring has minimal impact on flame temperatures but can have a pronounced impact on FEGT, thereby reducing NO_X emissions. A third influence is the reduction in fuel nitrogen content when a low-nitrogen fuel such as sawdust is used.

Process Optimization Several software packages have been developed or are under development that apply optimization procedures to the distributed control system of the boiler to provide tighter control of plant operation parameters [30]. The combustion process is optimized, resulting in lower NO_X emissions and improved boiler efficiency while maintaining safe, reliable, and consistent unit operation. Also, combustion optimization approaches have been developed where advanced computational and experimental approaches are used to make design and operational modifications to the process equipment and boiler as a whole [46].

The main software packages are the ULTRAMAX Method, Generic NO_X Control Intelligent System (GNOCIS/GNOCIS Plus), Boiler OP, QuickStudy, and Smart Burn [30,46]. The use of these packages has resulted in NO_X reductions of 10 to 40%, reduced unburned carbon levels by 25 to 50%, increased boiler efficiencies by 1 to 3%, and increased heat rates by 0.5 to 5%.

Flue Gas Treatment Flue gas treatment technologies are post-combustion processes to convert NO_X to molecular nitrogen or nitrates. The two primary strategies that have been developed for post-combustion control and are commercially available are selective catalytic reduction (SCR) and selective non-catalytic reduction (SNCR). Additional concepts are under development, including combining SCR and SNCR technologies (known as hybrid SCR/SNCR) and rich reagent injection; however, these are not extensively used at this time. Of these technologies, SCR is being identified by utilities as the strategy to meet stringent NO_X requirements. These technologies are discussed in the following sections, with an emphasis on SCR.

Selective Catalytic Reduction Selective catalytic reduction of NO_X using ammonia (NH_3) as the reducing gas was patented in the United States by Englehard Corporation in 1957 [47]. This technology can achieve NO_X reductions in excess of 90% and is widely used in commercial applications in Western Europe and Japan, which have stringent NO_X regulations, and is becoming the post-combustion technology of choice in the United States. Stringent NO_X regulations in Western Europe essentially mandate the installation of SCR, and approximately 40 GW of generating capacity are fitted with secondary NO_X reduction systems, the majority of which utilize SCR with only a few boilers using the SNCR process [48]. Similarly, SCR technology was introduced into commercial service in Japan in 1980 and has been applied to more than 23 GW of coal-fired generating capacity in 61 plants. U.S. utilities initially deployed SCR for coal-fired units for new and retrofit applications in 1991 and 1993, respectively [49]. SCR units have been installed on ~26 GW of generating capacity in the United States, but by the year 2007 more than 200 SCR installations with overall capacity greater than 100 GW are anticipated to be in place to meet NO_X targets mandated by the SIP-Call [48,49].

The SCR process uses a catalyst at approximately 570 to 750°F to facilitate a heterogeneous reaction between NO_X and an injected reagent, vaporized ammonia, to produce nitrogen and water vapor. Ammonia chemisorbs onto the active sites on the catalyst. The NO_X in the flue gas reacts with the adsorbed ammonia to produce nitrogen and water vapor. The principal reactions are [50]:

$$4NO + 4NH_3 + O_2 \longrightarrow 4N_2 + 6H_2O \tag{6-73}$$

$$2NO_2 + 4NH_3 + O_2 \longrightarrow 3N_2 + 6H_2O \tag{6-74}$$

A small fraction of the sulfur dioxide is oxidized to sulfur trioxide over the SCR catalyst. In addition, side reactions may produce the undesirable by-products ammonium sulfate ($(NH_4)_2SO_4$) and ammonium bisulfate (NH_4HSO_4), which can cause plugging and corrosion of downstream equipment. These side reactions are [47]:

$$SO_2 + \tfrac{1}{2}O_2 \longrightarrow SO_3 \tag{6-2}$$

$$2NH_3 + SO_3 + H_2O \longrightarrow (NH_4)_2\,SO_4 \tag{6-75}$$

$$NH_3 + SO_3 + H_2O \longrightarrow NH_4HSO_4 \tag{6-76}$$

The three SCR system configurations for coal-fired boilers are high-dust, low-dust, and tail-end, which are shown schematically in Figure 6-14 [50]. In a high-dust configuration, the SCR reactor is placed upstream of the particulate removal device between the economizer and the air preheater. This configuration (also referred to as hot-side, high-dust) is the most commonly used, particularly with dry-bottom boilers [30], and is the principle type planned for U.S. installations [48]. In this configuration, the catalyst is exposed to the fly ash and chemical compounds present in the flue gas that have the potential to degrade the catalyst by ash erosion and chemical reactions (*i.e.*, poisoning); however, these can be addressed by proper design as evidenced by the extensive use of this configuration.

In a low-dust installation, the SCR reactor is located downstream of the particulate removal device. This configuration (also referred to as hot-side, low-dust) reduces the degradation of the catalyst by fly ash erosion; however,

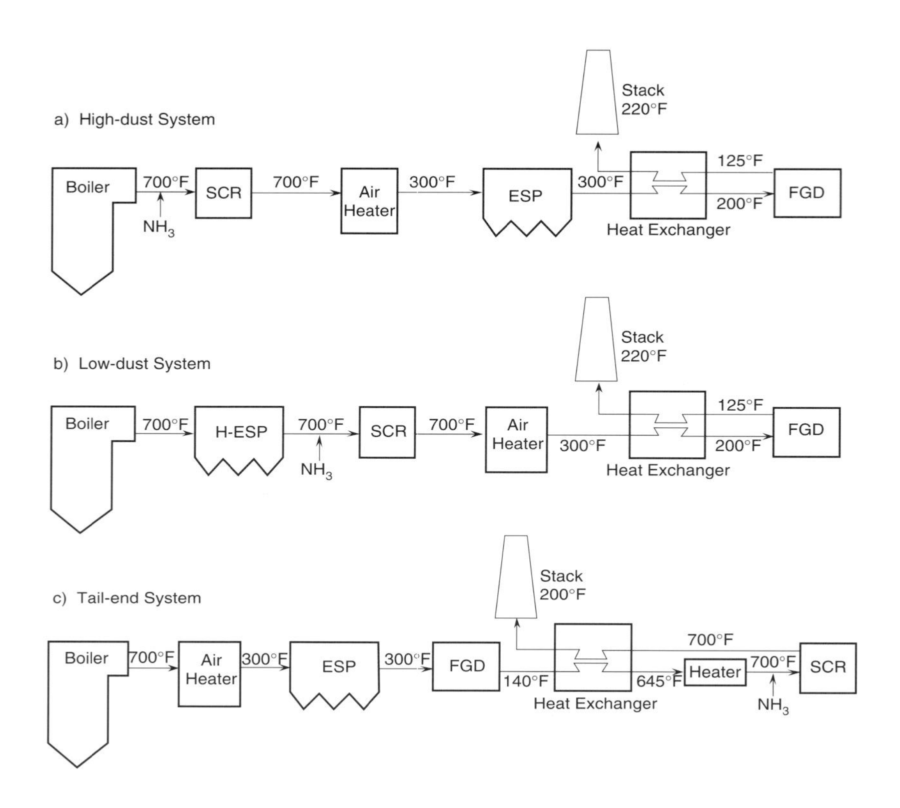

FIGURE 6-14. SCR configurations with typical system temperatures. (From EPA, *Performance of Selective Catalytic Reduction on Coal-Fired Steam Generating Units*, Office of Air and Radiation, U.S. Environmental Protection Agency, U.S. Government Printing Office, Washington, D.C., June 25, 1997.)

this configuration requires a costly hot-side ESP or a flue gas reheating system to maintain the optimum operating temperature.

In tail-end systems (also referred to as cold-side, low-dust), the SCR reactor is installed downstream of the FGD unit. It may be used mainly in wet-bottom boilers and also on retrofit installations with space limitations [30]; however, this configuration is typically more expensive than the high-dust configuration due to flue gas reheating requirements. This configuration does have the advantage of longer catalyst life and the use of more active catalyst formulations to reduce overall catalyst cost.

Several issues need to be considered in the design and operation of SCR systems, including coal characteristics, catalyst and reagent selections, process conditions, ammonia injection, catalyst cleaning and regeneration, low-load operation, and process optimization [30]. Coals with high sulfur in combination with significant quantities of alkaline, alkaline earth, arsenic, or phosphorus in the ash can severely deactivate a catalyst and reduce its service life. In addition, the SO_3 can react with residual ammonia, resulting in ammonium sulfate deposition in the air preheater and loss of performance.

The two leading geometries of SCR catalysts are honeycomb and plate [47]. The honeycomb form usually is an extruded ceramic, with the catalyst either incorporated throughout the structure (homogenous) or coated on the substrate. In the plate geometry, the support material is generally coated with catalyst. The catalyst commonly consists of a vanadium pentoxide active material on a titanium dioxide substrate.

For optimum SCR performance, the reagent must be well mixed with the flue gas and in direct proportion to the amount of NO_x reaching the catalyst. Anhydrous ammonia has been commonly used as the reagent, accounting for over 90% of current-world SCR applications [30]. It dominates planned installations in the United States, although numerous aqueous systems will be installed. Urea-based processes are being developed to address utilizing anhydrous ammonia, which is a hazardous and toxic chemical. When urea $CO(NH_2)_2$ is used, it produces ammonia, which is the active reducing agent, by the following reactions:

$$NH_2\text{–}CO - NH_2 \longrightarrow NH_3 + HNCO \tag{6-77}$$

$$HNCO + H_2O \longrightarrow NH_3 + CO_2 \tag{6-78}$$

During the operation of the SCR, the catalyst is deactivated by fly ash plugging, catalyst poisoning, and/or the formation of binding layers. The most common method of catalyst cleaning has been the installation of steam sootblowers, although acoustic cleaners have been successfully tested. Once the catalyst has been severely deactivated, it is conventional practice to add additional catalyst or replace it; however, several regeneration techniques have evolved over the last few years, providing extended service life for catalysts [30]. Low-load boiler operation can be problematic with SCR operation, specifically with high-sulfur coals. There is a minimum temperature below which the SCR should not be operated; therefore, system modifications, such as economizer bypass, to raise the SCR temperature during low-load operation may be required [30].

Selective Non-Catalytic Reduction Selective non-catalytic reduction is a proven, commercially-available technology that has been applied since 1974; over 300 systems are installed worldwide on various combustion sources, including utility applications [30]. The SNCR process involves injecting nitrogen-containing chemicals into the upper furnace or convective pass of a boiler within a specific temperature window without the use of an expensive catalyst. Various chemicals can be used that selectively react with NO in the presence of oxygen to form molecular nitrogen and water, but the two most common are ammonia and urea. Other chemicals that have been tested in research include amines, amides, amine salts, and cyanuric acid. In recent years, urea-based reagents such as dry urea, molten urea, or urea solution have been increasingly used, replacing ammonia at many plants because anhydrous ammonia is the most toxic and requires strict transportation, storage, and handling procedures [30]. The main reactions when using ammonia or urea are, respectively:

$$4NO + 4NH_3 + O_2 \longrightarrow 4N_2 + 6H_2O \quad (6\text{-}73)$$

$$4NO + 2CO\,(NH_2)_2 + O_2 \longrightarrow 4N_2 + 2CO_2 + 4H_2O \quad (6\text{-}79)$$

A critical issue is finding an injection location with the proper temperature window for all operating conditions and boiler loads. The chemicals then need to be adequately mixed with the flue gases to ensure maximum NO_x reduction without producing too much ammonia. Ammonia slip from an SNCR can affect downstream equipment by forming ammonium sulfates.

The temperature window varies for most of the reducing chemicals used but generally is between 1650 and 2100°F. Ammonia can be formed below the temperature window, and the reducing chemicals can actually form more NO_x above the temperature window. Ammonia has a lower operating temperature than urea: 1560 to 1920°F vs. 1830 to 2100°F, respectively. Enhancers such as hydrogen, carbon monoxide, hydrogen peroxide (H_2O_2), ethane (C_2H_6), light alkanes, and alcohols have been used in combination with urea to reduce the temperature window [51]. Several processes use proprietary additives with urea in order to reduce NO_x emissions [52].

The efficiency of reagent utilization is significantly less with SNCR than with SCR. In commercial SNCR systems, the utilization is typically between 20 and 60%; consequently, usually three to four times as much reagent is required with SNCR to achieve NO_x reductions similar to those of SCR. SNCR processes typically achieve 20 to 50% NO_x reduction with stoichiometric ratios of 1.0 to 2.0.

The major operational impacts of SNCR include air preheater fouling, ash contamination, N_2O emissions, and minor increases in heat rate. A major plant impact of SNCR is on the air preheater, where residual ammonia reacts with the SO_3 in the flue gas to form ammonium sulfate and bisulfate (see Reactions (6-2), (6-75), (6-76)), causing plugging and downstream corrosion. High levels of ammonia slip can contaminate the fly ash and reduce its sale or disposal. Significant quantities of N_2O can be formed when the reagent is injected into areas of the boiler that are below the SNCR optimum operating temperature range. Urea injection tends to produce a higher level of N_2O compared to ammonia. The unit heat rate is increased slightly due to the latent heat losses from vaporization of injected liquids and/or increased power requirements for high-energy injection systems. The overall efficiency and power losses normally range from 0.3 to 0.8% [30].

Hybrid SNCR/SCR Selective catalytic reduction generally represents a relatively high capital requirement, whereas selective non-catalytic reduction has a high reagent cost. A hybrid SNCR/SCR system balances these costs over the life cycle for a specific NO_x reduction level, provides improvements in reagent utilization, and increases overall NO_X reduction [30]. However, experience with these hybrid systems is limited, as full-scale power plant operation to date has only been in demonstrations. They are discussed here because they have demonstrated NO_X reductions as high as 60 to 70%.

In a hybrid SNCR/SCR system, the SNCR operates at lower temperatures than stand-alone SNCRs, resulting in greater NO_X reduction but also higher ammonia slip. The residual ammonia feeds a smaller-sized SCR reactor, which removes the ammonia slip and decreases NO_X emissions further. The SCR component may achieve only 10 to 30% NO_X reduction, with reagent utilization being as high as 60 to 80% [30]. Hybrid SNCR/SCR systems can be installed in various configurations, including [30]:

- SNCR with conventional reactor-housed SCR;
- SNCR with in-duct SCR, which uses catalysts in existing or expanded flue gas ductwork;
- SNCR with catalyzed air preheater, where catalytically active heat transfer elements are used;
- SNCR with a combination of in-duct SCR and catalyzed air heater.

Rich Reagent Injection Cyclone burners, with their turbulent and high-temperature environment, are conducive for NO_X production. Methods that cost less than installing SCRs to reduce NO_X production in cyclone-fired boilers have been tested, such as CWSF or biomass cofiring, while others are under development. One such process currently under development is the rich reagent injection (RRI) process, which involves injection of amine reagents in the fuel-rich zone above the main combustion zone at temperatures of 2370 to 3100°F. NO_X in the flue gas is converted to molecular nitrogen, and reductions of 30% have been achieved. The capital costs for an RRI system are consistent with those of SNCR; however, the operating costs are expected to be 2 to 3 times that of SNCR due to increased reagent usage.

NO_X Control in Fluidized-Bed Combustion

The fluidized-bed combustion (FBC) process described in Chapter 5 (Technologies for Coal Utilization) inherently produces lower NO_X emissions due to its lower operating temperature (*i.e.*, bed temperature of ~1600°F). Also, the bed is a reducing region where available oxygen is consumed by carbon, thereby reducing ionization of nitrogen. Additional combustion modifications or flue gas treatment for NO_X control, discussed previously in this chapter, can also be employed. Techniques currently used for FBC include reducing the peak temperature by flue gas recirculation (FGR), natural gas reburning (NGR), overfire air (OFA), fuel reburning, low excess air (LEA), and reduced air preheat [23]. Post-combustion control is also used, including SCR and SNCR, which achieve 35 to 90% NO_X reductions. Also, low nitrogen fuel can be used (*e.g.*, sawdust), thereby reducing the amount of fuel nitrogen available. Injecting sorbents into the combustion chamber or in the ducts can reduce NO_X by 60 to 90% [23].

NO_X Control in Stoker-Fired Boilers

Control of NO_X in stokers (specifically, traveling-grate and spreader stokers) include abatement methods to reduce the peak temperature, to reduce the residence time at peak temperature, and to chemically reduce the NO_X, in addition to using low-nitrogen fuels and injecting a sorbent [23]. In traveling-grate stokers, the peak temperature can be reduced by FGR, NGR, combustion optimization, OFA, LEA, water or steam injection, and reduced air preheat, thereby achieving 35 to 50% NO_X reduction. Air or fuel staging, which reduces the residence time at peak temperature, can achieve 50 to 70% NO_X reduction, while using SCR, SNCR, or fuel reburning technologies can achieve 55 to 80% NO_X reduction. Sorbent injection, which can achieve 60 to 90% NO_X reduction, and the use of fuels with low nitrogen content are technologies also employed. NO_X technologies used for spreader stokers are similar to traveling-grate stokers but achieve slightly different results. FGR, natural gas reburning, low-NO_X burners, combustion optimization, OFA, LEA, water or steam injection, and reduced air preheat temperature are control options to reduce peak temperatures that can achieve 50 to 65% NO_X reductions. Air or fuel staging or steam injection, which reduces the residence time at peak temperature, can achieve 50 to 65% NO_X reductions, while using SCR, SNCR, or fuel reburning technologies achieves 35 to 80% NO_X reductions. Additional NO_X reduction technologies include sorbent injection, which can achieve 60 to 90% reductions, and using lower nitrogen fuels.

Economics of NO_X Reduction/Removal

The costs for NO_X reduction/removal techniques are site and performance specific, thus making it difficult to compare generalized system costs. These techniques depend on several factors, including degree of retrofit difficulty, unit size, uncontrolled NO_X levels, and required NO_X reduction [30]. This section summarizes costs for the various systems using published data.

Low-NO_X Burners Wu [30] reported that the capital costs for a low-NO_X burner are in the range of \$650 to \$8300/MM Btu. The operating costs can range from \$340 to \$1500/MM Btu. The levelized costs can vary from \$240 to \$4300/short ton of NO_X removed, with the average cost being closer to the lower end of the range [53].

Furnace Air Staging The costs for furnace air staging are similar to those for low-NO_X burners [30]. The capital costs range from ~\$8 to \$23/kW, and the levelized costs range from $110 to $210/short ton of NO_x removed. If furnace air staging is combined with low-NO_x burners, the capital costs can increase to $15 to $30/kW, while the levelized costs remain relatively unchanged. Retrofits of furnace air staging in tangentially-fired boilers are generally more expensive than those in wall-fired boilers: $11 to $23/kW and $5 to $11/kW, respectively.

Flue Gas Recirculation The capital costs for conventional flue gas recirculation is similar to that for low-NO_x burners and overfire air: $8 to $35/kW [53]; however, capital costs of induced FGR, a design derivative of conventional forced flue gas desulfurization, have been reduced to $1 to $3/kW.

Fuel Staging (Reburn) The capital costs for reburn technology depend on the size of the unit, ease of retrofit, control system upgrade requirements, and, for natural gas reburn, availability of natural gas at the plant [30]. The retrofit costs are typically about $15 to $20/kW for natural gas, coal, or oil reburn, excluding the cost of any natural gas pipeline. The operating costs for a reburn retrofit are mainly due to the differential cost of the reburn fuel over the main fuel. For coal reburn, this cost is zero, but reburn fuels such as natural gas or oil are usually more expensive than the main fuel. This differential, however, can be offset by reductions in SO_2 emissions, ash remediation and disposal, and pulverizer power. The levelized cost for reburn is ~$110 to $210/short ton of NO_x removed [16].

Cofiring Cofiring of CWSF is not commercially used at this time. Biomass cofiring, on the other hand, is currently being demonstrated at several plants and commercial operations are being performed at seven utilities. The capital costs for biomass cofiring range from $175 to $250/kW [43].

Process Optimization The total turnkey installation cost for an advanced combustion control system ranges from $150,000 to $500,000 [30]. It is possible to achieve moderate cost reductions on a per-unit basis for similar units at the same power plant site. The size of the unit typically has little impact on the cost of a system.

Selective Catalytic Reduction (SCR) The capital costs for an SCR system depend on the level of NO_x removal and other site-specific conditions, such as inlet NO_x concentration, unit size, and ease of retrofit and range from $80 to $160/kW [54]. The capital costs of an SCR system include [30]:

- Catalyst and reactor system;
- Flow control skid and valving system;
- Ammonia injection grid;

- Ammonia storage;
- Piping;
- Ducts, expansion joints, and dampers;
- Fan upgrades/booster fans;
- Air preheater changes;
- Foundations, structural steel, and electricals;
- Installation.

The operating costs can vary from $1500 to $5800/MM Btu, and the levelized cost can range from $1800 to $10,900/MM Btu [30]. The operating costs include [30]:

- Ammonia usage;
- Pressure drop changes;
- Excess air change;
- Unburned carbon change;
- Ash disposal;
- Catalyst replacement;
- Vaporization/injection energy requirements;
- Other auxiliary power usage.

Selective Non-Catalytic Reduction Selective non-catalytic reduction is less capital intensive than SCR. The cost of an SNCR retrofit is $10 to $20/kW, whereas incorporating SNCR into a new boiler typically costs $5 to $10/kW [30]. The difference is due to the costs associated with modifying the existing boiler to install the reagent injection ports. The operating costs associated with the reagent, auxiliary power, and potential adverse plant impacts are of the order of $1 to $2 mills/kWh. The levelized costs average ~$1000/short ton of NO_X removed. A new, single-level approach to SNCR—SNCR trim—offers 20 to 30% NO_X reduction at about half the cost of conventional SNCR and is being tested by the EPRI [55]. SCNR trim has low operating costs, equivalent to only about $850/short ton of NO_X removed.

Other Flue Gas Treatment Processes Limited data are available for hybrid SNCR/SCR systems as they are still in the demonstration phase. A levelized cost estimate for a 500 MW boiler with 50% NO_X reduction is ~$5800/short ton of NO_X removed [30]. Similarly, cost data on the rich reagent injection process, which is under development, are not available.

Hybrid Flue Gas Treatment and Combustion Modifications A combination of flue gas treatment with combustion modification is increasingly being used. This technology provides higher overall NO_X reductions and can be more cost effective than stand-alone technology for the same level of NO_X control [30]. The costs of SCR can be reduced when it is used in combination with combustion modifications such as low-NO_X burners and overfire air [30]. Capital costs are lowered because combustion modifications lower the inlet NO_X concentration, which reduces the catalyst volume, support systems, and installation cost of SCR. In addition, operating costs are lower due to reductions in catalyst replacement and reagent consumption. SNCR can be combined with low-NO_X burners or gas reburn. SNCR and gas reburn have comparable economics at the same level of NO_X reduction; however, combining the two technologies considerably lowers costs while achieving a slightly higher NO_X reduction. An example of annual costs, reported by Wu [30], are ~\$1140, \$1120, and \$730 per short ton NO_X removed, respectively, for urea SNCR, gas reburn, and urea SNCR/gas reburn.

Particulate Matter

Particulate matter (PM) emissions from coal-fired electric utility boilers in the United States have decreased significantly since implementation of the 1970 Clean Air Act Amendments. In 2001, ~23 million short tons of particulate matter, reported as PM_{10} (*i.e.*, particles with an aerodynamic diameter ≤ 10 μm), were emitted from inventoried point and area sources, of which ~190,000 short tons (or ~1.6% of the total) were emitted by coal-fired electric utility boilers [56]. This is a substantial decrease from a total of ~1.7 million short tons of PM_{10} being emitted from coal-fired power plants in 1970, especially as coal consumption for electricity generation has increased more than 150% over this period, and the reduction is due to the application of particulate control technologies. Similarly, in 2001, annual emissions of particulates smaller than 2.5 μm (*i.e.*, $PM_{2.5}$), which is a subset of PM_{10}, were 100,000 short tons, or less than 0.8% of the total primary $PM_{2.5}$ emitted from all sources.

The application of control technologies to combustion sources is illustrated in Figures 6-1 and 6-2, which show the improvements in emissions rates from coal-fired power plants since 1970 as well as near-term projected emissions rates. As of 2000, 1020 coal-fired electric generators were equipped with particulate collectors and represented a total of more than 321,000 MW generating capacity [3].

Several particulate control technologies are available for coal-fired power plants, including electrostatic precipitators, fabric filters (baghouses), wet particulate scrubbers, mechanical collectors (cyclones), and hot-gas particulate filtration [57]. Of these, ESPs and fabric filters are currently the technologies of choice as they can meet current and pending legislation PM levels. While cleaning large volumes of flue gas, they achieve very high collection efficiencies and can remove fine particles. When operating properly, ESPs and baghouses can achieve overall collection efficiencies of 99.9% of primary particulates (over 99% control of PM_{10} and 95% control of $PM_{2.5}$), thereby achieving the 1978 New Source Performance Standards required limit of 0.03 lb PM per million Btu [58]. The primary particulate matter collection devices used in the power generation industry—ESPs and fabric filters (baghouses)—are discussed in this section. In addition, hybrid systems under development that combine ESPs and fabric filters in a single, overall system are presented.

Electrostatic Precipitators

Particulate and aerosol collection by electrostatic precipitation is based on the mutual attraction between particles of one electrical charge and a collection electrode of opposite polarity. This concept was pioneered by F. G. Cottrell in 1910 [6]. The advantages of this technology include the ability to handle large gas volumes (ESPs have been built for volumetric flow rates up to 4,000,000 ft^3/min), achieve high collection efficiencies (which vary from 99 to 99.9%), maintain low pressure drops (0.1–0.5 inH_2O), collect fine particles (0.5–200 μm), and operate at high gas temperatures (up to 1200°F). In addition, the energy expended in separating particles from the gas stream acts solely on the particles and not on the gas stream.

Electrostatic precipitators have been utilized to control particulate emissions from coal-fired boilers used for steam generation for about 60 years [7]. Initially, all ESPs were installed downstream of the air preheaters at temperatures of 270 to 350°F and are referred to as cold-side ESPs. ESPs installed upstream of air preheaters, where temperatures range from 600 to 750°F, are referred to as hot-side ESPs and use low-sulfur fuels with lower fly ash resistivity. In the early 1970s, ESPs were the preferred choice for high-efficiency particulate control devices [7]. Nearly 90% of U.S. coal-based electric utilities use ESPs to collect fine particles [59].

Operating Principles Several basic geometries are used in the design of ESPs, but the common design used in the power-generation industry is the plate-and-wire configuration. In this design (shown in Figure 6-15), the ESP consists of a large hopper-bottomed box containing rows of plates forming passages through which the flue gas flows. Centrally located in each passage are electrodes energized with high-voltage (45–70 kV), negative-polarity, direct current (dc) provided by a transformer–rectifier set [8]. Examples of various designs of rigid discharge electrodes are shown in Figure 6-16 [8]. The discharge electrode most commonly used in the United States is the weighted-wire electrode, while the rigid-frame electrode is commonly used in Europe [8]. The flow is usually horizontal, and the passageways are typically 8 to 10 inches wide. The height of a plate varies from 18 to 40 feet, and the length varies from 25 to 30 feet. The ESP is designed to reduce the flow of flue gas from 50 to 60 ft/sec to less than 10 ft/sec as it enters the ESP so the particles can be effectively collected.

The electrodes discharge electrons into the flue gas stream, ionizing the gas molecules. These gas molecules, with electrons attached, form negative ions. The gas is heavily ionized in the vicinity of the electrodes,

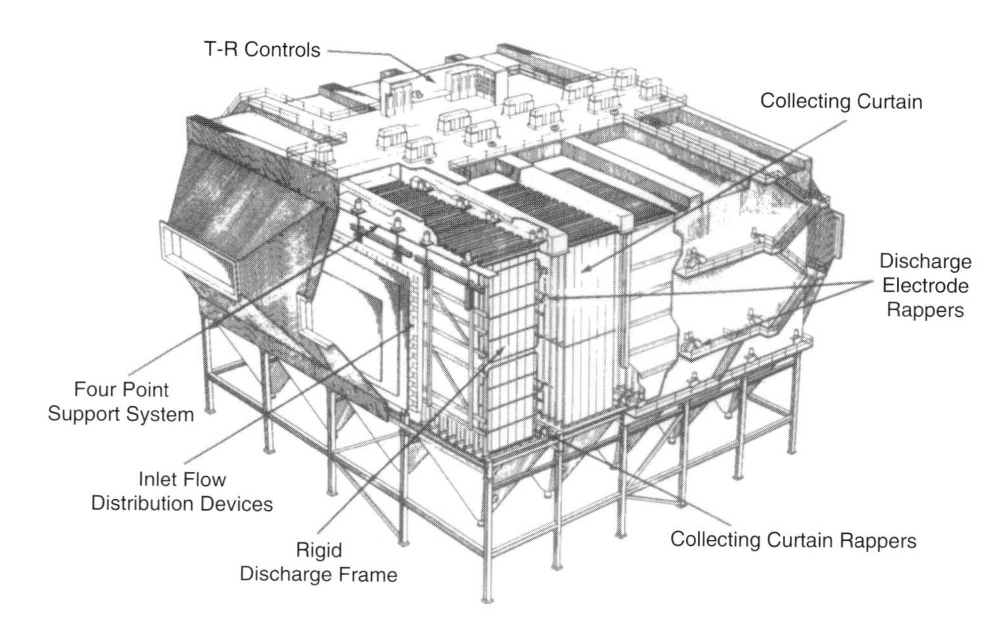

FIGURE 6-15. Electrostatic precipitator. (From B&W, Electrostatic Precipitator Product Sheet PS151 2M A 12/82, Babcok & Wilcox Co., Barberton, OH, December 1982.)

resulting in a visible blue corona effect. The fine particles are then charged through collisions with the negatively charged gas ions, resulting in the particles becoming negatively charged. Under the large electrostatic force, the negatively charged ash particles migrate out of the gas stream toward the grounded plates, where they collect and form an ash layer. These plates are periodically cleaned by a rapping system to release the layer into the ash hoppers as an agglomerated mass.

The speed at which the migration of the ash particles takes place is known as the migration or drift velocity. It depends upon the electrical force on the charged particle as well as the drag force developed as the particle attempts to move perpendicular to the main gas flow toward the collecting electrode [6]. The drift velocity, w, is defined as:

$$w = \frac{2.95 \times 10^{-12} p E_c E_p d_p}{\mu_g} K_C \tag{6-80}$$

where w is in meters per second, p is the dielectric constant for the particles (which typically lies between 1.50 and 2.40), E_c is the strength of the charging field (V/m), E_p is the collecting field strength (V/m), d_p is the particle diameter (μm), K_C is the Cunningham correction factor for particles with a diameter less than roughly 5 μm (dimensionless), and μ_g is the gas viscosity (kg/m/sec).

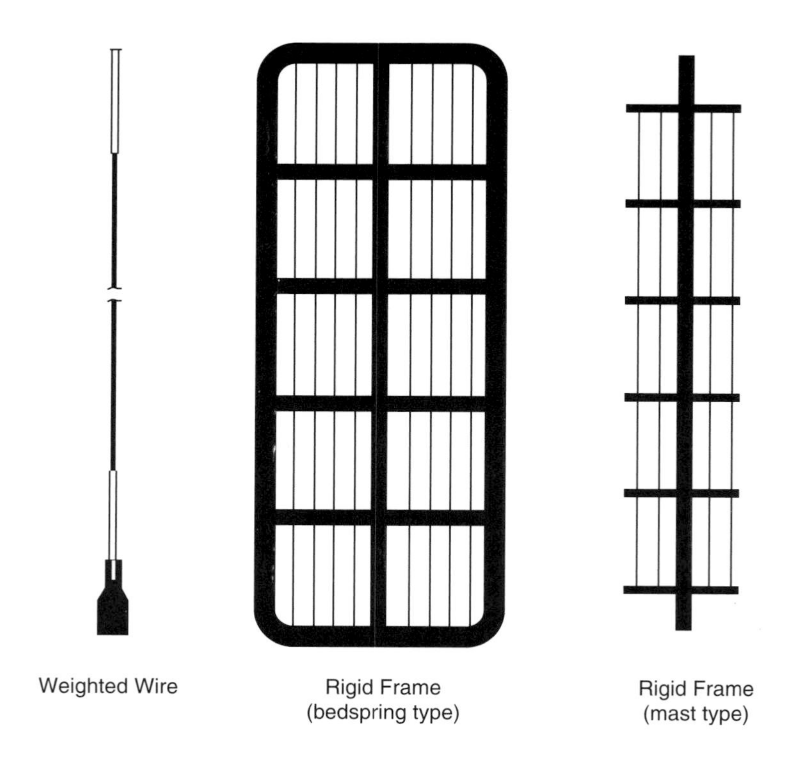

FIGURE 6-16. Rigid discharge electrode designs. (From Elliot, T. C., Ed., *Standard Handbook of Powerplant Engineering,* McGraw-Hill, New York, 1989. With permission.)

The Cunningham correction factor in Equation (6-80) is defined as [6]:

$$K_C = 1 + \frac{2\lambda}{d_p}\left[1.257 + 0.400e^{\left(\frac{-0.55d_p}{\lambda}\right)}\right] \tag{6-81}$$

where λ is the mean free path of the molecules in the gas phase. This quantity is given by:

$$\lambda = \frac{\mu_g}{0.499\rho_g u_m} \tag{6-82}$$

where u_m is the mean molecular speed (m/sec) and ρ_g is the gas density (kg/m^3). From the kinetic theory of gases, u_m is given by:

$$u_m = \left[\frac{8R_u T}{\pi M}\right]^{1/2} \tag{6-83}$$

where M is the molecular weight of the gas, T is temperature (K), and R_u is the universal gas constant (8.31×10^3 m^2/sec^2/mol K).

The drift velocity is used to determine collection efficiency using the Deutsch–Anderson equation:

$$\eta = 1 - e^{\left(-\frac{wA}{Q}\right)} \tag{6-84}$$

where w is the drift velocity, A is the area of collection electrodes, and Q is the volumetric flow rate. The units of w, A, and Q must be consistent because the factor wA/Q is dimensionless.

The ratio A/Q is often referred to as the specific collection area (SCA) and is the most fundamental ESP size descriptor [8]. Collection efficiency increases as SCA and w increase. The value of w increases rapidly as the voltage applied to the emitting voltage is increased; however, the voltage cannot be increased above that level at which an electric short circuit, or arc, is formed between the electrode and ground.

Factors that Affect ESP Performance Several factors affect ESP performance; of these, fly ash resistivity is the most important.

Fly Ash Resistivity Fly ash resistivity plays a key role in dust-layer breakdown and the ESP performance. Resistivity is dependent on the flue gas temperature and chemistry and on the chemical composition of the ash itself. Electrostatic precipitation is most effective in collecting dust in the resistivity range of 10^4 to 10^{10} ohm-cm [6]. In general, resistivities above 10^{11} ohm-cm are considered to be a problem because the maximum operating field strength is limited by the fly ash resistivity. Back corona, the migration of positive ions generated in the fly ash layer toward the emitting electrodes, which neutralizes the negatively charged particles, will result if the ash resistivity is greater than 10^{12} ohm-cm. If the fly ash resistivity is below 2×10^{10} ohm-cm, it is not considered to be a problem because the maximum operating field strength is limited by factors other than resistivity.

Examples of low- and high-resistivity fly ashes are shown in Figure 6-17, where resistivity is plotted as a function of temperature for two U.S. lignite samples from North Dakota and two subbituminous samples from the Powder River Basin [61]. The differences in fly ash resistivity are due to variations in ash composition. The low-resistivity fly ashes were produced from coals that contained higher levels of sodium in the coal ash. Higher sodium levels result in lower resistivity. Similarly, higher concentrations of iron lower resistivity. Higher levels of calcium and magnesium have the opposite effect on resistivity. This is illustrated in Figure 6-18, where the fly ash resistivities of two Texas lignites are shown along with the fly ash resistivities from the same two coals when injecting limestone for SO_2 control [62]. The addition of calcium through sorbent injection resulted in increasing the fly ash resistivities.

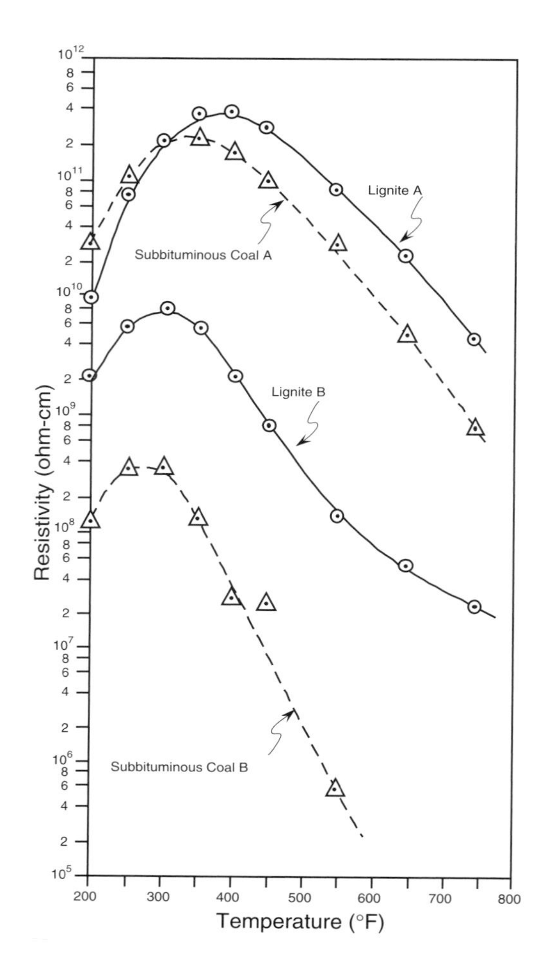

FIGURE 6-17. Illustration of effect of ash composition on fly ash resistivity for coals from the same geographical location. (Miller, B. G., unpublished data, 1986.)

Flue gas properties also affect fly ash resistivity. The two properties that have the most influence on ash resistivity are temperature and humidity. The effects of temperature can be observed in Figures 6-17 and 6-18. Similarly, as moisture content in the flue gas is increased, the fly ash resistivity decreases. The dome-shaped curves shown in Figures 6-17 and 6-18 are typical of fly ashes. The shape of the curves is due to a change in the mechanism of conduction through the bulk layer of particles as the temperature is varied [6]. The predominant mechanism below 300°F is surface conduction, where the electric charges are carried in a surface film adsorbed on the particle. As the temperature is increased above 300°F, the phenomenon of adsorption becomes less effective, and the predominant mechanism is

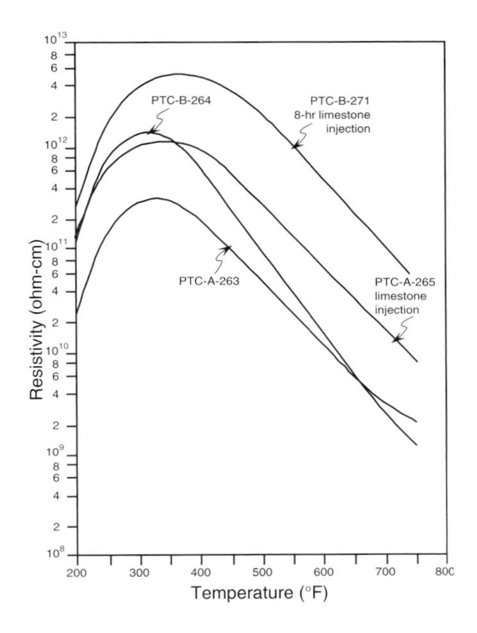

FIGURE 6-18. The effect of limestone addition on fly ash resistivity. (From Miller, B. G. *et al.*, Sulfur Capture by Limestone Injection During Combustion of Pulverized Panola County Texas Lignite, in *Proc. of the Gulf Coast Lignite Conference*, 1984.)

volume or intrinsic conduction. Volume conduction involves passage of an electric charge through the particles.

Other Factors The three primary mechanical deficiencies in operating units are gas sneakage, fly ash re-entrainment, and flue gas distribution [8]. Flue gas sneakage (*i.e.*, flue gas that is bypassing the effective region of the ESP) increases the outlet dust loading. Re-entrainment occurs when individual dust particles are not collected in the hoppers but are caught up in the gas stream, thus increasing dust loading in the ESP and resulting in higher outlet dust loadings. Nonuniform flue gas distribution throughout the entire cross section of the ESP decreases the collection ability of the unit.

Many additional factors can affect the performance of an ESP, including the quality and type of fuel. Changes in coal and ash composition, grindability, and the burner/boiler system are important. Fly ash resistivity increases with decreasing sulfur content, an issue that must be considered when switching to lower sulfur coals. Moisture content and ash composition affect resistivity, as discussed earlier. Changes in coal grindability can affect pulverizer performance by altering particle size distribution, which in turn can impact combustion performance and ESP performance. Modifications to the boiler system can affect temperatures or combustion performance and thereby impact ESP performance.

Methods to Enhance ESP Performance Difficulties in collecting high-resistivity fly ash and fine particulates have led to very large units being specified, unacceptable increases in ESP power consumption, and, in extreme cases, the use of fabric filters in lieu of ESPs [8]. As a result, concepts have been developed to overcome the technical limitations and maintain competitiveness with fabric filters, including [8]:

- Pulse energization, where a high-voltage pulse is superimposed on the base voltage to enhance ESP performance during operation under high-resistivity conditions;
- Intermittent energization, where the voltage to the ESP is turned off during selected periods to provide a longer period between each energization cycle and reduce the potential for back corona;
- Wide plate spacing, which reduces capital and maintenance costs and allows for thicker discharge electrodes and increased current density.

Another approach to achieving electrical resistivities in the desired range is the addition of conditioning agents to the flue gas stream. This technique is applied commercially to both hot-side and cold-side ESPs. Conditioning modifies the electrical resistivity of the fly ash and/or its physical characteristic by changing the surface electrical conductivity of the dust layer deposited on the collecting plates, increasing the space charge on the gas between the electrodes, and/or increasing dust cohesiveness to enlarge particles and reduce rapping re-entrainment losses [8]. Over 200 utility boilers are equipped with some form of conditioning in the United States [8].

The most common conditioning agents are sulfur trioxide (SO_3), ammonia (NH_3), and compounds related to them, as well as sodium compounds. Sulfur trioxide is most widely applied for cold-side ESPs, while sodium compounds are used for hot-side ESPs [8]. Although results vary between coal and system, the injection of 10 to 20 ppm of SO_3 can reduce the resistivity to a value that will permit good collection efficiencies. In select cases, SO_3 injection of 30 to 40 ppm has resulted in reductions of fly ash resistivity of 2 to 3 orders of magnitude (*e.g.*, from 10^{11} to $\sim 10^8$ ohm-cm) [6]. Disadvantages of SO_3 injection systems include the possibility of plume color degradation. Disadvantages of sodium compounds include potential problems with increased deposition and interference from certain fuel constituents, which affect the economics of the injection [8]. Combined SO_3–NH_3 conditioning is used in which the SO_3 adjusts the resistivity downward and the NH_3 modifies the space-charge effect, improves agglomeration, and reduces rapping re-entrainment losses [8].

Wet ESPs Dry ESPs, which have been discussed up to this point, have been successfully used for many years in utility applications for coarse and fine particulate removal. Dry ESPs can achieve a 99+% collection efficiency for particles 1 to 10 μm in size; however, dry ESPs cannot remove toxic gases and vapors that are in a vapor state at 400°F, cannot efficiently collect very small fly ash particles, and cannot handle moist or sticky particulate that would stick to the collection surface; they also require considerable space for multiple fields due to re-entrainment of particles and rely on mechanical collection methods to clean the plates that require maintenance and periodic shutdowns [63].

Wet electrostatic precipitators (WESPs) address these issues and are a viable technology to collect finer particulate than existing technology while also collecting aerosols. WESPs have been commercially available since their first introduction by F. G. Cottrell in 1907 [64]; however, they have primarily been used in small, industrial-type settings as opposed to utility power plants. WESPs have been in service for nearly 100 years in the metallurgical industry and in many other applications. They are used to control acid mists, submicron particulates (as small as 0.01 μm with 99.9% removal), mercury, metals, and dioxins/furans when installed as the final polishing device within a multipollutant control system [63]. When integrated with upstream air pollution control equipment, such as an SCR, dry ESP, and wet scrubber, multiple pollutants can be removed when the WESP serves as the final polishing device.

Wet electrostatic precipitators operate in the same three-step process as dry ESPs: charging, collecting, and cleaning of the particles from the collecting electrode [65]. However, cleaning of the collecting electrode is performed by washing the collection surface with liquid, rather than by mechanically rapping the collection plates. WESPs operate in a wet environment in order to wash the collection surface; therefore, they can handle a wider variety of pollutants and gas conditions than dry ESPs [65]. WESPs find their greatest use where:

- The gas in question has a high moisture content;
- The gas stream includes sticky particulate;
- The collection of submicron particulate is required;
- The gas stream has acid droplets of mist;
- The temperature of the gas stream is below the moisture dew point.

WESPs continually wet the collection surface and create a dilute slurry that flows down the collecting wall to a recycle tank, never allowing a layer of particulate cake to build up [65]. As a result, captured particulate is never re-entrained. Also, when firing low-sulfur coal, which produces a high resistivity dust, the electrical field does not deteriorate, and power levels within a WESP can be dramatically higher than in a dry ESP: 2000 W/1000 scfm versus 100 to 500 W/1000 scfm, respectively. Similar to a dry ESP, WESPs can be configured either as tubular precipitators (*i.e.*, the charging electrode is located down the center of a tube) with vertical gas flow or as plate precipitators with horizontal gas flow [66]. For a utility application, tubular WESPs are appropriate as a mist eliminator above a flue gas desulfurization scrubber, while the plate type can be employed at the back end of a dry ESP train for final polishing of the gas.

Fabric Filters
Historically, ESPs have been the principle control technology for fly ash emissions in the electric power industry. Small, relatively inexpensive ESPs could be installed to meet early federal and state regulations; however, as particulate control regulations have become more stringent, ESPs have become larger and more expensive. Also, increased use of low-sulfur coal has resulted in the formation of fly ash with higher electric resistivity, which is more difficult to collect; consequently, ESP size and cost have increased to maintain high collection efficiency [67]. As a result, interest in baghouses has increased. Baghouses offer extremely high collection efficiency (*i.e.*, *99.9* to 99.99+%) and are capable of filtering large volumes of flue gas, and their size and efficiency are relatively independent of the type of coal burned [67]. Baghouses are essentially huge vacuum cleaners consisting of a large number of long, tubular filter bags arranged in parallel flow paths. As the ash-laden flue gas passes through these filters, the particulate is removed. Advantages of fabric filters include high collection efficiency over a broad range of particle sizes; flexibility in design provided by the availability of various cleaning methods and filter media; wide range of volumetric capacities in a single installation, which may range from 100 to 5 million ft^3/min; reasonable operating pressure drops and power requirements; and the ability to handle a variety of solid materials [6]. Disadvantages of baghouses include their large footprints, the possibility of an explosion or fire if sparks are present in the vicinity of a baghouse, and difficulties encountered when handling hydroscopic materials due to cloth cleaning problems.

The first utility baghouse in the United States was installed on a coal-fired boiler in 1973 by the Pennsylvania Power and Light Company at its Sunbury Station [67]. This baghouse, as well as the next several baghouses installed, were small, and it was not until 1978 that the first large baghouse was installed on a utility boiler. This baghouse serviced a 350 MW pulverized coal-fired boiler at the Harrington Station of the Southwestern Public Service Company. Beginning in 1978, there has been a steady increase in the installation of utility commitments to baghouse technology, and currently more than 110 baghouses are in operation on utility boilers in the United States and service more than ~22,000 MW of generating capacity [67].

Filtration Mechanisms Filtration occurs when the particulate-laden flue gas is forced through a porous, solid medium, which captures the particles. In a baghouse, this solid medium is the filter bag and/or the residual dust cake on the bag. The important filtering mechanisms are three aerodynamic capture mechanisms: direct interception, inertial impaction, and diffusion.

Electrostatic attraction may also play a role with certain types of dusts/fiber combinations [6].

Direct interception occurs if the gas streamlines carrying the particles are close to the filter elements for contact. Inertial impaction occurs when the particles have sufficient momentum and cannot follow the gas stream when the stream is diverted by the filter element and the particles strike the filter. Diffusion results when the particle mass is very low and Brownian diffusion superimposes random motion on the streamline trajectory, thereby increasing the probability of the particle contacting and being captured by the filter [67]. Particles may be attracted to or repulsed by filters due to a variety of Coulombic and polarization forces. Particles larger than 1 μm are removed by impaction and direct interception, whereas particles from 0.001 to 1 μm are removed mainly by diffusion and electrostatic separation [6].

The effectiveness of a filter in capturing particles is reported in terms of collection efficiency or particle penetration. Particle penetration, P, is defined as the ratio of the particle concentration (mass or number of particles per unit volume of gas), also referred to as dust loading, on the outlet of the filter (*i.e.*, cleaned flue gas stream) to that on the inlet side of the filter (*i.e.*, dirty flue gas stream). Collection efficiency, η, is defined as:

$$\eta = 1 - P \tag{6-85}$$

Typically, both penetration and collection efficiency are multiplied by 100 and reported as a percent.

The filtration process can be divided into three distinct time regimes: (1) filtration by a clean fabric, which occurs only once in the life of a bag; (2) establishment of a residual dust cake, which occurs after many filtering and cleaning cycles; and (3) steady-state operation, in which the quantity of particulate matter removed during the cleaning cycles equals the amount collected during each filter cycle [67]. In general, the initial collection efficiency of new filters is quite low (<99% and as low as 75–90%), whereas a conditioned bag (*i.e.*, a bag that has retained residual particles in the fibers of the filter that cannot be removed by cleaning) may have a collection efficiency of 99.99+%. A dust cake will form on the filters, where the adhesive and cohesive forces acting between the particles and filter elements and among the particles, respectively, are sufficiently strong to allow particulate agglomerates to bridge the filter pores. The accumulated dust cake forms a secondary filter of much higher efficiency than the clean fabric. On a seasoned bag, residual dust cakes generally weigh 10 to 20 times as much as the ash deposited during an average cleaning cycle [67].

Operating Principles Baghouses remove particles from the flue gas within compartments arranged in parallel flow paths, with each compartment containing several hundred large, tube-shaped filter bags. Figure 6-19 is a cutaway view of a typical 10-compartment baghouse [67]. A baghouse on

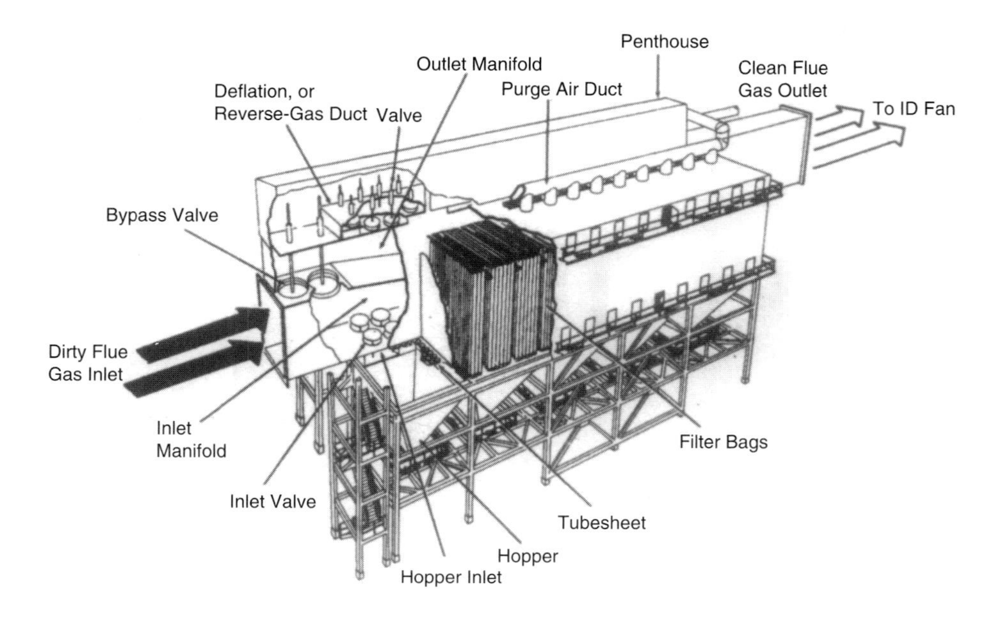

FIGURE 6-19. Cutaway view of a typical 10-compartment baghouse. (*Source:* Bustard, C. J. *et al.*, *Fabric Filters for the Electric Utility Industry*, Vol. 1, *General Concepts*, Electric Power Research Institute, Palo Alto, CA, 1988.)

a 500 MW coal-fired unit may be required to handle in excess of 2 million ft^3/min of flue gas at temperatures of 250 to 350°F. From an inlet manifold, the dirty flue gas, with typical dust loadings from 0.1 to 10 gr/ft^3 of gas (0.23 to 23 g/m^3), enters hopper inlet ducts that route it into individual compartment hoppers. From each hopper, the gas flows upward through the bags, where the fly ash is deposited. The clean gas is drawn into an outlet manifold, which carries it out of the baghouse to an outlet duct. Periodic operation requires shutdown of portions of the baghouse at regular intervals for cleaning. Cleaning is accomplished in a variety of ways, including mechanical vibration or shaking, pulse jets of air, and reverse air flow.

The two fundamental parameters in sizing and operating baghouses are the air-to-cloth (A/C) ratio and pressure drop across the filters. Other important factors that affect the performance of the fabric filter include the flue gas temperature, dew point, and moisture content, as well as particle size distribution and composition of the fly ash [68]. The A/C ratio, which is a fundamental fabric filter descriptor denoting the ratio of the volumetric flue gas flow (ft^3/min) to the amount of filtering surface area (ft^2), is reported in units of ft/min [8]. For fabric filters, it has been generally observed that the overall collection efficiency is enhanced (as the A/C ratio) that is, superficial filtration velocity decreases. Factors to be considered with the A/C ratio include type of filter fabric, type of coal and firing method, fly ash properties, duty cycle of the boiler, inlet fly ash loading, and cleaning method [68]. The A/C ratio determines the size of the baghouse and hence the capital cost.

Pressure drop is a measure of the energy required to move the flue gas through the baghouse. Factors affecting pressure drop are boiler type (which influences the fly ash particle size), filtration media, fly ash properties, and flue gas composition [68]. The pressure drop is an important parameter, as it determines the capital cost and energy requirements of the fans.

As the filter cake accumulates on the supporting fabric, the removal efficiency typically increases; however, the resistance to flow also increases. For a clean filter cloth, the pressure drop is about 0.5 inH_2O and the removal efficiency is low. After sufficient filter cake buildup, the pressure drop can increase to 2 to 3 inH_2O with a removal efficiency of 99+% [6]. When the pressure drop reaches 5 to 6 inH_2O, it is usually necessary to clean the filters.

The pressure drop for both the cleaned filter and the dust cake, ΔP_T, may be represented by Darcy's equation [6]:

$$\Delta P_T = \Delta P_R + \Delta P_C = \frac{\mu_g x_R V}{K_R} + \frac{\mu_g x_C V}{K_C} \tag{6-86}$$

where ΔP_R is the conditioned residual pressure drop; ΔP_C is the dust cake pressure drop; K_R and K_C are the filter and dust cake permeabilities, respectively; V is the superficial velocity; μ_g is the gas viscosity; and x_R and x_C are the filter and dust cake thicknesses, respectively. The permeabilities K_R and K_C are difficult quantities to predict with direct measurements as they are functions of the properties of the filter and dust such as porosity, pore size distribution, and particle size distribution. Therefore, in practice ΔP_R is usually measured after the bags are cleaned and ΔP_C is determined using the equation:

$$\Delta P_C = K_2 C_i V^2 t \tag{6-87}$$

where C_i is the dust loading and, along with V, is assumed constant during the filtration cycle; t is the filtration time; and K_2, the dust resistance coefficient, is estimated from:

$$K_2 = \frac{0.00304}{(d_{g,mass})^{1.1}} \left(\frac{\mu_g}{\mu_{g,70^\circ F}}\right) \left(\frac{2600}{\rho_p}\right) \left(\frac{V}{0.0152}\right)^{0.6} \tag{6-88}$$

where d_g is the geometric mass median diameter (m), μ_g is the gas viscosity (kg/m/sec), ρ_p is the particle density (kg/m^3), and V is the superficial velocity (m/sec).

Basic Types of Fabric Filters The three basic types of baghouses are reverse-gas, shake-deflate, and pulse-jet. They are distinguished by the cleaning mechanisms and by their A/C ratios. The A/C ratios for fabric filters range from a low of 1.0 to 12.0 ft/min depending on the type of cleaning mechanism used and characteristics of the fly ash [6]. Ash that accumulates on the bags in excess of the desired residual dust cake must be removed by periodic bag cleaning to reduce the gas flow resistance (and, hence, induced draft fan power requirements) and to reduce bag weight. In U.S. utility baghouses, cleaning is done off-line by isolating individual compartments for cleaning.

Reverse-Gas Fabric Filters Reverse-gas fabric filters are generally the most conservative design of the fabric filter types. They typically operate at low A/C ratios ranging from 1.5 to 3.5 ft/min [6,68]. Fly ash collection occurs on the inside of the bags, because the flue gas flow is from the inside of the bags to the outside, as illustrated in Figure 6-20 [67]. Reverse-gas baghouses use off-line cleaning, where compartments are isolated and cleaning air is passed from the outside of the bags into the inside, causing the bags to partially collapse and release the collected ash. The dislodged ash falls into the hopper. A variation of the reverse-gas cleaning method is the use of sonic energy for bag cleaning. With this method, low-frequency (<250–300 Hz), high-sound-pressure (0.3–0.6 inH_2O) pneumatic horns are sounded simultaneously, and the normal reverse-gas flow adds energy to the cleaning process. Reverse-gas fabric filters are widely used in the United States; approximately 90% of the utility baghouses employ this reverse-gas cleaning process [67].

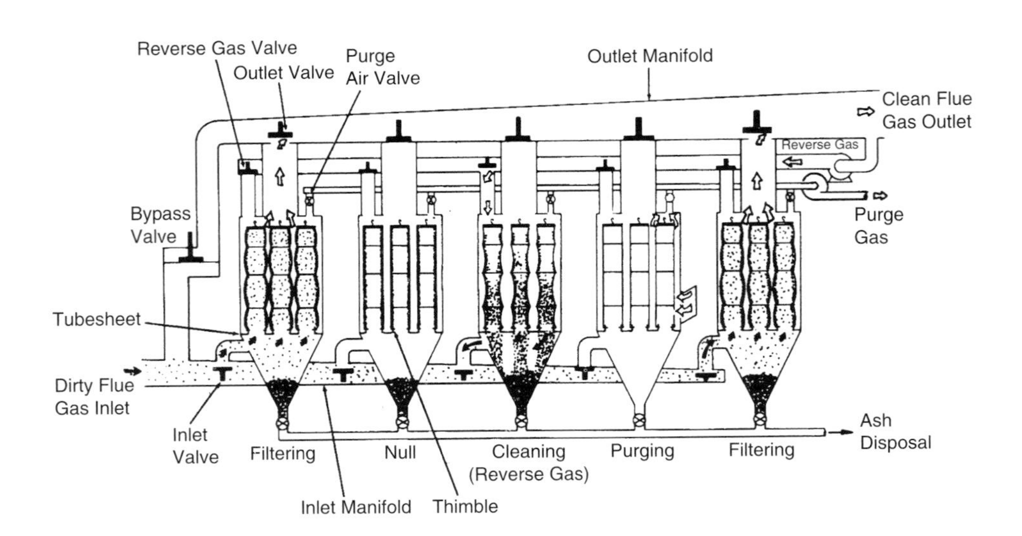

FIGURE 6-20. Schematic diagram of the compartments in a reverse-gas baghouse illustrating the flue gas and cleaning air flows during the various cycles of operation. (*Source:* Bustard, C. J. *et al.*, *Fabric Filters for the Electric Utility Industry*, Vol. 1, *General Concepts*, Electric Power Research Institute, Palo Alto, CA, 1988.)

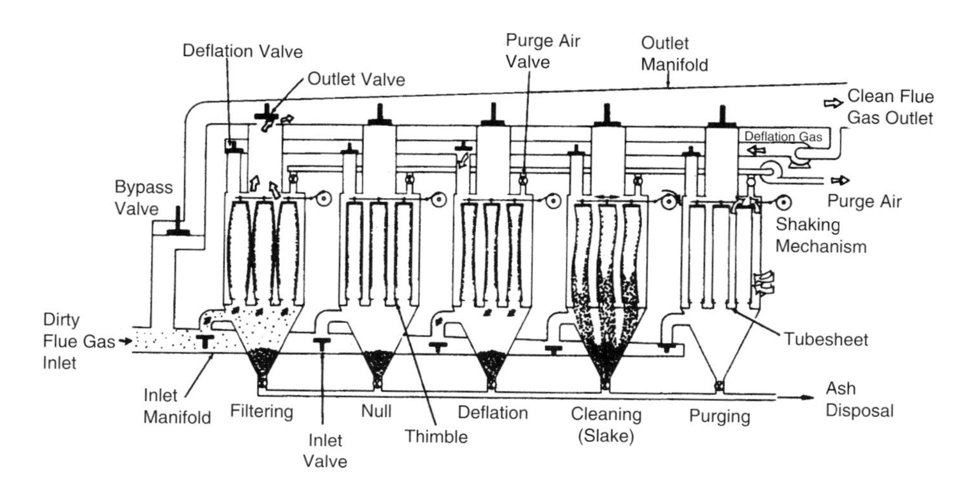

FIGURE 6-21. Schematic diagram of the compartments in a shake-deflate baghouse illustrating the flue gas and cleaning air flows during the various cycles of operation. (*Source:* Bustard, C. J. *et al.*, *Fabric Filters for the Electric Utility Industry*, Vol. 1, *General Concepts*, Electric Power Research Institute, Palo Alto, CA, 1988.)

Shake-Deflate Fabric Filters Shake-deflate baghouses are another low A/C type system (2 to 4 ft/min), and they collect dust on the inside of the bags similar to the reverse-gas systems [6]. With shake-deflate cleaning, a small quantity of filtered gas is forced backward through the compartment being cleaned, which is done off-line. The reversed filtered gas relaxes the bags but does not completely collapse them. As the gas is flowing or immediately after it is shut off, the tops of the bags are mechanically shaken for 5 to 20 sec at frequencies ranging from 1 to 4 Hz and at amplitudes of 0.75 to 2 in. [67]. The operating cycles of a shake-deflate baghouse are illustrated in Figure 6-21 [67]. Operating experience with shake-deflate baghouses in utility service has been good [8].

Pulse-Jet Fabric Filters In pulse-jet fabric filters, the flue gas flow is from the outside of the bag inward. This is illustrated in Figure 6-22 [67]. The A/C ratio is higher than reverse-air units and is typically 3 to 4 ft/min allowing for a more compact installation, but the ratio can vary from 2 to 5 ft/min [6]. Cleaning is performed with a high-pressure burst of air into the open end of the bag. Pulse-jet systems require metal cages on the inside of the bags to prevent bag collapse. Bag cleaning can be performed on-line by pulsing selected bags while the remaining bags continue to filter the flue gas. Three cleaning methods have evolved for the pulse-jet systems [68]:

- High-pressure (40–100 psig), low-volume pulse;
- Intermediate pressure (15–30 psig) and volume pulse;
- Low-pressure (7.5–10 psig), high-volume pulse.

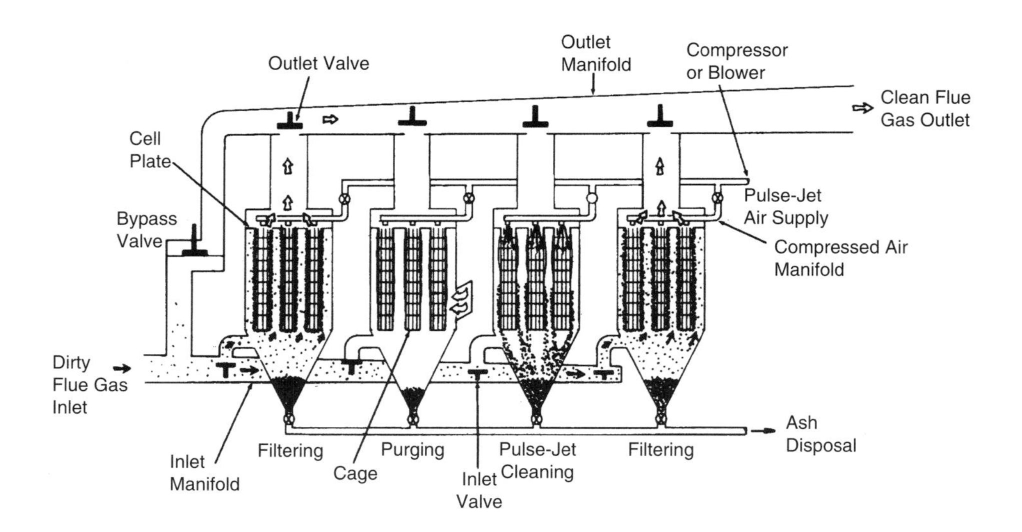

FIGURE 6-22. Schematic diagram of the compartments in a pulse-jet baghouse illustrating the flue gas and cleaning air flows during the various cycles of operation. (*Source:* Bustard, C. J. *et al.*, *Fabric Filters for the Electric Utility Industry*, Vol. 1, *General Concepts*, Electric Power Research Institute, Palo Alto, CA, 1988.)

The first method is used mainly in the United States, while the latter two methods are used primarily in larger boilers in Australia, Canada, and Western Europe [68].

Pulse-jet cleaning results in lower resistance to gas flow than the other two baghouse types, thus allowing smaller baghouses to filter the same volume of flue gas. Despite this, pulse-jet cleaning is not the preferred choice in the United States for utility boilers because of concerns that the more rigorous cleaning method results in lower particulate collection efficiency and shorter bag life. Pulse-jet baghouses are used in the United States, as well as Japan, for industrial boilers [68]. In Canada and Europe, pulse-jet systems are used in industrial plants and some large-sized utility plants. Much work has been done on improving fabrics for the filters, and the pulse-jet technology is becoming more attractive to utilities.

Fabric Filter Characteristics Fabric filters are made from woven, felted, and knitted materials with filter weights that generally range from as low as 5 oz/yd^2 to as high as 25 oz/yd^2 [6]. Filtration media are selected depending on the type of baghouse, their efficiency in capturing particles, system operating temperature, physical and chemical nature of the fly ash and flue gas, durability for a long bag life, and the cost of the fabric. Tables 6-9 and 6-10 provide some general data of the most commonly used fabrics and the criteria to select them, respectively [57]. Currently, there is a tendency toward using

TABLE 6-9
Fabric Filter Cloth Characteristics

Fiber	*Maximum Operating Temperature (°F)*	*Acid Resistance*	*Alkali Resistance*	*Dry Heat Resistance*	*Wet Heat Resistance*	*Flex and Abrasion Resistance*
Cotton	180	Poor	Good	Fair	Fair	Good
Polypropylene (Propex®)	200	Excellent	Excellent	Fair	Fair	Very good
Nylon (Neotex®)	250	Fair	Good	Good	Good	Excellent
Acrylic (Dratex®)	260	Excellent	Fair	Good	Good	Fair
Polyester (Terytex®)	300	Good	Fair	Good	Good	Very good
Ryton®	375	Excellent	Excellent	Very good	Very good	Very good
Aramid (Nomex®)	400	Fair	Good	Excellent	Excellent	Very good
Glass	500	Very good	Fair	Excellent	Excellent	Poor
P-84®	500	Good	Fair	Excellent	Excellent	Very good
Teflon®	500	Excellent	Excellent	Excellent	Excellent	Excellent
Tefari®	500	Very good	Excellent	Excellent	Excellent	Excellent

Source: Soud, H. N. and S. C. Mitchell, *Particulate Control Handbook for Coal-Fired Plants,* IEA Coal Research, London, 1997. With permission.

TABLE 6-10
Factors for Selecting Fabrics

Cotton

Cotton fabrics have good abrasion resistance and mechanical strength but are subject to rot, mildew, and shrinkage. Maximum operating temperature is 180°F.

Polypropylene (Propex®)

Polypropylene fabrics offer good tensile strength and abrasion resistance and perform well in organic and mineral acids, solvents, and alkalis. Polypropylene is attacked by nitric and chlorosulfonic acids and by sodium and potassium hydroxide at high temperatures and concentrations. Maximum operating temperature is 200°F.

Nylon (Neotex®)

Nylon fabrics have good tensile strength and alkali resistance. Nylon is degraded by mineral acids and oxidizing agents; this reaction is accelerated at high concentrations and temperatures. Maximum operating temperature is 250°F.

Acrylic (Dratex®)

The resistance of homopolymer acrylic fibers is excellent in organic solvents, good in oxidizing agents and mineral and organic acids, and fair in alkalis. They dissolve in sulfuric acid concentrations. Maximum operating temperature is 260°F.

(continued)

TABLE 6-10
(continued)

Polyester (Terytex®)
Polyester fabrics offer good resistance to most acids, oxidizing agents, and organic solvents. Concentrated sulfuric and nitric acids are the exception. Polyesters are dissolved by alkalis at high concentrations. Maximum operating temperature is 270°F.

Aramid (Nomex®)
Nomex® fabrics resist attack by mild acids, mild alkalis, and most hydrocarbons. Resistance to sulfur oxides above the acid dew point at temperatures above 150°F is better than polyester. Flex resistance of Nomex® is excellent. Maximum continuous operating temperature is 400°F.

Ryton®
Ryton® fabrics offer exceptional chemical resistance through the pH range. They resist thermal oxidation and are affected by concentrated nitric, sulfuric, and chromic acids. Maximum continuous operating temperature is 375°F.

Glass
Glass fabrics offer outstanding performance in high-heat applications. In general, by using a proprietary finish they become resistant to acids, except by hydrofluoric and hot phosphoric acid in their most concentrated forms. They are attacked by strong alkalis at room temperature and weak alkalis at higher temperatures. Glass is vulnerable to damage caused by abrasion and flex; however, the proprietary finishes can lubricate the fibers and reduce the internal abrasion caused by flexing. Maximum operating temperature is 500°F.

P-84®
P-84® fabrics resist common organic solvents and avoid high pH levels. They provide good acid resistance. P-84® offers superior collection efficiency due to irregular fiber structure. Maximum continuous operating temperature is 500°F.

Teflon®
Teflon® has excellent chemical resistance throughout the pH range, high particulate collection efficiency, and excellent abrasion resistance. Maximum continuous operating temperature is 500°F.

Tefair®
Tefair® has excellent chemical resistance throughout the pH range, excellent abrasion resistance, and high degree of efficiency. It is affected by concentrations of hydrofluoric acid and high concentrations of salts. Maximum continuous operating temperature is 500°F.

Source: Soud, H. N. and S. C. Mitchell, *Particulate Control Handbook for Coal-Fired Plants*, IEA Coal Research, London, 1997. With permission.

needle felts or polytetrafluorethylene (PTFE) membranes on woven glass, due to their ability to withstand higher temperatures (during system upsets which result in temperature excursions) and improved bag performance [57]. To protect bags against chemical attack, the fabrics are usually coated with other materials such as Teflon®, silicone, graphite, and Gore-Tex® [67].

Bags generally fall into two size categories: 30 to 36 ft in length by 1 ft in diameter or 20 to 22 ft in length by 8 in. in diameter [67]. Bag fabrics are constructed using combinations of texturized and untexturized yarns. Texturized yarns contain many broken filaments and are used to create fabric surfaces with properties suitable for retaining residual dust cakes to yield high collection efficiencies without excessive pressure drops. Smooth yarns are made of continuous, unbroken filaments and are stronger than texturized yarns.

Factors that Affect Baghouse Performance Key factors in proper baghouse design and operation are flue gas flow and properties, fly ash characteristics, and coal composition [8]. The baghouse must minimize pressure drop, maintain appropriate temperature and velocity profiles, and distribute the ash-laden flue gas evenly to the individual compartments and bags.

Particle size distribution of the fly ash and loading of the flue gas varies with type of combustion system [57]. Stoker-fired units produce ash with high carbon content, moderate loading, and large particle size distribution (compared to other combustion systems). Pulverized coal-fired systems produce ash with low carbon content, high loading, and fine particle size distribution. Cyclone-fired units produce ash with low carbon content, moderate loadings, and very fine particle size distribution. Fluidized-bed systems generally produce ash with high carbon content, high loading, and fine particle size distribution [8].

The sulfur content of the coal has been correlated to fabric filter operation. The cohesiveness of ash produced from high-sulfur coals is greater than from Western low-sulfur coals [8]. Also, maintaining the baghouse above the acid dew point is critical in high-sulfur coal applications. The fly ash properties are important because they affect the adhesion and cohesion characteristics of the dust cake which, in turn, affect the properties of the residual dust cake, collection efficiency, and cleanability of the bags.

Methods to Enhance Filter Performance The most recognized method to enhance fabric filter performance is the application of sonic energy, which was discussed previously. Virtually all reverse-gas baghouses have included sonic horns [8]. Gas conditioning has been explored for improving filter performance, although this technique is not performed commercially [8]. Low concentrations of ammonia and/or sulfur trioxide have been added in test programs to control fine particulate emissions and reduce pressure drop when firing low-rank fuels.

Hybrid Systems

Although the discussions of technologies in this chapter have mainly focused on commercial systems, this section briefly discusses two concepts that are under development for improving particulate capture. They are considered here because these technologies are expected to become commercial in the very near future, especially as particulate emissions become more stringent. Hybrid systems have been under development for over 10 years, as utilities are required to meet increasingly tighter emissions regulations for particulate matter as well as sulfur dioxide. Fly ash resistivity and dust loadings are affected by switching to low sulfur coals or injecting sorbents for sulfur dioxide control, which in turn can reduce ESP efficiency. The desire to reduce fine particulate emissions is also leading to innovative technologies. Two such systems, the compact hybrid particulate collector (COHPAC) and the advanced hybrid particulate collector (AHPC) have been developed to address these issues.

Compact Hybrid Particulate Collector The COHPAC, developed by the EPRI, involves the installation of a pulse-jet baghouse downstream of the ESP or retrofitted into the last field of an ESP [69,70]. Because the pulse-jet collector is operating as a polisher for achieving lower particulate emissions, the low dust loading to the baghouse allows the filter to be operated at higher A/C ratios (8 to 20 ft/min) without increasing the pressure drop. This system allows for the ability to retrofit existing units and achieve high efficiencies at relatively low cost. The COHPAC technology has been demonstrated at the utility scale, including full-scale operation at Alabama Power's E.C. Gaston Station (272 MW) and TU Electric's Big Brown Plants (two units, each 575 MW) [69,71]. Results from COHPAC operation have been positive. For example, at E.C. Gaston Station, the COHPAC has been operated with both on-line cleaning and long filter bags (*i.e.*, 23 feet) at filtration rates of 8.5 ft/min while providing low outlet emissions levels (<0.01 lb/MM Btu) and reduced pressure drops, even with occasional high inlet dust loadings. COHPAC is a promising technology for polishing particulate emissions and is expected to help utilities meet the more stringent particulate emissions standards.

Advanced Hybrid Particulate Collector Another hybrid system under development is the advanced hybrid particulate collector (AHPC). This technology, developed by the University of North Dakota Energy and Environmental Research Center (EERC) and being demonstrated by EERC, DOE, W. L. Gore & Associates, and Otter Tail Power Company, is unique because, instead of placing the ESP and fabric filter in series, the filter bags are placed directly between ESP collection plates [72]. A schematic diagram of the AHPC is shown in Figure 6-23 [73]. The collection plates are perforated with 45% open area to allow dust to reach the bags; however, because the particles become charged before they pass through the plates, over 90% of the particulate mass is collected on the plates before it ever reaches the bags [74]. The low dust loading to the bags allows them to be operated at a high filtration velocity (*i.e.*, smaller device as 65 to 75% fewer bags are needed)

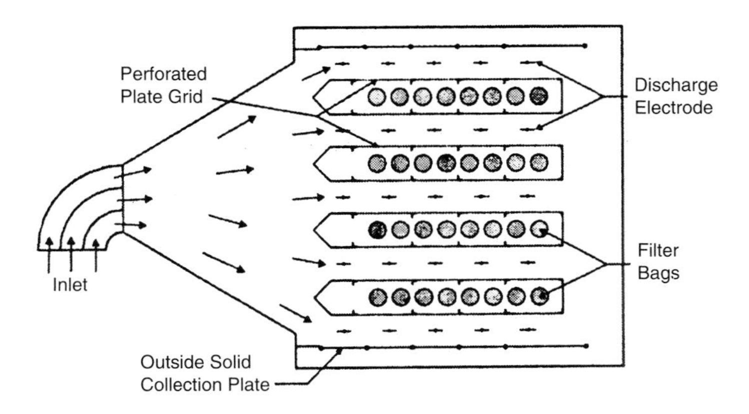

FIGURE 6-23. Schematic diagram of the Advanced Hybrid Particulate Collector. (From DOE, *Advanced Hybrid Particulate Collector Fact Sheet*, Office of Fossil Energy, U.S. Department of Energy, Washington, D.C., 2001.)

and to be cleaned without the normal concern for dust re-entrainment [75]. When pulses of air are used to clean the filter bag, the dislodged particles are injected into the ESP fields where they have another opportunity to be collected on the plates. Because these bags will not need to be cleaned as often as in typical baghouse operation, they are expected to have excellent performance over a long operating life, thereby leading to lower operating costs.

Particulate capture efficiencies of greater than 99.99% have been achieved in a 2.5 MW slipstream demonstration [76]. The AHPC technology is expected to increase fine particulate ($PM_{2.5}$) collection efficiency by one or two orders of magnitude (*i.e.*, 99.99 to 99.999%) [76]. A 450 MW demonstration is currently being conducted in the Big Stone cyclone-fired power plant, operated by Otter Tail Power Company and co-owned by Montana-Dakota Utilities, Northwestern Public Service, and Otter Tail Power Company, and burning coal from Wyoming's Powder River Basin.

Economics of Particulate Matter Control

As with other pollution control technology costs, the costs for particulate control systems are site specific and vary from country to country. They are influenced by the required emission limit and type of coal. This section summarizes the costs for ESPs, fabric filters, and hybrid systems using published data.

Electrostatic Precipitators The capital costs for a new ESP are between $40 and $60 kW, with the higher costs being associated with higher collection efficiencies [16]. Because most coal-fired power plants are already fitted with

ESPs, much of the published data relate to costs for upgrading existing ESPs. ESP rebuilds are less costly today due to greater market competition, the emergence of new construction techniques, and the use of wide plate spacing requiring less collecting plates. Wide plate spacing is one of the most economic and effective approaches to replacing internals. The cost benefits result from the need for fewer internal elements and materials, erection savings due to reuse of part of the original casing, and weight-savings effects on the existing support structures and foundations [16]. Costs for upgrading ESPs have been estimated at about $12/kW per field for a 500 MW unit, with the increased operating costs estimated to be $100,000 per year [16].

Flue gas conditioning has proved to be more cost effective than adding new fields. With difficult-to-collect fly ashes, conditioning allows operation without adding new fields. The reduction in ESP size with conditioning also lowers the operating costs because fewer fields and hoppers are used, thus decreasing the number of heaters and consequently the power consumption required [16]. A native SO_3 conditioning system for a 500 MW power plant requires a capital cost of $4.50/kW. Adding an anhydrous ammonia conditioning system to an existing SO_3 system would cost about $1/kW for a 50 MW unit, with the operating costs increasing by $50,000/year [16].

Staehle *et al.* [64] performed an economic analysis for using WESPs for SO_3 control at three different levels of control: 50, 80, and 95%. The capital costs for the three levels of control were $10, $15, and $20/kW, respectively. The total operating costs, based on 8000 hours of operation per year, were $120,000, $160,000, and $200,000 per year, respectively.

Fabric Filters Fabric filters are reported to cost between $50 and $70/kW [16]. Reverse-gas baghouses have higher capital and operating costs than pulse-jet baghouses because reverse-gas baghouses operate at a lower A/C ratio. Fabric filters are generally more expensive than ESPs for collection efficiencies up to 99.5%; however, baghouses become more cost effective for higher collection efficiencies. In addition, high resistivity fly ashes need to be upgraded to achieve high collection efficiencies, and baghouses have economic advantages over ESPs for fly ash resistivity greater than 10^{13} to 10^{14} ohm-cm. Operating costs for baghouses are also higher than ESPs due to bag replacement and auxiliary power requirements.

Hybrid Systems A cost analysis that was performed for a COHPAC that uses a pulse-jet bag filter following an ESP estimated that the capital costs have varied from $57 to $70/kW and operating costs from $320,000 to $570,000 per year, both depending on the unit size [16]. The analysis was performed for upgrading ESPs at a few coal-fired units ranging in size from 150 to 300 MW. Less information is available for the AHPC because it is in the early stages of commercialization. According to Gebert *et al.* [72], retrofit or ESP conversion jobs have been quoted in North America and Europe comparing the AHPC to a COHPAC design; in those cases where the ESP was old and required significant upgrades for the hybrid filter system to function well, the AHPC had the economic advantage. For example, the 450 MW Big Stone power plant conversion has a project cost for the overall filter system of $25/kW. Gebert *et al.* [72] anticipate that these costs will decline further as more systems are built and the design is further refined and optimized.

Pollutants with Pending Compliance Regulation

As discussed in Chapter 4 (Coal-Fired Emissions and Legislative Action in the United States), controls for mercury emissions will be in place by the end of 2007. A rule was proposed in December 2003 that will be finalized by December 2004, with initial compliance required by the end of 2007. In anticipation of the regulation, which has been expected by the industry since 1999 when the EPA initiated the mercury Information Collection Request (ICR; see Chapter 4 for a discussion of the ICR), a number of companies, government agencies, and institutions have been working on developing mercury control technologies while others have been working on developing monitoring instruments. This section discusses some of the leading options for mercury control. This section is not inclusive, as many technologies are being investigated at the bench- and pilot-scale; however, it does discuss several of the options closest to commercialization.

Mercury

Mercury exists in trace amounts in fossil fuels, vegetation, crustal material, and waste products [77]. Mercury vapor can be released to the atmosphere through combustion or natural processes where it can drift for a year or more, spreading over the globe. It has been estimated that 5500 short tons of mercury were emitted globally in 1995 from both natural and anthropogenic sources, with coal-fired power plants in the United States contributing about 48 short tons, or <1% of the total [77]. The complexity of the mercury control issue is illustrated in a simple example from the DOE: If the Houston Astrodome were filled with ping-pong balls representing the quantity of flue gas emitted from coal-fired power plants in the United States each year, 30,000,000,000 ping-pong balls would be required. Mercury emissions would be represented by 30 colored ping-pong balls, and the challenge by industry is to remove 21 of the 30 colored balls (for 70% compliance) from among the 30,000,000,000 balls. Technologies under development or being demonstrated that involve the removal of mercury from flue gas include sorbent injection, particulate collection systems, catalysts, or chemical additives to promote the oxidation of elemental mercury and facilitate its capture in particulate and sulfur dioxide control systems, as well as fixed structures in flue gas ducts that adsorb mercury.

Mercury in U.S. Coal

Over 40,000 fuel samples were analyzed as part of the ICR, and a summary of the ICR coal data, by point of origin for six regions and corresponding coal rank, is provided in Table 6-11 [78]. Appalachian bituminous coal and Western subbituminous coal accounted for ~75% of U.S. coal production in 1999 and over 80% of the mercury entering coal-fired power plants. The composition of these coals is quite different, which can affect their mercury emissions. Appalachian coals typically have high mercury, chlorine, and sulfur contents and low calcium content, resulting in a high percentage of oxidized mercury (*i.e.*, Hg^{2+}); in contrast, Western subbituminous coals typically have low concentrations of mercury, chlorine, and sulfur contents and high calcium content, resulting in a high percentage of elemental mercury (*i.e.*, Hg°).

Emissions from Existing Control Technologies from Coal-Fired Power Plants

Estimates for mercury emissions from coal-fired power plants with various control technologies, based on the 1999 ICR data, are given in Table 6-12 [78]. These data show that mercury emissions were estimated to be ~49 short tons in 1999. This estimate is based on 84 units tested in the third phase of the ICR (out of more than 1100 units in the United States); a question of bias has been raised based on the number of samples from Eastern versus Western coal-fired boilers, so the various estimates of mercury emissions range from 40 to 52 short tons/year [78]. The ICR data indicate that the speciation of mercury exiting the stack of the boilers is primarily gas-phase oxidized (43%) or elemental (54%) mercury, with some particulate-bound (3%) mercury present [77]. Table 6-12 provides information on the influence of various existing air pollution control devices (APCDs) on mercury removal; however, mercury capture across the APCDs can vary significantly based on coal properties, fly ash properties (including unburned carbon), specific APCD configurations, and other factors [77]. Mercury removals across cold-side ESPs averaged 27%, compared to 4% for hot-side ESPs [78]. Removals for fabric filters were higher, averaging 58% due to additional gas–solid contact time for oxidation. Both wet and dry FGD systems removed 80 to 90% of the gaseous oxidized mercury, but elemental mercury was not affected. High mercury removals (*i.e.*, 86%) in fluidized-bed combustors with fabric filters were attributed to mercury capture on high carbon content fly ash.

Pavlish *et al.* [78] provide an in-depth review of mercury emissions from existing control technologies, but the differences by coal rank are among the most significant findings of the ICR; specifically, units burning subbituminous coal and lignite frequently demonstrate worse mercury capture. For example, removal across a cold-side ESP averaged 35% for bituminous coal compared to 10% for low-rank coal. This is further illustrated in Figure 6-24, which shows the range of removal efficiencies across cold-side ESPs for

TABLE 6-11
Summary of ICR Data on Mercury in Coal

	Coal Rank						*Totals*
	Bituminous			*Subbituminous*	*Lignite*		
	Appalachian Region	*Interior Region*	*Western Region*	*Western Region*	*Fort Union Region*	*Gulf Coast Region*	
No. of samples	19,530	3763	1471	7989	424	623	—
	Average ICR coal analysis (dry basis)						
Hg (ppm)	0.126	0.09	0.049	0.068	—	0.119	—
Cl (ppm)	948	1348	215	124	—	221	—
S (%)	1.7	2.5	0.6	0.5	—	1.4	—
Ash (%)	11.7	10.4	10.5	7.9	—	23.6	—
Btu/lb	13,275	13,001	12,614	11,971	—	9646	—
	Other coal-related factors						
Ca (ppm, dry basis)	2700	6100	7000	14,000	—	33,000	—
Fe (ppm, dry basis)	16,000	23,000	4200	10,000	—	20,000	—
Moisture (% as received)	2.5	6.6	4.2	19.4	—	34.5	—
Typical heat rate (Btu/kWh)	10,002	10,067	10,047	10,276	10,805	10,769	—
	Regional coal production for utility use						
Million short tons (as received)	342	67	75	336	23	57	900
Million short tons (dry coal)	333	63	72	271	14	37	790
	Mercury in coal used by utilities						
Short tons of Hg	42.1	5.4	3.5	18.4	1.3	4.5	75.1
Pounds of Hg/10^{12} Btu	9.5	6.6	3.9	5.7	8.3	12.5	—
Pounds of Hg/GWh	0.0951	0.07	0.039	0.0584	0.09	0.134	—

Source: Pavlish, J. J. *et al.*, Status Review of Mercury Control Options for Coal-Fired Power Plants, *Fuel Processing Technology*, Vol. 82, 2003, pp. 89–165. With permission.

TABLE 6-12
Estimated Mercury Removal by Various Control Technologies

Control Technology	*Short Tons of Mercury Entering*	*No. of U.S. Power Plants*	*Number of ICRs for Part III Test Sites*	*Estimated Mercury Removals (%)*[a]	*Mercury Emission Calculation, EPRI ICRs (short tons)*
ESP cold	39.4	674	18	27	28.8
ESP cold + FGD wet	16.8	117	11	49	8.6
ESP hot	5.5	120	9	4	5.3
Fabric filter	2.9	58	9	58	1.2
Venturi particulate scrubber	2.2	32	9	18	1.8
Spray dryer + fabric filter	1.6	47	10	38	1
ESP hot + FGD wet	1.6	20	6	26	1.2
Fabric filter + FGD wet	1.5	14	2	88	0.2
Spray dryer + ESP cold	0.3	5	3	18	0.2
FBC + fabric filter	3.4	39	5	86	0.5
Integrated gasification combined cycle	0.07	2	2	4	0.1
FBC + ESP cold	0.02	1	1	—	0.1
Totals	75.3	1128	84	—	48.8

[a]Removals as percentage of mercury in coal calculated by the Electric Power Research Institute (EPRI).

Note: EPRI, Electric Power Research Institute; ICRs, Information Collection Requests; ESP, electrostatic precipitator; FGD, flue gas desulfurization; FBC, fluidized-bed combustion (FBC).

Source: Adapted from Pavlish *et al.* [78].

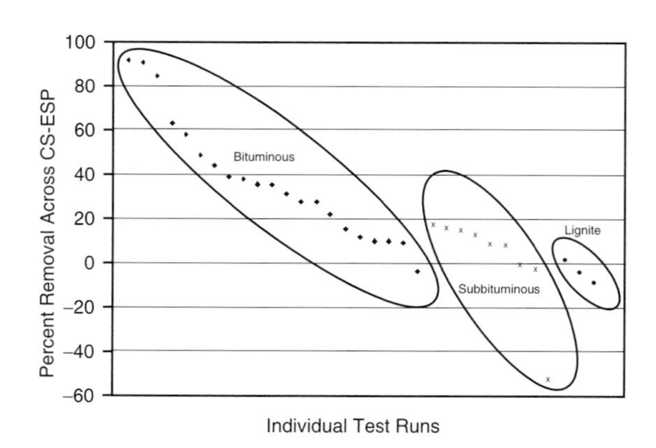

FIGURE 6-24. Mercury capture across cold-side ESPs. (From Feeley, T. J. *et al.*, *A Review of DOE/NETL's Mercury Control Technology R&D Program for Coal-Fired Power Plants*, DOE/NETL Hg R&D Review (National Energy Technology Laboratory, U.S. Department of Energy, Washington, D.C.), www.netl.doe.gov/, 2003.

bituminous coal, subbituminous coal, and lignite [77]. Mercury removal across a cold-side ESP followed by a WFGD averaged 65% for bituminous coal compared to 35% for low-rank coal [78]. The ICR data indicate that, for pulverized coal-fired units, the greatest co-benefit for mercury control is obtained for bituminous coal-fired units equipped with a fabric filter for particulate matter control and either a WFGD or spray dryer absorber for sulfur dioxide control. The worst performing pulverized bituminous coal-fired units were equipped with only a hot-side ESP [77].

The rank-dependency on mercury removal is due to the speciation of the mercury in the flue gas, which can vary significantly among power plants depending on coal properties. Power plants that burn bituminous coal typically have higher levels of oxidized mercury than power plants that burn subbituminous coal or lignite, possibly due to the higher chlorine and sulfur content of the bituminous coal. The oxidized mercury, as well as the particulate mercury, can be effectively captured in some conventional control devices such as an ESP, fabric filter, or FGD system, while elemental mercury is not as readily captured. The oxidized mercury can be more readily adsorbed onto fly ash particles and collected with the ash in either an ESP or fabric filter. Also, because the most likely form of oxidized mercury present in the flue gas—mercuric chloride ($HgCl_2$)—is water soluble, it is more readily absorbed in the scrubbing slurry of plants equipped with wet FGD systems compared to elemental mercury, which is not water soluble [77]. It has been speculated that the installation of SCRs or SNCRs could significantly increase oxidation and improve removal of mercury. This suggestion is based on European reports and testing performed at the University of North Dakota Energy and Environmental Research Center [78].

Near-Term Control Technologies

Many research organizations, federal agencies, technology vendors, and utilities are actively in the process of identifying, developing, and demonstrating cost-effective mercury control technologies for the electric utility industry. Many technology options are available at various levels of testing, demonstration, and commercialization but, based on the current state of development, sorbent injection, FGD, and coal cleaning represent the best potential for reducing mercury emissions and meeting the future mercury regulations.

Sorbent Injection Injection of activated carbon upstream of either an ESP or a baghouse is the retrofit technology that offers the greatest potential for controlling mercury emissions in plants that are not equipped with FGD scrubbers, which includes 75% of all U.S. power plants [78–80]. This is a challenging technology due to the low concentrations of mercury in the flue gas, the wide range of concentrations of acid gases and chlorine species that are present, and the relatively short gas residence time upstream of the particulate control device. Contact time, though, can be increased by using fabric filters. In addition, carbon injection upstream of a COHPAC fabric filter offers one of the most efficient and cost-effective approaches for reducing mercury emissions from coal-fired boilers. This combination of activated-carbon injection and COHPAC represents EPRI's patented TOXE-CON process and has the additional benefit of minimizing the impact on fly ash and its subsequent reuse because most of the fly ash is removed upstream by the ESP [81].

Wet Flue Gas Desulfurization Wet FGD systems are currently installed on about 25% of the electric-power-generating capacity in the United States. Although the primary function of wet scrubbers is to reduce sulfur dioxide emissions, bench- to large-scale testing has indicated that oxidized mercury (80–95%) can also be effectively captured in wet scrubbers. They are not effective, though, in capturing elemental mercury, and there is evidence that a portion of the oxidized mercury can be reduced to elemental mercury within the wet FGD system and emitted from the stack [77]. Bituminous coal, which typically has high concentrations of oxidized mercury, has the potential for achieving high overall mercury reduction. Low-rank coals also exhibit high capture of the oxidized mercury but because the concentration of oxidized mercury is low, they have low overall mercury reduction. Techniques to oxidize the vapor-phase elemental mercury prior to the wet scrubber are being aggressively studied. Also, methods to prevent the reduction of oxidized mercury to elemental mercury are also being investigated.

Coal Cleaning Coal cleaning is an option for removing mercury from the coal prior to utilization. Of the over 1 billion short tons of coal mined each year in the United States, about 600 to 650 million short tons are processed to some degree [82]. Coal cleaning removes pyritic sulfur and ash. Mercury tends to have a strong inorganic association (*i.e.*, it is associated with the pyrite), especially for Eastern bituminous coals, but mercury removal efficiencies reported for physical coal cleaning vary considerably. Physical coal cleaning is effective in reducing the concentration of many trace elements, especially if they are present in the coal in relatively high concentrations. The degree of reduction achieved is coal specific, relating in part to the degree of mineral association of the specific trace element and the degree of liberation of the trace element-bearing mineral. High levels of mercury removal (up to ~80%) have been demonstrated with advanced cleaning techniques such as column flotation and selective agglomeration [83], while conventional cleaning methods, such as heavy media cyclone, combined water-only cyclone/spiral concentrators, and froth flotation, have been shown to remove up to 62% of the mercury [84]. In both the conventional and advanced cleaning techniques, the results varied widely and were coal dependent.

TABLE 6-13
Estimates of Current and Projected Annualized Operating Cost for Mercury Emissions Control Technology

Coal		*Existing*	*Retrofit*	*Current Cost*	*Projected Cost*
Type	*S (%)*	*Controls*	*Control*	*(million$/kWh)*	*(million)$/kWh*
Bit	3	ESP cold + FGD	PAC	0.727–1.197	0.436–0.718
Bit	3	Fabric filter + FGD	PAC	0.305–0.502	0.183–0.301
Bit	3	ESP hot + FGD	PAC + PFF	1.501–NA	0.901–NA
Bit	0.6	ESP cold	SC + PAC	1.017–1.793	0.610–1.076
Bit	0.6	Fabric filter	SC + PAC	0.427–0.753	0.256–0.452
Bit	0.6	ESP hot	SC + PAC + PFF	1.817–3.783	1.090–2.270
Subbit	0.5	ESP cold	SC + PAC	1.150–1.915	0.690–1.149
Subbit	0.5	Fabric filter	SC + PAC	0.423–1.120	0.254–0.672
Subbit	0.5	ESP hot	SC + PAC + PFF	1.419–2.723	0.851–1.634

Note: Bit, bituminous coal; Subbit, subbituminous coal; ESP, electrostatic precipitator; PAC, powdered activated carbon; PFF, polishing fabric filter; SC, spray cooling; NA, not available.
Source: [85].

Cost Estimates to Control Mercury Emissions

Cost estimates for mercury compliance are currently very approximate and vary from $5000 to $70,000 per pound of mercury removed, from 0.03 to 0.8¢/kWh, and from $1.7 to $7 billion annually for the total national cost, depending on technical advances [78]. A breakdown of costs by various technology options is provided in Table 6-13 [85]. The costs of cleaning Eastern bituminous coals for mercury removal range from no additional cost (for coals already washed for sulfur removal) to a cost of $33,000 per pound of mercury removed [86]. The costs for cleaning Powder River Basin subbituminous coals are higher and approach $58,000 per pound of mercury removed [78]; however, mercury reductions from washing methods currently being applied are already built into the ICR mercury data for delivered coal; consequently, to realize a benefit from coal cleaning, higher levels of coal cleaning must be employed. Advanced cleaning methods can remove additional mercury, but they are generally not economical.

Potential Future Regulated Emissions

As discussed in Chapter 4, the increased CO_2 concentration in the atmosphere from fossil fuel combustion is causing concerns for global warming. The capture and sequestration of CO_2 from stationary combustion sources is considered an important option for the control of CO_2 emissions. Currently, however, there are no cost-effective technologies for coal-fired power plants available. It is estimated that the costs for capture and separation of CO_2 from flue gas comprises ~75% of the total costs of ocean or geologic sequestration. Consequently, DOE has a carbon sequestration program that is aggressively exploring technologies for CO_2 capture, as well as the subsequent CO_2 sequestration. The Carbon Sequestration Program, established in 1998, directly implements President Bush's Global Climate Change Initiative (announced on February 14, 2002), which has the goal of 18% reduction in greenhouse gas intensity of the United States by 2012 [87]. By 2018, the goal is to develop to the point of commercial deployment, systems for direct capture and sequestration of greenhouse gas and criteria pollutant emissions from fossil fuel conversion processes that result in near-zero emissions. This section contains a brief overview of the technologies being investigated for CO_2 capture and sequestration in anticipation of future CO_2 emissions legislation. The status and types of technologies under development are continually updated by the DOE on their Carbon Sequestration Program website (www.netl.doe.gov/coalpower/sequestration).

Carbon Dioxide

For a given energy content, coal, being primarily carbon, produces the most CO_2; oil produces less and natural gas, which derives a significant amount of its energy content from the hydrogen component of methane, produces the least. The EPA has published the following CO_2 emissions factors for fossil fuel combustion: 207 lb/MM Btu for coal, 168 lb/MM Btu for oil, and 117 lb/MM Btu for natural gas [88]. Also, lower rank coals, such as lignite and low-sulfur subbituminous coal, which is commonly used to replace high-sulfur bituminous coal, produce more carbon dioxide per unit of heat than the higher rank bituminous coal; consequently, average yearly emissions of CO_2 from electric utilities in the United States have been steadily increasing—from 206.7 to 208.2 lb of CO_2 per MM Btu from 1980 to 1997—due to switching from high-rank to low-rank coals [88].

Switching from a high-carbon fuel to a low-carbon fuel, such as from coal to natural gas, for electricity generation would greatly reduce CO_2 emissions; however, this strategy is dependent upon abundant, affordable natural gas. It is apparent that the United States does not have such a supply, as natural gas availability and affordability have been extremely volatile, as evidenced by each fall/winter since 1999. Switching the U.S. electricity-generating capacity from coal to natural gas is not sound energy policy; hence, options for capturing and sequestering CO_2 from coal-fired generators are being developed and are reviewed in the following sections.

Improving energy efficiency is considered a third method to reduce CO_2 emissions. A measure of efficiency of electricity generation is the heat rate, or the Btu consumed per kilowatt-hour (kWh) generated. A generating plant operating at 33% efficiency would have a heat rate of 10,400 Btu/kWh. The U.S. electric power industry has been on a trend toward increased efficiency since 1949, but that trend has been very slight in the past two decades [88]. Improving power plant efficiency is a major goal of not only the DOE but also industry, a topic that is discussed in more detail in Chapter 7 (Future Power Generation).

CO_2 Capture

In general, CO_2 can be separated, recovered, and purified from concentrated CO_2 sources by chemical and physical methods such as absorption, adsorption, or membrane separation. These separation and purification steps can produce pure CO_2 from power plant flue gas but they add considerable cost to the CO_2 conversion or sequestration system. Industrially, separation of CO_2 is usually performed utilizing the amine absorption process with monoethanol amine (MEA) [89]. The main reaction responsible for CO_2 chemical interaction with amine (*i.e.*, chemical adsorption) is believed to be the carbamate formation:

$$CO_2 + 2R_2NH \longleftrightarrow R_2NH_2^+ + R_2NCOO^- \qquad (6\text{-}89)$$

where R is an alkyl group. The Fluor Daniel Econamine FG CO_2 Recovery process that was developed by Dow Chemical is a widely used commercial process that uses an amine solution, containing a proprietary additive, to remove CO_2 economically from low-pressure, oxygen-containing gas streams similar to flue gas [90]. Some large-scale designs have been developed for CO_2 recovery from flue gas for use in CO_2-enhanced oil recovery [91]. Activated carbons and carbon molecular sieves are readily available commercially, and many studies have been conducted on CO_2 adsorption using such materials as well as other adsorbents such as zeolites, pillard clays, and metal oxides [15].

The DOE, through its Carbon Sequestration Program, is performing cost-shared CO_2 capture research and development covering a wide range of technology areas, including amine adsorbents, carbon adsorbents, membranes, sodium and other metal-based sorbents, electrochemical pumps, hydrates, and mineral carbonation [92]. The objective is to dramatically lower the capital cost and energy penalty associated with capturing dilute concentrations of CO_2 from large point sources such as power plants. Approximately one-third of the U.S. carbon emissions come from power plants.

A strong synergistic link exists between improved efficiency of fossil fuel conversion systems and carbon capture; the cost of capture per unit of product is less for a more efficient process. For example, heat and pressure integration between CO_2 capture and the other fossil fuel conversion systems can reduce parasitic steam and CO_2 recompression load [92]. Also, integrating CO_2 capture with SO_2/SO_3, NO_x, and mercury control can reduce or eliminate the need for scrubbers and other emissions control systems.

Advanced fuel conversion technologies such as gasification, oxygen (versus air) combustion, electrochemical cells, advanced steam reforming, and chemical looping produce a CO_2-rich exhaust stream that is ready for transport, storage, and utilization. Advanced fuel technologies are discussed in detail in Chapter 7. The CO_2 capture technologies that are being pursued by the DOE are categorized as pre-combustion decarbonization, oxygen-fired combustion, post-combustion capture, and advanced conversion processes [92].

Pre-Combustion Decarbonization Currently, 10 oxygen-fired gasifiers are in operation in the United States. Syngas from an oxygen-fired gasifier can be shifted to provide a stream of primarily H_2 and CO_2 at 400 to 800 psig. Regenerable glycol solvents can capture CO_2 and be regenerated to produce pure CO_2 at 15 to 25 psig. Supporting projects include [92]: (1) developing a high-temperature CO_2-selective membrane that will be suited to integrated gasification combined-cycle (IGCC) power generation systems and will enhance the efficiency of the water–gas shift reaction while recovering CO_2 for sequestration; (2) developing a process that captures CO_2 by combining it with water at low temperature and high pressure, thereby forming CO_2/water hydrates (ice-like macromolecular structures of CO_2 and water); (3) manufacturing a thermally-optimized membrane with better separation capabilities than current polymer members, with a focus on separation of CO_2, methane, and nitrogen gases in the range of 210 to 750°F; and (4) developing engineering evaluations of technologies for the capture, use, and disposal of CO_2 emphasizing IGCC power systems that produce both merchant hydrogen and electricity but also investigating CO_2 retrofit options for pulverized coal-fired boilers.

Oxygen-Fired Combustion Using oxygen instead of air during the combustion process produces a flue gas that is 90% pure CO_2 and is already at low pressure (10–15 psig); however, oxygen combustion requires three times more oxygen per kilowatt-hour of electricity generation than gasification. Currently, no oxygen-fired pulverized coal power plants are in commercial operation in the United States. Projects supporting this technology area include [92]: (1) designing a novel oxy-fuel boiler that incorporates a membrane to separate oxygen from the air, which is then used for combustion; (2) building on international work in advanced combustion in mixtures of oxygen and recycled flue gas; and (3) conducting tests of oxygen-enhanced combustion with the objective of lowering the cost of retrofit systems.

Post-Combustion Capture Flue gas from the 300 GW of U.S. pulverized coal-fired boiler capacity contains 12 to 18% CO_2. Currently, amine scrubbing with CO_2 compression to 1200 psig costs approximately $2000/kW and reduces the net power plant output by 12.5% [92]. Post-combustion capture supporting projects include [92]: (1) developing a CO_2 separation technology that uses a regenerable sodium-based sorbent to capture CO_2 from flue gas; (2) developing cost-effective electrochemical devices for the separation (electrochemical pump) and detection (sensors) of CO_2; (3) developing integrated collaborative technologies to prove the feasibility of advanced CO_2 separation and capture; (4) developing amine-enriched adsorbents prepared by chemical treatment of high surface area oxide materials with various amine compounds; (5) developing simultaneous removal of CO_2 and SO_2 by ammonia solution to recover the CO_2 by converting ammonium carbonate to ammonium bicarbonate solution and recycle ammonia; (6) predicting the performance of absorption/stripping of CO_2 with aqueous K_2CO_3; (7) developing an integrated modeling framework for evaluating alternative carbon sequestration technologies; (8) modifying the design of a tubular solid oxide fuel cell (SOFC) module to incorporate an afterburner stack of tubular oxygen transport membranes, thus oxidizing SOFC-depleted fuel in the anode exhaust to CO_2 that can then be easily separated; (9) developing viable technological solutions to safe and economic capture and storage of CO_2 underground; and (10) developing sorbents for CO_2 separation and removal via pressure (PSA) and/or temperature swing adsorption (TSA).

Work is also being performed on CO_2 capture that is outside of the DOE's Carbon Sequestration Program. One example is the development of a high-capacity, highly selective CO_2 adsorbent based on a novel "molecular basket" concept consisting of a mesoporous molecular sieve (MCM-41) and an immobilized branched-chain polymer (sterically branched polyethyenimine) with CO_2-capturing sites [93,94]. This adsorbent has progressed from the bench scale to pilot testing and, similar to the other projects summarized here, is undergoing further research and development to determine its commercial feasibility.

Advanced Conversion Currently, only a limited number of promising ideas have emerged in this area, none of which is at the commercial or demonstration phase. A supporting program in the DOE's Carbon Sequestration Program is the development of a method to use gasified coal or natural gas to reduce a metal oxide sorbent, thereby producing steam and high-pressure CO_2 [92].

CO_2 Sequestration

Although sequestration is not considered a power plant emissions control technology, it is discussed in this chapter because sequestration is vital to the overall power plant system in that options for disposal or use of the captured CO_2 are necessary. Sequestration encompasses all forms of carbon storage, including storage in terrestrial ecosystems, geologic formations, and perhaps oceans. Through the development of optimized field practices and technologies, the DOE's Carbon Sequestration Program seeks to quantify and improve the storage capacity of all potential reserves [92].

Geologic Sequestration There are several types of geologic formations in which CO_2 can be stored, including depleting oil reservoirs, depleting gas reservoirs, unmineable coal seams, saline formations, shale formations with high organic content, and others. Many power plants are located near geologic formations that are amenable to CO_2 storage. DOE supporting projects include the following goals [92]:

- Assess the feasibility of co-optimization of CO_2 sequestration and enhanced oil and gas recovery from oil reservoirs; currently, about 32 million short tons of CO_2 per year are injected into depleting oil reservoirs in the United States as part of enhanced oil operations, 10% of which is from anthropogenic sources. Current practices are not directed at optimizing CO_2 storage, and a typical storage rate is 2000 scf CO_2 per barrel of oil recovered;
- Acquire technical information for assessing the effects of CO_2–CH_4–N_2 mixing;
- Perform a demonstration of 4 million scf CO_2 per day using existing recovery technology to evaluate the viability of storing CO_2 in deep unmineable coal seams in the San Juan Basin in northwest New Mexico/Southwestern Colorado;
- Demonstrate a coal bed methane production technology (*i.e.*, slant hole) to drain natural gas from unmineable coal seams and inject CO_2 in the wells;
- Quantify the CO_2 storage potential of the Black Warrior coalbed methane region in Alabama;
- Investigate the ability of coal to enhance coalbed methane production while sequestering CO_2 by adsorption on the surface of various United States coals;
- Study deep saline formation in the Colorado Plateau and Rocky Mountain region to determine volume, fate, and transport of stored CO_2;
- Evaluate factors that affect chemical reactions that convert CO_2 to a stable solid in underground saline formations;
- Explore the use of hydraulic fracturing to improve the permeability of saline formations, thereby lowering the cost of CO_2 injection;
- Develop analytical tools to investigate the solution kinetics of CO_2 brines at temperatures and pressures appropriate for deep aquifer CO_2 sequestration;
- Develop and apply the criteria for characterizing optimal saline reservoirs for very-long-term sequestration of CO_2;
- Investigate the concept that Devonian shales, like coal, could serve as geologic sinks for CO_2;
- Investigate the technical and economic viability of sequestration of CO_2 from point sources in Texas low-rank coals, as well as the potential for enhanced coalbed gas recovery.

Terrestrial Sequestration The United States has vast agricultural and forest resources, and policymakers have looked to terrestrial sequestration as an option for reducing net greenhouse gas emissions. Currently, terrestrial uptake of CO_2 offsets approximately one-third of global anthropogenic CO_2 emissions. The uptake from domestic terrestrial ecosystems is expected to decrease 13% over the next 20 years as Northeast forests mature. Opportunities for enhanced terrestrial uptake include 1.5 million acres of land damaged by past mining practices, 32 million acres of Conservation Resource Protection (CRP) farmland, and 120 million acres of pastureland [92]. DOE supporting projects have the following goals [92]:

- Restore sustainable forests on Appalachian mined lands by demonstrating terrestrial sequestration for wood products, renewable energy, carbon storage, and other ecosystem services on three 30- to 40-hectare sites in West Virginia and Virginia;
- Enhance terrestrial carbon sinks through reclamation and reforestation of abandoned mine lands in the Appalachian region;
- Use low compaction reclamation techniques to facilitate reforestation using three 150- to 200-acre demonstrations;
- Determine the best way to increase carbon sequestration potential of land previously disturbed by mining, highway construction, or poor land management practices;
- Demonstrate and assess the life-cycle costs of integrating electricity production with enhanced terrestrial carbon sequestration at coal-mine spoil land at Tennessee Valley Authority's 2558 MW bituminous coal-burning Paradise Station;
- Explore terrestrial sequestration opportunities in degraded and rangeland in Southwestern United States.

Ocean Sequestration Oceans absorb, release, and store large amounts of CO_2 from the atmosphere. The two approaches to enhancing oceanic carbon sequestration both take advantage of the ocean's natural processes [92]. One approach is to enhance the productivity of the ocean biological systems through fertilization or other means. Another approach is to inject CO_2 into the deep ocean. Compared to terrestrial ecosystems and geologic formations, the concept of ocean sequestration is in a much earlier stage of development. Experimental results and observed surges in phytoplankton growth after dust clouds pass over certain ocean regions indicate that increasing the concentration of iron and other macronutrients in certain ocean waters can greatly increase the growth of phytoplankton and thus CO_2 uptake [92]. Ocean fertilization remains highly controversial because of uncertainty surrounding other changes it may cause. The DOE is not supporting any research projects in this area, although opportunities exist in establishing the scientific knowledge base needed to understand, assess, and optimize ocean fertilization; develop effective macronutrient seeding methodologies; and assess long-term CO_2 fate and flux. Although no pilot or commercial applications of ocean injection have been performed, small-scale experiments have been conducted and the DOE is supporting projects in their Carbon Sequestration Program, which include [92]:

- Collaborating in an international effort to understand two-phase CO_2 plumes;
- Providing logistical and technical support to the international collaboration;
- Studying the feasibility of large-scale CO_2 ocean sequestration;
- Synthesizing CO_2/water hydrates to study their physical properties;
- Studying deep-ocean liquid CO_2/water/calcium carbonate storage.

Multipollutant Control

The concept of controlling or removing more than one pollutant using a single control device has been of interest to the coal-fired industry for many years. Initially, technologies to simultaneously control NO_x and SO_2, particulate matter and NO_x, or all three pollutants were developed and tested at various scales in the DOE's Clean Coal Technology program beginning in the late 1980s (see Chapter 7 for details of this program). Simultaneous NO_x and SO_2 control was demonstrated using the SNOX™, SNRB™, and integrated dry NO_x/SO_2 emissions control systems [95]. Each of these demonstration projects involved a unique combination of control technologies to achieve reduction of NO_x and SO_2 emissions. The SNOX™ process uses an SCR, catalytic SO_2 converter (to SO_3), and a wet-gas sulfuric acid tower for removing 93 to 94% and 95% of the NO_x and SO_2, respectively. The SNRB™ process combines the removal of SO_2, NO_x, and particulates in one unit, a high-temperature baghouse (fitted with a catalyst in the bag cages for NO_x reduction) located between the economizer and the combustion air preheater with a calcium- or sodium-based sorbent injected for SO_2 control. It achieved 80%, 50 to 95%, and 99+% control of SO_2, NO_x, and particulate matter, respectively. The integrated dry NO_x/SO_2 emissions control system is comprised of low-NO_x burners, overfire air, SNCR, and duct sorbent injection and has achieved up to 80 and 70% reductions in NO_x and SO_2, respectively. These three technologies are potentially applicable to flue gas cleaning for all types of conventional coal-fired units, including stoker, cyclone, and pulverized coal-fired boilers. Capital costs for these three systems are estimated at \$305, \$253, and \$190/kW, respectively, for the SNOX, SNRB, and integrated dry NO_x/SO_2 emissions control systems. The operating costs are estimated at \$12.1 mills/kWh for the SNRB process, and the SNOX process generates a \$6.1 mills/kWh credit [95]. No data are available for the integrated process.

Similarly, technologies to simultaneously remove NO_x and particulate matter have also been developed and tested. These technologies, which consisted of coating substrates such as ceramic membrane filters with SCR catalysts, were initially being developed for high-pressure, high-temperature atmospheres found in gasification systems. This work was applied under conventional pulverized coal-fired conditions to investigate simultaneous NO_X and particulate removal [15,94,96]. The technology removed NO_X and particulate matter (at PM removal efficiencies greater than high-efficiency P-84 polyimide bags) and had the added benefit of removing a significant amount of mercury, as 79% of the mercury was removed across a 6000-acfm ceramic membrane filter system [96]. However, unacceptable pressure drops across the ceramic filters have not been resolved (*i.e.*, fine particles embedded into the ceramic substrate), thus hindering the commercialization of this process [97].

The interest in multipollutant control, or integrated emissions control, has intensified in the United States over the last several years primarily due to impending multipollutant legislation such as the Clear Skies Act. The pollutants of interest include SO_2, NO_X, mercury, SO_3, and fine particulate matter. The list of multipollutant technologies is long and continues to grow; however, very few of the integrated emissions control technologies have advanced to the point where reliable cost data are available, although several are beginning to generate this information through pilot- and demonstration-scale testing [98]. Similarly, Canada is evaluating multipollutant control for reducing SO_X, NO_X, particulate, and mercury emissions levels comparable to those of a natural gas combined-cycle plant with SCR [99].

Several promising technologies are currently being tested, and a few of those are reviewed here. Several of these evaluations are cofunded by various DOE programs, state agencies, and industry. Note that several of the technologies discussed in previous sections are also considered multipollutant control options.

ECO Process

Powerspan Corporation has developed an integrated air pollution control technology that achieves major reductions in emissions of NO_X (90%), SO_2 (98%), fine particulate matter (95%), and mercury (80–90%) from coal-fired power plants; it has been tested in a 1-MW slipstream [100]. The patented technology, Electro-Catalytic Oxidation (ECO), also reduces emissions of air toxic compounds such as arsenic and lead as well as acid gases such as hydrochloric acid (HCl).

In commercial operation, the ECO process is to be installed downstream of a power plant's existing ESP or baghouse [100]. It treats the flue gas in three steps to achieve multipollutant removal. In the first process step, a barrier discharge reactor oxidizes gaseous pollutants to higher oxides (*i.e.*, nitric oxide to nitrogen dioxide, a portion of the sulfur dioxide to sulfuric acid, and mercury to mercuric oxide). Following the barrier discharge reactor is the ammonia scrubber, which removes unconverted sulfur dioxide and nitrogen dioxide produced in the barrier discharge. A WESP follows the scrubber and it, along with the scrubber, captures acid aerosols produced by the discharge reactor, fine particulate matter, and oxidized mercury. The WESP also captures aerosols generated in the ammonia scrubber. Liquid effluent from the ammonia scrubber contains dissolved sulfate and nitrate salts, along with mercury and captured particulate matter. It is sent to a by-product recovery system, which includes filtration to remove ash and activated carbon adsorption for mercury removal. The treated by-product stream, free of mercury and ash, can be processed to form ammonia sulfate/nitrate fertilizer. Powerspan's capital cost estimate is ~\$200/kW, including balance of plant modifications [100]. The levelized operating and maintenance costs are estimated to be \$2.0 to \$2.5 mills/kWh.

Airborne Process

Airborne Pollution Control, in cooperation with LG&E Energy Corporation, The Babcock & Wilcox Company, and USFilter HPD Systems, has developed an emerging multipollutant, post-combustion control system [101]. The technology combines the use of dry sodium bicarbonate injection coupled with enhanced wet sodium carbonate scrubbing to provide SO_X, NO_X, mercury, and other heavy metal reductions. Although sodium bicarbonate scrubbing is well known as an effective flue gas cleanup process, commercial application has been prevented by the high cost of sodium bicarbonate, the limited economic value of the scrubber product (*i.e.*, sodium sulfate), and the economic and environmental issues associated with sodium sulfate disposal [98]. Airborne Pollution Control has developed a recycling process that will regenerate sodium sulfate back into sodium bicarbonate and a sulfate-based fertilizer product that may eliminate the financial and disposal barriers. Testing has reached the 5-MW level.

LoTOx Process

British Oxygen Corporation's low-temperature oxidation process, LoTOx, was originally developed for NO_X control but has shown multipollutant reduction capabilities, as well [98]. The LoTOx process injects ozone into the flue gas at temperatures below 300°F to react with NO and NO_2 to form soluble higher oxides that can be removed with a wet scrubber. NO_X reduction efficiencies greater than 90% have been achieved and, because the ozone also oxidizes elemental mercury to soluble oxide species, mercury removal is also possible, with reductions greater than 90% achieved in laboratory tests. Capital costs for LoTOx are estimated at \$289/kW [98]. Levelized operating and maintenance costs are estimated at \$12.96 mills/kWh.

Mobotec Systems

Mobotec USA, Inc., is exploring furnace sorbent injection using limestone or trona in combination with rotary opposed fire air (ROFA) and ROTAMIX (a second-generation SNCR and sorbent injection process) for SO_2 and mercury removal. Mixing in the ROFA and ROTAMIX systems creates optimal conditions for achieving multipollutant reduction by providing ample turbulence and residence time within a specific temperature window [98]. Tests with trona and limestone have achieved reductions in SO_2 of 69 and 64%, respectively; SO_3, 90 and 90%; HCl, 0 and 75%; mercury, 89 and 67%; NO_x, 4 and 11%; and particulate matter, 18 and 80% [102].

Others

Combining sorbent injection for mercury control with other technologies for NO_x and/or SO_x removal represents another multipollutant control option [98]. Many companies are exploring sorbent and chemical injection techniques that can remove mercury at reasonable costs, including Sorbent Technologies Corporation, ADA-Environmental Solutions, URS Corporation, EPRI, and Alstom Power; universities such as the University of North Dakota Energy and Environmental Research Center and Penn State University; and federal agencies such as the DOE's National Energy Technology Laboratory.

References

1. DOE, *National Energy Technology Laboratory Accomplishments FY 2002* (Office of Fossil Energy, U.S. Department of Energy, Washington, D.C., August 2003).
2. DOE, *Clean Coal Technology—Environmental Benefits of Clean Coal Technologies,* Topical Report No. 18 (U.S. Department of Energy, U.S. Government Printing Office, Washington, D.C., April 2001).
3. U.S. Energy Information Administration, *Electric Power Annual 2000,* Vol. II (Office of Coal, Nuclear, Electric, and Alternate Fuels, U.S. Department of Energy, U.S. Government Printing Office, Washington, D.C., November 2002).
4. Soud, H. N., *Developments in FGD* (IEA Coal Research, London, 2000).
5. Srivastava, R. K., C. Singer, and W. Jozewicz, SO_2 Scrubbing Technologies: A Review, in *Proc. of the AWMA 2000 Annual Conference and Exhibition* (Air and Waste Management Association, Pittsburgh, PA, 2000).
6. Wark, K., C. F. Warner, and W. T. Davis, *Air Pollution: Its Origin and Control,* Third ed. (Addison-Wesley Longman, Menlo Park, CA, 1998).
7. Davis, W. T. (editor), *Air Pollution Engineering Manual,* Second ed. (John Wiley & Sons, New York, 2000).

8. Elliot, T. C. (editor), *Standard Handbook of Powerplant Engineering* (McGraw-Hill, New York, 1989).
9. U.S. Energy Information Agency, *U.S. Coal Reserves: 1997 Update* (U.S. Department of Energy, Office of Coal, Nuclear, Electric and Alternate Fuels, U.S. Government Printing Office, Washington, D.C., February 1999).
10. Harrison, C. D., Fuel Options to Mitigate Emissions Reduction Costs, in *Proc. of the 28th International Technical Conference on Coal Utilization and Fuel Systems* (Coal & Slurry Technology Association, Washington, D.C., 2003).
11. Stultz, S. C. and J. B. Kitto (editors), *Steam: Its Generation and Use*, 40th ed. (Babcock & Wilcox Co., Barberton, OH, 1992).
12. Radcliffe, P. T., *Economic Evaluation of Flue Gas Desulfurization Systems* (Electric Power Research Institute, Palo Alto, CA, 1991).
13. Rhudy, R., M. McElroy, and G. Offen, Status of Calcium-Based Dry Sorbent Injection SO_2 Control, in *Proc. of the Tenth Symposium on Flue Gas Desulfurization*, November 17–21, 1986, pp. 9-69–9-84.
14. Bland, V. V. and C. E. Martin, *Full-Scale Demonstration of Additives for* NO_2 *Reduction with Dry Sodium Desulfurization*, EPRI GS-6852 (Electric Power Research Institute, Palo Alto, CA, June 1990).
15. Miller, B. G., A. L. Boehman, P. Hatcher, H. Knicker, A. Krishnan *et al.*, *The Development of Coal-Based Technologies for Department of Defense Facilities: Phase II Final Report*, DE-FC22-92PC92162, prepared for the U.S. Department of Energy Federal Energy Technology Center, Pittsburgh, PA, July 31, 2000, 784 pages.
16. Wu, Z. *Air Pollution Control Costs for Coal-Fired Power Stations* (IEA Coal Research, London, 2001).
17. Smith, D. J., Cost of SO_2 Scrubbers Down to $100/kW, *Power Engineering*, September 2001, pp. 63–68.
18. Blythe G., B. Horton, and R. Rhudy, EPRI FGD Operating and Maintenance Cost Survey, in *Proc. of the EPRI–DOE–EPA Combined Utility Air Pollution Control Symposium: The MEGA Symposium*, Vol. I, *SO_2 Controls*, 1999, pp. 1-21–1-34.
19. EPA, *EPA Acid Rain Program: 2001 Progress Report* (Office of Air and Radiation, U.S. Environmental Protection Agency, U.S. Government Printing Office, Washington, D.C., November 2002).
20. EPA, *Latest Findings on National Air Quality: 2002 Status and Trends* (Office of Air Quality Planning and Standards, U.S. Environmental Protection Agency, U.S. Government Printing Office, Washington, D.C., August 2003).
21. Mitchell, S. C., *NO_X in Pulverized Coal Combustion* (IEA Coal Research, London, 1998).
22. Davidson, R. M., *How Coal Properties Influence Emissions* (IEA Coal Research, London, 2000).
23. EPA, *Technical Bulletin, Nitrogen Oxides (NO_X): Why and How They Are Controlled* (Office of Air Quality Planning and Standards, U.S. Environmental Protection Agency, U.S. Government Printing Office, Washington, D.C., November 1999).

24. Lawn, C. J. (editor), *Principles of Combustion Engineering for Boilers* (Academic Press, London, 1987).
25. Tsiou, C., H. Lin, S. Laux, and J. Grusha, Operating Results from Foster Wheeler's New Vortex Series Low-NO_X Burners, in *Proc. of Power Gen 2000*, 2000.
26. Steitz, T. H. and R. W. Cole, Field Experience in Over 30,000 MW of Wall-Fired Low NO_X Installations, in *Proc. of Power Gen 1996*, 1996.
27. Steitz, T. H., J. Grusha, and R. Cole, Wall-Fired Low NO_X Burner Evolution for Global NO_X Compliance, in *Proc. of the 23rd International Technical Conference on Coal Utilization and Fuel Systems* (Coal & Slurry Technology Association, Washington, D.C., 1998).
28. Patel, R. L., D. E. Thornock, R. W. Borio, B. G. Miller, and A. W. Scaroni, Firing Micronized Coal with a Low NO_X RSFC Burner in an Industrial Boiler Designed for Oil and Gas, in *Proc. of the Thirteenth Annual International Pittsburgh Coal Conference*, 1996.
29. Borio, R. W., R. L. Patel, D. E. Thornock, B. G. Miller, A. W. Scaroni, and J. G. McGowan, *Task 5—Final Report: One Thousand Hour Demonstration Test in the Penn State Boiler*, DE-AC22-91PC91160, prepared for the U.S. Department of Energy, Federal Energy Technology Center, Pittsburgh, PA, March 1998.
30. Wu, Z., *NO_X Control for Pulverized Coal Fired Power Stations* (IEA Coal Research, London, 2002).
31. U.S. Energy Information Administration, *Net Generation by Energy Source: Electric Utilities, 1990 through July 2003* (U.S. Energy Information Agency, U.S. Department of Energy, Washington, D.C.), www.eia.doe.gov/cneaf/electricity/epm (released November 2003).
32. EPA, *Control of NO_X Emissions by Reburning* (Office of Research and Development, U.S. Environmental Protection Agency, U.S. Government Printing Office, Washington, D.C., February 1996).
33. Wendt, J. O. L., C. V. Sternling, and M. A. Matovich, Reduction of Sulfur Trioxide and Nitrogen Oxides by Secondary Fuel Injection, in *Proc. of the 14th Symposium (International) on Combustion* (Combustion Institute, Pittsburgh, PA, 1973), pp. 897–904.
34. Myerson, A. L., F. R. Taylor, and B. G. Faunce, Ignition Limits and Products of the Multistage Flames of Propane–Nitrogen Dioxide Mixtures, in *Proc. of the 6th Symposium (International) on Combustion* (Combustion Institute, Pittsburgh, PA, 1957), pp. 154–163.
35. Stoesssner, R. D. and E. Zawadzki, Coal Water Slurry Dual Firing Project for Homer City Station: Phase I Test Results, in *Proc. of the 16th International Technical Conference on Coal Utilization and Fuel Systems* (Coal & Slurry Technology Association, Washington, D.C., 1991), pp. 599–608.
36. Falcone Miller, S., B. G. Miller, A. W. Scaroni, S. A. Britton, D. Clark, W. P. Kinneman, S. V. Pisupate, R. Poe, R. Wasco, and R. T. Wincek, *Coal-Water Slurry Fuel Combustion Program* (Pennsylvania Electric Company, Erie, PA, 1993), 98 pp.

37. Morrison, J. L., B. G. Miller, and A. W. Scaroni, *Determining Coal Slurryability: A UCIG/Penn State Initiative*, WO3852-06 (Electric Power Research Institute, Palo Alto, CA, January 1998).
38. Ashworth, R. A. and T. M. Sommer, Economical Use of Coal Water Slurry Fuels Produced from Impounded Coal Fines, in *Proc. of Effects of Coal Quality on Power Plants* (Electric Power Research Institute, Palo, Alto, CA, 1997).
39. Falcone Miller, S., J. L. Morrison, and A. W. Scaroni, The Effect of Cofiring Coal-Water Slurry Fuel Formulated from Waste Coal Fines with Pulverized Coal on NO_X Emissions, in *Proc. of the 21st International Technical Conference on Coal Utilization and Fuel Systems* (Coal & Slurry Technology Association, Washington, D.C., 1996).
40. Miller, B. G., S. Falcone Miller, J. L. Morrison, and A. W. Scaroni, Cofiring Coal-Water Slurry Fuel with Pulverized Coal as a NO_X Reduction Strategy, in *Proc. of the 14th International Pittsburgh Coal Conference*, 1997.
41. Battista, J. J., personal communication, 1997.
42. Tillman, D. A., NO_X Reduction Achieved Through Biomass Cofiring, in *Proc. of the 20th Annual International Pittsburgh Coal Conference*, September 2003.
43. Tillman, D A., personal communication, November 2003.
44. Tillman, D. A. and N. S. Harding, *Fuels of Opportunity: Characteristics and Uses in Combustion Systems* (Elsevier, London, 2004).
45. Tillman, D. A., B. G. Miller, and D. Johnson, Analyzing Opportunity Fuels for Firing in Coal-Fired Boilers, in *Proc. of the 20th Annual International Pittsburgh Coal Conference*, September 2003.
46. Vasquez, E. R., H. Gadalla, K. McQuistan, F. Iman, and R. E. Sears, NO_X Control in Coal-Fired Cyclone Boilers Using SmartBurn Combustion Technology, in *Proc. of the EPRI–DOE–EPA Combined Power Plant Air Pollution Control: The MEGA Symposium*, 2003.
47. DOE, *Clean Coal Technology—Control of Nitrogen Oxide Emissions: Selective Catalytic Reduction (SCR)*, Topical Report No. 9 (U.S. Department of Energy, U.S. Government Printing Office, Washington, D.C., July 1997).
48. McIlvaine, R. W., H. Weiler, and W. Ellison, SCR Operating Experience of German Powerplant Owners as Applied to Challenging U.S. High-Sulfur Service, in *Proc. of the EPRI–DOE–EPA Combined Power Plant Air Pollution Control: The MEGA Symposium*, 2003.
49. Cichanowicz, J. E., L. L. Smith, L. J. Muzio, and J. Marchetti, 100 GW of SCR: Installation Status and Implications of Operating Performance on Compliance Strategies, in *Proc. of the EPRI–DOE–EPA Combined Power Plant Air Pollution Control: The MEGA Symposium*, 2003.
50. EPA, *Performance of Selective Catalytic Reduction on Coal-Fired Steam Generating Units* (Office of Air and Radiation, U.S. Environmental Protection Agency, U.S. Government Printing Office, Washington, D.C., June 25, 1997).
51. Lodder, P. and J. B. Lefers, Effect of Natural Gas, C_2H_6, and CO on the Homogenous Gas Phase Reduction of NO_X by NH_3, *Chemical Engineering Journal*, Vol. 30, No. 3, 1985, p. 161.

52. Ciarlante, V. and M. A. Zoccola, Conectiv Energy Successfully Using SNCR for NO_X Control, *Power Engineering*, Vol. 105, No. 6, June 2001, pp. 61–62.
53. Frederick, N., R. K. Agrawai, and S. C. Wood, NO_X Control on a Budget: Induced Flue Gas Recirculation, *Power Engineering*, Vol. 107, No. 7, July 2003, pp. 28–32.
54. Hoskins, B., Uniqueness of SCR Retrofits Translates into Broad Cost Variations, *Power Engineering*, Vol. 107, No. 5, May 2003, pp. 25–30.
55. EPRI, *EPRI 2002 Annual Report* (Electric Power Research Institute, Palo Alto, CA, 2003), pp. 11–12.
56. EPA, *National Emission Inventory (NEI) Air Pollution Emission Trends, Updated August 2000: Average Annual Emissions, All Criteria Pollutants Years Including 1997–2001* (U.S. Environmental Protection Agency, Washington, D.C.), www.epa.gov/ttn/chief/trends/ (last updated September 4, 2003).
57. Soud, H. N. and S. C. Mitchell, *Particulate Control Handbook for Coal-Fired Plants* (IEA Coal Research, London, 1997).
58. DOE, *Description—PM Emissions Control* (U.S. Department of Energy, Washington, D.C.), www.netl.doe.gov/coalpower/environment/pm/description.html (last updated December 2, 2003).
59. DOE, *Controlling Air Toxics with Electrostatic Precipitators Fact Sheet* (Office of Fossil Energy, U.S. Department of Energy, Washington, D.C., 1997).
60. B&W, *Electrostatic Precipitator Product Sheet*, PS151 2M A 12/82 (Babcok & Wilcox Co., Barberton, OH, December 1982).
61. Miller, B. G., unpublished data, 1986.
62. Miller, B. G., S. J. Miller, G. P. Lamb, and J. A. Luppens, Sulfur Capture by Limestone Injection During Combustion of Pulverized Panola County Texas Lignite, in *Proc. of Gulf Coast Lignite Conference*, 1984.
63. Buckley, W. and I. Ray, Application of Wet Electrostatic Precipitation Technology in the Utility Industry for PM2.5 Control, in *Proc. of the EPRI–DOE–EPA Combined Power Plant Air Pollution Control: The MEGA Symposium*, 2003.
64. Staehle, R. C., R. J. Triscori, G. Ross, K. S. Kumar, and E. Pasternak, The Past, Present and Future of Wet Electrostatic Precipitators in Power Plant Applications, in *Proc. of the EPRI–DOE–EPA Combined Power Plant Air Pollution Control: The MEGA Symposium*, 2003.
65. Altman, R., G. Offen, W. Buckley, and I. Ray, Wet Electrostatic Precipitation Demonstrating Promise for Fine Particulate Control, Part I, *Power Engineering*, Vol. 105, No. 1, January 2001, pp. 37–39.
66. Altman, R., W. Buckley, and I. Ray, Wet Electrostatic Precipitation Demonstrating Promise for Fine Particulate Control, Part II, *Power Engineering*, Vol. 105, No. 2, February 2001, pp. 42–44.
67. Bustard, C. J., K. M. Cushing, D. H. Pontius, W. B. Smith, and R. C. Carr, *Fabric Filters for the Electric Utility Industry*, Vol. 1, *General Concepts* (Electric Power Research Institute, Palo Alto, CA, 1988).
68. Soud, H. N., *Developments in Particulate Control for Coal Combustion* (IEA Coal Research, London, 1995).

69. Miller, R. L., W. A. Harrison, D. B. Prater, and R. L. Chang, Alabama Power Company E.C. Gaston 272 MW Electric Steam Plant—Unit No. 3 Enhanced COHPAC I Installation, in *Proc. of the EPRI–DOE–EPA Combined Utility Air Pollution Control Symposium: The MEGA Symposium*, Vol. III, *Particulates and Air Toxics*, 1997.
70. *Proc. of the EPRI–DOE–EPA Combined Utility Air Pollution Control Symposium: The MEGA Symposium*, Vol. III, *Particulates and Air Toxics*, 1997.
71. Cushing, K. M., W. A. Harrison, and R. L. Chang, Performance Response of COHPAC I Baghouse During Operation with Normal and Artificial Changes in Inlet Fly Ash Concentration and During Injection of Sorbents for Control of Air Toxics, in *Proc. of the EPRI–DOE–EPA Combined Utility Air Pollution Control Symposium: The MEGA Symposium*, Vol. III, *Particulates and Air Toxics*, 1997.
72. Gebert, R., C. Rinschler, D. Davis, U. Leibacher, P. Studer, W. Eckert, W. Swanson, J. Endrizzi, T. Hrdlicka, S. J. Miller, M. L. Jones, Y. Zhuang, and M. Collings, Commercialization of the Advanced Hybrid Filter Technology, in *Proc. of the Conference on Air Quality III: Mercury, Trace Elements, and Particulate Matter* (University of North Dakota, Grand Forks, 2002).
73. DOE, *Advanced Hybrid Particulate Collector Fact Sheet* (Office of Fossil Energy, U.S. Department of Energy, Washington, D.C., June 2001).
74. DOE, Control Technology Advanced Hybrid Particulate Collector (U.S. Department of Energy, Washington, D.C.), www.netl.doe.gov/coalpower/environment/pm/con_tech/hybrid.html (last updated December 2, 2003).
75. Blankinship, S., Hybrid Filter Technology Weds ESPs with Bag Filters, *Power Engineering*, Vol. 106, No. 2, February 2002, p 9.
76. DOE, *Demonstration of a Full-Scale Retrofit of the Advanced Hybrid Particulate Collector (AHPC) Collector Fact Sheet* (Office of Fossil Energy, U.S. Department of Energy, Washington, D.C., February 2003).
77. Feeley, T. J., J. Murphy, J. Hoffman, and S. A. Renninger, *A Review of DOE/NETL's Mercury Control Technology R&D Program for Coal-Fired Power Plants*, DOE/NETL Hg R&D Review (National Energy Technology Laboratory, U.S. Department of Energy, Washington, D.C.), www.netl.doe.gov/ (April 2003).
78. Pavlish, J. J., E. A. Sondreal, M. D. Mann, E. S. Olson, K. C. Galbreath, D. L. Laudal, and S. A. Benson, Status Review of Mercury Control Options for Coal-Fired Power Plants, *Fuel Processing Technology*, Vol. 82, 2003, pp. 89–165.
79. Sjostrom, S., J. Bustard, and R. Chang, Mercury Removal Trends and Options for Coal-Fired Power Plants with Full-Scale ESPs and Fabric Filters, in *Proc. of the Nineteenth Annual International Pittsburgh Coal Conference*, 2002.
80. Starns, T., J. Bustard, M. Durham, C. Martin, R. Schlager, S. Sjostrom, C. Lindsey, B. Donnelly, R. Afonso, R. Chang, and S. Renninger, Results of Activated Carbon Injection Upstream of Electrostatic Precipitators for Mercury Control,

in *Proc. of the EPRI–DOE–EPA Combined Power Plant Air Pollution Control: The MEGA Symposium*, 2003.

81. Bustard, C. J., M. Durham, C. Lindsey, T. Starns, C. Martin *et al.*, Results of Activated Carbon Injection for Mercury Control Upstream of a COHPAC Fabric, in *Proc. of the EPRI–DOE–EPA Combined Power Plant Air Pollution Control: The MEGA Symposium*, 2003.
82. National Research Council, *Coal Waste Impoundments: Risks, Responses, and Alternatives* (National Academy Press, Washington, D.C., 2002).
83. Jha, M. C., F. J. Smit, G. L. Shields, and N. Moro, *Engineering Development of Advanced Physical Fine Coal Cleaning for Premium Fuel Applications: Project Final Report*, DOE Contract No. DE-AC22-92PC92208 (U.S. Department of Energy, Washington, D.C., September 1997).
84. Akers, D. J. and C. E. Raleigh, The Mechanism of Trace Element Removal During Coal Cleaning, *Coal Preparation*, Vol. 19, No. 3, 1998, pp. 257–269.
85. Kilgroe, J. D. and R. K. Srivastava, *Technical Memorandum: Control of Mercury Emissions from Coal-Fired Electric Utility Boilers* (U.S. Environmental Protection Agency, Washington, D.C., September 2000).
86. Akers, D. J. and B. Toole-O'Neil, Coal Cleaning for HAP Control: Cost and Performance, in *Proc. of the 23rd International Technical Conference on Coal Utilization and Fuel Systems* (Coal & Slurry Technology Association, Washington, D.C., 1998).
87. DOE, *Carbon Sequestration Technology Roadmap and Program Plan* (National Energy Technology Laboratory, U.S. Department of Energy, Washington, D.C., March 12, 2003).
88. Los Alamos, *The Products of Coal: Electricity and Carbon Dioxide* (Clean Coal Compendium, Los Alamos National Laboratory, Los Alamos, NM), www.lanl.gov/projects/cctc/climate/Coal_CO2.html (last modified, December 17, 1999).
89. Chakma, A., CO_2 Capture Processes: Opportunities for Improved Energy Efficiencies, *Energy Conversion and Management*, Vol. 38, 1997, pp. S51–S56.
90. Sander, M. T. and C. L. Mariz, The Fluor Daniel Econamine FG Process: Past Experience and Present-Day Focus, *Energy Conversion and Management*, Vol. 33, 1995, pp. 813–818.
91. Mariz, C. L., Carbon Dioxide Recovery: Large Scale Design Trends, *Journal of Canadian Petroleum Technology*, Vol. 37, No. 7, 1998, pp. 42–47.
92. DOE, *Carbon Sequestration: Why Sequestration?* (National Energy Technology Laboratory, U.S. Department of Energy, Washington, D.C.), www.netl.doe.gov/coalpower/sequestration (last updated July 25, 2003).
93. Song, C. X. Xu, J. M. Andresen, B. G. Miller, and A. W. Scaroni, Novel "Molecular Basket" Adsorbent for CO_2 Capture, in *Proc. of the Seventh International Conference on Carbon Dioxide Utilization* (October 2003).
94. Miller, B. G., S. Falcone Miller, S. V. Pisupati, C. Song, R. S. Wasco *et al.*, *The Development of Coal-Based Technologies for Department of Defense Facilities: Phase III Final Report*, DE-FC22-92PC92162, prepared for the U.S. Department of Energy Federal Energy Technology Center, Pittsburgh, PA, January 31, 2004, 600 pages.

95. DOE, *Clean Coal Technology—Technologies for the Combined Control of Sulfur Dioxide and Nitrogen Oxides Emissions from Coal-Fired Boilers*, Topical Report No. 13. (U.S. Department of Energy, U.S. Government Printing Office, Washington, D.C., May 1999).
96. Miller, B. G., S. Falcone Miller, R. T. Wincek, and A. W. Scaroni, A Demonstration of Fine Particulate and Mercury Removal in a Coal-Fired Industrial Boiler Using Ceramic Membrane Filters and Conventional Fabric Filters, in *Proc. of the EPRI–DOE–EPA Combined Utility Air Pollutant Symposium: The Mega Symposium*, 1999.
97. Miller, B. G., S. Falcone Miller, R. T. Wincek, and A. W. Scaroni, A Preliminary Evaluation of Ceramic Filters Including the Use of Nondestructive X-Ray Computerized Tomography, in *Proc. of the 26th International Technical Conference on Coal Utilization and Fuel Systems*, 2001.
98. Schimmoller, B. K., Lack of Environmental Certainty Renews Emphasis on Low-Cost Emissions Control, *Power Engineering*, Vol. 107, No. 9, September 2003, pp. 32–38.
99. Cameron, D. H., C. E. Martin, W. A. Campbell, and R. A. Stobbs, The Future of Multi-Pollutant Control for Coal-Fired Boilers: A Canadian Perspective, in *Proc. of the EPRI–DOE–EPA Combined Power Plant Air Pollution Control: The MEGA Symposium*, 2003.
100. McLarnon, C. R. and D. Steen, Combined SO_2, NO_x, PM, and Hg Removal from Coal Fired Boilers, in *Proc. of the EPRI–DOE–EPA Combined Power Plant Air Pollution Control: The MEGA Symposium*, 2003.
101. Mortson, M. E. and F. C. Owens II, Multi Pollutant Control with the Airborne Process, in *Proc. of the EPRI–DOE–EPA Combined Power Plant Air Pollution Control: The MEGA Symposium*, 2003.
102. Haddad, E., J. Ralson, G. Green, and S. Castagnero, Full-Scale Evaluation of a Multi-Pollutant Reduction Technology: SO_2, Hg, and NO_x, in *Proc. of the EPRI–DOE–EPA Combined Power Plant Air Pollution Control: The MEGA Symposium, 2003.*

CHAPTER 7

Future Power Generation (Near-Zero Emissions During Electricity Generation)

President George W. Bush has pledged that the United States will be a leader in the long-term effort to achieve energy security [1]. To do this, a balanced and diversified portfolio of energy resources is required, which includes fossil, nuclear, and renewable energy sources. Clean coal is recognized as a crucial element in the United States' overall policy, as recently evidenced by President Bush's commitment to a 10-year, $2 billion clean coal research initiative (which is discussed later in this chapter). Coal is a popular fuel choice because worldwide reserves could last two to three centuries, it is widely dispersed throughout the world, and it is among the most economic of energy resources. However, coal is among the most environmentally problematic of all energy resources, and much effort has gone into addressing this over the last two decades, with the U.S. Department of Energy (DOE) spearheading the development of clean coal technologies.

In the foreseeable future, the energy needed to sustain economic growth in the United States (as well as most countries of the world, for that matter) will continue to come largely from fossil fuels, with coal playing a leading role. In supplying this energy need, however, the United States must address growing global and regional environmental concerns and energy prices [2]. Maintaining low-cost electricity while demand grows and environmental pressures increase requires new technologies. These technologies must allow the United States to use its indigenous resources wisely, cleanly, and efficiently.

The existing fleet of coal-fired power plants in the United States is faced with increasingly stringent environmental regulations, of which air emissions are the primary focus. The Clean Air Act, particularly the 1990 amendments (see Chapter 4, Coal-Fired Emissions and Legislative Action in the United States), addresses the environmental performance of coal-based power systems specifically targeting emissions of sulfur dioxide, nitrogen oxides, hazardous air pollutants (including mercury), and fine particulate matter. Carbon dioxide is also of concern because of its potential to contribute to global climate change.

As discussed in Chapters 4 and 6 (Emissions Control Strategies for Power Plants), much progress has been made in reducing emissions from coal-fired power plants; however, with more stringent regulations anticipated in the near future, as well as the DOE's goal to develop near-zero emissions coal-fired power plants, major research and development activities are necessary to develop advanced coal power-generation technologies, with full-scale demonstrations a key to realizing the benefits of these programs [3]. This chapter discusses programs that have been implemented by the United States over the last 20 years to develop new technologies, which are also being exported or utilized by other countries. The various programs, under the umbrella of the DOE's Coal Power Program, address near- and long-range needs and include developing cost-effective environmental control technologies to comply with current and emerging regulations and developing technologies for near-zero emissions power and clean fuels plants with carbon dioxide management capability, respectively [3]. The DOE programs discussed in this chapter include the Clean Coal Technology (CCT) Demonstration Program, Power Plant Improvement Initiative (PPII), Clean Coal Power Initiative (CCPI), Vision 21, and FutureGen.

The Clean Coal Technology Program is providing a portfolio of technologies that will ensure that the U.S. recoverable coal reserves of 274 billion short tons can continue to supply the country's energy needs economically and in an environmentally sound manner. Under the Clean Coal Technology Program, cost-effective environmental control devices were developed for existing power plants. In addition, a new generation of technologies that can produce electricity and other commodities as well as provide the efficiencies and environmental performance responsive to global climate change concerns were developed [4]. The PPII projects focus on technologies enabling existing coal-fired power plants to meet increasingly stringent environmental regulations at the lowest possible cost [5]. The CCPI Program is an innovative technology demonstration program that fosters more efficient clean coal technologies for use in existing and new power generation facilities [6]. Vision 21 is the DOE's initiative to effectively remove environmental concerns associated with the use of fossil fuels for producing electricity and transportation fuels through better technology; the design basis is scheduled to be completed by 2015 with plant deployment by 2020 [5]. The DOE is particularly interested in coal-based energy plants. A specific integrated co-production plant program is FutureGen. FutureGen is an integrated carbon dioxide capture and sequestration and hydrogen research initiative to design, build, and operate a nearly emission-free, coal-fired electric power and hydrogen production plant [7].

Clean Coal Technology Demonstration Program

Nearly 20 years ago it was recognized that, given the need to respond to environmental objectives, new technologies would be necessary if coal was to continue as a viable source of secure energy [2]. In 1985, the Clean Coal Technology Demonstration Program was initiated with the objective to demonstrate a new generation of advanced coal utilization technologies. This investment in technology forms a solid foundation for addressing global and regional environmental concerns while providing low-cost energy that can compete in a deregulated electric power marketplace.

Clean Coal Technology Program Evolution

During the 1970s and early 1980s, many of the government-sponsored technology demonstrations focused on synthetic fuels production technology. The Synthetic Fuels Corporation (SFC) was established under the Energy Security Act of 1980 to reduce U.S. vulnerability to disruptions of crude oil imports [4]. The purpose of the SFC was accomplished by encouraging the private sector to build and operate synthetic fuel production facilities that used abundant domestic energy resources, specifically coal and oil shale. The strategy was for the SFC to be primarily a financier of pioneer commercial and near-commercial scale facilities [4]. By 1985, the market drivers for synthetic fuels dissolved as oil prices declined, world oil prices stabilized, and a short-term buffer was provided by the Strategic Petroleum Reserve. Congress responded to the decline of private-sector interest in the production of synthetic fuels and abolished the SFC in 1986.

The CCT Program was initiated in October 1984 through Public Law 98-473, Joint Resolution Making Continuing Appropriations for Fiscal Year 1985 and Other Purposes [4]. The United States moved from an energy policy based on synthetic fuels production to a more balanced policy. This policy established that the United States have an adequate supply of energy maintained at a reasonable cost and consistent with environmental, health, and safety objectives. Energy stability, security, and strength were the foundations for this policy. Coal was recognized as an essential element in this energy policy for the foreseeable future because the following existed [4]:

- Well-understood coal resource base;
- Available technology and skilled labor base to safely and economically extract, transport, and use coal;
- Existing multibillion-dollar infrastructure to gather, transport, and deliver coal to serve the domestic and international marketplace;
- Secure and abundant energy resource within the country's borders which is relatively invulnerable to disruptions because the coal production is dispersed and flexible, the delivery network is vast, and the stockpiling capability is great;

- Potential for export opportunities of U.S.-developed, coal-based technologies due to coal being the fuel of necessity in many lesser developed economies.

Congress recognized that the continued viability of coal as an energy resource was dependent on the demonstration and commercial application of a new generation of advanced coal-based technologies with improved operational, economic, and environmental performance; consequently, the CCT Program was established [4]. The DOE issued the first solicitations (CCT-I) in 1986 for clean coal technology projects and selected a broad range of projects in four major product markets: environmental control devices, advanced electric power generation, coal processing for clean fuels, and industrial applications. In February 1988, the second solicitation (CCT-II) was issued and provided for the demonstration of technologies that were capable of achieving significant reductions in SO_2, NO_x, or both, from existing power plants that were to be more cost effective than current technologies and capable of commercial deployment in the 1990s. The emphasis of the solicitation was on SO_2 and NO_x, precursors of acid rain, as a major presidential initiative was launched to address acid rain. The DOE issued a third solicitation (CCT-III) in May 1989 with essentially the same objective as the second but which also encouraged technologies that would produce clean fuels from run-of-mine coal. The next two solicitations recognized emerging energy and environmental issues, such as global climate change and capping of SO_2 emissions, and focused on seeking highly efficient, economically competitive, and low-emissions technologies. Specifically, the fourth solicitation (CCT-IV), released in January 1991, had as its objective the demonstration of energy-efficient, economically competitive technologies capable of retrofitting, repowering, or replacing existing facilities while achieving significant reductions in SO_2 and NO_x emissions. The fifth solicitation (CCT-V) was released in July 1992 to provide for demonstration projects that significantly advanced the efficiency and environmental performance of technologies applicable to new or existing facilities.

CCT Program Funding and Costs

The five CCT Program solicitations resulted in the demonstration of 38 projects as of 2003. Several additional projects were started; however, for a variety of reasons (primarily financial), these projects were terminated. In fact, two advanced power generation projects (*i.e.*, City of Lakeland, Department of Water & City Utilities projects; summarized in the next section as part of the portfolio of 38 projects) were active for approximately a decade but were terminated in the summer of 2003 due to financial issues. Of the 38 projects, in addition to the two recently terminated, 31 demonstration projects have been completed, two are in operation, one is in construction, and two are in design [8,9]. The 38 projects have resulted in a combined commitment by the federal government and the private sector of $5.2 billion. The DOE's cost-share for these projects is ~$1.7 billion, or ~34% of the total [4]. The project participants (*i.e.*, non-federal government participants) are providing the remaining $3.4 billion, or 66%, of the total. Table 7-1 summarizes the costs, cost-sharing, and application categories of the CCT projects.

CCT Program Projects

The CCT Program projects provide a portfolio of technologies that will enable coal to continue to provide low-cost, secure energy vital to the U.S. economy while satisfying energy and environmental goals. The projects are spread across the country in 18 states, as shown in Figure 7-1 [4]. Table 7-2 lists each project, non-federal government participant, location of the project, solicitation under which the projects were awarded, and status of

TABLE 7-1
CCT Project Costs and Cost-Sharing (in Thousands of Dollars)

	Total Project Costs	*Percent (%)*	*Cost-Share Dollars*		*Cost-Share Percent*	
			DOE	*Participant*	*DOE*	*Participant*
Subprogram						
CCT-I	844,363	16	236,640	604,723	28	72
CCT-II	318,577	6	139,229	179,348	44	56
CCT-III	1,325,329	26	576,918	748,411	44	56
CCT-IV	950,429	18	439,063	511,366	46	54
CCT-V	1,765,009	34	360,982	1,404,027	20	80
Total	5,203,707	100	1,775,832	3,447,875	34	66
Application category						
Advanced Electric Power Generation	2,864,284	55	1,118,865	1,745,419	39	61
Environmental Control Devices	620,110	12	252,866	367,244	41	59
Coal Processing for Clean Fuels	431,810	8	192,029	239,781	44	56
Industrial Applications	1,287,503	25	192,072	1,095,431	15	85
Total	5,203,707	100	1,755,832	3,447,875	34	66

Source: DOE, *Clean Coal Technology Demonstration Program: Program Update 2001 Including Power Plant Improvement Initiative Projects*, Office of Fossil Energy, U.S. Department of Energy, Washington, D.C., July 2002.

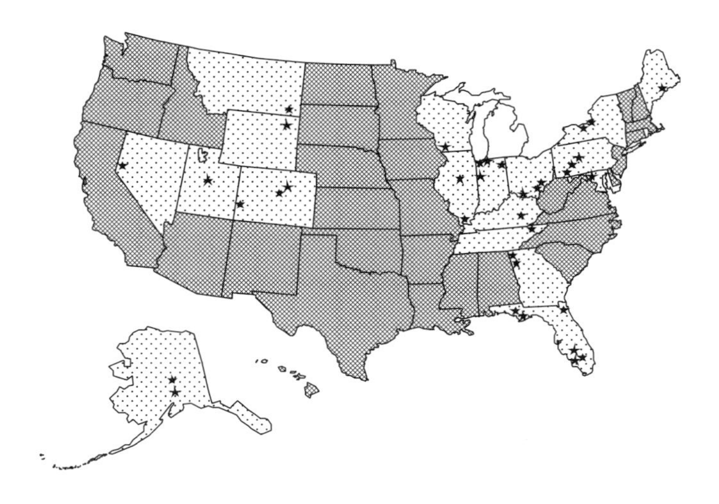

FIGURE 7-1. Location of the CCT program projects. (From DOE, *Clean Coal Technology Demonstration Program: Program Update 2001 Including Power Plant Improvement Initiative Projects*, Office of Fossil Energy, U.S. Department of Energy, Washington, D.C., July 2002.)

TABLE 7-2
CCT Program Demonstration Projects

Project	*Participant*	*Solicitation/Status*
Environmental Control Devices		
SO_2 Control Technologies		
10 MW demonstration of gas suspension absorption	AirPol, Inc.	CCT-III/completed 03/94
Confined zone dispersion flue gas desulfurization (FGD) demonstration	Bechtel Corp.	CCT-III/completed 06/93
LIFAC sorbent injection desulfurization demonstration project	LIFAC-North America	CCT-III/completed 06/94
Advanced flue gas desulfurization (FGD) demonstration project	Pure Air on the Lake, LP	CCT-II/completed 06/95
Demonstration of innovative applications of technology for the CT-121 flue gas desulfurization (FGD) process	Southern Company Services, Inc.	CCT-II/completed 12/94
NO_x Control Technologies		
Demonstration of advanced combustion techniques for a wall-fired boiler	Southern Company Services, Inc.	CCT-II/completed 04/03

(continued)

TABLE 7-2
(continued)

Project	Participant	Solicitation/Status
Demonstration of coal reburning for cyclone boiler NO_X control	Babcock & Wilcox Co.	CCT-II/completed 12/92
Full-scale demonstration of low-NO_X cell burner retrofit	Babcock & Wilcox Co.	CCT-III/completed 04/93
Evaluation of gas reburning and low-NO_X burners on a wall-fired boiler	Energy and Environmental Research Corporation	CCT-III/completed 01/95
Micronized coal reburning demonstration for NO_X control	New York State Electric & Gas Corp.	CCT-IV/completed 04/99
Demonstration of selective catalytic reduction technology for the control of NO_X emissions from high-sulfur, coal-fired boilers	Southern Company Services, Inc.	CCT-II/completed 07/95
180-MW demonstration of advanced tangentially fired combustion techniques for the reduction of NO_X emissions from coal-fired boilers	Southern Company Services, Inc.	CCT-II/completed 12/92
Combined SO_2/NO_X Control Technologies		
SNOX™ flue gas cleaning demonstration project	ABB Environmental Systems	CCT-II/completed 12/94
LIMB demonstration project extension and cool-side demonstration	Babcock & Wilcox Co.	CCT-I/completed 08/91
SOx-NOx-Rox-Box™ flue gas cleanup demonstration project	Babcock & Wilcox Co.	CCT-II/completed 05/93
Enhancing the use of coals by gas reburning and sorbent injection	Energy and Environmental Research Corporation	CCT-I/completed 10/94
Milliken clean coal technology demonstration project	New York State Electric & Gas Corp.	CCT-IV/completed 06/98
Integrated dry NO_X/SO_2 emissions control system	Public Service of Colorado	CCT-III/completed 12/96
Advanced Electric Power Generation		
Fluidized-Bed Combustion		
McIntosh unit 4A pressurized circulating fluidized-bed (PCFB) demonstration project	City of Lakeland, Lakeland Electric	CCT-III/terminated 2003

(continued)

TABLE 7-2
(*continued*)

Project	*Participant*	*Solicitation/Status*
McIntosh unit 4B topped pressurized circulating fluidized-bed (PCFB) demonstration project	City of Lakeland, Lakeland Electric	CCT-IV/terminated 2003
JEA large-scale circulating fluidized-bed (CFB) combustion demonstration project	JEA	CCT-I/operating
Tidd pressurized fluidized-bed combustion (PFBC) demonstration project	The Ohio Power Co.	CCT-I/completed 03/95
Nucla circulating fluidized-bed (CFB) demonstration project	Tri-State Generation and Transmission Association, Inc.	CCT-I/completed 01/91
Integrated Gasification Combined Cycle		
Kentucky Pioneer Energy integrated gasification combined cycle (IGCC) demonstration project	Kentucky Pioneer Energy, LLC	CCT-V/under construction
Tampa Electric integrated gasification combined cycle (IGCC) project	Tampa Electric Company	CCT-III/completed 10/01
Piñon Pine integrated gasification combined cycle (IGCC) power project	Sierra Pacific Power Co.	CCT-IV/completed 01/01
Wabash River coal gasification repowering project	Wabash River Coal Gasification Repowering Project Joint Venture	CCT-IV/completed 12/99
Advanced Combustion/Heat Engines		
Clean coal diesel demonstration	TIAX (formerly Arthur D. Little, Inc.)	CCT-V/under construction
Healy clean coal project	Alaska Industrial Development and Export Authority	CCT-III/completed 12/99
Coal Processing for Clean Fuels		
Commercial-scale demonstration of the liquid-phase methanol (LPMEOH™) process	Air Products Liquid Phase Conversion Company, LP	CCT-III/completed 12/02
Development of the Coal Quality Expert™	ABB Combustion Engineering, Inc.; CQ, Inc.	CCT-I/completed 12/95
ENCOAL® mild coal gasification project	ENCOAL Corp.	CCT-III/completed 07/97

(*continued*)

TABLE 7-2
(*continued*)

Project	*Participant*	*Solicitation/Status*
Advanced coal conversion process development	Western SynCoal, LLC	CCT-I/completed 01/01
Industrial Applications		
Clean Power from Integrated Coal/Ore Reduction (CPICOR™)	CPICOR™ Management Company, LLC	CCT-V/design
Blast furnace granular-coal injection system demonstration project	Bethlehem Steel Corp.	CCT-III/completed 11/98
Advanced cyclone combustor with internal sulfur, nitrogen, and ash control	Coal Tech Corp.	CCT-I/completed 05/90
Cement kiln flue gas recovery scrubber	Passamaquoddy Tribe	CCT-II/completed 09/93
Pulse combustor design qualification test	ThermoChem, Inc.	CCT-IV/completed 09/01

Source: Data from DOE [4,9].

the project. The participants listed in Table 7-2 are the primary non-federal government companies, although each project had several supporting team members [4]. The projects are listed in Table 7-2 by four basic market sectors. A synopsis of the projects is provided in the following text, with an emphasis on emissions achievements.

Environmental Control Devices

The initial thrust of the CCT Program addressed acid rain, and 18 projects have been completed involving SO_2 and NO_x control for coal-fired boilers. The technologies demonstrated provide a suite of cost-effective control options for the full range of boiler types. The projects included seven NO_x emission control systems installed on more than 1750 MW of utility generating capacity, five SO_2 emission control systems installed on ~770 MW, and six combined SO_2/NO_x emission control systems installed on more than 665 MW of capacity [4].

SO_2 Control Technologies The CCT Program successfully demonstrated two sorbent injection systems, one spray dryer system, and two advanced flue gas desulfurization (AFGD) systems. Sulfur dioxide reductions varying from 50 to 90+% were demonstrated. AirPol, Inc., demonstrated that FLS milfo, Inc.'s gas suspension absorption system was an economic option for achieving Phase II 1990 Clean Air Act Amendments SO_2 compliance in coal-fired boilers using high-sulfur coal [10]. The demonstration was performed using a vertical, single-nozzle reactor (*i.e.*, spray dryer) with integrated sorbent particulate recycle in a 10 MW equivalent slipstream of flue gas from a Tennessee Valley Authority (West Paducah, Kentucky) 175 MW wall-fired boiler. Sulfur dioxide reductions of 60 to 90% were obtained firing 2.7 to 3.5% sulfur coal [4].

Bechtel Corporation demonstrated SO_2 removal capabilities of in-duct confined zone dispersion (CZD)/flue gas desulfurization (FGD) technology—specifically, to define the optimum process operating parameters and to determine the operability, reliability, and cost-effectiveness of CZD/FGD during long-term testing and its impact on downstream operations and emissions [11]. The demonstration was performed using half of the flue gas from the Pennsylvania Electric Company Seward Station 147 MW tangentially fired boiler. Sulfur dioxide reductions of 50% were achieved firing 1.5 to 2.5% sulfur coal [4].

LIFAC-North America (a joint venture partnership between Tampella Power Corporation and ICF Kaiser Engineers, Inc.) demonstrated the LIFAC sorbent injection process, with furnace sorbent injection and sulfur capture occurring in a vertical activation reactor [12]. The LIFAC process was shown to be easily retrofitted to power plants with space limitations and burning high-sulfur coals. The 60 MW demonstration was performed on the Richmond (Indiana) Power & Light Whitewater Valley Station and achieved 70% SO_2 removal firing 2.0 to 2.9% sulfur coal [4].

Pure Air on the Lake, L.P. (a subsidiary of Pure Air, which is a general partnership between Air Products and Chemicals, Inc., and Mitsubishi Heavy Industries America, Inc.) demonstrated Pure Air's AFGD process to reduce SO_2 emissions by 95% or more at approximately one-half the cost of conventional scrubbing technology, to significantly reduce space requirements, and to create no new waste streams [13]. A single SO_2 absorber that was built for Baily (Indiana) Generating Station Unit Nos. 7 and 8 (*i.e.*, 528 MW) achieved 95% SO_2 capture firing 2.3 to 4.7% sulfur coal [4].

Southern Company Services, Inc., demonstrated Chiyoda Corporation's Chiyoda Thoroughbred-121 AFGD process for combined particulate and SO_2 capture with high reliability [14]. Testing performed at the Georgia Power Company Plant Yates, Unit No. 1 (100 MW), achieved over 90% SO_2 removal efficiency at SO_2 inlet concentrations of 1000 to 3000 ppm with >97% limestone utilization, 97.7 to 99.3% particulate removal, >95% HCl and HF capture, 80 to 98% capture of most metals, <50% capture of mercury and cadmium, and <70% capture of selenium [4].

NO_x Control Technologies Under the CCT Program, seven NO_x control technologies were assessed encompassing low-NO_x burners (LNBs), advanced overfire air (AOFA), reburning, selective catalytic reduction (SCR), selective non-catalytic reduction (SNCR), and combinations of them.

NO_X reductions varying from 37 to 80% were demonstrated. Southern Company Services, Inc., performed a demonstration using Foster Wheeler's LNB with AOFA and the Electric Power Research Institute's (EPRI's) Generic NO_X Control Intelligent System (GNOCIS) computer software to achieve 50% NO_X reduction; to determine the contributions of AOFA and LNB to NO_X reduction and the parameters for optimal LNB/AOFA performance; and to assess the long-term effects of LNB, AOFA, combined LNB/AOFA, and the GNOCIS advanced digital controls on NO_X reduction, boiler performance, and auxiliary components [9,15]. The demonstration was performed on the Georgia Power Plant Hammond, Unit No. 4, which is a 500 MW wall-fired boiler, and achieved 68% NO_X reduction with fly ash loss-on-ignition (LOI) increasing from a baseline of 7 to 8–10% [4].

The Babcock & Wilcox Company (B&W) demonstrated the technical and economic feasibility of their coal reburning system to achieve greater than 50% NO_X reduction with no serious impact on cyclone combustor operation, boiler performance, or other emission streams [16]. The demonstration was performed on the Wisconsin Power and Light Company 100 MW Nelson Dewey Station, Unit No. 2, and achieved 52 to 62% NO_X reduction with 30% heat input from the coal [4]. B&W also demonstrated the cost-effective reduction of NO_X from a large, base-loaded, coal-fired utility boiler with their Low-NO_X Cell Burner (LNCB®) system to achieve at least 50% NO_X reduction without degradation of boiler performance at less cost than that of conventional low-NO_X burners [17]. The demonstration was performed on the Dayton (Ohio) Power and Light Company 605 MW J.M. Stuart Plant, Unit No. 4, and achieved 48 to 58% NO_X reduction, experienced average CO emissions of 28 to 55 ppm, increased fly ash production without affecting the electrostatic precipitator (ESP) performance, and increased unburned carbon (UBC) losses by $\sim$28% [4].

The Energy and Environmental Research Corporation (EERC), currently GE Energy and Environmental Research, performed a demonstration to attain up to a 70% decrease in NO_X emissions from an existing wall-fired utility boiler firing low-sulfur coal using both natural gas reburning (GR) and LNBs, as well as to assess the impact of GR-LNB technology on boiler performance [18]. The demonstration was performed on a 172 MW wall-fired boiler (Public Service Company of Colorado Cherokee Station, Unit No. 3). It achieved 37 to 65% NO_X reduction, with 13 to 18% of the heat input coming from the natural gas; reduced SO_2 and particulate loadings by the percentage heat input by natural gas reburning; and resulted in acceptable carbon-in-ash and CO levels with GR/LNB operation [4].

New York State Electric & Gas Corporation (NYSEG) demonstrated micronized coal reburning with the objective to achieve at least 50% NO_X reduction on a cyclone burner and 25 to 35% NO_X reduction on a tangentially fired boiler [19]. Demonstrations were performed on the NYSEG Milliken Station (Lansing, New York) Unit No. 1, which is a 148 MW tangentially-fired boiler, and Eastman Kodak Company's Kodak Park

(Rochester, New York) Power Plant Unit No. 1, which is a 60 MW cyclone boiler. Nitrogen oxide reductions of 59 and 28% were achieved in the cyclone- and tangentially fired units, respectively. The micronized coal consisted of 17 and 14% of heat input, respectively. LOI was maintained at <5% at Milliken Station but increased from baseline levels of 10–15 to 40–50% at Kodak Park [4].

Southern Company Services, Inc., evaluated the performance of eight SCR catalysts with different shapes and chemical compositions when applied to operating conditions found in U.S. pulverized coal-fired utility boilers firing U.S. high-sulfur coal under various operating conditions, while achieving as much as 80% NO_x removal [20]. In this demonstration project, the SCR facility consisted of three 2.5 MW equivalent SCR reactors supplied by separate flue gas streams and six 0.20 MW equivalent reactors for a total of 8.7 MW equivalent using flue gas from the Gulf Power Company Plant Crist (Pensacola, Florida), Unit No. 4. The reactors were sized to produce data that will allow the SCR process to be scaled up to commercial size. Nitrogen oxide reductions of over 80% were achieved at an ammonia slip well under the 5 ppm level deemed acceptable for commercial operation [4].

Southern Company Services, Inc., also demonstrated short- and long-term NO_x reduction capabilities of ABB Combustion Engineering Inc.'s (now Alstom Power, Inc.) Low-NO_x Concentric Firing System (LNCFS™) at various combinations of OFA and coal nozzle positioning [21]. The demonstration was performed on Gulf Power Company's 180 MW tangentially fired Plant Lansing Smith (Lynn Haven, Florida), Unit No. 2. Reductions in NO_x of 37 to 45% were achieved [4].

Combined SO_2/NO_x Control Technologies Six combined SO_2/NO_x control technologies were assessed under the CCT Program. These technologies used various combinations of technologies and demonstrated NO_x and SO_2 reductions of 40–94 and 50–95%, respectively. ABB Environmental Systems demonstrated Haldor Topsoe's SNOX™ catalytic advanced flue gas cleanup system at an electric power plant using U.S. high-sulfur coals with the objective to remove 95% of the SO_2 and more than 90% of the NO_x from the flue gas, as well as produce a salable by-product of concentrated sulfuric acid [22]. In the SNOX™ process, particulate is removed using a high-efficiency baghouse, NO_x is reduced in a catalytic reactor using ammonia, and SO_2 is oxidized to SO_3 in a second catalytic reactor and subsequently hydrolyzed to concentrated sulfuric acid. Testing performed in a 35 MW equivalent slipstream from the Ohio Edison Niles (Ohio) Station, Unit No. 2 (108 MW) achieved SO_2 reductions in excess of 95% and NO_x reductions averaging 94%, produced a sulfuric acid that exceeded federal specifications for a Class I acid, eliminated CO and hydrocarbon emissions due to the presence of the SO_2 catalyst, and exhibited high capture efficiency of most air toxics (except for mercury) in the high-efficiency baghouse [4].

B&W demonstrated that their limestone injection multistage burner (LIMB) process can achieve up to 50% NO_x and SO_2 reductions and that Consolidated Coal Company's Coolside duct injection of lime sorbents can achieve removal of up to 70% SO_2 [23]. The testing was performed at the Ohio Edison 105 MW Edgewater Station (Lorain, Ohio), Unit No. 4. The LIMB process reduces SO_2 by injecting dry sorbent into the boiler above the burners—in this case, B&W's DRB-XCL® low-NO_x burners; SO_2 removal efficiencies varying from 45 to 60% with lime-based products and 22 to 40% with limestone were achieved, while nitrogen oxide reductions of 40 to 50% were obtained. The Coolside process, which is a humidified duct injection process, achieved 70% SO_2 reduction [4].

B&W also demonstrated their SOx-NOx-Rox Box™ (SNRB™) process with the objective of achieving greater than 70% SO_2 removal and 90% or higher reduction in NO_x emissions while maintaining particulate emissions below 0.03 lb/MM Btu [24]. The SNRB™ process combines the removal of SO_2, NO_x, and particulates in one unit: a high-temperature baghouse. Sulfur dioxide is removed using sorbent injection, NO_x is reduced by injecting ammonia in the presence of an SCR catalyst inside the bags, and particulate is removed using high-temperature fiber filter bags. The testing was performed in a 5 MW equivalent slipstream from Ohio Edison Company's 156 MW R.E. Burger Plant (Dilles Bottom, Ohio), Unit No. 5, and SO_2 and NO_x reductions of 80–90 and 90%, respectively, were achieved. In addition, air toxic removal efficiency was comparable to that of the ESP at the plant, except that HCl and HF were reduced by 95 and 84%, respectively [4].

Energy and Environmental Research Corporation performed a demonstration in which natural gas reburning was combined with in-furnace sorbent injection with the objective of reducing NO_x by 60% and SO_2 by at least 50% in two different boiler configurations—tangentially and cyclone-fired units—while burning high-sulfur Midwestern coal [25]. Testing was performed on the Illinois Power Company 71 MW Hennepin Plant, Unit No. 1 (tangentially fired boiler), and on the City Water, Light and Power (Springfield, Illinois) 40 MW Lakeside Station, Unit No. 7 (cyclone-fired boiler). Nitrogen oxide reductions averaged 67 and 66%, respectively, for the tangentially and cyclone-fired boilers, while SO_2 reductions averaged 53 and 58%, respectively [4].

New York State Electric & Gas Corporation performed a demonstration using Saarberg-Hölter-Umwelttechnik (S-H-U), GmbH's formic acid-enhanced, wet limestone scrubber technology; ABB Combustion Engineering's LNCFS™ process; Stebbins Engineering and Manufacturing's split-module absorber; ABB Air Preheater's heat-pipe air preheater; and NYSEG's plant emissions optimization advisor (PEOA) with the objective of achieving high-sulfur capture efficiency and NO_x and particulate control at minimum power requirements, zero wastewater discharge, and the production of by-products instead of wastes from the scrubber [26].

The flue gas from NYSEG's Milliken Station (Lansing, New York), Unit Nos. 1 and 2 (300 MW), was used in the project, and sulfur dioxide removals of 98 and 95% were demonstrated with and without formic acid, respectively, and 39% NO_x reduction was achieved [4].

Public Service Company of Colorado demonstrated the integration of five technologies—B&W's DRB-XCL® low-NO_x burners with OFA, in-duct sorbent injection, flue gas humidification, and furnace (urea) injection—with the objective of achieving 70% reduction in NO_x and SO_2 emissions and, more specifically, to assess the integration of a down-fired low-NO_x burner with in-furnace urea injection and dry sorbent in-duct injection with humidification for SO_2 removal [27]. Testing performed on the Public Service Company of Colorado Arapahoe Station (Denver, Colorado), 100 MW Unit No. 4, demonstrated 70% SO_2 removal and 62 to 80% NO_x reduction [4].

Advanced Electric Power Generation Technology

The CCT Program provides a range of advanced electric power generation options for both repowering and new power generation in response to the need for load growth as well as environmental concerns. The emphasis of this program category included technologies that could effectively repower aging power plants faced with the need to both control emissions and respond to growing power demands. Repowering is an important option because existing power generation sites have significant value and warrant investment because the infrastructure is in place and siting new plants represents a major undertaking.

These advanced systems offer greater than 20% reductions in greenhouse gas emissions; SO_2, NO_x, and particulate emissions far below New Source Performance Standards (NSPSs); and salable solid and liquid by-products [4]. Over 1800 MW of capacity are represented by 11 projects, including five fluidized-bed combustion systems (two completed, one ongoing, and two terminated in June 2003 after completing designs), four integrated gasification combined cycle systems (three completed and one ongoing), and two advanced combustion/heat engine systems (one completed and one delayed). The advanced electric power generation technology projects selected under the CCT Program are characterized by high thermal efficiency, very low pollutant emissions, reduced CO_2 emissions, few solid waste problems, and enhanced economics. Five generic advanced electric power generation technologies are demonstrated in the CCT Program: fluidized-bed combustion, integrated gasification combined cycle, integrated gasification fuel cell, coal-fired diesel, and slagging combustion.

Fluidized-Bed Combustion City of Lakeland (Florida), Lakeland Electric was selected by the DOE for two CCT Program projects in 1989 and 1993; however, in 2003 these projects were terminated due to economic issues [9,28]. The first project, a pressurized circulating fluidized-bed (PCFB) project, was to demonstrate Foster Wheeler Corporation's PCFB technology coupled with Siemens Westinghouse's ceramic candle-type, hot-gas cleanup system and power generation technologies, which were to represent a cost-effective, high-efficiency, low-emissions means of adding generating capacity at greenfield sites or in repowering applications [4]. The second project, to be performed on the same boiler, was to demonstrate topped PCFB technology in a fully commercial power generation setting, thereby advancing the technology for future plants that will operate at higher gas turbine inlet temperatures and will be expected to achieve cycle efficiencies in excess of 45%.

JEA (formerly Jacksonville (Florida) Electric Authority) is demonstrating atmospheric circulating fluidized-bed (ACFB) combustion at a scale larger than previously operated. The objective of the project is to demonstrate ACFB combustion at 297.5 MW, which represents a scale up from previously constructed facilities; to verify expectations of the technology's economic, environmental, and technical performance; to provide potential users with the data necessary for evaluating a large-scale ACFB combustion as a commercial alternative; to accomplish greater than 90% SO_2 removal; and to reduce NO_X emissions by 60% when compared with conventional technology [4]. The CFB boiler has operated at full load and achieved rated output and the demonstration test program has begun, but no published results are available [29].

The Ohio Power Company performed a pressurized fluidized-bed (bubbling) combustion (PFBC) demonstration to verify expectations of PFBC economic, environmental, and technical performance in a combined-cycle repowering application at utility scale; to accomplish greater than 90% SO_2 removal; and to achieve an NO_X emission level of 0.3 lb/MM Btu at full load [30]. The demonstration was performed at the Ohio Power Company 70 MW Tidd Plant (Brilliant, Ohio), Unit No. 1, and was the first large-scale operational demonstration of PFBC in the United States. Sulfur dioxide removal efficiency of 90 to 95% was achieved at full load with calcium-to-sulfur (Ca/S) ratios of 1:1.4 and 1:5, respectively [4]. NO_X emissions were 0.15 to 0.33 lb/MM Btu, CO emissions were less than 0.01 lb/MM Btu, and particulate emissions were less than 0.02 lb/MM Btu. Operationally, the PFBC boiler demonstrated commercial readiness.

Tri-State Generation and Transmission Association, Inc., demonstrated the feasibility of ACFB technology at the utility scale and evaluated the economic, environmental, and operational performance at that scale [31]. Three small, coal-fired stoker boilers at the Nucla Station (Nucla, Colorado) were replaced with a new 110 MW atmospheric CFB boiler. Environmentally, SO_2 capture efficiencies of 70 and 95% were achieved at Ca/S ratios of 1.5 and 4.0, respectively; NO_X emissions averaged 0.18 lb/MM Btu; CO emissions ranged from 70 to 140 ppm; particulate emissions ranged from 0.0072 to 0.0125 lb/MM Btu (or 99.9% removal efficiency); and solid waste was essentially benign and showed potential as an agricultural solid amendment, soil/roadbed stabilizer, or landfill cap [4].

Integrated Gasification Combined Cycle The integrated gasification combined cycle (IGCC) process has four basic steps: (1) fuel gas is generated from a gasifier; (2) either the fuel gas is passed directly to a hot-gas cleanup system to remove particulates, sulfur, and nitrogen compounds or the gas is first cooled to produce steam and then cleaned conventionally; (3) the clean fuel gas is combusted in a gas turbine generator to produce electricity; and (4) the residual heat in the hot exhaust from the gas turbine generator is recovered in a heat-recovery steam generator, and the steam is used to produce additional electricity in a steam turbine generator. IGCC systems are among the cleanest and most efficient of the emerging clean coal technologies [4]. Sulfur, nitrogen compounds, and particulate matter are removed before the fuel is combusted (*i.e.*, before combustion air is added), resulting in a much lower volume of gas to be treated in a post-combustion scrubber. With hot-gas cleanup, IGGC systems have the potential for efficiencies of over 50%. An example of an IGCC system is shown in Figure 7-2 [32].

In a coal gasifier, the sulfur in the coal is released in the form of hydrogen sulfide (H_2S) rather than SO_2 as in a combustion process. Several commercial processes are capable of removing H_2S; more than 99% of the H_2S can be removed from the gas, making it as clean as natural gas. Energy conversion in fuels cells is more efficient than traditional energy conversion devices and can be as high as 60%. A typical fuel cell system using coal as a fuel includes a coal gasifier with a gas cleanup system, a fuel cell that uses the coal gas to generate electricity (direct current) and heat, an inverter to convert direct current to alternating current, and a heat recovery system that can be used to produce additional electric power in a bottoming steam cycle [4].

Fuel cells do not rely on combustion; instead, an electrochemical reaction generates electricity. Electrochemical reactions release the chemical energy that bonds atoms together—in this case, the atoms of hydrogen and oxygen [33]. The fuel cell is extremely clean and highly efficient. In a clean coal technology application, the fuel cell is fueled either by hydrogen extracted from the coal gas or a mixture of synthesis gas (low-Btu gas consisting of CO and H_2). In a coal gasification/fuel cell application, coal gas is supplied to the anode, and air and CO_2 are supplied to the cathode to produce electricity and heat. The principal waste product from the fuel cell is water.

Fuel cells are often categorized by the material used to separate the electrodes, which is termed the electrolyte. The most mature fuel cell concept is the phosphoric acid fuel cell [33]. Other concepts include the molten carbonate fuel cell (MCFC), which uses a hot mixture of lithium and potassium carbonate as the electrolyte, and the solid oxide fuel cell, which uses a hard ceramic material instead of a liquid electrolyte. The MCFC is integrated with one of the Clean Coal Technology IGCC projects described below.

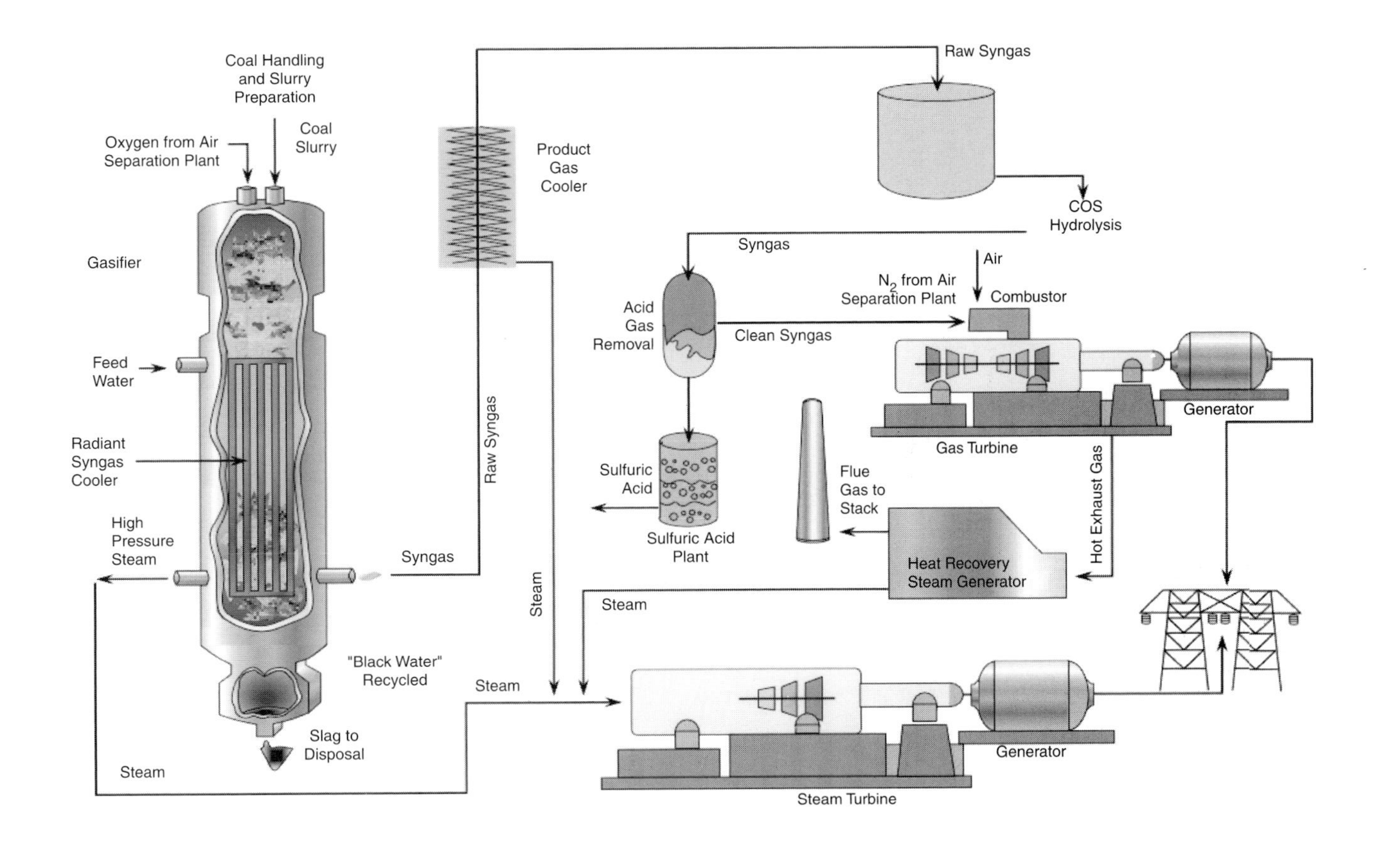

FIGURE 7-2. Schematic diagram of an IGCC system. (From DOE, *Clean Coal Technology, Tampa Electric Integrated Gasification Combined-Cycle Project: An Update*, Office of Fossil Energy, U.S. Department of Energy, Washington, D.C., July 2000.)

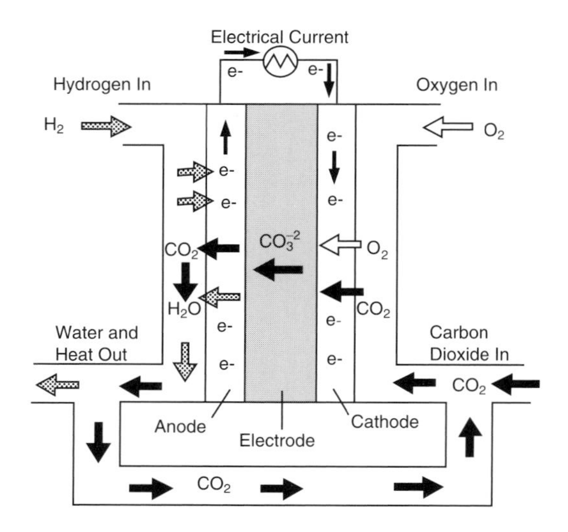

FIGURE 7-3. Schematic diagram of a molten carbonate fuel cell. (From DOE, Energy Efficiency and Renewable Energy—Hydrogen, Fuel Cells, and Infrastructure Technologies Program, U.S. Department of Energy, Washington, D.C., www.eere.energy.gov/hydrogenandfuelcells/fuelcells/types.html#mcfc. Last updated January 27, 2003.)

The MCFC evolved from work in the 1960s aimed at producing a fuel cell which would operate directly on coal [34]. While direct operation on coal seems less likely today, operation on coal-derived fuel gases is both technically and economically viable. The MCFC, shown schematically in Figure 7-3, uses a molten carbonate salt mixture as its electrolyte. The composition of the electrolyte varies but usually consists of lithium carbonate and potassium carbonate. At an operating temperature of about 1200°F, the salt mixture is liquid and a good ionic conductor. The MCFC reactions that occur are [34]:

$$\text{Anode reactions: } H_2 + CO_3^{2-} \longrightarrow H_2O + CO_2 + 2e^- \tag{7-1}$$

$$CO + CO_3^{2-} \longrightarrow 2CO_2 + 2e^- \tag{7-2}$$

$$\text{Cathode reaction: } O_2 + 2CO_2 + 4e^- \longrightarrow 2CO_3^{2-} \tag{7-3}$$

The anode process involves a reaction between hydrogen and carbon monoxide and the carbonate ions from the electrolyte which produces water and carbon dioxide and releases electrons to the anode [34]. The cathode process combines oxygen and carbon dioxide from the oxidant stream with electrons from the cathode to produce carbonate ions that enter the electrolyte. The use of carbon dioxide in the oxidant stream requires a system for collecting carbon dioxide from the anode exhaust and mixing it with the cathode feed stream. Of the four IGCC projects, three have completed operations and one recently broke ground [4,9]. The project that broke ground on August 13, 2003, will incorporate an MCFC with a coal gasifier.

Tampa Electric Company demonstrated an advanced IGCC system using Texaco's (now ChevronTexaco) pressurized, oxygen-blown, entrained-flow gasifier technology [36]. The objective was to demonstrate IGCC technology in a greenfield commercial electric utility application at the 250 MW size using an entrained-flow, oxygen-blown gasifier with full heat recovery, conventional coal-gas cleanup, and an advanced gas turbine with nitrogen injection for power augmentation and NO_X control [4]. The IGCC system shown in Figure 7-2 is that of the Polk system [32]. The demonstration was performed at the Tampa Electric Company Polk Power Station (Mulberry, Florida) and achieved greater than 98% sulfur capture, while NO_X emissions were reduced by over 90% compared with a conventional pulverized coal-fired power plant, particulate matter was well below the regulatory limits set for the Polk plant site, and carbon burnout exceeded 95% [32]. The plant is currently in commercial operation.

Sierra Pacific Power Company tested IGCC using the KRW air-blown, pressurized fluidized-bed coal gasification system [37]. The objective was to demonstrate air-blown, pressurized fluidized-bed IGCC technology incorporating hot-gas cleanup; evaluate a low-Btu gas combustion turbine; and assess long-term reliability, availability, maintainability, and environmental performance at a scale sufficient to determine commercial potential. The emission targets were to remove more than 95% of the sulfur in the coal and emit less than 70% NO_X and 20% less CO than in a comparable conventional coal-fired plant [4]. The 107 MW demonstration (shown in the block diagram in Figure 7-4), performed at the Sierra Pacific Power Company Tracy Station (Reno, Nevada), experienced many operational difficulties, and steady-state operation was not reached in the course of the testing; therefore, environmental performance could not be evaluated. The project did succeed in identifying and working through a number of problems, made possible only through a full-scale demonstration, and positioned the technology for commercialization. In addition, the testing proved the ability of the KRW gasifier to produce coal-derived synthesis gas of design quality [4].

The Wabash River Coal Gasification Repowering Project Joint Venture—a joint venture of Dynegy, Inc. (formerly Destec Energy, Inc.) and PSI Energy, Inc.—demonstrated IGCC using Global Energy's two-stage, pressurized, oxygen-blown, entrained-flow gasification system (*i.e.*, E-Gas Technology^{TM}) [39]. A schematic diagram of the system is shown in Figure 7-5. The objective was to demonstrate utility repowering with the E-Gas Technology^{TM}, including advancements in the technology relevant to the use of high-sulfur bituminous coal, and to assess the long-term reliability, availability, and maintainability of the system in a commercial-scale unit [4]. The 296 MW demonstration was successfully performed at PSI Energy's Wabash River Generating Station (Terre Haute, Indiana) and achieved sulfur capture efficiency greater than 99%. The sulfur-based pollutants were

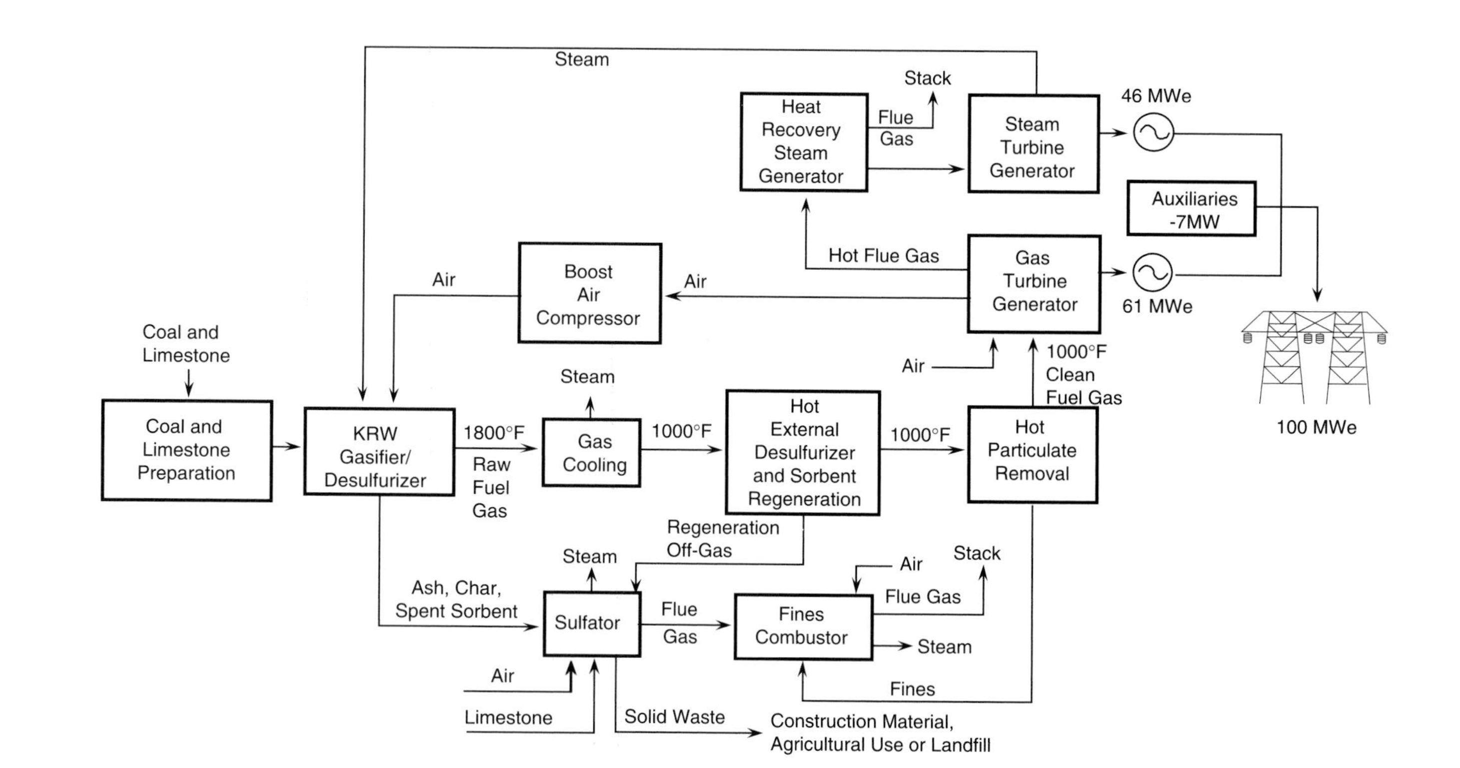

FIGURE 7-4. Block diagram of the Piñon Pine IGCC system. (From DOE, *Clean Coal Technology: The Piñon Pine Power Project,* Technical Report No. 6, Office of Fossil Energy, U.S. Department of Energy, Washington, D.C., December 1996.)

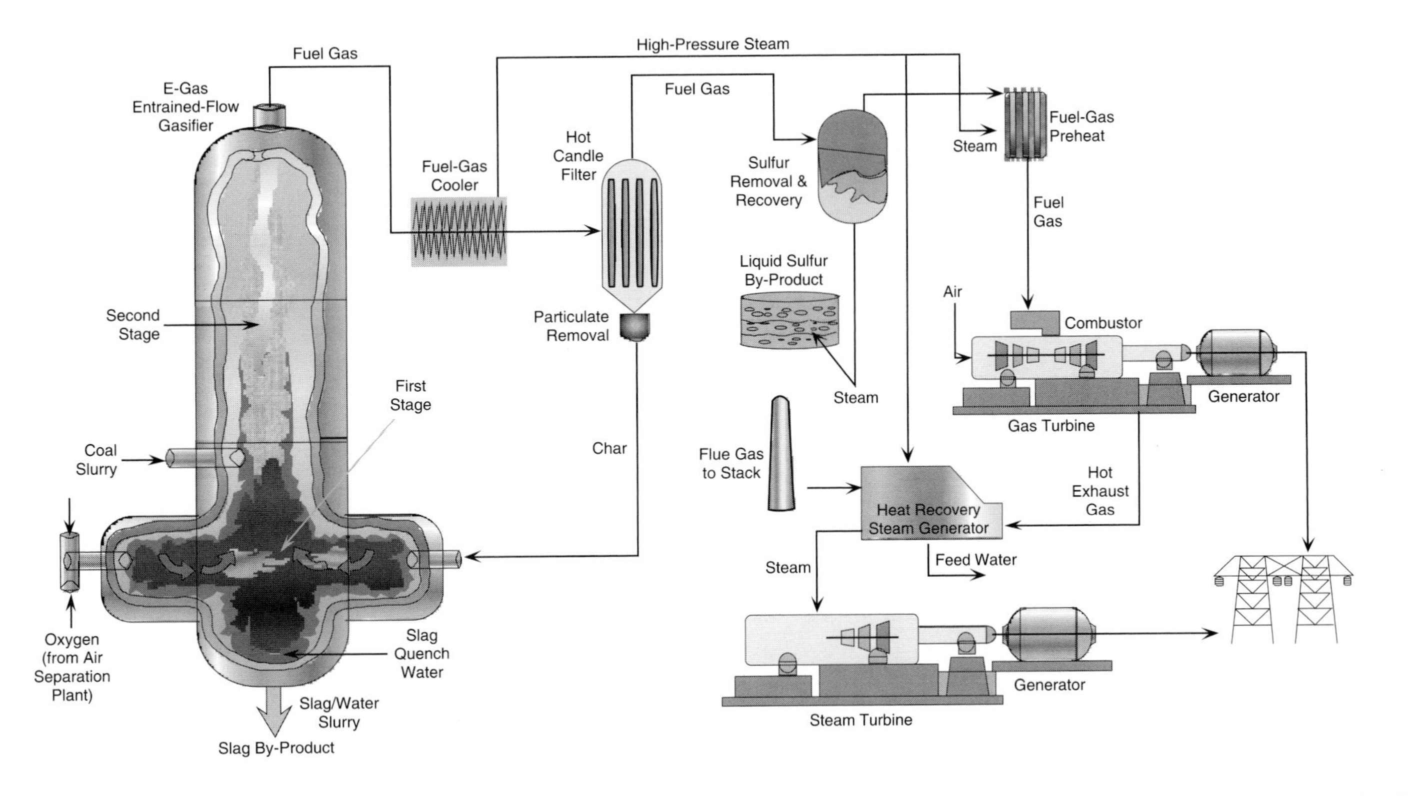

FIGURE 7-5. Schematic diagram of the Wabash River coal gasification system. (From DOE, *Clean Coal Technology, The Wabash River Coal Gasification Repowering Project: An Update*, Office of Fossil Energy, U.S. Department of Energy, Washington, D.C., September 2000.)

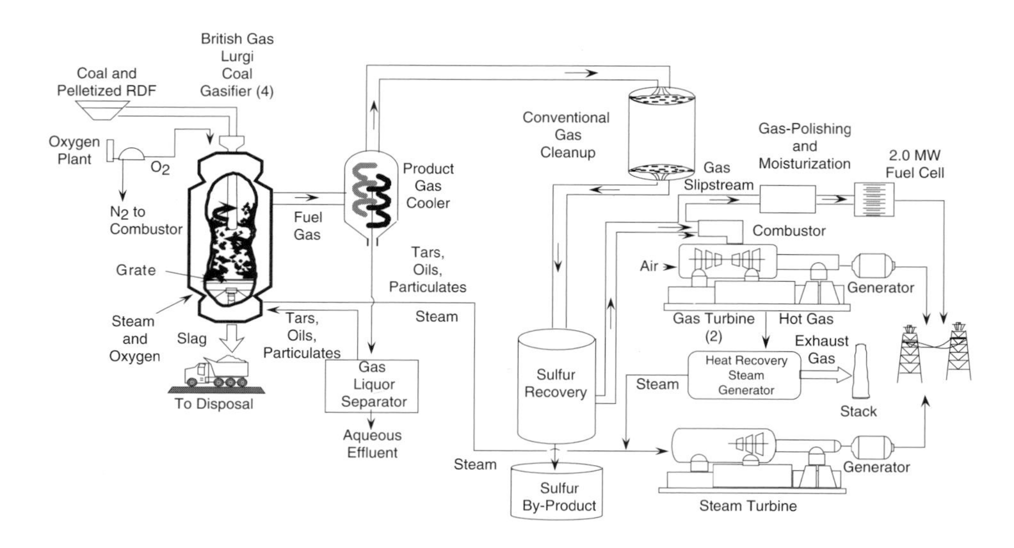

FIGURE 7-6. Schematic diagram of the Kentucky Pioneer Energy IGCC demonstration project (with MCFC slipstream shown). (From DOE, *Clean Coal Technology Demonstration Program: Program Update 2001 Including Power Plant Improvement Initiative Projects*, Office of Fossil Energy, U.S. Department of Energy, Washington, D.C., July 2002.)

converted into 99.99% pure sulfur, NO_X emissions were 0.15 lb/MM Btu (thus meeting 2003 target emission limits for ozone non-attainment areas), particulate emissions were below detectable limits, CO emissions averaged 0.05 lb/MM Btu, and coal ash was converted to a low-carbon vitreous slag valued as an aggregate in construction or as grit for abrasives and roofing metals [4]. The plant is currently in commercial operation.

Kentucky Pioneer Energy, LLC, was awarded a CCT Program project to demonstrate and assess the reliability, availability, and maintainability of a utility-scale IGCC system using a high-sulfur bituminous coal and refuse-derived fuel blend in oxygen-blown, fixed-bed, BGL slagging gasifiers, and the operability of an MCFC fueled by coal gas [4]. The IGCC system, shown in Figure 7-6, is located at the East Kentucky Power Cooperative Smith site (Trapp, Kentucky), and its capacity is 580 MW. The MCFC portion of the project, which is a slipstream of fuel gas fed to the gas turbine (and rated at 2.0 MW), has been moved to the Wabash River site [9]. The IGCC system to be demonstrated in this project is suitable for both repowering applications and new power plants. Permitting for the IGCC system is still under way. The fuel cell portion of the project broke ground in August 2003 [9].

Advanced Combustion/Heat Engines Two projects are demonstrating advanced combustion/heat engine technology. One project has been completed, and the other is delayed. Alaska Industrial Development and Export

Authority demonstrated TRW's clean coal combustion system integrated with B&W's spray dryer absorber (SDA) with sorbent recycle [41]. The demonstration was performed adjacent to Healy Unit No. 1. The objective was to demonstrate an innovative new power plant design featuring integration of an advanced combustor coupled with both high- and low-temperature emissions control processes. Emissions were controlled using TRW's advanced entrained/slagging combustors through staged fuel and air injection for NO_x control and limestone injection for SO_2 control. Additional SO_2 control was accomplished using B&W's activated recycle SDA. Carbon burnout goals of greater than 99% were achieved and emissions were successfully controlled: NO_x emissions averaged 0.245 lb/MM Btu, SO_2 removal efficiencies in excess of 90% were achieved with typical emissions of 0.038 lb/MM Btu, particulate matter emissions were 0.0047 lb/MM Btu, and CO emissions were less than 130 ppm at 3.0% O_2 [4].

TIAX (formerly Arthur D. Little, Inc.) is demonstrating a coal-fired diesel engine operation with the objective to prove the design, operability, and durability of the coal diesel engine during 4000 hours of operation and to test a coal slurry in the diesel [4]. A Fairbanks Morse 18-cylinder, heavy-duty engine (6.4 MW) modified to operate on Alaskan subbituminous coal made into a low-rank coal-water fuel is expected to have very low NO_x and SO_2 emission levels (50–70% below current NSPSs). In addition, the demonstration plant, located at the University of Alaska (Fairbanks), is expected to achieve 41% efficiency, with future plant designs expected to reach 48% efficiency, which will result in a 25% reduction in CO_2 emissions compared with conventional coal-fired plants. Testing has been delayed due to TIAX's reorganization [9].

Coal Processing for Clean Fuels Technology

The CCT Program also addresses approaches to converting raw run-of-mine coals to high-energy density, low-sulfur products. Four projects completed in the category of coal processing for clean fuels represent a diversified portfolio and include two projects that produced high-energy density solid fuels (see discussion in Chapter 5, Technologies for Coal Utilization), one of which also produced a liquid product equivalent to No. 6 fuel oil; one project that demonstrated a new methanol production process; and one project that complemented the process demonstrations by providing an expert computer model that enables a utility to assess the environmental, operational, and cost impact of utilizing coals not previously burned at a facility, including upgraded coal and coal blends [4].

ENCOAL Corporation demonstrated SGI International's Liquids-From-Coal (LFC®) process at Triton Coal Company's Buckskin Mine (located near Gillette, Wyoming) [42]. The project objective was to demonstrate the integrated operation of a number of novel processing steps to produce two higher heating value fuel forms with lower sulfur contents from mild gasification of low-sulfur subbituminous coal and to provide sufficient products for potential end users to conduct burn tests. The process, described in Chapter 5, produces a Process-Derived Fuel (PDF®) and Coal-Derived Liquid (CDL®). The LFC® process consistently produced 250 short tons/day of PDF® and 250 barrels/day of CDL® from 500 short tons of run-of-mine coal per day. The PDF® contains 0.26% sulfur with a heat content of 11,100 Btu/lb (compared with 0.45% sulfur and 8300 Btu/lb for the feed coal) [4]. The CDL® contains 0.6% sulfur and has a heating value of 140,000 Btu/gallon (compared with 0.8% sulfur and 150,000 Btu/gallon for No. 6 fuel oil) [4].

Western SynCoal LLC (formerly Rosebud SynCoal Partnership, a subsidiary of Montana Power Company's Energy Supply Division) demonstrated their advanced coal conversion process (ACCP) of upgrading low-rank subbituminous coal and lignite [43]. The process, described in Chapter 5, was performed to demonstrate the potential of ACCP to produce a stable coal product having a moisture content as low as 1%, sulfur content as low as 0.3%, and heating value up to 12,000 Btu/lb [4]. The ACCP project processed over 2.8 million short tons of raw coal at Colstrip, Montana, to produce nearly 1.9 million short tons of SynCoal® products that were shipped to utility and industrial users. Lower emissions of SO_2 and NO_x were reported in addition to increased power plant output due to the higher grade of fuel burned.

Air Products Liquid Phase Conversion Company, LP (a limited partnership between Air Products and Chemicals, Inc., the general partner, and Eastman Chemical Company), demonstrated Air Products and Chemicals' liquid-phase methanol process [44]. The objective was to demonstrate, on a commercial scale, the production of methanol from coal-derived synthesis gas using the LPMEOH™ process; to determine the suitability of methanol produced during this demonstration for use as a chemical feedstock or as a low SO_2- and NO_x-emitting alternative fuel in stationary and transportation applications; and to demonstrate, if practical, the production of dimethyl ether (DME) as a mixed co-product with methanol [4]. The LPMEOH™ process, illustrated in Figure 7-7, was successfully operated for 69 months, and the demonstration ended in December 2002 [45]. Over the entire operating period, the demonstration facility (located at Kingsport, Tennessee) operated at an on-stream availability of 97.5% and produced nearly 104 million gallons of methanol, all of which was accepted by Eastman Chemical Company for use in downstream chemical processes. The facility is currently being operated in a commercial mode by Eastman Chemical Company [45]. The process was developed to enhance IGCC power generation by producing a clean-burning, storable liquid fuel from clean coal-derived gas. Methanol contains no sulfur and has exceptionally low NO_x characteristics when burned.

The final project in this category was the development of CQ, Inc.'s (Homer City, Pennsylvania) EPRI Coal Quality Expert™ (CQE™) computer software [47]. The objective of the project was to provide the utility industry with a PC software program it could use to confidently and inexpensively

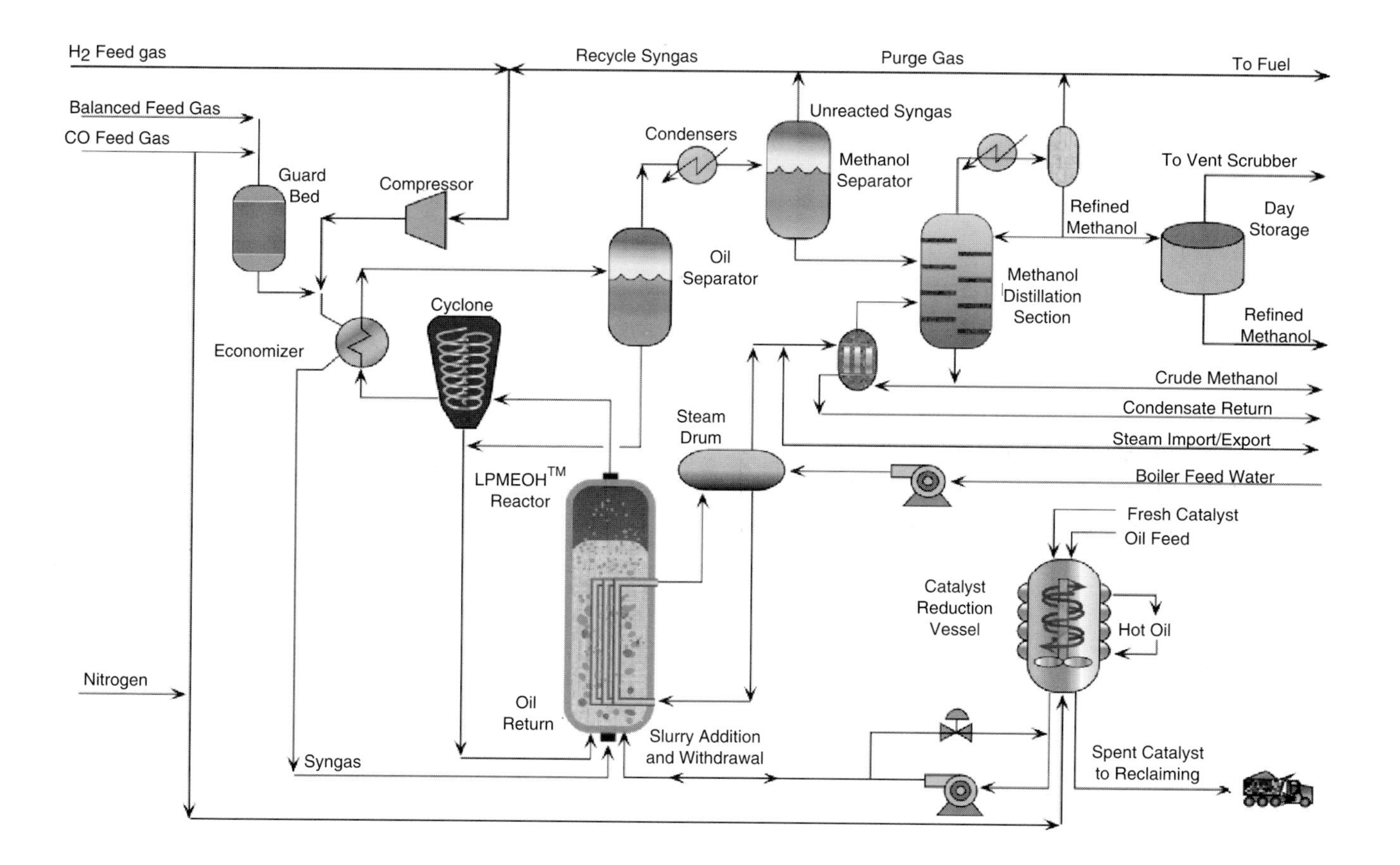

FIGURE 7-7. LPMEOH™ demonstration unit process flow diagram. (From DOE, *Clean Coal Technology: Commercial-Scale Demonstration of the Liquid Phase Methanol (LPMEOH™) Process*, Technical Report No. 11, Office of Fossil Energy, U.S. Department of Energy, Washington, D.C., April 1999.)

evaluate the potential for coal cleaning, blending, and switching options to reduce emissions while producing the lowest cost electricity [4]. Specifically, the project was intended to (1) enhance the existing Coal Quality Information Systems (CQIS™) database and Coal Quality Impact Model (CQIM™) to allow assessment of the effects of coal cleaning on specific boiler costs and performance; and (2) develop and validate CQE™, a model that allows accurate and detailed prediction of coal quality impacts on total power plant operating cost and performance. The model that was developed evaluates the impacts of coal quality, capital improvements, operational changes, and environmental compliance alternatives on power plant emissions, performance, and production costs [4].

Industrial Applications Technology

Projects were also undertaken to address pollution problems associated with using coal in the industrial sector. Five projects encompass substitution of coal for 40% of the coke in iron making (completed), integration of a direct iron-making process with the production of electricity (ongoing), reduction of cement kiln emissions and solid-waste generation (completed), demonstration of an industrial-scale slagging combustor (completed), and demonstration of a pulse combustor system (completed). Although electricity can be produced at the industrial scale, these projects are not discussed in this chapter as they have limited or no applicability in power generation and, except for the Blast Furnace Granular-Coal Injection System Demonstration Project [48] and Passamaquoddy Technology Recovery Scrubber™, have not been considered commercial successes in that no domestic or international sales have been made of the demonstrated technologies nor are they in continued operation at the demonstration site [49].

CCT Program Accomplishments

Over the past 15 to 20 years, the Clean Coal Technology Demonstration Program has successfully demonstrated technologies that [50]:

- Increase efficiency and reduce emissions from coal-fired power plants and industrial facilities;
- Expand the number of options, such as fluidized-bed boilers and gasifiers, available for the clean use of coal;
- Produce coal-based fuels that burn cleaner and help reduce emissions.

Many of the technologies that have been demonstrated under this Program are now in commercial use. Table 7-3 lists the 38 projects discussed in the previous section and those projects considered commercial successes to date are noted. Commercial success is considered if domestic or international

TABLE 7-3
Clean Coal Technology Program Commercial Successes to Date

Project	*Participant*	*Location*	*Commercial Status*
Gas suspension absorption	AirPol, Inc.	West Paducah, Kentucky	Domestic and international sales
Confined zone dispersion	Bechtel Corp.	Seward, Pennsylvania	—[a]
LIFAC sorbent injection	LIFAC–North America	Richmond, Indiana	Domestic and international sales; continued operation
Advanced flue gas desulfurization	Pure Air	Chesterton, Indiana	Continued operation
CT-121 flue gas scrubber	Southern Company Services	Newnan, Georgia	International sales; continued operation
NO_X control, wall-fired boiler	Southern Company Services	Coosa, Georgia	Domestic and international sales; continued operation
Coal reburning	B&W Co.	Cassville, Wisconsin	Continued operation
Low-NO_X cell burner	B&W Co.	Aberdeen, Ohio	Domestic sales; continued operation
Gas reburning/ low-NO_X burners	EERC	Denver, Colorado	Domestic and international sales; continued operation
Micronized coal reburning	NYSEG	Lansing, New York	Continued operation
Selective catalytic reduction	Southern Company Services	Pensacola, Florida	Domestic and international sales
NO_X control, tangentially fired boiler	Southern Company Services	Lynn Haven, Florida	Domestic and international sales; continued operation
SNOX™ flue gas cleaning	ABB	Niles, Ohio	—
LIMB SO_2/NO_X control	B&W Co.	Lorain, Ohio	Domestic and international sales
SNRB™ process	B&W Co.	Dilles Bottom, Ohio	—
Gas reburning, sorbent injection	EERC	Hennepin and Springfield, Illinois	Continued operation
Milliken clean coal	NYSEG	Lansing, New York	Domestic sales
Dry NO_X/SO_X control system	Public Service of Colorado	Denver, Colorado	Domestic sales; continued operation

(continued)

TABLE 7-3
(continued)

Project	*Participant*	*Location*	*Commercial Status*
McIntosh 4A pressurized fluidized-bed combustion (PFBC)	City of Lakeland	Lakeland, Florida	—
McIntosh 4B pressurized fluidized-bed combustion (PFBC)	City of Lakeland	Lakeland, Florida	—
JEA fluidized-bed combustion (FBC)	JEA	Jacksonville, Florida	—
Tidd pressurized fluidized-bed combustion (PFBC)	Ohio Power Company	Brilliant, Ohio	International sales
Nucla circulating fluidized-bed (CFB)	Tri-State	Nucla, Colorado	Domestic and international sales
Kentucky Pioneer	Kentucky Pioneer	Trapp, Kentucky	—
Tampa Electric integrated gasification combined cycle (IGCC)	Tampa Electric	Mulberry, Florida	Domestic and international sales
Piñon Pine Power	Sierra Pacific	Reno, Nevada	—
Wabash River Repowering	Wabash River Coal Gasification J.V.	West Terre Haute, Indiana	Continued operation
Clean coal diesel	TIAX	Fairbanks, Alaska	—
Healy clean coal	Alaska Industrial Development and Export Authority	Healy, Alaska	—
LPMEOH™ process	Air Products	Kingsport, Tennessee	Domestic sales; continued operation
Coal Quality Expert™	ABB and CQ, Inc.	Multiple sites	Domestic and international sales
ENCOAL® mild gasification	ENCOAL Corp.	Gillette, Wyoming	Domestic and international sales pending

(continued)

TABLE 7-3
(continued)

Project	*Participant*	*Location*	*Commercial Status*
Advanced coal conversion process	Western SynCoal, LLC	Colstrip, Montana	Extended continued operation
Integrated coal/ore reduction	CPICOR™ Management Co.	Vineyard, Utah	—
Blast furnace coal injection	Bethlehem Steel Corp.	Burns Harbor, Indiana	Domestic sales; continued operation
Advanced cyclone combustor	Coal Tech Corporation	Williamsport, Pennsylvania	—
Cement kiln scrubber	Passamaquoddy Tribe	Thomaston, Maine	Continued operation
Pulse combustor	ThermoChem, Inc.	Baltimore, Maryland	—

[a]Nothing reported.

sales are made or if the technology continues to operate commercially at the demonstration site. Others have identified success based on patents and awards granted to Clean Coal Technology Program projects [51].

Commercial sales (as of 2000), domestic and international, resulting from the CCT Program projects include approximately 130 gasifiers, 160 fluidized-bed units, 2900 NO_X reduction units, and 200 SO_2 removal units [4,51]. In addition, approximately 30 utilities have the model for coal processing for clean fuels.

Prior to the CCT Program, scrubbers capable of high SO_2 removal were costly to build and difficult to maintain, placed a significant parasitic energy load on the plant output, and produced a sludge waste requiring disposal [52]. The demonstration projects conducted under the CCT Program have cut operating and capital costs in half, provided SO_2 removal efficiencies of 95 to 98%, produced valuable by-products, mitigated plant efficiency losses, and captured multiple air pollutants. If CCT-developed technologies were applied to all U.S. coal-fired boilers at an average efficiency of 90%, total SO_2 emissions could be further reduced by ~10 million short tons/year [52]. Currently, about one-fourth of the total U.S. coal-fired capacity has FGD units installed. The United States has about 260 units with a total capacity of 85,000 MW, which is the largest number of FGD installations in the world.

Prior to the CCT Program, NO_X control technology proven in U.S. utility service was essentially nonexistent. The CCT Program has met the regulatory challenge by developing and incorporating emerging NO_X control technologies into a portfolio of cost-effective compliance options for the full range of boiler types being used commercially [52]. Products of the CCT Program for NO_X control include:

- Low-NO_X burners, overfire air, and reburning systems that modify the combustion process to limit NO_X formation;
- Post-combustion control options using SCR and SNCR;
- Artificial-intelligence-based control systems that effectively handle numerous dynamic parameters to optimize operational and environmental performance of boilers.

As a result, over three-fourths of U.S. coal-fired power plants have installed low-NO_X burners. Reburning and artificial intelligence systems have made significant market penetration as well. All sites that developed these NO_X control technologies have retained them for commercial use. In addition, several commercial installations of SCR and, to some extent, SNCR have been installed with many planned for installation in the near future.

The CCT Program has provided the foundation for power in the twenty-first century through successful demonstration of FBC and IGCC projects on a commercial scale. These technologies are inherently clean, producing negligible emissions of SO_2, NO_X, and particulate matter. The IGCC demonstration projects have achieved excellent environmental performance, with emissions as low as 0.02 lb SO_2 per MM Btu and 0.08 lb NO_X per MM Btu. In addition, the higher thermal efficiency processes result in significant reductions in CO_2 emissions.

Power Plant Improvement Initiative (PPII)

The success of the CCT Program serves as a model for other cooperative government/industry programs aimed at introducing new technologies into the commercial marketplace. Two follow-up programs have been developed that build on the successes of the CCT Program: the Power Plant Improvement Initiative (PPII) and the Clean Coal Power Initiative (CCPI). The PPII, established by the Department of the Interior and Related Agencies Appropriations for Fiscal Year 2001 (Public Law 106-291), is a cost-shared program, patterned after the CCT Program and directed toward improved reliability and environmental performance of the nation's existing coal-burning power plants [29]. Authorized by the U.S. Congress in 2001, the PPII originally involved eight projects; two projects withdrew after being selected, four are in progress, and two are under negotiation. The four projects that are under way have a total cost of more than \$41 million, of which the private sector is contributing nearly \$24 million, exceeding the 50% private sector cost share mandated by Congress.

PPII Projects

The PPII projects focus on technologies enabling coal-fired power plants to meet increasingly stringent environmental regulations at the lowest possible cost. With many plants threatened with shutdowns because of environmental concerns, more effective and lower cost emission controls can keep generators operating while improving the quality of the nation's air and water [5]. A brief description of the ongoing projects and their objectives is provided here.

Sunflower Electric Power Company, with GE Energy and Environmental Research (formerly Energy and Environmental Research Corporation), will be demonstrating a unique combination of high-tech combustion modifications and sophisticated control systems with the goal of reducing NO_X emissions to 0.15 to 0.22 lb/MM Btu [53]. The project will be performed on Holcomb Station (Garden City, Kansas), a 360 MW, pulverized coal-fired boiler firing subbituminous coal. The reduction in NO_X emissions will be accomplished by modifying existing low-NO_X burners, separating the overfire air, installing fuel flow measurement transducers, balancing the combustion air, and using neural network controls.

Otter Tail Power Company, with Montana Dakota Utilities, North-Western Public Service, W.L. Gore & Associates, Inc., and the University of North Dakota Energy and Environmental Research Center, will be demonstrating, in a full-scale application, a hybrid technology that increases the particulate matter capture in coal plants up to 99.99% by integrating fabric filtration and electrostatic precipitation [54]. The advanced hybrid particulate collector (AHPC), discussed in Chapter 6, combines the best features of an electrostatic precipitator (ESP) and a baghouse in the same housing, providing major synergism between the two methods to overcome the problem of excessive fine particulate emissions that escape collection in an ESP and the re-entrainment of dust in a baghouse. The demonstration is being performed on Big Stone Power Plant's (Big Stone City, South Dakota) 450 MW cyclone-fired boiler and is a scale-up from a 2.5 MW slipstream test program that was performed at the plant.

Tampa Electric Company and Pegasus Technology, Inc., will be demonstrating control of boiler fouling on Big Bend Power Station's (Apollo Beach, Florida) 445 MW unit using a neural-network soot-blowing system in conjunction with advanced controls and instruments [55]. Ash and slag deposition compromise plant efficiency by impeding heat transfer to the working fluid, leading to higher fuel consumption and higher emissions. The process optimization is targeted to reduce total NO_X generation by 30% or more, improve heat rate by 2%, and reduce particulate matter emissions by 5%. As compared to competing technologies, this system could be an extremely cost-effective technology that has the ability to be readily adapted to virtually any pulverized coal-fired boiler.

In the fourth PPII project that is under way, Universal Aggregates LLC (a joint venture between CONSOL Energy, Inc., and SynAggs, Inc.)

will design, build, and operate an aggregate manufacturing plant that converts 115,000 short tons/year of spray dryer by-products into 150,000 short tons/year of lightweight masonry blocks or lightweight concrete [56]. Only ~18% of flue gas desulfurization residue in the United States is recycled, with the remainder landfilled. This process will reduce plant disposal costs and the environmental drawbacks of landfilling by producing a salable by-product. The demonstration will be located by the Birchwood Power Facility in King George County, Virginia.

The two projects under negotiation with the DOE also pertain to emissions reduction. In one project, CONSOL Energy, Inc., is proposing to demonstrate a multipollutant control system that can cost-effectively reduce NO_X, SO_2, acid gases (*i.e.*, hydrochloric and hydrofluoric acids), and mercury from smaller coal-fired power plants using single-bed, in-duct SCR combined with low-NO_X combustion technology [57]. In the second project, proposed by TIAX, a hybrid system composed of lower cost components from three established NO_X reduction systems (*i.e.*, fuel-lean gas reburn, SNCR, and SCR) will be developed and demonstrated to reduce NO_X emissions to 0.15 lb/MM Btu at lower costs than conventional SCR [58].

Benefits of the PPII

The PPII, a precursor to CCPI, is designed to establish commercial-scale demonstrations of coal technologies to ensure energy supply reliability [3]. The PPII is poised to make near-term contributions to air quality improvements and focuses on technology that can be commercialized over the next few years.

Clean Coal Power Initiative (CCPI)

The second follow-on program to the CCT Program is the Clean Coal Power Initiative (CCPI), which was initiated by President Bush in 2002 and is an innovative technology demonstration program that fosters more efficient clean coal technologies for use in existing and new power generation facilities in the United States [6]. Candidate technologies are demonstrated at a significant scale to ensure proof-of-operation prior to widespread commercialization. Technologies emerging from this program will help to meet the president's new environmental objectives for the United States, as detailed in the Clear Skies Initiative, Global Climate Change Initiative, and FutureGen, and to advance pollution control and coal utilization, both in the United States and abroad. Early demonstrations emphasize technologies that are applicable to existing power plants and include construction of new plants. Later demonstrations will include systems comprising advanced turbines, membranes, fuel cells, gasification technologies, and hydrogen production [6]. CCPI is a multiyear program funded at a total federal cost of up to $2 billion, with the private sector cost-share being at least 50%. CCPI responds to President Bush's commitment to clean coal technology development as part of his National Energy Policy [59]. Priorities covered by the National Energy Policy include increasing the country's domestic energy supply, protecting its environment, ensuring a comprehensive energy delivery system, and enhancing national energy security [3].

The Clean Coal Power Initiative will be conducted over four solicitations (Rounds 1 through 4). Round 1 proposals were submitted in 2002, with eight projects selected in January 2003 out of 36 proposals submitted (although one project has subsequently withdrawn). The projects selected are comprised of power generation, co-production, multipollutant emissions control, advanced control systems, and by-product utilization. Negotiations are under way, and the projects total $1.23 billion, of which the DOE is providing $286 million or 23% of the total [60]. Round 2 planning activities are under way; the solicitation is to be released in 2004 and project selections are planned for 2005. The emphasis of the second solicitation will be on efficiency improvements and advanced multipollutant (including mercury) controls for Clear Skies technologies. Round 3 (tentative award date 2007) will emphasize co-production, membranes, fuel cells, and energy systems with efficiencies greater than 50%. Near-zero emissions, hydrogen production and transportation, sequestration, and efficiencies greater than 55% (Vision 21 modules) will be emphasized in Round 4 (tentative award date 2009).

Program Importance

The strength and security of the U.S. economy are closely linked to the availability, reliability, and cost of electric power. Economic growth is linked to reliable and affordable electric power. Electricity requirements for the United States are steadily increasing, and coal will play a significant role in satisfying U.S. energy needs. CCPI will help meet these energy electricity demands by demonstrating new generation technologies [6]. CCPI will also enable effective use of existing facilities and prepare for their retirement by demonstrating technological improvements in efficiency; advanced low-cost, high-performance emissions control technologies; and reliability at new and existing plants. CCPI is closely aligned with research, development, and demonstration activities being performed under the DOE's Coal and Power Systems core research and development programs that are working toward ultra-clean fossil-fuel-based energy systems in the twenty-first century [2,3]. CCPI technologies will address existing and new regulatory requirements and complement the goals of the FutureGen Project, which is an initiative to create the world's first coal-based, zero-emission electricity and hydrogen plant (see later discussion). CCPI will help the United States achieve improved power plant performance and near-zero emissions and is integral to achieving new plant performance targets identified in the

DOE's roadmap for existing and future energy plants [3,7,61]. Existing plant roadmap performance objectives include reducing costs for NO_X and high-efficiency mercury control and achieving particulate matter targets in 2010 of 99.99% capture of 0.1- to 10-μm particulates [3]. The long-term roadmap goals are aimed at achieving near-zero emissions power and clean fuels plants with CO_2 management capability. The long-term new plant performance targets are presented in Table 7-4 [3,61].

TABLE 7-4
New Plant Performance Targets (representing best integrated plant technology capability)

Air Emissions	*Reference Plant[a] 98% SO_2 Removal 0.15 lb NO_X per MM Btu 0.01 lb PM per MM Btu Mercury (Hg)*	*2010 99% SO_2 Removal 0.05 lb NO_X per MM Btu[b] 0.005 lb PM per MM Btu[c] 90% Hg Removal[d]*	*2020 (Vision 21) >99% SO_2 Removal <0.01 lb NO_X per MM Btu 0.002 lb PM per MM Btu 95% Hg Removal*
By-product utilization	30%	50%	Near 100%
Plant efficiency (HHV)[e]	40%	45–55%	50–60%
Availability[f]	>80%	>85%	≥90%
Plant capital cost ($/kW)[g]	900–1300	900–1000	800–900
Cost of electricity (¢/kWh)[h]	3.5	3.0–3.2	<3.0

[a]Plant that can be built using current state-of-the-art technology; plant meets New Source Performance Standards; NO_X levels below 0.15 lb/MM Btu can be achieved with a combination of advanced combustion and SCR technologies; some mercury (Hg) reduction is achieved as a co-benefit with existing environmental control technologies. By-product utilization represents an average for existing plant locations, as actual plant utilization ranges from essentially zero to near 100%. No carbon capture and sequestration; figures reflect current cooling tower technology or use.
[b]For NO_X, reduce cost for achieving <0.10 and 0.05 lb/MM Btu to three-fourths that of SCR by 2005 and 2010, respectively.
[c]Achieve particulate matter (PM) targets for existing plant in 2010 and 99.99% capture of 0.10–10 μm particles.
[d]Achieve 50–70% Hg reduction at less than three-fourths the cost of activated carbon injection by 2005.
[e]HHV, based on fuel higher heating value.
[f]Percent of time capable of generating power.
[g]Range reflects projection for different plant technologies that will achieve environmental performance and energy cost targets.
[h]Bus-bar cost of electricity.
Source: Data from Eastman [3] and DOE [61].

Round 1 CCPI Projects

Early demonstrations of CCPI will emphasize advanced technologies that are applicable to existing power plants but will also include construction of new, advanced, clean coal power plants. Figure 7-8 shows the locations of the seven projects selected and the amount of funding, both DOE and private sector, for each project. Two of the projects are directed at new ways to comply with the Clear Skies Initiative, which calls for dramatic reductions in air pollutants from power plants over the next 15 years [60]. Three projects are expected to contribute to the Climate Change Initiative to reduce greenhouse gases. The final two projects will use abundant waste materials that resulted from earlier coal mining activities while reducing air pollution through coal gasification and multipollutant control systems. Brief descriptions of the technologies being demonstrated in each project are provided here.

New Technologies for Clear Skies Initiative

Wisconsin Electric Power Company will design, install, operate, and evaluate the TOXECON process as an integrated emissions control system for mercury, particulate matter, SO_2, and NO_x [3,62]. A TOXECON unit, shown schematically in Figure 7-9 [3], will be installed on the combined flue gas of Wisconsin Electric Power Company's Presque Isle Power Plant Units 7, 8, and 9 for a total of 270 MW. TOXECON represents an option for greater than 80% mercury control for coal-fired power plants, may be the primary mercury control option for western coals, and may be the only choice for units with hot-side ESPs [62]. The TOXECON configuration allows for separate treatment or disposal of the fly ash collected in the primary particulate control device. Powdered activated carbon will be injected for mercury control, and sodium-based additives will be injected for reducing SO_2 and NO_x emissions.

Colorado Springs Utilities (CSU) and Foster Wheeler are demonstrating an advanced coal-fired power plant that will have unprecedented low emissions levels using advanced low-cost emission control systems [3,62]. Fully integrated, multilayered emission controls are being combined with CFB combustion to produce what is expected to be the cleanest coal-fired unit in the world while maintaining cost competitiveness and high unit reliability. Control of NO_x will be performed using an advanced staged-combustion process coupled with an advanced SNCR system. Sulfur dioxide removal of 96 to 98% is expected using a three-stage approach, and an integrated trace metal control system will remove up to 90% of the mercury, lead, and other metals, as well as virtually all acid gases. The demonstration will be performed at full scale using a 150 MW unit at the CSU Ray D. Nixon Power Plant located near Colorado Springs.

New Technologies to Meet Climate Change Goals

Great River Energy will demonstrate a process for reducing the moisture content of lignite, thereby increasing the value of high-moisture fuels in

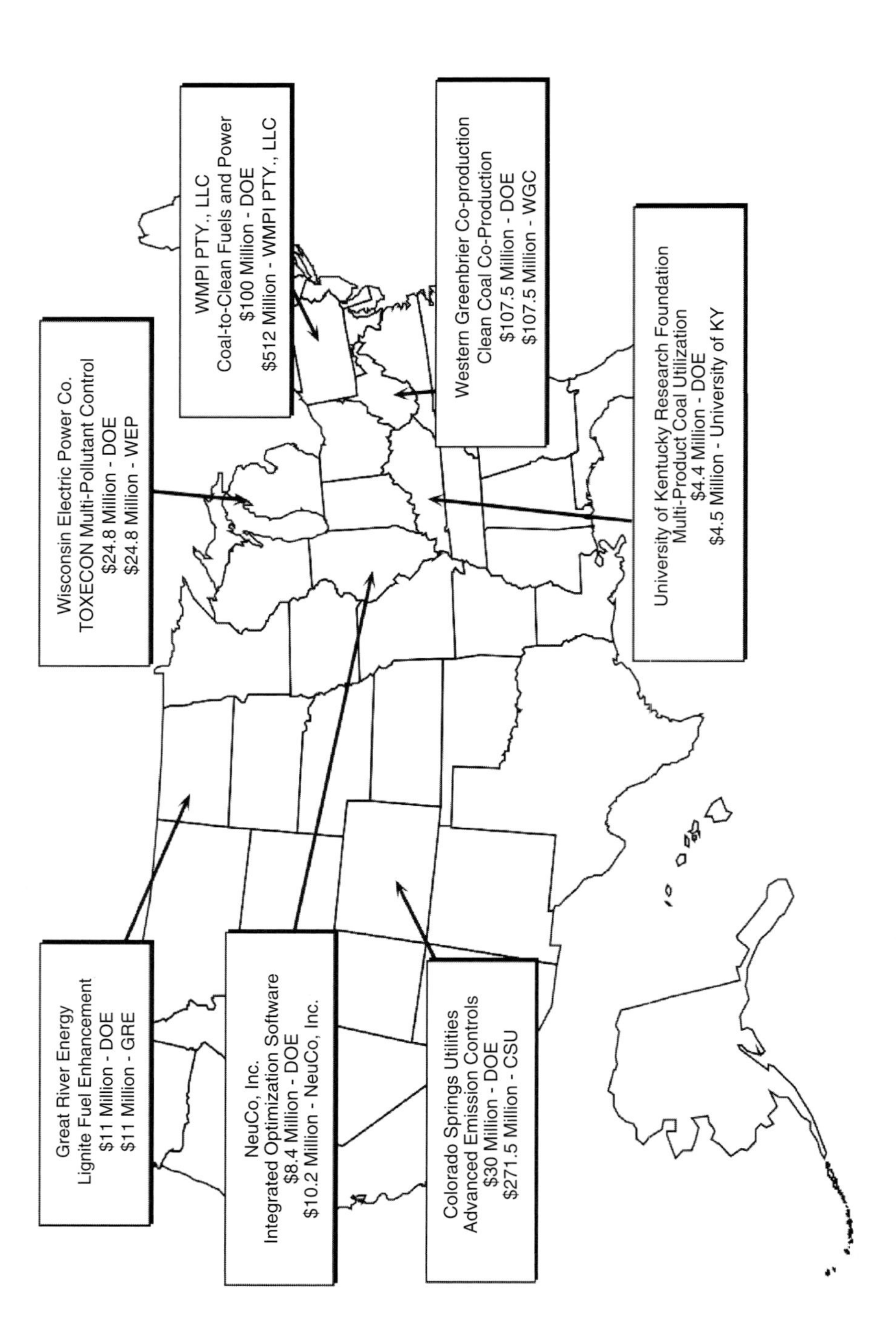

FIGURE 7-8. Clean Coal Power Initiative Round 1 projects. (From Sarkus, T. A., *Clean Coal Power Initiative: Tackling Coal's Environmental Challenges*, presentation to the National Coal Council, www.netl.doe.gov/coalpower/ccpi/main.html, December 4, 2003.)

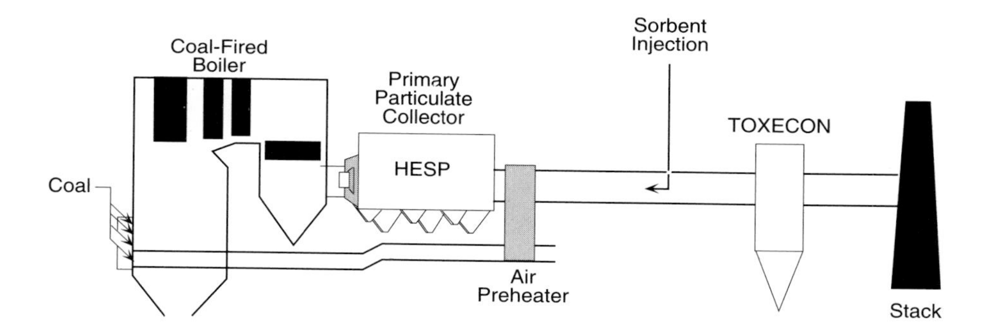

FIGURE 7-9. TOXECON configuration for mercury control in a coal-fired boiler. (From Eastman, M. L., *Clean Coal Power Initiative*, presented at the Clean Coal Power Conference, www.netl.doe.gov/coalpower/ccpi/program_info.html, November 18, 2003.)

electrical generation plants; increasing the net generating capacity of units that burn high-moisture coal; increasing the energy supply of units that burn high-moisture coal; increasing the cost-effectiveness of the country's electrical generation industry; improving the environment by reducing emissions from coal-fired power plants; and increasing the value of the country's lignite reserves [3,62]. The demonstration will be performed at Great River Energy's Coal Creek Station (*i.e.*, two 546 MW pulverized coal-fired units located at Underwood, North Dakota) and will use waste heat in two fluidized-bed driers to reduce the lignite moisture content by 10 percentage points, which is typically about 40%.

NeuCo, Inc. (Boston, Massachusetts) will design, develop, and demonstrate integrated on-line optimization systems at Dynegy Midwest Generation's Baldwin Energy Complex (three 600 MW coal-fired units) located in Baldwin, Illinois [3,62]. The modules will address sootblowing, SCR operations, overall unit thermal performance, and plant-wide profit optimization by reducing NO_X, increasing fuel efficiency, and increasing reliability. The increases in fuel efficiency (*i.e.*, heat rate production) will provide commensurate reductions in greenhouse gases, mercury, and particulate matter. The technology platform, consisting of neural networks, algorithms, and fuzzy logic techniques, will be used to comprehensively apply optimization techniques to a variety of systems within coal power plants using existing control technologies and then will link these systems to each other.

The University of Kentucky Research Foundation of Lexington, Kentucky, in partnership with LG&E Energy Corporation of Louisville, Kentucky, will design, construct, and demonstrate an advanced coal ash beneficiation processing plant at the LG&E 2200 MW Ghent Power Plant in Ghent, Kentucky [3,62]. The process, based on hydraulic classification and froth flotation technology, separates the UBC from power plant ash or from ash disposal ponds and recycles it for use as fuel. In addition, the process upgrades the ash to make it suitable for producing a high-strength alternative to portland cement and uses a beneficiated coarse ash to produce lightweight aggregate suitable for use in concrete masonry materials such as building blocks or to be used as sand for construction applications. The new process has the potential to reduce the manufacture of portland cement, one of the highest generators of CO_2 of any industrial process.

New Technologies to Produce Clean Energy from Waste Coal Piles

Waste Management and Processors, Inc. (WMPI PTY., LLC), of Gilberton, Pennsylvania, has assembled a leading technology and engineering team (*i.e.*, Nexant, Inc., Shell Global Solutions B.V. U.S., Uhde GmbH, and SASAOL Technology, Ltd.) to design, engineer, construct, and demonstrate the first clean coal power facility in the United States using waste coal gasification as the basis for the clean power, thermal energy, and clean liquid fuels production [3,62]. The demonstration will be performed at Gilberton, Pennsylvania, and will convert wastes generated from anthracite cleaning into electric power and high-value, premium, ultra-clean transportation fuels. The plant will gasify 4700 short tons/day of coal wastes to produce a synthesis gas of hydrogen and carbon monoxide, of which most will be combusted to produce electricity (*i.e.*, 41 MW), with a portion of the synthesis gas being converted into about 5000 barrels/day of synthetic hydrocarbon liquids via Fischer–Tropsch (F-T) synthesis. The Gilberton project is depicted in the block diagram in Figure 7-10. The ultra-clean transportation fuels that will be produced include naphtha, kerosene, and diesel fuel, which will be in the form of an ultra-clean, high-cetane diesel fuel that contains no sulfur

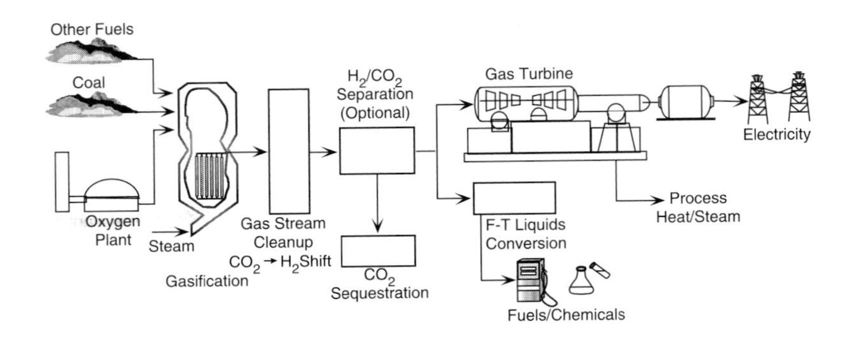

FIGURE 7-10. Block diagram of the WMPI Pty., LLC, process. (From DOE, *Clean Coal Technology Demonstration Program: Program Update 2001 Including Power Plant Improvement Initiative Projects*, Office of Fossil Energy, U.S. Department of Energy, Washington, D.C., July 2002.)

or aromatic hydrocarbons. The F-T naphtha formed during the F-T synthesis can be upgraded to clean-burning reformulated gasoline. F-T naphtha is also an excellent feedstock for steam cracking or olefin production or an on-board reforming feed for fuel-cell-powered vehicles. The low-smoke-point kerosene is a niche-market jet fuel. Approximately 1300 and 3800 barrels/day, respectively, of F-T naphtha and ultra-clean diesel and jet fuel will be produced.

Western Greenbrier Co-Generation, LLC (a public service entity formed to serve the interests of three municipalities in West Virginia—Rainelle, Rupert, and Quinwood), will demonstrate an 85 MW, clean coal, co-production process in Rainelle, West Virginia [3,62]. The power plant will be an innovative Alstom Power, Inc., CFB boiler system utilizing bituminous coal waste and integrated with an advanced, multipollutant control system for SO_X, NO_X, particulate matter, and mercury. An integrated co-production facility will manufacture structural bricks using ash from the boiler and wood waste from an adjacent industrial process.

CCPI Benefits

The Clean Coal Power Initiative bridges the gap between the CCT Program and PPII and implementation of Vision 21 systems, ensuring ongoing development of advanced systems for power production emerging from the DOE's core fossil-fuel research programs. Successful completion of the CCPI Program will introduce technologies to the U.S. marketplace that can achieve compliance with emerging air regulations and National Energy Policy (NEP) priorities. CCPI provides an important platform to implement the NEP recommendation to increase investment in clean coal technology [3]. The program will also help to ensure that upcoming regulations are science and engineering based and exploit emerging technologies developed under CCPI. CCPI will mitigate costs and reduce the technical and environmental risks associated with advanced technology development and will serve as a proving ground in the United States to speed technologies to market both at home and abroad, thereby ensuring the realization of early environmental benefits [3].

The CCPI Program benefits are expected to be substantial when compared to the investment costs. Unless advanced technologies achieve widespread commercial use, which must occur through demonstrations, the projected benefits will not be achieved. Benefits expected from the program include [6]:

- Reduced fuel costs due to higher plant efficiencies;
- Lower capital costs for construction of new plants and repowered facilities;
- Lower capital and operating costs for existing plants;

- Reduced costs of environmental compliance;
- Avoided environmental costs (*e.g.*, health, infrastructure, agriculture);
- Enhanced industrial competitiveness leading to increased domestic and international sales;
- Additional jobs.

Vision 21

The DOE is providing the foundation needed to build a future generation of fossil energy-based power systems capable of meeting the energy and environmental demands of the twenty-first century. This initiative—Vision 21—is the DOE's approach for developing the technology needed for ultra-clean twenty-first-century energy plants. The overall goal of Vision 21 is to effectively remove, at competitive costs, all of the environmental concerns associated with the use of fossil fuels for producing electricity and transportation fuels [63]. Achieving this goal will require an intensive, long-range (*i.e.*, 15–20 years) research and development effort that emphasizes innovation and commercialization of revolutionary technologies. First-generation systems emerging from the CCT Program, PPII, and CCPI are providing or will provide the basis for Vision 21, including: (1) a knowledge base from which to launch commercial systems that will experience increasingly improved cost and performance over time through design refinement, and (2) platforms on which to test new components that will result in improvements in cost and performance.

Vision 21 is based on three premises: The United States will rely on fossil fuels for a major share of its energy needs well into the twenty-first century; a diverse mix of energy resources, including coal, gas, oil, biomass, and other renewables, as well as nuclear, must be used for strategic and security reasons, rather than using a limited subset of these resources; and research and development directed at resolving energy and environmental issues can find affordable ways to make energy conversion systems meet ever stricter environmental standards [63]. Vision 21 plants will effectively remove environmental constraints as an issue in the use of fossil fuels; emissions of traditional pollutants, including smog- and acid-raining species, will be near zero; and the greenhouse gases (carbon dioxide) will be reduced 40 to 50% by efficiency improvements and reduced to zero if combined with sequestration. In addition, Vision 21 plants will address water use, by-product utilization, sustainability (*i.e.*, no future legacies), timely deployment of new technology, and affordable, competitive systems with other energy options. The Vision 21 energy plant performance targets are listed in Table 7-5 [63,64], and the technology concept is shown in Figure 7-11 [65].

computational modeling and development of virtual demonstration capability, advanced controls and sensors, environmental control technology, and advanced manufacturing and modularization [66].

Systems integration in the energy plants will use "smart" systems integration techniques to combine high-performance subsystems into very clean and efficient low-cost plants. Sub-elements include systems engineering, dynamic response and control, and industrial ecology.

Plant designs are the major product of the Vision 21 Program. Sub-elements include designs for components and subsystems, prototype plants, and commercial plants. In addition, a virtual demonstration capability will be developed [66].

Significant activity is under way to address these program elements, such as in-house research projects performed by the DOE; demonstrations being conducted through the CCT, PPII, and CCPI programs; and research and development projects funded by the DOE through numerous solicitations specifically targeted for the Vision 21 Program.

Vision 21 Benefits

Many benefits will be realized by Vision 21 successes. Environmental barriers to fossil fuel use will be removed, including advances in control of smog- and acid-rain-forming pollutants and particulate and hazardous air pollutants, capture and sequestration of carbon dioxide, and minimization and utilization of solid wastes. Affordable energy costs will be maintained by using a wide range of low-cost fossil fuel options. Useful co-products, including transportation fuels, will be produced in the energy plant that will reduce reliance on imported oil, stabilize oil prices, and improve the U.S. balance of trade. The United States will continue its leadership role in clean energy technology by promoting export of fossil energy and environmental technology, equipment, and sales.

FutureGen

On February 27, 2003, President Bush announced plans for the United States to build a prototype of the fossil fuel power plant of the future: FutureGen. In a statement released by the president, he stated: "Today I am pleased to announce that the United States will sponsor a $1 billion, 10-year demonstration project to create the world's first coal-based zero-emissions electricity and hydrogen power plant. This project will be undertaken with international partners and power and advanced technology providers to dramatically reduce air pollution and capture and store emissions of greenhouse gases. We will work together on this important effort to meet the world's growing energy needs, while protecting the health of our people and environment" [67]. FutureGen is a cost-shared venture with private-sector and

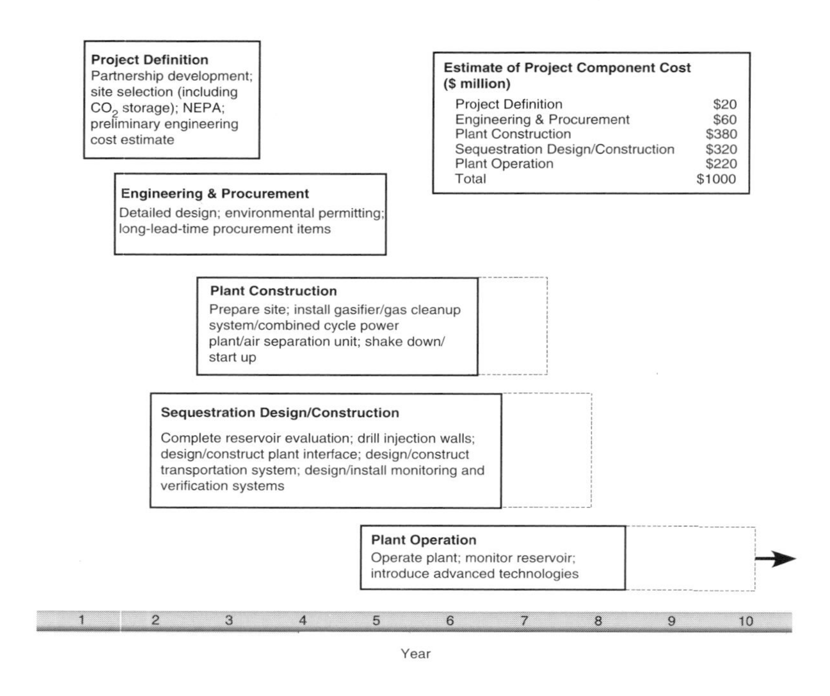

FIGURE 7-12. FutureGen project timelines, components, and estimated costs. (From DOE, *FutureGen—A Sequestration and Hydrogen Research Initiative Fact Sheet*, Office of Fossil Energy, U.S. Department of Energy, Washington, D.C., February 2003.)

international partners that will combine electricity and hydrogen production with the virtual total elimination of harmful emissions, including greenhouse gases, through sequestration [64,68]. The FutureGen plant will serve as the test bed for demonstrating the best technologies the world has to offer.

The power industry will be asked to organize a consortium to manage the project and share in the project costs. Current plans call for the plant to be built over the next five years and operated for at least five years beyond that [64]. A generic timeline showing the approximate period of performances for the five project components is provided in Figure 7-12 [69]. Figure 7-12 also lists the broad activities to be performed as well as the estimated costs for each project component.

Nearly every aspect of the prototype plant will employ cutting-edge technology. Rather than using traditional coal combustion technology, the plant will be based on coal gasification technology, with a hydrogen-rich gas being produced. The hydrogen could then be combusted in a turbine, used in fuel cells to produce clean electricity, or fed to a refinery to upgrade petroleum products [64]. In the future, the plant could become a model hydrogen-production facility to supply a fleet of hydrogen-powered vehicles.

Pollutants such as SO_2 and NO_x will be cleaned from the coal gases and converted to usable by-products such as fertilizers and soil enhancers [64]. Mercury pollutants will be removed, and CO_2 will be captured and sequestered in deep underground geologic formations. Candidate reservoirs include depleted oil and gas reservoirs, unmineable coal seams, deep saline aquifers, and basalt formations [69]. The reservoirs will be intensively monitored to verify the permanence of the CO_2 storage.

The prototype plant will be sized to generate approximately 275 MW of electricity, *i.e.*, equivalent to an average mid-size, coal-fired power plant [64]. The plant would be a stepping stone toward a future coal-fueled power plant that not only would be emissions free but would also operate at unprecedented fuel efficiencies.

The goals of the project include [69]:

- Design, construct, and operate a nominal 275 MW (net equivalent output) prototype plant that produces electricity and hydrogen with near-zero emissions. The size of the plant is dictated by the need for producing commercially relevant data, including the requirement for producing one million metric tons per year of CO_2 to adequately validate the integrated operation of the gasification plant and the receiving geologic formation;
- Sequester at least 90% of the CO_2 emissions from the plant with the future potential to capture and sequester nearly 100%;
- Prove the effectiveness, safety, and permanence of CO_2 sequestration;
- Establish standardized technologies and protocols for CO_2 measuring, monitoring, and verification;
- Validate the engineering, economic, and environmental viability of advanced coal-based, near-zero emission technologies that by 2020 will (1) produce electricity with less than a 10% increase in cost compared to non-sequestered systems, and (2) produce hydrogen at $4.00/MM Btu (wholesale), which is equivalent to $0.48/gallon of gasoline, or $0.22/gallon less than the current wholesale price of gasoline.

Benefits of the DOE's Clean Coal Power Program/Demonstrations

The DOE's coal-based research programs and associated demonstrations offer many benefits. Demonstrations are necessary for advancing technologies and achieving widespread commercial use. The federal government's support of the Clean Coal Power Program, the DOE's leadership role in performing the program and demonstrations, and industry's participation in the program and demonstrations are crucial to bringing these technologies to market and ensuring the continued use of coal in an environmentally acceptable manner. In addition to the environmental benefit of achieving near-zero-emission coal-based plants, economic and security benefits are also realized.

A study by the Southern Company Services, Inc., estimated that the DOE coal-based research programs related to large-scale power generation are estimated to provide over $100 billion in benefits to the U.S. economy through 2020 at a cost to the federal budget of less than $4 billion [70]. Sarkus and Smouse [70] also report that, in a study conducted by EPRI, the benefits of coal research and development to consumers are estimated to be $1380 billion for the period 2007 to 2050.

In the benefits study forecasted to 2020, five savings categories were selected [63]: (1) savings in fuel costs (based on higher efficiency systems); (2) savings due to the reduced capital cost of building new plants (due to lower capital cost of plants using advanced technology); (3) savings in the cost of control technology used on existing plants (due to lower capital and operating costs to achieve environmental regulations); (4) savings from avoided environmental costs due to the reduction in emissions achieved by advanced technology (estimated credit for avoided environmental costs for health, infrastructure, and agriculture realized); and (5) increased technology export resulting from more competitive U.S. technology (clean coal technology could increase U.S. sales of technology abroad by 10–15%). Benefits that are not included in the study are those gained from utilizing these technologies beyond 2020 (*e.g.*, repowering existing plants to realize higher efficiency and lower fuel costs), savings in business sectors due to the implementation of advanced coal-processing technology (*e.g.*, freeing natural gas use for other sectors), and potential savings if carbon dioxide regulations are enacted. The economic benefits are listed in Table 7-6 for the five categories [63].

The projected cumulative benefits to 2020 of approximately $100 billion represent a significant return on a forecasted investment of $11 billion,

TABLE 7-6
Economic Benefits of DOE's Clean Coal Power Program

Savings Categories	*Cumulative Benefits ($ billions, 2003–2020)*
Fuel cost	10
Capital cost (new plants)	12
Control technology cost (existing plants)	32
Avoided environmental costs	10
Technology export	36
Total benefit	100

Source: DOE, *Vision 21*, National Energy Technology Laboratory, U.S. Department of Energy, Washington, D.C., www.netl.doe.gov/coalpower/vision21/main.html (last modified January 5, 2004).

of which about $4 billion will come from the federal government [63]. For perspective, the end-use price of electricity in the United States is greater than $230 billion/year, and fuel cost estimates for U.S. coal-fired power generation range from $25 to $30 billion/year. The capital cost savings reflect savings of $100 and $200/kW for new plants built in 2010 and 2020, respectively. The savings in control technology include savings resulting from increased by-product utilization. Avoided environmental costs considered only SO_2 (at $200/ton SO_2) and NO_x (at $800/ton NO_x) emissions. The estimate for increased exports is based on current estimates for power-generating machinery and equipment (*i.e.*, $30–$35 billion/year) and for increased market penetration for clean coal technology primarily in developing countries. Other benefits include increased jobs from technology export, estimated to be 75,000 new jobs per year in 2010, increasing to 200,000 per year in 2020.

References

1. Abraham, S., Remarks to the Clean Coal and Power Conference (November 17, 2003), www.fossil.energy.gov/news/speeches/03/03_sec_cleancoal_111703.html.
2. DOE, *Coal & Power Systems Strategic and Multi-Year Program Plans* (Office of Fossil Energy, U.S. Department of Energy, Washington, D.C., February 2001).
3. Eastman, M. L., Clean Coal Power Initiative, presented at the Clean Coal Power Conference (November 18, 2003), www.netl.doe.gov/coalpower/ccpi/program_info.html.
4. DOE, *Clean Coal Technology Demonstration Program: Program Update 2001 Including Power Plant Improvement Initiative Projects* (Office of Fossil Energy, U.S. Department of Energy, Washington, D.C., July 2002).
5. DOE, *Power Plant Improvement Initiative (PPII)* (National Energy Technology Laboratory, U.S. Department of Energy, Washington, D.C.), www.netl.doe.gov/coalpower/ccpi/ppii_projects.html (last modified January 5, 2004).
6. DOE, *Clean Coal Technology Demonstrations Program Facts: Clean Coal Power Initiative (CCPI)* (Office of Fossil Energy, U.S. Department of Energy, Washington, D.C., July 2003).
7. DOE, *Clean Coal Technology Roadmap: "CURC/EPRI/DOE Consensus Roadmap"* (National Energy Technology Laboratory, U.S. Department of Energy, Washington, D.C.), www.netl.doe.gov/coalpower/ccpi/main.html (last modified January 6, 2004).
8. DOE, *Clean Coal Technology Compendium CCT Program* (U.S. Department of Energy, Washington, D.C.), www.lanl.gov/projects/cctc/programs/program.html (last modified January 22, 2003).
9. DOE, *Clean Coal Today*, DOE/FE-0215P-54 (Office of Fossil Energy, U.S. Department of Energy, Washington, D.C., Issue No. 54, Summer 2003).

10. Hsu, F. E., *10 MW Demonstration of Gas Suspension Absorption Final Project Performance and Economics Report*, DE-F22-90PC90542, June 1995.
11. Bechtel Corp., *Confined Zone Dispersion Project: Final Technical Report*, DE-FC22-90PC90546, June 1994.
12. LIFAC North America, *LIFAC Demonstration at Richmond Power and Light Whitewater Valley Unit No. 2—Project Performance and Economics: Final Report*, DE-FC22-9090548, April 1988.
13. Pure Air on the Lake, LP, *Advanced Flue Gas Desulfurization (AFGD) Demonstration Project: Final Technical Report*, DE-FC22-90PC89660, April 1996.
14. Southern Company Services, Inc., *Demonstration of Innovative Applications of Technology for Cost Reductions to the CT-121 FGD Process: Final Report*, DE-FC22-90PC89650, January 1997.
15. Southern Company Services, Inc., *Demonstration of Advanced Combustion NO_X Control Techniques for a Wall-Fired Boiler: Project Performance Summary*, DE-FC22-90PC89651, January 2001.
16. Babcock & Wilcox Co., *Demonstration of Coal Reburning for Cyclone Boiler NO_X Control: Final Project Report*, DE-FC22-90PC89659, February 1994.
17. Babcock & Wilcox Co., *Full-Scale Demonstration of Low-NO_X Cell™ Burner Retrofit: Final Report*, DE-FC22-90PC90545, July 1994.
18. Energy and Environmental Research Corporation (EERC), *Evaluation of Gas Reburning and Low-NO_X Burners on a Wall-Fired Boiler: Final Report*, DE-FC91-PC90547, July 1998.
19. CONSOL, Inc., *Milliken Clean Coal Technology Demonstration Project: Final Report*, DE-FC22-93PC92642, October 1999.
20. Southern Company Services, Inc., *Demonstration of Selective Catalytic Reduction (SCR) Technology for the Control of Nitrogen Oxide (NO_X) Emissions from High-Sulfur Coal-Fired Boilers: Final Report*, DE-FC22-90PC89652, October 1996.
21. Energy Technologies Enterprises Corp., *180 MW Demonstration of Advanced Tangentially Fired Combustion Techniques for the Reduction of Nitrogen Oxide (NO_X) Emissions from Coal-Fired Boilers*, DE-FC22-90PC89653, December 1993.
22. Asea Brown Boveri Environmental Systems, *SNOX Demonstration Project, Project Performance and Economics: Final Report*, DE-FC22-90PC89655, July 1996.
23. Goots, T. R., M. J. DePero, and P. S. Nolan, *LIMB Demonstration Project Extension and Coolside Demonstration*, DE-FC22-87PC79798, November 1992.
24. Babcock & Wilcox Co., *SOx-NOx-Rox Box™ Flue Gas Cleanup Demonstration Project Performance Summary*, DE-FC22-90PC89656, June 1999.
25. Energy and Environmental Research Corp., *Enhancing the Use of Coals by Gas Reburning—Sorbent Injection Final Report*, DE-FC22-87PC79796, February 1997.

26. New York State Electric & Gas Corp., *Milliken Clean Coal Technology Demonstration Project, Project Performance and Economics: Final Report*, DE-FC22-93PC92642, April 1999.
27. Hunt, T. and T. J. Hanley, *Integrated Dry NO_x/SO_2Emissions Control System: Final Report*, DE-FC22-91PC90550, November 1997.
28. Raskin, N., Foster Wheeler Power Group, Inc., personal communication, January 8, 2004.
29. DOE, *Clean Coal Technology: The JEA Large-Scale CFB Combustion Demonstration Project*, Technical Report No. 22 (Office of Fossil Energy, U.S. Department of Energy, Washington, D.C., March 2003).
30. Ohio Power Company, *Tidd PFBC Demonstration Project: Final Report*, DE-FC21-87MC24132, August 1995.
31. Colorado-Ute Electric Association, Inc., *NUCLA Circulating Atmospheric Fluidized-Bed Demonstration Project: Final Report*, DE-FC21-89MC25137, October 1991.
32. DOE, *Clean Coal Technology: Tampa Electric Integrated Gasification Combined-Cycle Project—An Update* (Office of Fossil Energy, U.S. Department of Energy, Washington, D.C., July 2000).
33. DOE, *Clean Coal Technology: The New Coal Era* (Office of Fossil Energy, U.S. Department of Energy, Washington, D.C., June 1990).
34. DOD, *Fuel Cell Information Guide: Molten Carbonate Fuel Cells* (Engineer Research and Development Center, U.S. Corps of Engineers, U.S. Department of Defense, Washington, D.C.), www.dodfuelcell.com/molten.html (last updated January 6, 2004).
35. DOE, *Hydrogen, Fuel Cells, and Infrastructure Technologies Program* (Energy Efficiency and Renewable Energy, U.S. Department of Energy, Washington, D.C.), www.eere.energy.gov/hydrogenandfuelcells/fuelcells/types.html#mcfc (last updated January 27, 2003).
36. Hornick, M. J. and J. E. McDaniel, *Tampa Electric Polk Power Station Integrated Gasification Combined Cycle Project*, DE-FC21-91MC27363, August 2002.
37. Cargill, P., G. DeJonghe, T. Howsley, B. Lawson, L. Leighton, and M. Woodward, *Piñon Pine IGCC Project: Final Technical Report*, DE-FC21-92MC29309, January 2001.
38. DOE, *Clean Coal Technology: The Piñon Pine Power Project*, Technical Report No. 6. (Office of Fossil Energy, U.S. Department of Energy, Washington, D.C., December 1996).
39. Wabash River Energy, Ltd., *Wabash River Coal Gasification Repowering Project: Final Technical Report*, DE-FC21-91MC29310, August 2000.
40. DOE, *Clean Coal Technology: The Wabash River Coal Gasification Repowering Project—An Update* (Office of Fossil Energy, U.S. Department of Energy, Washington, D.C., September 2000).
41. Alaska Industrial Development and Export Authority, *Healy Clean Coal Project: Project Performance and Economics Final Report*, DE-FC22-91PC90544, April 2001.

42. ENCOAL, *ENCOAL Mild Gasification Project,* DE-FC21-90MC27339, September 1997.
43. DOE, *Clean Coal Technology Compendium CCT Program* (U.S. Department of Energy, Washington, D.C.), www.lanl.gov/projects/cctc/factsheets/rsbud/adcconvdemo.html (last modified December 2, 2002).
44. Heydorn, E. C., B. W. Diamond, and R. D. Lilly, *Commercial-Scale Demonstration of the Liquid Phase Methanol (LPMEOH^{TM}) Process: Final Report,* DE-FC22-92PC90543, June 2003.
45. DOE, *Clean Coal Today,* DOE/FE-0215P-53 (Office of Fossil Energy, U.S. Department of Energy, Washington, D.C., Issue No. 53, Spring 2003).
46. DOE, *Clean Coal Technology: Commercial-Scale Demonstration of the Liquid Phase Methanol (LPMEOH^{TM}) Process,* Technical Report No. 11. (Office of Fossil Energy, U.S. Department of Energy, Washington, D.C., April 1999).
47. CQ, Inc., *Development of a Coal Quality Expert,* DE-FC22-90PC89663, June 20, 1998.
48. Bethlehem Steel Corp., *Blast Furnace Granular Coal Injection System,* DE-FC21-MC27362, October 1999.
49. NMA, *Clean Coal Technology: Current Progress, Future Promise* (National Mining Association, Washington, D.C.), www.nma.org (last updated March 2003).
50. DOE, *Institutional Plan FY2003–2007* (Office of Fossil Energy, U.S. Department of Energy, National Energy Technology Laboratory, Washington, D.C., October 2002).
51. Energetics, Inc., *Clean Coal Technology Commercial Successes,* draft report for the U.S. DOE (Office of Fossil Energy, U.S. Department of Energy, Washington, D.C.), January 2000.
52. DOE, *Clean Coal Technology: Environmental Benefits of Clean Coal Technology,* Technical Report No. 18. (Office of Fossil Energy, U.S. Department of Energy, Washington, D.C., April 2001).
53. DOE, *Integration of Low-NO_x Burners with an Optimization Plan for Boiler Combustion,* Power Plant Improvement Initiative Project Fact Sheet (Los Alamos National Laboratory, U.S. Department of Energy, Washington, D.C.), www.lanl.gov/projects/ppii/factsheets/sunflower/sunflower_demo.html (last modified May 1, 2003).
54. DOE, *Demonstration of a Full-Scale Retrofit of the Advanced Hybrid Particulate Collector Technology,* Power Plant Improvement Initiative Project Fact Sheet (Los Alamos National Laboratory, U.S. Department of Energy, Washington, D.C.), www.lanl.gov/projects/ppii/factsheets/otter/otter_demo.html (last modified February 20, 2003).
55. DOE, *Big Bend Power Station: Neural Network-Sootblower Optimization,* Power Plant Improvement Initiative Project Fact Sheet (Los Alamos National Laboratory, U.S. Department of Energy, Washington, D.C.), www.lanl.gov/projects/ppii/factsheets/neural/neural_demo.html (last modified October 31, 2002).
56. DOE, *Commercial Demonstration of the Manufactured Aggregate Processing Technology Utilizing Spray Dry Ash,* Power Plant Improvement

Initiative Project Fact Sheet (Los Alamos National Laboratory, U.S. Department of Energy, Washington, D.C.), www.lanl.gov/projects/ppii/factsheets/aggregate/aggregate_demo.html (last modified February 27, 2003).

57. DOE, *Greenidge Multi-Pollutant Control Project*, Power Plant Improvement Initiative Project Fact Sheet (Los Alamos National Laboratory, U.S. Department of Energy, Washington, D.C.), www.lanl.gov/projects/ppii/factsheets/greenidge/greenidge_demo.html (last modified October 23, 2002).
58. DOE, *Development of Hybrid FLGR/SNCR/SCR Advanced NO_X Control Technology*, Power Plant Improvement Initiative Project Fact Sheet (Los Alamos National Laboratory, U.S. Department of Energy, Washington, D.C.), www.lanl.gov/projects/ppii/factsheets/advnox/advnox_demo.html (last modified October 23, 2002).
59. NEPD Group, *National Energy Policy* (National Energy Policy Development, U.S. Government Printing Office, Washington, D.C., May 2001).
60. Sarkus, T. A., *Clean Coal Power Initiative: Tackling Coal's Environmental Challenges*, presentation to the National Coal Council (National Energy Technology Laboratory, U.S. Department of Energy, Washington, D.C.), www.netl.doe.gov/coalpower/ccpi/main.html (accessed December 4, 2003).
61. DOE, *Clean Coal Technology Roadmap: "CURC/EPRI/DOE Consensus Roadmap" Background Information* (National Energy Technology Laboratory, U.S. Department of Energy, Washington, D.C.), www.netl.doe.gov/coalpower/ccpi/main.html (accessed January 6, 2004).
62. DOE, *Techline: Secretary of Energy Announces First Projects to Meet President's Commitment to New Clean Coal Technologies* (National Energy Technology Laboratory, U.S. Department of Energy, Washington, D.C.), www.netl.doe.gov/publications/press/2003/tl_ccpi_2003sel.html (accessed January 15, 2003).
63. DOE, *Vision 21* (National Energy Technology Laboratory, U.S. Department of Energy, Washington, D.C.), www.netl.doe.gov/coalpower/vision21/main.html (last modified January 5, 2004).
64. DOE, *Clean Coal Technology Demonstration Program: Program Update 2003—Includes Clean Coal Technology Demonstration Program, Power Plant Improvement Initiative, and Clean Coal Power Initiative Projects* (Office of Fossil Energy, U.S. Department of Energy, Washington, D.C., December 2003).
65. DOE, *Vision 21: Clean Energy for the 21st Century* (Office of Fossil Energy, U.S. Department of Energy, Washington, D.C., November 1998).
66. DOE, *Vision 21 Program Plan: Clean Energy for the 21st Century* (Office of Fossil Energy, U.S. Department of Energy, Washington, D.C., April 1999).
67. The White House, Statement by the President, www.whitehouse.gov/news/releases/2003/02/20030227-11.html, February 27, 2003.
68. DOE, *A Prospectus for Participation by Foreign Governments in FutureGen* (Office of Fossil Energy, U.S. Department of Energy, Washington, D.C., June 20, 2003).

69. DOE, *FutureGen: A Sequestration and Hydrogen Research Initiative Fact Sheet* (Office of Fossil Energy, U.S. Department of Energy, Washington, D.C., February 2003).
70. Sarkus, T. A. and S. M. Smouse, *Implementing the U.S. Department of Energy's Power Plant Improvement and Clean Coal Power Initiatives: Coal-Tech 2002: Mine Mouth Power Plant* (Los Alamos National Laboratory, U.S. Department of Energy, Washington, D.C., August 21, 2002), www.lanl.gov/projects/ppii/news/whatsnew.html.

CHAPTER 8

Coal's Role in Providing United States Energy Security

America's economic engine is fueled primarily by fossil fuels, a trend that is expected to continue for several decades. Coal, oil, and natural gas supplied 85% of the nation's total energy, 69% of its electricity, and nearly all of its transportation fuels in 2002 [1]. The contribution of fossil fuels to the U.S. energy supply is expected to increase; the U.S. Department of Energy (DOE) forecasts that fossil fuels will be supplying 90% of the nation's energy by 2020 because of projected growth in natural gas consumption. Of the ~97 quadrillion Btu of energy consumed in the United States in 2002, 23% (22.18 quadrillion Btu) was attributed to coal, while natural gas, petroleum, nuclear power, and renewable energy consumption was 24, 39, 8, and 6%, respectively [1]. The important role of fossil fuels, in general, and coal, in particular, in the U.S. energy picture is further illustrated in Figure 8-1, which depicts the contribution of coal to the various end-use sectors—transportation, residential, commercial, and industrial—along with the energy consumed to generate electricity. As discussed in Chapter 2 (Present, Past, and Future Role of Coal), coal is primarily used in the generation of electric power, and the contributions from coal as well as the various other energy sources for producing electricity are factored into the individual end-use sector graphs shown in Figure 8-1.

America's economic strength was established largely due to abundant and inexpensive energy. U.S. consumers benefit from some of the lowest electricity rates of any free-market economy, and coal has been a major contributor to the surge in electrification of the country; however, increasing electricity rates and natural gas prices are affecting most Americans. The volatility in natural gas prices and availability is becoming more troublesome as evidenced by the frequent episodes of low availability and high price. Rising crude oil prices are reflected in increased gasoline and heating oil prices, which create an economic strain on consumers. Conflicts in the

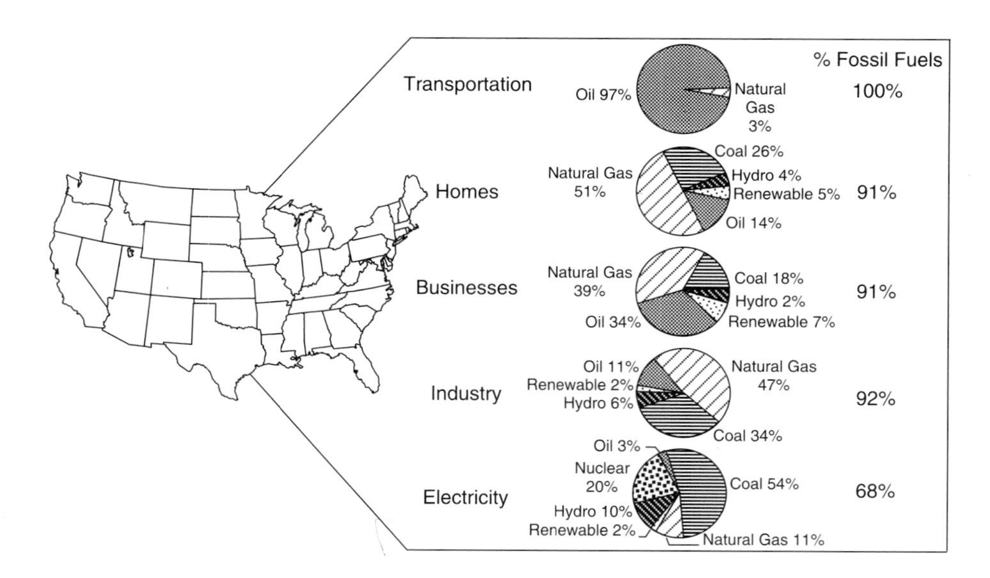

FIGURE 8-1. Distribution of energy consumption in the United States in 2000. (From DOE, *National Energy Technology Laboratory Accomplishments by 2000*, Office of Fossil Energy, Washington, D.C., September 2001.)

Middle East affect the world oil markets and prices and result in economic and emotional strains on U.S. citizens. Threats to the energy security of the United States include concerns of terrorist activities targeting nuclear power facilities and the oil and gas infrastructure, an ever-increasing dependency on oil and natural gas imports, an aging infrastructure (*e.g.*, power transmission lines) or an insufficient infrastructure (*e.g.*, natural gas pipelines), and low reserve capacity. Many of these issues can be addressed through the use of coal. This chapter discusses the contribution of coal to providing U.S. energy security. Specifically, it discusses the impact of volatile energy prices and fuel availability on the economy, the stability of coal prices and their economic impact, the use of natural gas rather than coal in power generation, the potential of coal to reduce the U.S. dependency on imported oil, the development of a coal-based hydrogen economy, and even the role of coal in providing security to the nation's food supply. The role of coal in international energy security and sustainable development is also discussed.

Overview of U.S. Energy Security Issues

Energy security is a complex issue and, in the case of the United States, is ensured when the nation can deliver energy economically, reliably, in an environmentally sound way, and safely in quantities sufficient to support the growing economy and defense needs [3]. This will require policies that support expansion of the energy supply and delivery infrastructure (with sufficient storage and generating reserves), diversity of fuels, and redundancy of infrastructure to meet the demands of economic growth. The United States will need increased contributions toward energy security from all available sources over the long term, including conservation, traditional sources of energy, renewable resources, and new energy sources, such as hydrogen. In order to enhance energy security, the United States must [3]:

- Encourage conservation and energy efficiency;
- Maintain diverse energy supplies while enhancing domestic production and delivery;
- Maximize economic efficiency;
- Accelerate research and development to create and deploy advanced energy technologies;
- Develop and implement effective contingency and emergency plans;
- Develop policies based on sound science and realistic economic, national security, and environmental needs in order to make decisions that are timely, consistent, and coordinated with energy security, economic, and environmental objectives.

U.S. energy security is closely linked to global energy markets and trends, North American energy resources, and energy production, transportation, and storage systems, as well as changing patterns of consumption [3]. One energy source of particular concern is petroleum; in 2002, more than 60% of the oil consumed in the United States was imported [1]. World oil demand is projected to increase by nearly 44 million barrels per day by 2020; ~60% of the increased demand will be in developing countries, and 60% of the total demand will be in the transportation sector. Approximately 63% of the oil will be supplied by the Organization of Petroleum Exporting Countries (OPEC), and 44% will come from the Persian Gulf region [3,4]. These trends have enormous implications for U.S. energy security, as global competition and potential disruptions to the United States will impact availability and cost to consumers.

Energy patterns in the United States are projected to shift dramatically by 2020 and also have energy security implications [3,5]. Total U.S. energy consumption is projected to increase from 97 quadrillion Btu in 2001 to ~127 quadrillion Btu in 2020. Petroleum is expected to remain the dominant fuel in U.S. markets, maintaining about a 40% market share, with domestic production remaining constant and the United States relying heavily on imported oil. The transportation sector, in which 96% of the energy is supplied by oil, is expected to grow more rapidly than any other sector, increasing from about 13 million barrels per day (bbl/day) in 2001 to ~19 million bbl/day in 2020. Natural gas consumption is projected to increase from 23 trillion cubic feet in 2001 to 34 trillion cubic feet in 2020, primarily as a result of rapid growth in demand for electricity generation. Domestic natural gas production and imports from Canada are projected to increase from 19 and 3.5 trillion cubic feet, respectively, in 2001 to 28.5 and 5.5 trillion cubic feet, respectively, in 2020. The market share of renewable energy is projected to remain constant, increasing from 5 quadrillion Btu in 2001 to 8.9 quadrillion Btu in 2020. Coal consumption is projected to increase from ~1 billion short tons in 2002 to ~1.4 billion short tons in 2020, with about 90% of the total coal demand being used for electricity generation. Total U.S. electricity demand is projected to increase at an average annual rate of 1.8% through 2020.

The energy security implications of imported oil and the use of natural gas in the power generation industry are discussed later in this chapter. The availability and price of natural gas, especially if it is used in large quantities in the power industry, and the use of natural gas as a feedstock in the chemical/fertilizer industry, for residential heating, and in the transportation sector are discussed here, as is the application of reforming natural gas to produce hydrogen.

The U.S. energy infrastructure is the keystone of the American way of life. It is the foundation for all other critical infrastructures, including information and telecommunications, postal service and shipping, public health, and agriculture [6]. Ensuring the safety, security, and reliability of the energy infrastructure is vital to homeland security and economic prosperity.

National Energy Plan and Coal Utilization

The U.S. government, aware of its growing demand for energy, is trying to make rational decisions to avoid any energy shortfalls. These shortfalls result in high prices at the gasoline pumps, high home heating bills due to high natural gas and fuel oil prices, rolling blackouts during periods of cold weather when natural-gas-fired power plants cannot obtain sufficient quantities of gas, and blackouts due to an aging electricity distribution infrastructure, to name a few. In the next 20 years, U.S. overall energy consumption is expected to increase by more than 30%; oil demand, 33%; natural gas consumption, 62%; and electricity demand, 45%, due at least in part to the growth in information technology [7].

President Bush's National Energy Policy makes 105 specific recommendations, including 42 on promoting conservation and protecting the environment, 35 on diversifying energy supply and modernizing antiquated infrastructure, and 25 on enhancing national energy security [8,9]. Energy security affects all nations, and President Bush has pledged that the United States will be a leader in the long-term effort to achieve that goal [10]. The policies needed for energy security differ among nations but the basics are the same and are identified as core themes in the U.S. National Energy Policy [7,10]:

- Energy conservation must be increased;
- A balanced and diversified portfolio of energy resources is required;
- Protection of the environment is necessary as growth in domestic energy production and consumption is increased;
- Technology innovations are essential to achieve energy security goals;
- Global alliances and markets must be strengthened.

Energy Conservation/Efficiency

Good energy policy begins with the efficient use of energy. If the energy intensity (*i.e.*, amount of energy required to generate a dollar of gross domestic product, or GDP) of the U.S. economy remains constant, the energy demand in 2020 will increase to 175 quadrillion Btu [7]. The energy plan and current policies are projected to improve energy efficiency, and the demand can be lowered to ~127 quadrillion Btu in 2020. This will require employing cutting-edge technology. The national energy plan focuses on developing fuel-efficient vehicles, higher efficiency appliance standards, and combined heat and power technologies, and on establishing efficiency-base tax credits. Energy efficiency is an important part of the national energy plan but cannot by itself close the gap between projected energy demand and projected domestic energy production.

Diversity of Fuel Sources

The national energy plan calls for a balanced and diversified portfolio of energy resources, which includes coal, oil, natural gas, nuclear power, and renewable and alternative energy sources such as biomass, solar, wind, geothermal, and hydropower. The national energy plan recognizes that it will likely take decades for renewable and alternative resources to make major contributions to our energy mix. At the same time, electricity demand in the United States is forecast to rise 45% by 2020 [7], which translates into adding the equivalent of 1300 to 1900 new power plants (*i.e.*, 60–90 new plants per year) to the existing 5000 power plants in service. If current policies and practices remain unchanged, most of those plants, possibly more than 90%, will be fired by natural gas [7]. This is especially disturbing because the current oil and gas delivery system is experiencing increased stress, and bottlenecks are developing. It is the consensus of the natural gas industry that the United States will need an additional 38,000 miles of major transmission pipelines and 263,000 miles of smaller distribution lines by 2015 to bring the necessary natural gas to homes and businesses [7]. Energy security dictates a more balanced approach to new power generation; consequently, one measure of the National Energy Plan established the 10-year, $2 billion

Clean Coal Power Initiative Program (discussed in Chapter 7, Future Power Generation) to develop improved clean coal power technologies.

Environmental Protection

The National Energy Policy recognizes the need to conserve and improve the quality of the environment by reducing emissions from energy production and consumption [10]. Environmental progress can be made without compromising the expansion of the U.S. economy, especially through development of better technology and establishment of greater regulatory certainty relating to coal electricity generation by implementing clear policies, easily applied to business decisions [7,8]. Specific proposals for coal call for mandatory reduction for emissions of sulfur dioxide, nitrogen oxides, and mercury from electricity generation. In addition, the National Energy Policy recommends streamlining the permitting process for power plant siting and allowing utilities to modify plants without fear of new litigation [8].

Technological Innovations

It is understood that scientific breakthroughs and technological innovations are essential to achieving the energy security goals of the United States, improving the use of energy resources, and developing future energy systems [10]. Clean coal is a crucial element in the overall energy policy, and the Clean Coal Power Initiative identified in the energy plan will assist in developing future energy systems such as Vision 21 energy plants.

Global Alliances and Markets

The national energy plan recognizes the need to strengthen global alliances and markets. The United States is committed to international cooperation to strengthen energy trade relationships, accelerate scientific/technological progress, and spread the benefits of energy and environmental advances throughout the world [10]. Developments in clean coal technology are especially important for marketing to the world.

The Role of Coal in the National Energy Plan

The Clean Coal Power Initiative is a major commitment by the United States to further develop clean coal power technologies. This is crucial, as the electric industry is expected to increase significantly by 2020 and natural gas is the fuel currently identified to meet much of that growth if circumstances do not change. In addition, environmental constraints are tightening with respect to the use of current coal-based plants and technology [7]. The National Energy Policy recognizes that there are several positive issues regarding the use of coal. Coal is an abundant and inexpensive resource.

Coal is the most abundant fuel source in the United States; the recoverable reserves of ~250 million short tons represent a 200- to 300-year supply at current rates of consumption. The abundance of coal helps to keep coal prices low and stable. The average delivered cost of coal in 2003 to U.S. power plants was about $1.25/million Btu compared to $5.80/million Btu for natural gas, although natural gas prices have a history of volatile price swings [11]. In 2002, the average cost of coal at the burner tip was about $13/MWh, while the average cost of natural gas was $37/MWh [11]. Coal-based industries employ many more workers than any other energy source [8]. Coal is an indigenous resource that is not subject to unreliable weather conditions, price volatility, or disruptions from foreign suppliers. Reliance on a single source such as natural gas can lead to constraints on supply or significant price increases. Technology developed by the partnership of industry and the government has lessened the environmental impact of coal technologies [7]. Coal can be burned cleanly using clean coal technology, and the more rapidly clean coal technology can be retrofitted onto older power plants the cleaner the air will become [8]. Also, future systems under development will improve system efficiencies and reduce emissions even further.

Energy and the Economy

Energy prices have a significant impact on the U.S. economy, as evidenced by the oil embargoes and the recent rising energy prices (*i.e.*, for natural gas and oil). Prior to the oil embargo of 1973/1974, total energy expenditures comprised 8% of the U.S. GDP; the share of petroleum expenditures was slightly less than 5%, and natural gas expenditures accounted for 1% [12]. The price shocks of the 1970s and early 1980s resulted in these rising dramatically to 14, 8, and 2%, respectively, by 1981. For the next two decades, the shares have decreased consistently to approximately pre-embargo levels; however, these shares are now beginning to increase again starting in the late 1990s due to higher natural gas and petroleum prices.

High energy prices result in increased inflation. Viewed from a long-term perspective, inflation, as measured by the rate of change in the consumer price index (CPI), tracks movements in the world oil price [12]. Oil and other energy prices constitute a portion of the actual CPI but also impact other (downstream) commodity prices that will have a lagged effect on the CPI inflation. Since the 1970s, observable and dramatic changes in GDP growth have occurred as the world oil price has undergone dramatic change [12]. The price shocks of 1973/1974, the late 1970s/early 1980s, and the early 1990s were all followed by recessions, and higher energy prices in 2003 contributed to the downturn in the U.S. economy.

The strength and security of the U.S. economy are closely linked to the availability, reliability, and cost of electric power. Since 1970, real GDP in the United States and electricity generation have been clearly linked, as

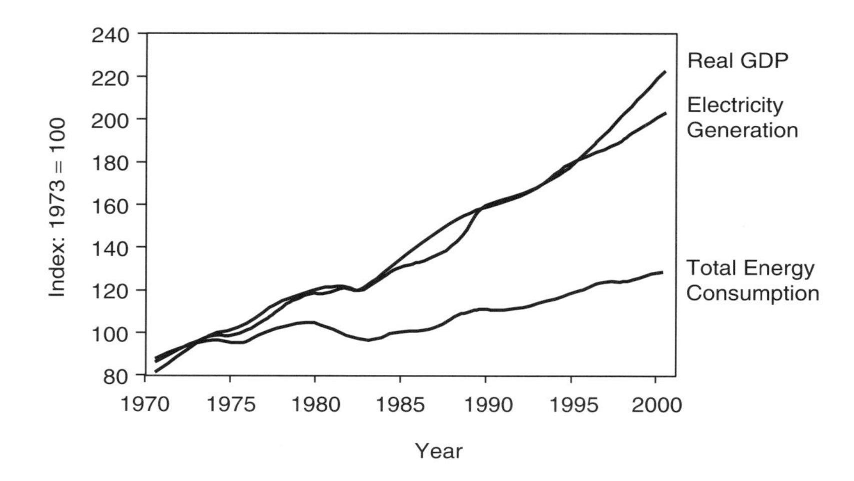

FIGURE 8-2. Economic growth and energy security are linked to electricity consumption. (From Eastman, M. L., *Clean Coal Power Initiative*, presented at the Clean Coal Power Conference, www.netl.doe/gov/coalpower/ccpi/program_info.html, November 18, 2003.)

illustrated in Figure 8-2 [13]. Because economic growth is linked to reliable and affordable electric power, continued use of domestic coal resources will play a significant role in satisfying the energy needs of the United States. This is likely to continue through the middle of the twenty-first century and beyond [13].

The lowest cost electrical generation plants are coal fired. In 2001, 20 of the 25 lowest cost steam-generating plants in the United States were fueled by coal [14]. Figure 8-3 shows that most states with low-cost electricity receive a large amount of their generation from coal, with the noted exception of the three Pacific Northwest states that have ample hydroelectric resources [15]. Conversely, states with the highest cost of electricity generate less than 30% of their electricity from coal. The five states with the lowest electricity costs all generate more than 94% of their electricity from coal (*i.e.*, Indiana, Kentucky, West Virginia, Wyoming, and Utah).

The trend of lower cost electricity from coal generation will continue to be more evident over the next 5 years [15]. For example, the three states with the lowest energy costs in 2000 were Washington, Oregon, and Idaho because of their hydroelectric power generation capabilities [15]; however, according to Williams [15], they have fully utilized their hydroelectric capabilities and now rely on the regional spot market, which is driven by natural gas for incremental power, and none of these three states is among the 10 lowest states for electricity prices in 2002. The abundance of coal results in low coal prices and keeps coal prices stable, both in long-term contracts and the spot market. States that rely on coal for most of their generation are insulated from wholesale power price spikes that have followed the volatility of the

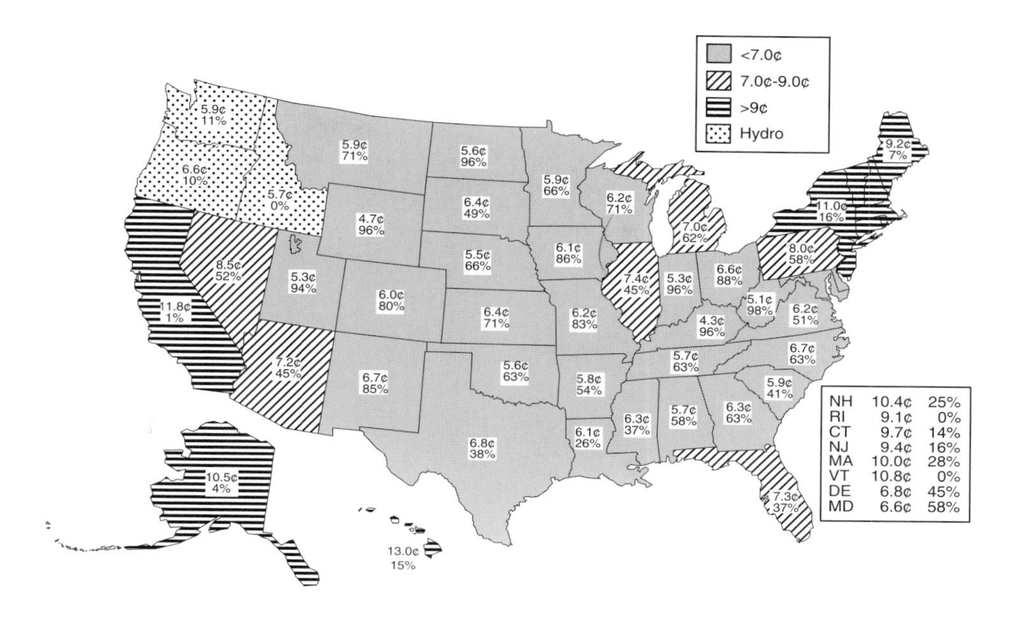

FIGURE 8-3. U.S. retail electricity costs for January to November 2002. Values shown for each state are the average retail price (¢/kWh) and the percent of total generation from coal for 2001. (From Williams, J., *Power Engineering*, Vol. 107, No. 6, June 2003, pp. 31–36. With permission.)

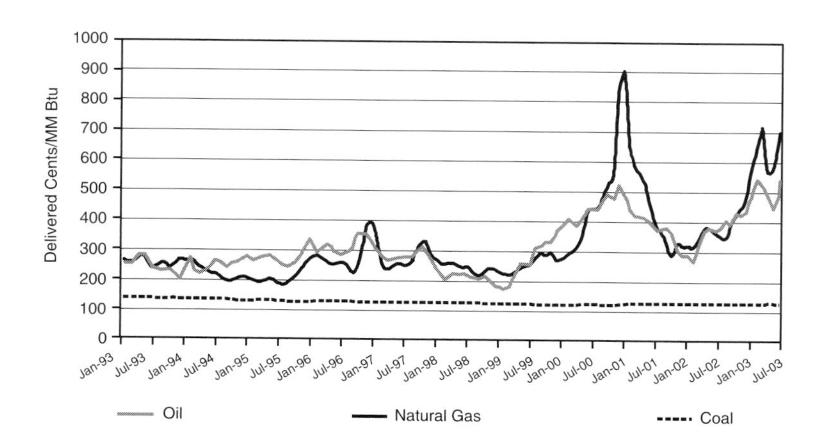

FIGURE 8-4. Delivered fuel prices for the period January 1993 through May 2003. (From Roberts, A., *Coal Age*, Vol. 108, No. 10, Nov./Dec. 2003, pp. 37–38. With permission.)

natural gas market. Figure 8-4 shows delivered fuel prices for coal, natural gas, and fuel oil for the period from January 1993 through May 2003 [11].

Natural Gas Use in Power Generation

Natural gas is a premium fossil fuel that is easily transported, can be used in many applications, often is the least expensive option from a capital investment viewpoint, and burns with low levels of emissions. Of the fossil fuels, natural gas releases the lowest quantity of carbon dioxide per million Btu of energy consumed. For these reasons, natural gas has become a popular choice among the residential, commercial, industrial, and electricity generation sectors and is also becoming increasingly popular in compressed natural gas vehicles.

With this popularity have come serious supply and demand issues. In 2002, approximately 24 quadrillion Btu of natural gas were consumed in the United States [1]. Of this, approximately 36, 25, 23, and 16% were consumed by the industrial, electric power, residential, and commercial sectors, respectively. Natural gas consumption is projected to increase to ~35 quadrillion Btu, with electric power generation being responsible for much of the increase. This is a concern, because supply is currently not keeping pace with demand as natural gas imports are increasing, prices are volatile, and shortages are occurring during periods of extreme weather.

Natural gas prices are increasing as demand exceeds supply, with electric power generation causing much of the current market strain. Over the past 4 years, new gas-fired plants generating about 200,000 MW have been built, significantly increasing demand for natural gas and affecting natural gas prices [16]. The supply of natural gas is limited, not only by production but also by distribution. The infrastructure to deliver natural gas is not expected to keep pace with demand [17]. Economic concerns, regulatory hurdles, and capital budget constraints are inhibiting proper development of the delivery infrastructure.

The natural gas crisis impacts all sectors that utilize gas; however, the crisis becomes especially newsworthy when residential and industrial users are severely impacted. In the recent past (*i.e.*, beginning with the winter of 2000/2001), natural gas price spikes and their impact on the residential and industrial sectors have received much attention, including interest from politicians. Natural gas prices have been rising for the past 5 years, beginning in January 1999, with spikes being observed during three of the last four winters (*i.e.*, winters of 2000/2001, 2002/2003, and 2003/2004). Winter heating bills tracked by the Energy Information Administration (EIA) showed that the average price of natural gas for residential consumers was \$9.53, \$7.38, and \$8.39 per 1000 cubic feet for the winters of 2000/2001, 2001/2002, and 2002/2003, respectively [18]. The EIA estimated the average natural gas price during the winter of 2003/2004 to be \$9.57 per 1000 cubic feet [18]. The cold weather that hit the northeast United States in January 2004 resulted in warnings to the public to conserve energy and to prepare for rolling blackouts because natural-gas-fired power plants were shutting down due to insufficient gas supply.

Industrial consumers, including chemical, fertilizer, and other process industries that use natural gas as a feedstock, are complaining that the high prices are making their products noncompetitive in international markets, thereby aggravating the economic downturn the United States was experiencing in 2003 [16]. This prompted U.S. DOE Secretary Spencer Abraham to write a letter to 30 senators in June 2003 calling for, among other steps, electric utilities to switch from natural gas to coal and other energy sources. Many industrial users, such as the chemical industry, are supporting any energy aid program or initiative, such as switching power plants away from natural gas, that takes away the non-industrial demand for natural gas—be it coal, nuclear, renewables, or conservation [19]. There have been discussions regarding building natural gas supplies, but these have a horizon of at least 10 years, which is too long for the chemical industry.

The natural gas crisis is being closely scrutinized by the U.S. Congress. House Speaker Dennis Hastert (Republican, Illinois) chaired a Task Force for Affordable Natural Gas and planned a series of fact-finding meetings in August and September 2003. The Task Force was charged with identifying the causes of the natural gas shortage, the impact of natural gas prices on the American economy, and short- and long-term plans to encourage a stable supply of natural gas to ease prices. U.S. Congressman John Peterson (Republican, Pennsylvania), who was one of 18 members appointed to the Task Force, held the first public field meeting at Penn State's University Park campus in August 2003 which brought in numerous witnesses from across Pennsylvania. A recurring theme throughout the testimony was the ever-increasing natural gas prices being experienced by homeowners, businesses, and others. Testimony ranged from homeowners on a fixed budget who could not afford their home heating bills to businesses such as a local hospital that upgraded an old and inefficient heating/cooling system to a natural gas system only to find that their natural gas bills nearly doubled from 1999 to 2003 [20]. A sobering testimony was provided by representatives of Columbia Gas of Pennsylvania, who stated that natural gas demand is growing faster than natural gas production, so consumers should expect higher prices as well as greater price volatility [21]. Solutions to the problem included opening areas closed to exploration and production, pipeline construction, increasing storage capacity (*i.e.*, storing natural gas underground during the summer for use in the winter months), increasing conservation efforts, and providing assistance to lower income households. Congress is currently deciding on a course of action to address the natural gas crisis.

A balanced approach to new power generation is needed. Natural gas is a premium fuel, and its use in base-loaded power generation needs to be questioned. Coal, with its abundance and stable prices, should be a fuel of choice for new power generation plants. Coal technologies are becoming more environmentally sound and, as discussed in Chapter 7, will be making even greater strides in efficiency and pollutant reduction in the future.

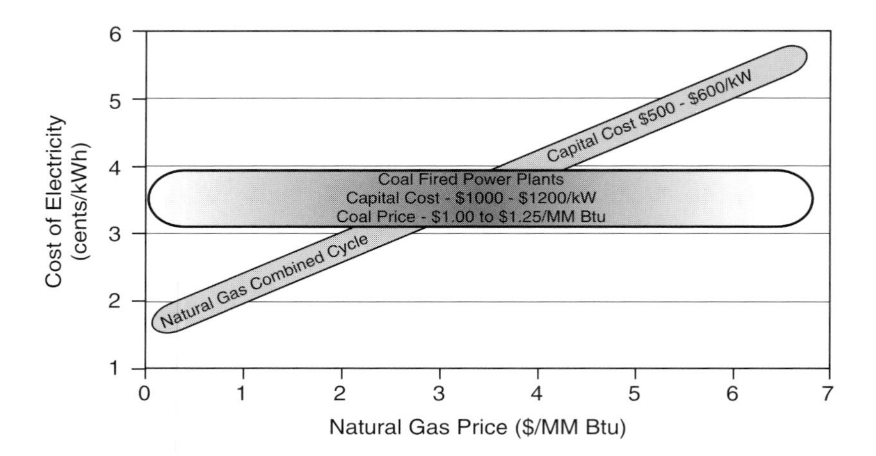

FIGURE 8-5. Comparison of coal technologies with natural gas combined cycle power plant for various natural gas prices. (From DOE, *Coal Technologies Are Cost Competitive*, Federal Energy Technology Center, U.S. Department of Energy, Washington, D.C., www.fetc.doe.gov/coalpower/powersystems/images/costcomp.jpg, accessed December 13, 2001.)

Coal technologies are cost competitive with natural gas (see Figure 8-5) and need to be seriously considered in lieu of natural-gas-fired power plants. A sound energy policy with the goal of energy security mandates this.

The Potential of Coal to Reduce U.S. Dependency on Imported Crude Oil

Inexpensive crude oil contributes to the U.S. economic prosperity; however, the increasing reliance upon imported crude oil makes the United States vulnerable to oil supply disruptions and threatens the nation's economic and energy security. As evidenced by the 1973 Arab oil embargo and the 1979 Iranian revolution, abrupt and prolonged losses of crude oil from the Persian Gulf region drastically affect the U.S. economy, increase unemployment, and boost inflation [23]. Even shorter periods of spiking prices, such as that experienced during the 1990 Gulf War, resulted in a downturn in the U.S. economy. In 1979, President Carter called this situation "a clear and present danger to our national security" [23]. Since the 1970s, there has been much discussion about energy security and national security by the U.S. Congress and White House; however, the United States is more dependent upon foreign oil today than it was in the 1970s. In 1973, the United States imported about 34% of its crude oil [24]. Currently, the United States imports approximately 60% of the oil it uses [1], and it is estimated that this could increase to 75% by 2010 [24]. The current U.S. dependency on imported crude oil is illustrated in Figure 8-6, which shows the petroleum flow in the United States in 2002 [1].

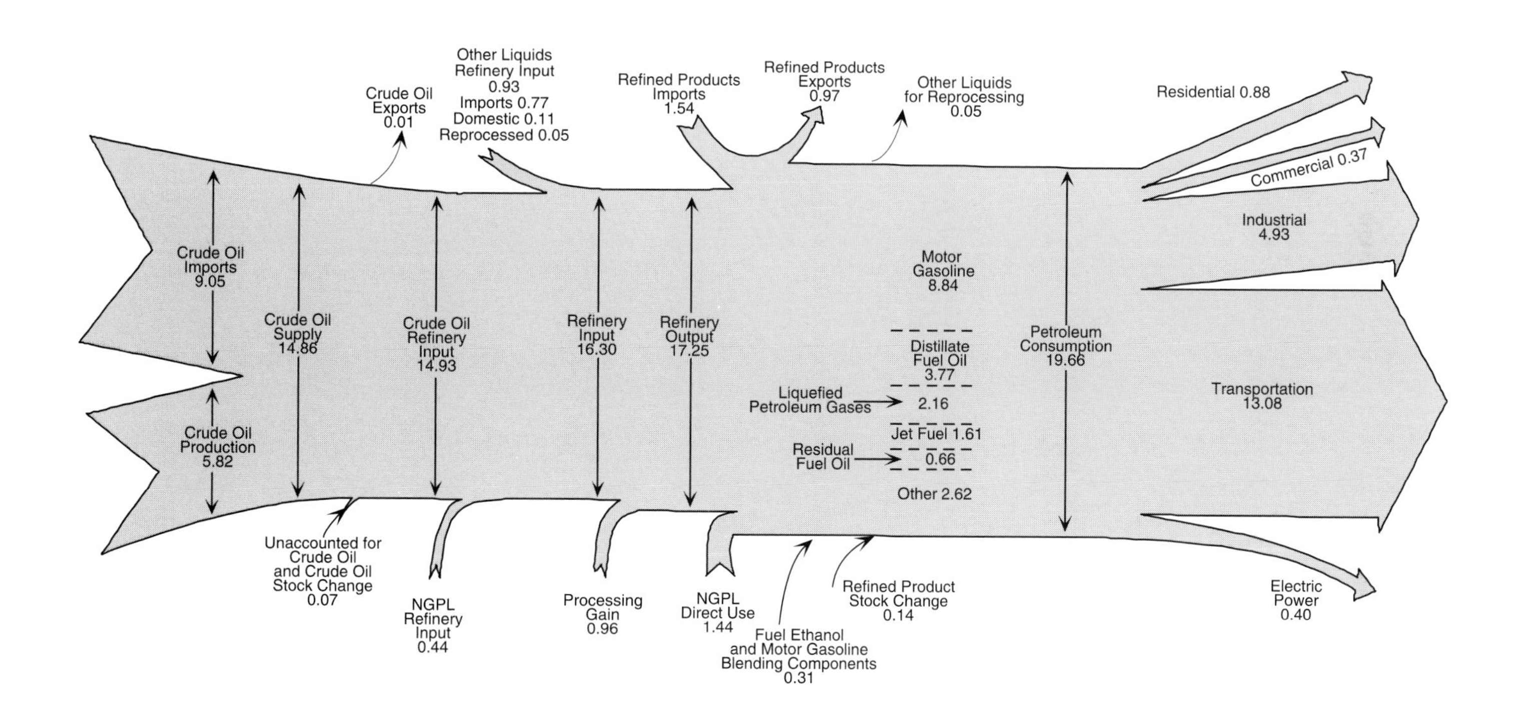

FIGURE 8-6. Petroleum flow in the United States in 2002 in million barrels per day. (From EIA, *Annual Energy Review 2002*, Energy Information Administration, U.S. Department of Energy, Washington, D.C., October 2003.)

The solution to achieving energy security in the United States does not involve isolationism or switching to importing crude oil only from non–Persian Gulf countries, even if either of these options could be realized. Taylor [25] points out that being energy independent, with respect to oil consumption, does not insulate oneself from a global crisis. The oil price spike of 1979 affected Great Britain, where all of the oil Great Britain consumed came from the North Sea, as much as it affected Japan, a country that imported all of its oil. The reason for this is the market for crude oil is global, not regional.

The United States is increasingly dependent on foreign crude oil and is very vulnerable to an oil supply disruption. Currently, four major producers provide over one-third of the U.S. oil supply: Canada, Mexico, Venezuela, and the Persian Gulf region (Bahrain, Iran, Iraq, Kuwait, Qatar, Saudi Arabia, and the United Arab Emirates) [26]; therefore, reducing the quantity of imported crude oil consumed in the United States is of utmost importance and is a crucial step to attain energy and economic security.

While the price of crude oil may be influenced by global events, the goal of reducing the total amount of imported crude oil should be pursued. Approximately two-thirds of the petroleum consumed in the United States is in the transportation sector, with another 25% being consumed in the industrial sector. Reducing the reliance of the transportation sector on oil is clearly a key to improving energy security. Options to lessen the dependency on imported crude oil include improving vehicle fuel efficiency and diversifying the feedstocks to produce transportation fuels, transportation fuel additives, and liquid fuels/feedstocks to the industrial sector. Feedstock diversification can be achieved by using biomass and coal. Utilizing biomass for producing biofuels and additives is important and should be pursued; however, coal can provide the greatest and quickest impact in reducing dependence on imported crude oil due to the vast coal resources and proven technological capability to produce liquid fuels from coal. Technologies exist to convert coal to liquid products, as discussed in Chapter 5 (Technologies for Coal Utilization) and Chapter 7 (Future Power Generation), with activities under way to further improve these processes. Gasification followed by Fischer–Tropsch synthesis is the leading processing candidate for producing the liquid fuels. The importance of coal to produce a variety of products, including transportation fuels, is evidenced by the direction of the DOE's research and development programs addressing future plants that produce power, fuels, and chemicals.

The Resurgence of Coal in Electric Power Generation

Fuel diversity for power generation is necessary for energy security, which is recognized by both industry and lawmakers. Tom Ridge, Secretary of the U.S. Department of Homeland Security, was a supporter of energy security while serving as governor of Pennsylvania; he led Pennsylvania toward competitive electricity markets and supported the development of new generating capacity in Pennsylvania with an emphasis on fuel diversity [27]. The first major coal plant to be built in Pennsylvania in 20 years was dedicated during his tenure and will be operational in the spring of 2004. A 561 MW, fluidized-bed power plant burning bituminous coal wastes has been constructed at the Seward Station to replace an aging 30 MW, pulverized-coal, wall-fired boiler. Pennsylvania lawmakers are not the only ones who realize that new capacity fueled by a stable coal supply is essential as opposed to making virtually all new capacity dependent on natural-gas-fired units in light of the inherent price volatility of natural gas. Regulators, lawmakers, and industry in Wisconsin, Illinois, Iowa, Kentucky, North Dakota, Wyoming, South Carolina, Louisiana, and elsewhere are aggressively promoting diversification with coal [27].

Much of the new capacity is being built on brownfield sites (*i.e.*, sites where industrial activities have been performed) because of existing permitting coupled with the presence of infrastructure such as rail or barge, water, and transmission access [27]. Some new capacity is being installed as replacements, such as at the Seward plant, while the rest is being installed at greenfield sites (*i.e.*, new industrial sites). The plants being installed are using advanced combustion and emissions technology. The Seward plant, constructed by Reliant Energy, is utilizing fluidized-bed technology with ultra-low emissions. Peabody Energy is in the process of installing two pairs of 1500 MW pulverized coal-fired boilers at the Thoroughbred Energy Campus (Muhlenberg County, Kentucky) and Prairie State Energy Campus (Washington County, Illinois), and these boilers will be among the cleanest coal plants east of the Mississippi River [27,28]. One, possibly two, 500 MW units are being planned in North Dakota as part of the state's Vision 21 program to install ultra-clean boiler systems firing lignite [29]. These are but a few of the projects under way, and more are anticipated as the DOE's Clean Coal Power Initiative and Vision 21 Program proceed. Approximately 61,000 MW of coal-fired generating capacity is currently planned [6]. Figure 8-7 shows the anticipated annual capacity additions, and Table 8-1 summarizes the capacity additions by state [6].

Production of Hydrogen from Coal

Chapter 7 provided a discussion of hydrogen production from coal gasification, along with the DOE's leading activities in this area; therefore, a detailed discussion of hydrogen production from coal is not presented in this section. Future energy systems will produce several products: electricity, chemicals, and other fuels such as hydrogen. Hydrogen is a clean burning fuel because the product of combustion is water; much research is currently under way in the United States exploring hydrogen production, storage, and utilization.

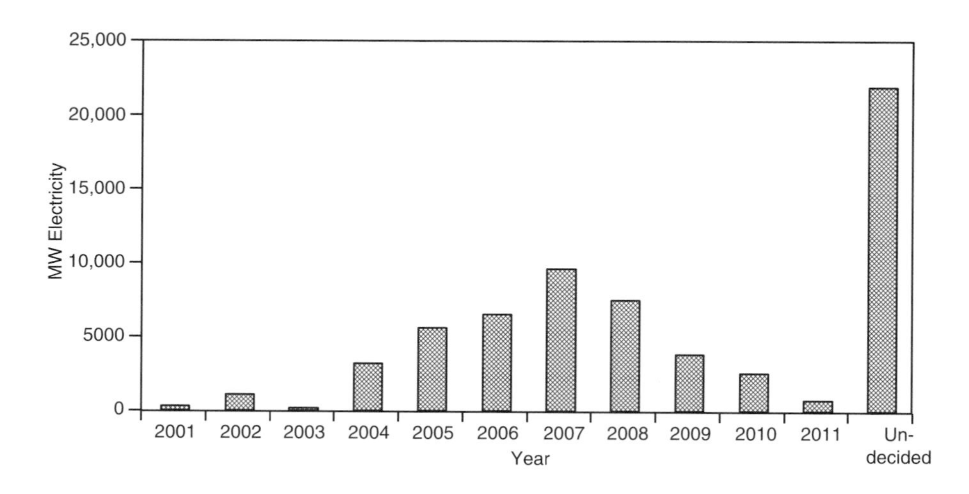

FIGURE 8-7. Annual capacity additions for coal-fired electric power generation. (From DOE, *Energy Security Technology*, National Energy Technology Laboratory, U.S. Department of Energy, Washington, D.C., www.netl.doe.gov/homeland/energy/security_main.html, last updated January 24, 2003.)

TABLE 8-1
Summary of Coal-Fired Electric Power Generation Additions by State

State	*Number of Plants*	*Capacity (MW)*
Alaska	1	200
Arizona	2	1600
Arkansas	1	1400
California	1	2500
Colorado	3	1430
Florida	3	1250
Georgia	1	1400
Illinois	10	8041
Indiana	1	500
Iowa	2	790
Kansas	1	660
Kentucky	8	5045
Louisiana	1	600
Maryland	1	180
Minnesota	3	1375
Mississippi	1	440
Missouri	2	1000
Montana	6	2263
Nebraska	2	820

(continued)

TABLE 8-1
(continued)

State	*Number of Plants*	*Capacity (MW)*
Nevada	1	500
New Mexico	2	1800
North Dakota	1	500
Ohio	3	2380
Oklahoma	2	1600
Oregon	1	500
Pennsylvania	3	1845
South Carolina	2	1440
Tennessee	1	1000
Texas	1	750
Utah	3	2700
Virginia	1	1600
Washington	1	2500
West Virginia	4	1435
Wisconsin	5	3000
Wyoming	3	740
Undecided	8	6000
Total	94	62,209

Source: DOE, *Energy Security Technology* (National Energy Technology Laboratory, U.S. Department of Energy, Washington, D.C.), www.netl.doe.gov/homeland/energy/security_main.html (last updated January 24, 2003).

Any discussion of energy security for the United States involving hydrogen utilization must include coal as the feedstock.

Currently, nearly all hydrogen production is based on fossil fuels. Worldwide, 48% of hydrogen is produced from natural gas, 30% from oil (and mostly consumed in refineries), 18% from coal, and the remaining 4% via water electrolysis [30]. Most of the hydrogen produced in the United States is done by steam reforming, or as a by-product of petroleum refining and chemicals production [30]. Steam reforming uses thermal energy to separate the hydrogen from the carbon components in methane (and sometimes methanol) and involves the reaction of the fuel with steam on catalytic surfaces. The first step of the reaction (when using natural gas) decomposes the natural gas into hydrogen and carbon monoxide and is then followed by the shift reaction that reacts the carbon monoxide with steam to produce carbon dioxide and hydrogen via the reactions:

$$CH_4 + H_2O \longrightarrow 2H_2 + CO \quad (8\text{-}1)$$

$$CO + H_2O \longrightarrow CO_2 + H_2 \quad (8\text{-}2)$$

The use of natural gas to produce hydrogen does not make any sense. Natural gas is a premium fuel for the residential and industrial sectors and possibly will have an increased role in the transportation sector in compressed natural gas mass-transit systems. Depending upon natural gas for hydrogen production has many of the same negative consequences to the economy as does using natural gas for power generation.

Hydrogen has many uses, including as a feedstock to fuel cells, as a fuel for power generation, and as a transportation fuel; however, much work needs to be done before most of these applications will become commercial, some of which will take two or three decades. In short, the reality of a full hydrogen economy is many years in the future. During the development of these hydrogen technologies, the DOE is aggressively pursuing "clean" technologies for hydrogen production using coal. This option provides the most energy and economic security to the United States because of the abundance and price of coal.

The Role of Coal in Providing Security to the U.S. Food Supply

When discussing U.S. energy/national security, one issue that has been raised over the last few years is the role of coal in providing security to the U.S. food supply. There has been concern among some individuals in the U.S. rendering industry and certain government agencies about ensuring a safe food supply from both natural and humanmade threats (*i.e.*, terrorism, disease outbreaks). The diseases of most concern include hoof-and-mouth disease, bovine spongiform encephalopathy (BSE, or "mad cow" disease), and chronic wasting disease (CWD) [31–36]. These concerns stem from the incident of BSE and hoof-and-mouth disease widely reported in Europe, the outbreaks of CWD in the United States (such as the 2004 incident in Wisconsin) [34,35], the first reports of BSE in North America which occurred in Canada, the detection of BSE in the United States in December 2003 [37], and the 2004 threat of acts of terrorism against the United States. Penn State has been working with various rendering companies, trade organizations, and government agencies for several years to evaluate the potential of using animal fats, proteins, and carcasses as boiler fuels [38–40].

It was noted in 2001 that BSE, which is a major health concern in Europe, had a very low risk potential for occurring in the United States, as determined by a landmark study by Harvard University [41]. The study did identify a potential pathway for the spread of BSE in the animal chain through deadstock (*i.e.*, dead and downer animals) on farms and ranches. This issue was being addressed by specific handling practices in the rendering industry and through regulatory options [31–33,36]. Interestingly, it was a paralyzed cow, a downer, that was detected in the U.S. food chain as having BSE that set off the scare that resulted in many changes in the rendering/packing industry, negatively affected the export market (resulting in layoffs of meat packing workers), and resulted in a downturn in the beef industry economy [36,42,43]. New regulations anticipated in Spring 2004 are expected to increase the amount of specified risk material from about 1.5 lb/beef carcass (brain and spinal cord) to 100 lb/carcass, forcing the industry to explore various disposal options [36].

One issue with regard to a major disease outbreak is how to dispose of a large number of carcasses in a short period of time, which will be necessary to limit proliferation of the disease. A recent example, although on a much smaller scale, was the need to identify disposal options for deer in Wisconsin infected with CWD in 2002 [34,35]. Approximately 10,000 to 25,000 deer in a 360-square-mile area were targeted to be killed to stop the spread of CWD [34]; however, identifying options for disposing of the carcasses was a major problem. Options identified included landfilling, rendering, and incineration/digestion [34]. Rendering was an option in this case because there were facilities that had the capability to process this quantity of material (each deer weighs 120 to 140 lb). There was a concern, though, about the wastewater that would be produced during the rendering. Preliminary tests were performed in which a few deer carcasses were disposed of at a foundry, and apparently the tests were successful (although the number of carcasses disposed of was small and no stack testing was performed to measure emissions). The decision was made to incinerate the carcasses [44]. This was an example where the number of carcasses (and total weight) could be handled with existing methods and infrastructure; however, an important concern is how to dispose of a large herd of cattle in an emergency situation and what options would be available. With proper planning and foresight, both utility and industrial boilers can be used to handle such an emergency. Doing so will involve a partnership of the packing/rendering industry, the federal government, and boiler owner/operators. The federal government's role would include emergency oversight and funding for modifying boilers (prior to the emergency) and handling/disposing of carcasses during the emergency.

The types of boilers that can be used in an emergency include boilers currently designed for liquid fuels as well as existing fluidized-bed boilers. It is understood that, should there be an outbreak of a highly contagious disease such as hoof-and-mouth disease, the herd is to be quarantined with severe restrictions on movement into and out of the infected area. The rendering industry can move a mobile system into this area via rail, and rendered fats can be utilized in the boiler, thus supplying steam to the rendering process as well as providing rendered fats, which have a heating value nearly as high as number 6 fuel oil (*i.e.*, ~18,000 Btu/lb), to other boilers in the immediate area. However, the number of carcasses that can be processed per day in this manner is limited, and other options should be explored, including cofiring the carcasses with coal in a fluidized-bed system or firing the carcasses in a boiler designed specifically for this feedstock. In the case of a hoof-and-mouth outbreak, the boiler would need to be nearby. For other cases, such as outbreaks of CWD, avian influenza (AI), or the identification of BSE cases, it will be easier to transport the animals or birds to the boilers and the number of candidate boilers would be larger.

In one example of planning for a national emergency, targeted fluidized-bed boilers throughout the country could be modified with emergency handling systems (although one hopes that they will never have to be used) that will take an entire carcass, grind it (as is currently done in the rendering industry), and feed the "chunked" cow directly into the fluidized-bed boiler. The chunked cow, with a heating value of about 4000 Btu/lb [36], can easily be cofired with coal and can even be a stand-alone fuel in a boiler designed for this feedstock. The first step in identifying candidate boilers will be to overlay maps of locations of fluidized-bed and other appropriate utility and industrial boilers with rendering and packing facilities as well as major herd concentrations (which tend to correspond to meat packing facilities). As an example, Figure 5-11 in Chapter 5 provides a map of the United States showing the locations of coal-fired, fluidized-bed boilers that could be utilized.

The rendering industry is currently interested in installing industrial-sized, coal-fired, fluidized-bed boilers to dispose of deadstock, and preliminary inquiries have been made of boiler vendors. The rendering industry is interested in this because it has the potential to solve the problem of disposal of deadstock, which can be 100,000 to 200,000 carcasses per year, while producing steam to meet their processing needs [45].

Similarly, an outbreak of AI in Virginia in 2002 was handled by disposing of approximately 16,000 tons of dead birds in landfills at a cost of $2.5 million [46]; however, Brglez [46] has shown that the birds could have been rendered into meal and utilized as a fuel at a nearby cement plant at a savings of $1 million to the industry. The concept of using animal meal to displace 10 to 15% of the coal (on a heat input basis) in cement kilns is currently being practiced in France and Italy [46]. Utilization of these types of materials as a cofire fuel with coal, instead of landfilling them, should become standard practice. This has three benefits: reduced disposal costs for the food industry, reduced fuel costs to the boiler operator, and enhanced national security to the U.S. food supply.

Coal's Role in International Energy Security and Sustainable Development

In order to preserve the option for utilizing coal in the future and to ensure that coal usage is performed in an environmentally acceptable manner, the Electric Power Research Institute (EPRI) undertook the Global Coal Initiative (GCI) in 2000 [47,48]. This initiative builds on and supplements DOE initiatives and worldwide coal combustion advances that are aimed at maintaining the strategic value of coal as a power-generating fuel worldwide.

The GCI involves a consortium of participants worldwide—coal suppliers, coal users, equipment manufacturers, and industry/government consortia—that are incorporating a long-term focus on development options for resolving carbon-energy conflicts and enabling the sustainable, competitive use of coal at near-zero emissions by 2020. The long-term viability of coal as a worldwide generating fuel depends on finding ways to further reduce or even eliminate the environmental impact of coal, including CO_2 emissions. The GCI is engaged in projects that focus on near-term operational movements and longer term coal-retention solutions [48]. The projects will help sustain a diversity of generating fuel resources as a hedge against price fluctuations, enhance the value of existing coal-fired power plants, and provide options for using coal in new plants [48].

One activity under the GCI is a new valuation framework to weigh private and public investment in advanced coal technologies to show the value to society of developing superior coal technologies to maintain fuel diversity [47]. Results suggest that the coal research and development in the United States will produce approximately $1.4 trillion in net benefits to U.S. consumers by 2050.

The essential six elements of the GCI, endorsed by the various factions of the global coal community, are (1) ultra-supercritical plant designs; (2) low-volatile coal combustion; (3) the value of real options for coal; (4) CO_2 control options; (5) advanced coal gasification, fluidized-bed, and other coal options; and (6) lignite and low-rank coal plant improvements [47,48]. Through ultra-supercritical plant designs, the initiative is evaluating and developing materials for advanced steam cycles in order to enhance efficiency and reduce emissions of pollutants and CO_2 from next-generation plants. The GCI is providing innovative solutions to operating problems with low-volatile coals, which are prevalent in China, India, South Africa, and Australia, to reduce fuel costs, maintain satisfactory performance during cycling and at low loads, and maintain NO_x emissions and unburned carbon levels at desired levels. Real options models are being developed and assessments performed so power producers can more effectively utilize coal as an option in their future asset base. Technological approaches and costs to design, procure equipment, construct, and operate power plants that separate and sequester CO_2 will be evaluated. Advanced approaches for utilizing coal will be developed and evaluated, including gasification, advanced fluidized-bed combustion, and hydrogen and chemical co-production, an effort that supports the DOE's various programs discussed in Chapter 7. The GCI is also developing solutions to fuel processing and boiler slagging in power plants that burn high-moisture lignite. There are vast lignite fields in the Dakotas, the Texas–Mississippi basin, Germany and Central Europe, Australia, China, and India; however, challenging issues related to the high moisture content need to be resolved.

The GCI complements the United Nations initiative to address coal and sustainable development [49]. Although coal and sustainable development may be contradictory terms to some, this was not the message that came from the World Summit on Sustainable Development held in Johannesburg, South Africa, in August/September 2002, where it was agreed that not only is coal compatible with sustainable development, but it is also essential [49]. Sustainable development was defined as "development that meets the needs of the present generation without undermining the capacity of future generations to meet their needs."

At the summit, the World Bank noted that improved energy services can enhance indoor air quality and reduce health hazards (by switching from traditional biomass and fossil fuels for cooking and heating), increase income (by reducing the time in collecting traditional biomass and implementing small-scale manufacturing and service activities), bring environmental benefits (by stopping deforestation and its subsequent soil degradation), and provide educational opportunities (by providing lighting or other services of direct educational relevance) [49]. At the summit, it was estimated that 1.6 billion people in developing countries have no access to electricity, while 2.4 billion rely on primitive biomass for cooking and heating and that, in 30 years, there would still be 1.4 billion people without electricity and 2.6 billion people still relying on traditional biomass. It was these challenges that the summit addressed and a plan of implementation was developed. One aspect of the plan asks governments to diversify energy supply by developing advanced, cleaner, more efficient, affordable, and cost-effective energy technologies, including fossil fuel technologies, and to transfer these technologies to developing countries.

At the summit, it was clear that coal has been, and must continue to be, a major contributor of sustainable development because coal can be found in most nations of the world [49]. On a global scale, coal accounts for ~25% of world primary energy consumption and fuels 35% of the world electricity generation, ~5 billion short tons/year are consumed, the international coal trade consists of ~600 million short tons/year, there are more than 200 years of reserves at current rates of consumption, coal is a "conflict-free" energy source, and there is a diversity of reserves and production. Coal must be used efficiently and cleanly, and it was recognized at the summit that coal technologies continue to improve and effective options are now available for countries of all levels of economic development with respect to emissions reduction. Coal continues to be the foundation for economic and social development of the world's largest economies in both the developed and developing world. Coal is a major component in securing an improved quality of life for billions of people worldwide who gain access to it through the energy services they need for daily life, industrial development, and social advancement. A sustainable energy policy can be achieved by encouraging electrification, establishing sound environmental regulations, safeguarding energy supplies through diversification, and continuing to support advanced coal technologies.

Concluding Statements

Consumers in the United States are feeling the impact of volatile foreign oil supplies with high prices at the gas pumps and through increased costs for goods and services that use oil and related fuels in their production processes. Volatile natural gas prices are also negatively impacting consumers. These issues are challenging U.S. policymakers to implement an energy policy that ensures greater independence from foreign energy interests and thereby greater energy security for the United States. Low-cost and abundant coal is a domestic resource of great strategic value.

Coal is the most abundant energy resource in the United States and fuels as much electricity generation as all other sources combined. The United States has more coal reserves than all of the oil reserves in the world. The energy value in Montana's coal reserves, for instance, are greater than the oil reserves in Saudi Arabia, Kuwait, Iran, and Iraq combined [14].

Coal's strategic value as a fuel source is due to its role in increasing economic growth and development and providing energy security at the local, national, and international levels [50]. The usage of coal provides jobs, supports infrastructure through taxes, supports economic growth and electrification, encourages productivity through electric technologies, provides a reliable energy source, is resistant to energy shocks, and stabilizes power prices. Coal-fired power plants provide system security through infrastructure reliability, which prevents sudden disruptions because [50]:

- There are many diverse sources of coal, thus the supply of coal is less vulnerable to disruption;
- Coal-fired plants have better storage capability near the power generation site because they are less vulnerable to transportation disruption;
- Coal has a more certain availability during peak demand;
- Coal-fired power plants are less vulnerable to outages because they are a mature, reliable technology;
- Coal-fired power plants are less vulnerable to terrorism compared to nuclear power plants, natural gas pipelines, or liquefied natural gas facilities.

Energy security in the United States cannot be overemphasized, and its importance is clearly evident in the National Energy Policy as well as in the DOE's Coal Power Program. One of the DOE's strategic goals is to protect national and economic security by promoting a diverse supply of reliable, affordable, and environmentally sound energy [51]. This goal is accomplished by developing technologies that foster a diverse supply of affordable and environmentally sound energy, improving energy efficiency, providing for reliable delivery of energy, exploring advanced technologies that make fundamental changes in energy options, and guarding against energy emergencies. Benefits of the Coal Power Program include reducing dependence on imported oil, which can be achieved by co-production of power and environmentally attractive fuels such as Fischer–Tropsch liquids and hydrogen. Additional benefits include maintaining diversity of energy resource options to avoid over-reliance on natural gas for power generation, encouraging economical use of natural gas in other sectors, and reducing energy price volatility and supply uncertainty. The Coal Power Program also retains domestic manufacturing capabilities and U.S. energy technology leadership to enhance economic growth and security. Technologies developed through the DOE's Coal Power Program can also be used to ensure a safe food source for the United States.

The energy security and sustainability of the United States depend on sufficient energy supplies to support U.S. and global economic growth, and coal is a major contributor to this security. Coal-fired electricity generation will enable and stimulate economic growth and social welfare. Diversifying energy production, through the use of coal in power generation and production of chemicals and hydrogen, provides the United States with energy security. Economic growth and energy security will enable cost-effective environmental controls and continued energy affordability.

References

1. EIA, *Annual Energy Review 2002* (Energy Information Administration, U.S. Department of Energy, Washington, D.C., October 2003).
2. DOE, *National Energy Technology Laboratory Accomplishments FY2000* (Office of Fossil Energy, Washington, D.C., September 2001).
3. USEA, *National Energy Security Post 9/11* (U.S. Energy Association, Washington, D.C., June 2002).
4. EIA, *International Energy Outlook 2002* (Energy Information Administration, U.S. Department of Energy, Washington, D.C., March 2002).
5. EIA, *Annual Energy Outlook 2003* (Energy Information Administration, U.S. Department of Energy, Washington, D.C., January 2003).
6. DOE, *Energy Security Technology* (National Energy Technology Laboratory, U.S. Department of Energy), www.netl.doe.gov/homeland/energy/security_main.html (last updated January 24, 2003).
7. Kripowicz, R. S., *Remarks to the Clean Coal Power Initiative Planning Workshop* (National Energy Technology Laboratory, U.S. Department of Energy, Washington, D.C., September 28, 2001), www.netl.doe.gov/coalpower/ccpi/main.html.
8. Caylor, B., Coal's Role in the National Energy Plan, *Energia*, Vol. 13, No. 3, 2001, p. 5.
9. NEPD Group, *National Energy Policy* (National Energy Policy Development, U.S. Government Printing Office, Washington, D.C., May 2001).

10. Abraham, S., Remarks to the Clean Coal and Power Conference (Office of Fossil Energy, U.S. Department of Energy, Washington, D.C., November 17, 2003), www.fossil.energy.gov/news/speeches/03/03_sec_cleancoal_111703.html.
11. Roberts, A., Four Common Sense Reasons to Burn Coal, and One More, *Coal Age*, Vol. 108, No. 10, Nov./Dec. 2003, pp. 37–38.
12. EIA, *Energy Price Impacts on the U.S. Economy* (Energy Information Administration, U.S. Department of Energy, Washington, D.C., April 2001), www.eia.doe.gov/oiaf/economy/energy_price.html.
13. Eastman, M. L., *Clean Coal Power Initiative*, presented at the Clean Coal Power Conference (National Energy Technology Laboratory, U.S. Department of Energy, Washington, D.C., November 18, 2003), www.netl.doe/gov/coalpower/ccpi/program_info.html.
14. Svec, V., *Again a Crisis with a Solution: Energy in America and Coal-Based Generation* (American Coal Council, Phoenix, AZ, 2003), pp. 15–19.
15. Williams, J., Transmission Upgrades: Key to Low-Cost Power and Greater Use of Coal-Fired Generation, *Power Engineering*, Vol. 107, No. 6, June 2003, pp. 31–36.
16. Smock, B., Natural Gas Crisis Revisited, *Power Engineering*, Vol., 107, No. 8, Aug. 2003, p. 5.
17. Silverstein, K., Pipeline Pressures on the Rise, *Energy Industry Issues Newsletter*, December 10, 2003, www.utilipoint.com.
18. EIA, *Short-Term Energy Outlook: January 2004* (Energy Information Administration, U.S. Department of Energy, Washington, D.C.), www.eia.doe.gov/emeu/steo/pub/contents.html#electric (last modified January 7, 2004).
19. Johnson, J., Congress Tackles Energy Bill Again, *Chemical & Engineering News*, Vol. 82, No. 3, Jan. 19, 2004, pp. 68–60.
20. McAleer, A. G., Testimony of A. Gordon McAleer (U.S. House Task Force for Natural Gas Affordability Public Hearing, State College, PA, August 8, 2003).
21. Murphy, T. J., Testimony of Terrence J. Murphy (U.S. House Task Force for Natural Gas Affordability Public Hearing, State College, PA, August 8, 2003).
22. DOE, *Coal Technologies are Cost Competitive* (Federal Energy Technology Center, U.S. Department of Energy, Washington, D.C.), www.fetc.doe.gov/coalpower/powersystems/images/costcomp.jpg (accessed December 13, 2001).
23. Coon, C., *Strengthening National Security Through Energy Security* (The Heritage Foundation, Washington, D.C., April 2002), www.heritge.org/Research/EnergyandEnvironment/WM94.cfm.
24. NRE, *Energy Security* (National Renewable Energy Laboratory, U.S. Department of Energy, Washington, D.C.), www.nrel.gov/clean_energy/security.html (accessed January 18, 2004).
25. Taylor, J., *Don't Worry About Energy Security* (Cato Institute, Washington, D.C., October 18, 2001), www.cato.org/cgi-bin/scripts/printtech.cgi/dailys/10-18-01.html.
26. DOE, *Biofuels and U.S. Energy Security* (Office of Energy Efficiency and Renewable Energy, U.S. Department of Energy, Washington, D.C.), www.ott.doe.gov/biofuels/energy_security.html (last updated December 12, 2003).

27. Anon., Homeland Security: U.S. Brownfield, *Power Engineering,* Vol. 106, No. 6, June 2002, pp. 28–34.
28. Anon., Peabody's Englehardt Says Two 1500-MW Coal-Burning Power Plants Will Be Built, *Coal Age,* Vol. 108, No. 8, Sept. 2003, p. 16.
29. Lignite Energy Council, www.lignitevision21.com (accessed February 2, 2004).
30. EERE, *Hydrogen* (Office of Energy Efficiency and Renewable Energy, U.S. Department of Energy, Washington, D.C., February 5, 2003), www.eere.energy.gov/hydrogenandfuelcells/hydrogen/production/html.
31. Mullane, D. K., Taylor By-Products, Inc., personal communications, 2002.
32. Detwiler, L., Animal and Plant Health Inspection Service, U.S. Department of Agriculture, personal communications, 2002.
33. Adams, J., Animal Health & Farm Services, National Milk Producers Federation, personal communication, March 2002.
34. Orlander, D., National Renders Association Contract Veterinarian, personal communication, May 5, 2002.
35. Anon., Deer Disease Raises Concerns About Venison, *Centre Daily Times,* June 2, 2002.
36. Harlan, D., Taylor-Excel Food Solutions Company, personal communication, March 2004.
37. C&EN, Mad Cow Disease Outbreak, *Chemical & Engineering News,* Vol. 82, No. 1, Jan. 5, 2004, p. 11.
38. Miller, B. G., D. A. Clark, M. A. Hill, J. Larsen, T. Clemens, and T. Wehr, A Demonstration of Pig Lard as an Industrial Boiler Fuel, in *Proc. of the 24th International Conference on Coal Utilization and Fuel Systems* (Coal & Slurry Technology Association, Washington, D.C., March 1999), pp. 743–754.
39. Miller, B. G., S. Falcone Miller, and A. W. Scaroni, Utilizing Agricultural By-Products in Industrial Boilers: Penn State's Experience and Coal's Role in Providing Security for Our Nation's Food Supply, in *Proc. of the Nineteenth Annual International Pittsburgh Coal Conference* (University of Pittsburgh, Pittsburgh, PA, September 2002).
40. Miller, B. G. and S. Falcone Miller, Utilizing Biomass in Industrial Boilers: The Role of Biomass and Industrial Boilers in Providing Energy/National Security, in *Proc. of the First CIBO Industrial Renewable Energy and Biomass Conference* (Council of Industrial Boiler Owners, Burke, VA, April 7–9, 2003).
41. USDA, *Evaluation of the Potential for Bovine Spongiform Encephalopathy in the United States* (Animal and Plant Health Inspection Service, U.S. Department of Agriculture, Washington, D.C., November 2001), www.aphis.usda.gov/oa/bse/.
42. Anon., Mad Cow Fears Leading to Layoffs, *Centre Daily Times,* January 24, 2004.
43. Anon., *'Mad Cow' Scare Could Cost Pennsylvania Millions,* Penn State Live website, http://live.psu.edu/story/5261, January 15, 2004.
44. Orlander, D., National Renders Association Contract Veterinarian, personal communication, March 13, 2003.
45. McMurtry, W. R., Darling International, personal communications, 2004.

46. Brglez, B. Major, U.S. Army Veterinary Corps, personal communication, February 14, 2003.
47. Armor, T., Coal-Fired Power Plants: Increasingly Lean and Green, *Power Engineering*, Vol. 105, No. 9, September 2001, pp. 40–43.
48. Anon., Bush Energy Policy Resonates with Global Coal Initiative, *EPRI Journal*, Vol. 26, No. 1, Summer 2001, pp. 21–23.
49. World Coal Institute, *Coal and Sustainable Development: The World Summit on Sustainable Development and Its Implications*, prepared for the United Nations Economic and Social Council Committee on Sustainable Energy, www.wci-coal.com, October 16, 2002.
50. Bruno, W. A., Coal's Role in International Energy Security and Sustainable Development, in *Proc. of the 2003 Conference on Unburned Carbon on Utility Fly Ash*, October 28, 2003.
51. DOE, *Clean Coal Technology Roadmap: "CURC/EPRI/DOE Consensus Roadmap" Background Information* (National Energy Technology Laboratory, U.S. Department of Energy, Washington, D.C., January 6, 2004), www.netl.doe.gov/coalpower/ccpi/main.html.

APPENDIX A

Coal-Fired Emission Factors

TABLE A-1
Emission Factors for SO_X, NO_X, and CO from Bituminous and Subbituminous Coal Combustion[a]

Firing Configuration	SO_X^b		NO_X^c		$CO^{d,e}$	
	Emission Factor (lb/ton)	*Emission Factor* Rating	*Emission Factor* (lb/ton)	*Emission Factor* Rating	*Emission Factor* (lb/ton)	*Emission Factor Rating*
PC, dry-bottom, wall-fired,[f] bituminous pre-NSPS[g]	38S	A	22	A	0.5	A
PC, dry-bottom, wall-fired,[f] bituminous pre-NSPS[g] with low-NO_X burner	38S	A	11	A	0.5	A
PC, dry-bottom, wall-fired,[f] bituminous NSPS[g]	38S	A	12	A	0.5	A
PC, dry-bottom, wall-fired,[f] subbituminous pre-NSPS[g]	35S	A	12	C	0.5	A
PC, dry-bottom, wall-fired,[f] subbituminous NSPS[g]	35S	A	7.4	A	0.5	A
PC, dry-bottom, cell-burner-fired, bituminous	38S	A	31	A	0.5	A

(continued)

TABLE A-1
(continued)

Firing Configuration	SO_x^b		NO_x^c		$CO^{d,e}$	
	Emission Factor (lb/ton)	*Emission Factor* Rating	*Emission Factor* (lb/ton)	*Emission Factor* Rating	*Emission Factor* (lb/ton)	*Emission Factor Rating*
PC, dry-bottom, cell-burner-fired, subbituminous	35S	A	14	E	0.5	A
PC, dry-bottom, tangentially fired, bituminous, pre-NSPS[g]	38S	A	15	A	0.5	A
PC, dry-bottom, tangentially fired, bituminous, pre-NSPS[g] with low-NO_x burner	38S	A	9.7	A	0.5	A
PC, dry-bottom, tangentially fired, bituminous, NSPS[g]	38S	—	10	A	0.5	A
PC, dry-bottom, tangentially fired, subbituminous, pre-NSPS[g]	35S	A	8.4	A	0.5	A
PC, dry-bottom, tangentially fired, subbituminous, pre-NSPS[g]	35S	A	7.2	A	0.5	A
PC, wet bottom, wall-fired,[f] bituminous, pre-NSPS[g]	38S	A	31	D	0.5	A
PC, wet bottom, tangentially fired, bituminous, NSPS[g]	38S	A	14	E	0.5	A
PC, wet bottom, wall-fired, subbituminous	35S	A	24	E	0.5	A
Cyclone furnace, bituminous	38S	A	33	A	0.5	A
Cyclone furnace, subbituminous	35S	A	17	C	0.5	A

(continued)

TABLE A-1
(continued)

Firing Configuration	SO_x^b		NO_x^c		$CO^{d,e}$	
	Emission Factor (lb/ton)	*Emission Factor* Rating	*Emission Factor* (lb/ton)	*Emission Factor* Rating	*Emission Factor* (lb/ton)	*Emission Factor Rating*
Spreader stoker, bituminous	38S	B	11	B	5	A
Spreader stoker, subbituminous	35S	B	8.8	B	5	A
Overfeed stoker[h]	38S	B	7.5	A	6	B
	(35S)	—	—	—	—	—
Underfeed stoker	31S	B	9.5	A	11	B
Hand-fed units	31S	D	9.1	E	275	E
FBC, circulating-bed	—[i]	E	5.0	D	18	E
FBC, bubbling-bed	—[i]	E	15.2	D	18	D

[a]Data from EPA, *Compilation of Air Pollutant Emission Factors*, AP-42, Fifth ed., U.S. Environmental Protection Agency, U.S. Government Printing Office, Washington, D.C., January 1995. Factors represent uncontrolled emissions, unless otherwise specified, and should be applied to coal feed, as fired. Tons are short tons.

[b]Expressed as SO_2, including SO_2, SO_3, and gaseous sulfates. The factor in parentheses should be used to estimate gaseous SO_x emissions for subbituminous coal. In all cases, S is weight percent (wt.%) sulfur content of coal as fired. Emission factor would be calculated by multiplying the wt.% sulfur in the coal by the numerical value preceding S. For example, if a fuel is 1.2% sulfur, then S = 1.2. On average, for bituminous coal, 95% of fuel sulfur is emitted as SO_2, and only about 0.7% of fuel sulfur is emitted as SO_3 and gaseous sulfate. An equally small percent of fuel sulfur is emitted as particulate sulfate. Small quantities of sulfur are also retained in bottom ash. With subbituminous coal, about 10% more fuel sulfur is retained in the bottom ash and particulate because of the more alkaline nature of the coal ash. Conversion to gaseous sulfate appears about the same as for bituminous coal.

[c]Expressed as NO_2. Generally, 95 vol% or more of NO_x present in combustion exhaust will be in the form of NO, the rest NO_2. To express factors as NO, multiply factors by 0.66. All factors represent emissions at baseline operations (*i.e.*, 60 to 110% load and no NO_x control measures).

[d]Nominal values achievable under normal operating conditions; values 1 or 2 orders of magnitude higher can occur when combustion is not complete.

[e]Emission factors for CO_2 emissions from coal combustion should be calculated using pounds CO_2 per ton coal = 72.6C, where C is the wt.% carbon content of the coal. For example, if the carbon content is 85%, then C equals 85.

[f]Wall-fired includes front and rear-wall-fired units, as well as opposed-wall-fired units.

[g]Pre-NSPS boilers are not subject to any NSPS. NSPS boilers are subject to Subpart D or Subpart D(a). Subpart D boilers are boilers constructed after August 17, 1971, and with a heat input rate greater than 250 million Btu per hour (MM Btu/hr). Subpart D(a) boilers are boilers constructed after September 18, 1978, and with a heat input rate greater than 250 MMBtu/hr.

[h]Includes traveling-grate, vibrating-grate, and chain-grate stokers.

[i]SO_2 emission factors for fluidized-bed combustion are a function of fuel sulfur content and calcium-to-sulfur ratio. For both bubbling-bed and circulating-bed design, use pounds SO_2/ton coal = $39.6(S)(Ca/S)^{-1.9}$. In this equation, S is the wt.% sulfur in the fuel and Ca/S is the molar calcium-to-sulfur ratio in the bed. This equation may be used when the Ca/S is between 1.5 and 7. When no calcium-based sorbents are used and the bed material is inert with respect to sulfur capture, the emission factor for underfeed stokers should be used to estimate the SO_2 emissions. In this case, the emission factor ratings are E for both bubbling- and circulating-bed units.

TABLE A-2
Emission Factors for CH_4, Total Non-Methane Organic Compounds (TNMOC), and N_2O from Bituminous and Subbituminous Coal Combustion[a]

Firing Configuration	CH_4^b		TNMOC[b,c]		N_2O	
	Emission Factor (lb/ton)	*Emission Factor* Rating	*Emission Factor* (lb/ton)	*Emission Factor* Rating	*Emission Factor* (lb/ton)	*Emission Factor Rating*
PC-fired, dry-bottom, wall-fired	0.04	B	0.06	B	0.03	B
PC-fired, dry-bottom, tangentially fired	0.04	B	0.06	B	0.08	B
PC-fired, wet-bottom	0.05	B	0.04	B	0.08	E
Cyclone furnace	0.01	B	0.11	B	0.09[c]	E
Spreader stoker	0.06	B	0.05	B	0.04	E
Spreader stoker with multiple cyclones, and reinjection	0.06	B	0.05	B	0.04	E
Spreader stoker with multiple cyclones, no reinjection	0.06	B	0.05	B	0.04	E
Overfeed stoker	0.06	B	0.05	B	0.04	E
Overfeed stoker with multiple cyclones	0.06	B	0.05	B	0.04	E
Underfeed stoker	0.8	B	1.3	B	0.04	E
Underfeed stoker with multiple cyclones	0.8	B	1.3	B	0.04	E
Hand-fed units	5	E	10	E	0.04	E
FBC, bubbling-bed	0.06	E	0.05	E	3.5	B
FBC, circulating-bed	0.06	E	0.05	E	3.5	B

[a]Data from EPA, *Compilation of Air Pollutant Emission Factors*, AP-42, Fifth ed., U.S. Environmental Protection Agency, U.S. Government Printing Office, Washington, D.C., January 1995. Tons are short tons. Factors represent uncontrolled emissions, unless otherwise specified, and should be applied to coal feed, as fired.

[b]Nominal values achievable under normal operating conditions; values 1 or 2 orders of magnitude higher can occur when combustion is not complete.

[c]TNMOC are expressed as C_2 to C_{16} alkane equivalents. Because of limited data, the effects of firing configuration on TNMOC emission factors could not be distinguished. As a result, all data were averaged collectively to develop a single average emission factor for pulverized coal units, cyclones, spreaders, and overfeed stokers.

TABLE A-3
Uncontrolled Emission Factors for PM and PM_{10} from Bituminous and Subbituminous Coal Combustion[a]

Firing Configuration	*Filterable PM*[b]		*Filterable PM_{10}*	
	Emission Factor (lb/ton)	*Emission Factor* Rating	*Emission Factor* (lb/ton)	*Emission Factor* Rating
PC-fired, dry-bottom, wall-fired	10A	A	2.3A	E
PC-fired, dry-bottom, tangentially fired	10A	B	2.3A[c]	E
PC-fired, wet-bottom	7A[d]	D	2.6A	E
Cyclone furnace	2A[d]	E	0.26A	E
Spreader stoker	66[e]	B	13.2	E
Spreader stoker with multiple cyclones, and reinjection	17	B	12.4	E
Spreader stoker with multiple cyclones, no reinjection	12	A	7.8	E
Overfeed stoker[f]	16[g]	C	6.0	E
Overfeed stoker with multiple cyclones[f]	9	C	5.0	E
Underfeed stoker	15[h]	D	6.2	E
Underfeed stoker with multiple cyclones	11	D	6.2[h]	E
Hand-fed units	15	E	6.2[i]	E
FBC, bubbling-bed	—[j]	E	—[j]	E
FBC, circulating-bed	—[j]	E	—[j]	E

[a]Data from EPA, *Compilation of Air Pollutant Emission Factors*, AP-42, Fifth ed., U.S. Environmental Protection Agency, U.S. Government Printing Office, Washington, D.C., January 1995. Factors represents uncontrolled emissions, unless otherwise specified, and should be applied to coal feed, as fired. Tons are short tons.

[b]Based on EPA Method 5 (front half catch). Where particulate is expressed in terms of coal ash content, the A factor is determined by multiplying weight percent (wt.%) ash content of coal (as fired) by the numerical value preceding the A. For example, if coal with 8% ash is fired in a PC-fired, dry-bottom unit, the particulate matter emission factor would be 10×8, or 80 lb/ton.

[c]No data found; emission factor for PC-fired, dry-bottom boilers used.

[d]Uncontrolled particulate emissions, when no fly ash reinjection is employed. When control device is installed, and collected fly ash is reinjected to boiler, particulate from boiler reaching control equipment can increase up to a factor of 2.

[e]Accounts for fly ash settling in an economizer, air heater, or breaching upstream of control device or stack. (Particulate directly at boiler outlet typically will be twice this level). Factor should be applied even when fly ash is reinjected to boiler form air heater or economizer dust hoppers.

[f]Includes traveling-grate, vibrating-grate, and chain-grate stokers.

[g]Accounts for fly ash settling in breaching or stack base. Particulate loadings directly at boiler outlet typically can be 50% higher.

[h]Accounts for fly ash settling in breaching downstream of boiler outlet.

[i]No data found; emission factor for underfeed stoker was used.

[j]No data found; emission factor for spreader stoker with multiple cyclones and reinjection was used.

TABLE A-4
Condensable Particulate Matter Emission Factors for Bituminous and Subbituminous Coal Combustion[a]

Firing Configuration[b]	*Controls*[c]	*CPM-TOT*[d,e]		*CPM-IOR*[d,e]		*CPM-ORG*[d,e]	
		Emission Factor (lb/MM Btu)	*Emission Factor Rating*	*Emission Factor* (lb/MM Btu)	*Emission Factor Rating*	*Emission Factor* (lb/MM Btu)	*Emission Factor Rating*
All pulverized coal-fired boilers	All PM controls (without FGD controls)	0.1S–0.03[f]	B	80% of CPM-TOT emission factor[e]	E	20% of CMP-TOT emission factor[e]	E
All pulverized coal-fired boilers	All PM controls combined with FGD controls	0.02	E	ND	—	ND	E
Spreader stoker, traveling-grate overfeed stoker, underfeed stoker	All PM controls or uncontrolled	0.04	C	80% of CPM-TOT emission factor[e]	E	20% CPM-TOT emission factor[e]	E

[a]Data from EPA, *Compilation of Air Pollutant Emission Factors*, AP-42, Fifth ed., U.S. Environmental Protection Agency, U.S. Government Printing Office, Washington, D.C., January 1995. All condensable PM is assumed to be less than 1.0 μm in diameter.
[b]No data are available for cyclone boilers or for atmospheric fluidized-bed combustion (AFBC) boilers. For cyclone boilers, use the factors provided for pulverized coal-fired boilers and applicable control devices. For AFBC boilers, use the factors provided for pulverized coal-fired boilers with PM and FGD controls.
[c]PM, particulate matter; FGD, flue gas desulfurization.
[d]CPM-TOT, total condensable particulate matter; CPM-IOR, inorganic condensable particulate matter; CPM-ORG, organic condensable particulate matter; ND, no data.
[e]Factors should be multiplied by fuel rate on a heat input basis (MM Btu), as fired. To convert to lb/ton of bituminous coal, multiply by 26 MM Btu/ton. To convert to lb/ton of subbituminous coal, multiply by 20 MM Btu/ton.
[f]S = coal sulfur percent by weight, as fired. For example, if the sulfur percent is 1.04, then $S = 1.04$. If the coal sulfur percent is 0.4 or less, use a default emission factor of 0.01 lb/MM Btu rather than the emission equation.

TABLE A-5
Emission Factors for Trace Elements, Polycyclic Organic Matter (POM), and HCOH from Uncontrolled Bituminous and Subbituminous Coal Combustion (Emission Factor Rating: E)[a]

Firing Configuration	*Emission Factor* (lb/10^{12} Btu)									
	As	*Be*	*Cd*	*Cr*	*Pb*[b]	*Mn*	*Hg*	*Ni*	*POM*	*HCOH*
Pulverized-coal, configuration unknown	ND	ND	ND	1922	ND	ND	ND	ND	ND	112[c]
Pulverized-coal, wet-bottom	538	81	44–70	1020–1570	507	808–2980	16	840–1290	ND	ND
Pulverized-coal, dry-bottom	684	81	44.4	1250–1570	507	228–2980	16	1030–1290	2.08	ND
Pulverized-coal, dry-bottom, tangential	ND	ND	ND	ND	ND	ND	ND	ND	2.4	ND
Cyclone furnace	115	<81	28	212–1502	507	228–1300	16	174–1290	ND	ND
Stoker, configuration unknown	ND	73	ND	19–300	ND	2170	16	775–1290	ND	ND
Spreader stoker	264–542	ND	21–43	942–1570	507	ND	ND	ND	ND	221[d]
Overfeed stoker, traveling-grate	542–1030	ND	43–82	ND	507	ND	ND	ND	ND	140[c]

[a]Data from EPA, *Compilation of Air Pollutant Emission Factors*, AP-42, Fifth ed., U.S. Environmental Protection Agency, U.S. Government Printing Office, Washington, D.C., January 1995. The emission factors in this table represent the ranges of factors reported in the literature. If only one data point was found, it is still reported in this table. To convert from lb/10^{12} Btu to pg/J, multiply by 0.43. ND, no data.
[b]Lead emission factors were taken directly from an EPA background document for support of the National Ambient Air Quality Standards.
[c]Based on two units; 133×10^6 Btu/hr and 155×10^6 Btu/hr.
[d]Based on one unit; 59×10^6 Btu/hr.

TABLE A-6
Cumulative Particle Size Distribution and Size-Specific Emission Factors for Dry Bottom Boilers Burning Pulverized Bituminous and Subbituminous Coal[a]

Particle Size[b] (μm)	*Cumulative Mass % ≤ Stated Size*					*Cumulative Emission Factor*[c] (lb/ton)				
	Uncontrolled	*Controlled*				*Uncontrolled*[d]	*Controlled*[e]			
		Multiple-Cyclone	*Scrubber*	*ESP*	*Baghouse*		*Multiple Cyclone*[f]	*Scrubber*[g]	*ESP*[g]	*Baghouse*[f]
15	32	54	81	79	97	3.2A	1.08A	0.48A	0.064A	0.02A
10	23	29	71	67	92	2.3A	0.58A	0.42A	0.054A	0.02A
6	17	14	62	50	77	1.7A	0.28A	0.38A	0.024A	0.02A
2.5	6	3	51	29	53	0.6A	0.06A	0.3A	0.024A	0.01A
1.25	2	1	35	17	31	0.2A	0.02A	0.22A	0.01A	0.006A
1.00	2	1	31	14	25	0.2A	0.02A	0.18A	0.01A	0.006A
0.625	1	1	20	12	14	0.10A	0.02A	0.12A	0.01A	0.002A
Total	100	100	100	100	100	10A	2A	0.6A	0.08A	0.02A

[a]Data from EPA, *Compilation of Air Pollutant Emission Factors*, AP-42, Fifth ed., U.S. Environmental Protection Agency, U.S. Government Printing Office, Washington, D.C., January 1995. Tons are short tons. To convert from lb/ton to kg/Mg, multiply by 0.5. Emission factors are pounds of pollutant per ton of coal combusted, as fired. ESP, electrostatic precipitator.
[b]Expressed as aerodynamic equivalent diameter.
[c]A = coal ash weight percent (wt.%) as fired; for example, if coal ash weight is 8.2%, then A = 8.2.
[d]Emission factor rating: C.
[e]Estimated control efficiency for multiple cyclones is 80%; for scrubber, 94%; for ESP, 99.2%; and for baghouse, 99.8%.
[f]Emission factor rating: E.
[g]Emission factor rating: D.

TABLE A-7
Cumulative Particle Size Distribution and Size-Specific Emission Factors for Wet-Bottom Boilers Burning Pulverized Bituminous Coal (Emission Factor Rating: E)[a]

Particle Size[b] (μm)	*Cumulative Mass % ≤ Stated Size*			*Cumulative Emission Factor*[c] (lb/ton)		
	Uncontrolled	*Controlled*		*Uncontrolled*	*Controlled*[d]	
		Multiple-Cyclone	*ESP*		*Multiple-Cyclone*	*ESP*
15	40	99	83	2.8A	1.38A	0.046A
10	37	93	75	2.6A	1.3A	0.042A
6	33	84	63	2.32A	1.18A	0.036A
2.5	21	61	40	1.48A	0.86A	0.022A
1.25	6	31	17	0.42A	0.44A	0.01A
1.00	4	19	8	0.28A	0.26A	0.004A
0.625	2	—[e]	—[e]	0.14A	—[e]	—[e]
Total	100	100	100	7.0A	1.4A	0.056A

[a]Data from EPA, *Compilation of Air Pollutant Emission Factors*, AP-42, Fifth ed., U.S. Environmental Protection Agency, U.S. Government Printing Office, Washington, D.C., January 1995. Tons are short tons. To convert from lb/ton to kg/Mg, multiply by 0.5. Emission factors are pounds of pollutant per ton of coal combusted as fired. ESP, electrostatic precipitator.
[b]Expressed as aerodynamic equivalent diameter.
[c]A = coal ash weight percent (wt.%), as fired; for example, if coal ash weight is 2.4%, then A = 2.4.
[d]Estimated control efficiency for multiple cyclones is 94%; for ESPs, 99.2%.
[e]Insufficient data.

TABLE A-8
Cumulative Size Distribution and Size-Specific Emission Factors for Cyclone Furnaces Burning Bituminous Coal (Emission Factor Rating: E)[a]

Particle Size[b] (μm)	*Cumulative Mass % ≤ Stated Size*			*Cumulative Emission Factor*[c] (lb/ton)		
	Uncontrolled	*Controlled*		*Uncontrolled*	*Controlled*[d]	
		Multiple-Cyclone	*ESP*		*Multiple-Cyclone*	*ESP*
15	33	95	90	0.66A	0.114A	0.013A
10	13	94	68	0.26A	0.112A	0.011A
6	8	93	56	0.16A	0.112A	0.009A
2.5	5.5	92	36	0.11A[c]	0.11A	0.006A
1.25	5	85	22	0.10A[e]	0.10A	0.04A
1.00	5	82	17	0.10A[e]	0.10A	0.004A
0.625	0	—[f]	—[f]	0	—[f]	—[f]
Total	100	100	100	2A	0.12A	0.016A

[a]Data from EPA, *Compilation of Air Pollutant Emission Factors*, AP-42, Fifth ed., U.S. Environmental Protection Agency, U.S. Government Printing Office, Washington, D.C., January 1995. Tons are short tons. To convert from lb/ton to kg/Mg, multiply by 0.5. Emission factors are pounds of pollutant per ton of coal combusted as fired. ESP, electrostatic precipitator.
[b]Expressed as aerodynamic equivalent diameter.
[c]A = coal ash weight percent (wt.%), as fired; for example, if coal ash weight is 2.4%, then A = 2.4.
[d]Estimated control efficiency for multiple cyclones is 94%; for ESPs, 99.2%.
[e]These values are estimates based on data from a controlled source.
[f]Insufficient data.

TABLE A-9
Cumulative Particle Size Distribution and Size-Specific Emission Factors for Overfeed Stokers Burning Bituminous Coal[a]

Particle Size[b] *(μm)*	*Cumulative Mass % ≤ Stated Size*		*Cumulative Emission Factor*(lb/ton)			
			Uncontrolled		*Multiple-Cyclone Controlled*[c]	
	Uncontrolled	*Multiple-Cyclone Controlled*	*Emission Factor*	*Emission Factor Rating*	*Emission Factor*	*Emission Factor Rating*
15	49	60	7.8	C	5.4	E
10	37	55	6.0	C	5.0	E
6	24	49	3.8	C	4.4	E
2.5	14	43	2.2	C	3.8	E
1.25	13	39	2.0	C	3.6	E
1.00	12	39	2.0	C	3.6	E
0.625	—[d]	16	—[d]	C	1.4	E
Total	100	100	16.0	C	9.0	E

[a]Data from EPA, *Compilation of Air Pollutant Emission Factors*, AP-42, Fifth ed., U.S. Environmental Protection Agency, U.S. Government Printing Office, Washington, D.C., January 1995. Tons are short tons. To convert from lb/ton to kg/Mg, multiply by 0.5. Emission factors are pounds of pollutant per ton of coal combusted as fired. ESP, electrostatic precipitator.
[b]Expressed as aerodynamic equivalent diameter.
[c]Estimated control efficiency for multiple cyclones is 80%.
[d]Insufficient data.

TABLE A-10
Cumulative Particle Size Distribution and Size-Specific Emission Factors for Underfeed Stokers Burning Bituminous Coal (Emission Factor Rating: C)[a]

Particle Size[b] (μm)	*Cumulative Mass % ≤ Stated Size*	*Uncontrolled Cumulative Emission Factor*[c] (lb/ton)
15	50	7.6
10	41	6.2
6	32	4.8
2.5	25	3.8
1.25	22	3.4
1.00	21	3.2
0.625	18	2.7
Total	100	15.0

[a]Data from EPA, *Compilation of Air Pollutant Emission Factors*, AP-42, Fifth ed., U.S. Environmental Protection Agency, U.S. Government Printing Office, Washington, D.C., January 1995. Tons are short tons. To convert from lb/ton to kg/Mg, multiply by 0.5. Emission factors are pounds of pollutant per ton of coal combusted, as fired.
[b]Expressed as aerodynamic equivalent diameter.
[c]May also be used for uncontrolled hand-fired units.

TABLE A-11
Cumulative Particle Size Distribution and Size-Specific Emission Factors for Spreader Stokers Burning Bituminous Coal[a]

Particle Size[b] (μm)	*Cumulative Mass % ≤ Stated Size*					*Cumulative Emission Factor (lb/ton)*				
		Controlled					*Controlled*			
	Uncontrolled	*Multiple-Cyclone*[c]	*Multiple-Cyclone*[d]	*ESP*	*Baghouse*	*Uncontrolled*[e]	*Multiple-Cyclone*[e,f]	*Multiple-Cyclone*[d,e]	*ESP*	*Baghouse*[e,g]
15	28	86	74	97	72	18.5	14.6	8.8	0.46	0.086
10	20	73	65	90	60	13.2	12	7.8	0.44	0.072
6	14	51	52	82	46	9.2	8.6	6.2	0.40	0.056
2.5	7	8	27	61	26	4.6	1.4	3.2	0.30	0.032
1.25	5	2	16	46	18	3.3	0.4	2.0	0.22	0.022
1.00	5	2	14	41	15	3.3	0.4	1.6	0.20	0.018
0.625	4	1	9	C[h]	7	2.6	0.2	1.0	C[h]	0.006
Total	100	100	100	100	100	66.0	17.0	12.0	0.48	0.12

[a]Data from EPA, *Compilation of Air Pollutant Emission Factors*, AP-42, Fifth ed., U.S. Environmental Protection Agency, U.S. Government Printing Office, Washington, D.C., January 1995. Tons are short tons. To convert from lb/ton to kg/Mg, multiply by 0.5. Emissions are pounds of pollutant per ton of coal combusted, as fired. ESP, electrostatic precipitator.
[b]Expressed as aerodynamic equivalent diameter.
[c]With flyash reinjection.
[d]Without flyash reinjection.
[e]Emission factor rating: C.
[f]Emission factor rating: E.
[g]Estimated control efficiency for ESP is 99.22%; for baghouse, 99.8%.
[h]Insufficient data.

TABLE A-12

Emission Factors for SO_X, NO_X, CO, and CO_2 from Uncontrolled Lignite Combustion (Emission Factor Rating: C, Except As Noted)[a]

Firing Configuration	SO_X *Emission Factor*[b] (lb/ton)	NO_X *Emission Factor* (lb/ton)	*CO Emission Factor* (lb/ton)	CO_2 *Emission Factor*[e] (lb/ton)	*TNMOC*[g,h,i] *Emission Factor* (lb/ton)
Pulverized-coal, dry-bottom, tangential	30S	7.1[f]	ND	72.6C	0.04
Pulverized-coal, dry-bottom, wall-fired,[c] pre-NSPS[d]	30S	13	0.25	72.6C	0.04
Pulverized-coal, dry-bottom, wall-fired,[c] NSPS[d]	30S	6.3	0.25	72.6C	0.04
Cyclone	30S	15	ND	72.6C	0.07
Spreader stoker	30S	5.8	ND	72.6C	0.03
Traveling-grate, overfeed stoker	30S	ND	ND	72.6C	0.03
Atmospheric, fluidized-bed combustor	10S[i]	3.6	0.15	72.6C	0.03

[a]Data from EPA, *Compilation of Air Pollutant Emission Factors*, AP-42, Fifth ed., U.S. Environmental Protection Agency, U.S. Government Printing Office, Washington, D.C., January 1995. Tons are short tons. To convert from lb/ton to kg/Mg, multiply by 0.5. To convert from lb/ton to lb/MM Btu, multiply by 0.0625. ND, no data.

[b]S = wt.% sulfur content of lignite, wet basis; for example, if the sulfur content equals 3.4%, then S = 3.4. For high-sodium ash ($Na_2O > 8\%$), use 22S. For low-sodium ash ($Na_2O < 2\%$), use 34S. If ash sodium content is unknown, use 30S.

[c]Wall-fired includes front- and rear-wall-fired units, as well as opposed-wall-fired units; NSPS, New Source Performance Standard.

[d]Pre-NSPS boilers are not subject to an NSPS. NSPS boilers are subject to Subpart D or Subpart D(a). Subpart D boilers are boilers constructed after August 17, 1971, with a heat input greater than 250 million Btu per hour (MM Btu/hr). Subpart D(a) boilers are boilers constructed after September 18, 1978, with a heat input rate greater than 250 MM Btu/hr.

[e]Emission factor rating: B. C = wt.% carbon of lignite, as-fired basis; for example, if carbon content equals 63%, then C = 63. If the %C value is not known, a default CO_2 emission value of 4600 lb/ton may be used.

[f]Emission factor rating: A.

[g]TNMOC: Total non-methane organic compounds. Emission factors were derived from bituminous coal data in the absence of lignite data assuming emissions are proportional to coal heating value. TNMOC are expressed as C_2 to C_{16} alkane equivalents. Because of limited data, the effects of firing configuration on TNMOC emission factors could not be distinguished. As a result, all data were averaged collectively to develop a single average emission factor for pulverized coal, cyclones, spreaders, and overfeed stokers.

[h]Nominal values achievable under normal operating conditions; values 1 or 2 orders of magnitude higher can occur when combustion is not complete.

[i]Using limestone bed material.

TABLE A-13
Emission Factors for NO_X and CO from Lignite Combustion with NO_X Controls[a]

Firing Configuration	*Control Device*	*NO_X*		*CO*	
		Emission Factor (lb/ton)	*Emission Factor Rating*	*Emission Factor* (lb/ton)	*Emission Factor Rating*
Subpart D boilers:[b] pulverized-coal, tangentially fired	Overfire air	6.8	C	ND	NA
Pulverized-coal, wall-fired	Overfire air and low NO_X burners	4.6	C	0.48	D
Subpart D(a) boilers:[b] pulverized coal, tangentially fired	Overfire air	6.0	C	0.1	D

[a]Data from EPA, *Compilation of Air Pollutant Emission Factors*, AP-42, Fifth ed., U.S. Environmental Protection Agency, U.S. Government Printing Office, Washington, D.C., January 1995. Tons are short tons. To convert from lb/ton to kg/Mg, multiply by 0.5. To convert from lb/ton to lb/MM Btu, multiply by 0.0625. ND, no data; NA, not applicable.

[b]Subpart D boilers are boilers constructed after August 17, 1971, with a heat input rate greater than 250 million Btu per hour (MM Btu/hr). Subpart D(a) boilers are boilers constructed after September 18, 1978, with a heat input rate greater than 250 MM Btu/hr.

TABLE A-14
Emission Factors for Polynuclear Organic Matter (POM) from Controlled Lignite Combustion (Emission Factor Rating: E)[a]

Firing Configuration	*Control Device*	*Emission Factor* ($lb/10^{12}$ Btu)
Pulverized-coal	High efficiency, cold-side ESP	2.3
Pulverized-coal, dry bottom	Multi-cyclone	1.8–18[b]
Pulverized-coal, dry bottom	ESP	2.6[b]
Cyclone furnace	ESP	0.11[c]–1.6[b]
Spreader stoker	Multi-cyclone	15[c]

[a]Data from EPA, *Compilation of Air Pollutant Emission Factors*, AP-42, Fifth ed., U.S. Environmental Protection Agency, U.S. Government Printing Office, Washington, D.C., January 1995. To convert from $lb/10^{12}$ Btu to pg/J, multiply by 0.43. ESP, electrostatic precipitator.
[b]Primarily trimethyl propenyl naphthalene.
[c]Primarily biphenyl.

TABLE A-15
Emission Factors for Filterable PM and N_2O from Uncontrolled Lignite Combustion (Emission Factor Rating: E, Except As Noted)[a]

Firing Configuration	*Filterable PM Emission Factor*[b] (lb/ton)	N_2O *Emission Factor*[c] (lb/ton)
Pulverized-coal, dry-bottom, tangentially fired	6.5A	ND
Pulverized-coal, dry-bottom, wall-fired	5.1A	ND
Cyclone	6.7A[c]	ND
Spreader stoker	8.0A	ND
Other stoker	3.4A	ND
Atmospheric fluidized-bed combustor	ND	2.5

[a]Data from EPA, *Compilation of Air Pollutant Emission Factors*, AP-42, Fifth ed., U.S. Environmental Protection Agency, U.S. Government Printing Office, Washington, D.C., January 1995. Tons are short tons. To convert from lb/ton to kg/Mg, multiply by 0.5. To convert from lb/ton to lb/MM Btu, multiply by 0.0625. ND, no data.
[b]A = wt.% ash content of lignite, wet basis; for example, if the ash content is 5%, then A = 5.
[c]Emission factor rating: C.

TABLE A-16
Emission Factors for Filterable Particulate Matter (PM) Emissions from Controlled Lignite Combustion (Emission Factor Rating: C, Except As Noted)[a]

Firing Configuration	*Control Device*	*Filterable PM Emission Factor* (lb/ton)
Subpart D boilers[b]	Baghouse wet scrubber	0.08A0.05A
Subpart D(a) boilers[b]	Wet scrubber	0.01A
Atmospheric fluidized-bed combustor[c]	ESP	0.07A

[a]Data from EPA, *Compilation of Air Pollutant Emission Factors*, AP-42, Fifth ed., U.S. Environmental Protection Agency, U.S. Government Printing Office, Washington, D.C., January 1995. Tons are short tons. A = wt.% ash content of lignite, wet basis; for example, if lignite is 2.3% ash, then A = 2.3. To convert from lb/ton to kg/Mg, multiply by 0.5. To convert from lb/ton to lb/MM Btu, multiply by 0.0625. ESP, electrostatic precipitator.
[b]Subpart D boilers are boilers constructed before August 17, 1971, with a heat input rate greater than 250 million Btu per hour (MM Btu/hr). Subpart D(a) boilers are boilers constructed after September 18, 1978, with a heat input rate greater than 250 MM Btu/hr.
[c]Emission factor rating: D.

TABLE A-17
Condensable Particulate Matter (PM) Emission Factors for Lignite Combustion[a]

Firing Configuration[b]	*Controls*[c]	*CPM-TOT*[d,e]		*CPM-IOR*[d,e]		*CPM-ORG*[d,e]	
		lb/MM Btu	*Rating*	lb/MM Btu	*Rating*	lb/MM Btu	*Rating*
All pulverized coal-fired boilers	All PM controls (without FGD controls)	0.1S–0.03[f]	C	80% of CPM-TOT emission factor[e]	E	20% of CPM-TOT emission factor[e]	E
All pulverized coal-fired boilers	All PM controls combined with FGD control	0.02[f]	E	ND	—	ND	—
Traveling-grate overfeed stoker, spreader stoker	All PM controls or uncontrolled	0.04	D	80% of CPM-TOT emission factor	E	20% of CPM-TOT emission factor	E

[a]Data from EPA, *Compilation of Air Pollutant Emission Factors*, AP-42, Fifth ed., U.S. Environmental Protection Agency, U.S. Government Printing Office, Washington, D.C., January 1995. All condensable PM is assumed to be less than 1.0 μm in diameter.

[b]No data are available for cyclone boilers; for cyclone boilers, use the factors provided for pulverized-coal-fired boilers and applicable controls.

[c]FGD, flue gas desulfurization.

[d]CPM-TOT, total condensable particulate matter; CPM-IOR, inorganic condensable particulate matter; CPM-ORG, organic condensable particulate matter; ND, no data.

[e]Factors should be multiplied by fuel rate on a heat input basis (MM Btu), as fired. To convert to lb/short ton of lignite, multiply by 16 MM Btu/short ton.

[f]S = wt.% coal sulfur, as fired; for example, if the sulfur percent is 1.04, then S = 1.04. If the coal sulfur percent is 0.4 or less, use a default emission factor of 0.01 lb/MM Btu rather than the emission equation.

TABLE A-18
Emission Factors for Trace Elements From Uncontrolled Lignite Combustion (Emission Factor Rating: E)[a]

Firing Configuration	*Emission Factor* ($lb/10^{12}$Btu)						
	As	*Be*	*Cd*	*Cr*	*Mn*	*Hg*	*Ni*
Pulverized-coal, wet-bottom	2730	131	49–77	1220–1880	4410–16250	21	154–1160
Pulverized-coal, dry-bottom	1390	131	49	1500–1880	16200	21	928–1160
Cyclone furnace	235–632	131	31	253–1880	3760	21	157–1160
Stoker, configuration unknown	ND	118	ND	ND	11800	21	ND
Spreader stoker	538–1100	ND	23–47	1130–1880	ND	ND	696–1160
Traveling-grate (overfed) stoker	1100–2100	ND	47–90	ND	ND	ND	ND

[a]Data from EPA, *Compilation of Air Pollutant Emission Factors*, AP-42, Fifth ed., U.S. Environmental Protection Agency, U.S. Government Printing Office, Washington, D.C., January 1995. To convert from $lb/10^{12}$ Btu to pg/J, multiply by 0.43. ND, no data.

TABLE A-19
Cumulative Particle Size Distribution and Size-Specific Emission Factors for Boilers Firing Pulverized Lignite (Emission Factor Rating: E)[a]

Particle Size[b] *(μm)*	*Cumulative Mass % ≤ Stated Size*		*Cumulative Emission*[c] (lb/ton)	
	Uncontrolled	*Multiple-Cyclone Controlled*	*Uncontrolled*	*Multiple-Cyclone Controlled*[d]
15	51	77	3.4A	1.0A
10	35	67	2.3A	0.88A
6	26	57	1.7A	0.75A
2.5	10	27	0.66A	0.36A
1.25	7	16	0.47A	0.21A
1.00	6	14	0.40A	0.19A
0.625	3	8	0.19A	0.11A
Total	—	—	6.6A	1.3A

[a]Data from EPA, *Compilation of Air Pollutant Emission Factors*, AP-42, Fifth ed., U.S. Environmental Protection Agency, U.S. Government Printing Office, Washington, D.C., January 1995. Tons are short tons. Based on tangentially fired units. For wall-fired units, multiply emission factors in the table by 0.79.

[b]Expressed as aerodynamic equivalent diameter.

[c]A = wt.% ash content of lignite, wet basis; for example, if lignite is 3.4% ash, then A = 3.4. To convert from lb/ton to kg/Mg, multiply by 0.5. To convert from lb/ton to lb/MM Btu, multiply by 0.0625.

[d]Estimated control efficiency for multiple cyclone is 80%, averaged over all particle sizes.

TABLE A-20
Cumulative Particle Size Distribution and Size-Specific Emission Factors for Lignite-Fired Spreader Stokers (Emission factor rating: E)[a]

Particle Size[b] (μm)	*Cumulative Mass % ≤ Stated Size*		*Cumulative Emission*[c] (lb/ton)	
	Uncontrolled	*Multiple-Cyclone Controlled*	*Uncontrolled*	*Multiple-Cyclone Controlled*[d]
15	28	55	2.2A	0.88A
10	20	41	1.6A	0.66A
6	14	31	1.1A	0.50A
2.5	7	26	0.56A	0.42A
1.25	5	23	0.40A	0.37A
1.00	5	22	0.40A	0.35A
0.625	4	—[e]	0.33A	—[e]
Total			8.0A	1.6A

[a]Data from EPA, *Compilation of Air Pollutant Emission Factors*, AP-42, Fifth ed., U.S. Environmental Protection Agency, U.S. Government Printing Office, Washington, D.C., January 1995. Tons are short tons.
[b]Expressed as aerodynamic equivalent diameter.
[c]A = wt.% ash content of lignite, wet basis; for example, if lignite is 5% ash, then A = 5. To convert from lb/ton to kg/Mg, multiply by 0.5. To convert from lb/ton to lb/MM Btu, multiply by 0.0625.
[d]Estimated control efficiency for multiple cyclone is 80%.
[e]Insufficient data.

TABLE A-21
Default CO_2 Emission Factors for U.S. Coals (Emission Factor Rating: C)[a]

Coal Type	*Average %C*[b]	*Conversion Factor*[c]	*Emission Factor*[d] (lb/ton coal)
Subbituminous	66.3	72.6	4810
High-volatile bituminous	75.9	72.6	5510
Medium-volatile bituminous	83.2	72.6	6040
Low-volatile bituminous	86.1	72.6	6250

[a]Data from EPA, *Compilation of Air Pollutant Emission Factors*, AP-42, Fifth ed., U.S. Environmental Protection Agency, U.S. Government Printing Office, Washington, D.C., January 1995. Tons are short tons. This table should be used only when an ultimate analysis is not available. If the ultimate analysis is available, CO_2 emissions should be calculated by multiplying the percent carbon (%C) by 72.6. This resultant factor would receive a quality rating of B.
[b]Based on average carbon contents for each coal type (dry basis) based on extensive sampling of U.S. coals.
[c]Based on the following equation:

$$\frac{44 \text{ ton } CO_2}{12 \text{ ton C}} \times 0.99 \times 2000 \frac{\text{lb } CO_2}{\text{ton } CO_2} \times \frac{1}{100\%} = 72.6 \frac{\text{lb } CO_2}{\text{ton } \%C}$$

where 44 = molecular weight of CO_2, 12 = molecular weight of carbon, and 0.99 = fraction of fuel oxidized during combustion.
[d]To convert from lb/ton to kg/Mg, multiply by 0.5.

TABLE A-22
Emission Factors for Various Organic Compounds from Controlled Coal Combustion[a]

Pollutant[b]	*Emission Factor* (lb/ton)[c]	*Emission Factor Rating*
Acetaldehyde	5.7E–04	C
Acetophenone	1.5E–05	D
Acrolein	2.9E–04	D
Benzene	1.3E–03	A
Benzyl chloride	7.0E–04	D
bis(2–Ethylhexyl)phthalate (DEHP)	7.3E–05	D
Bromoform	3.9E–05	E
Carbon disulfide	1.3E–04	D
2-Chloroacetophenone	7.0E–06	E
Chlorobenzene	2.2E–05	D
Chloroform	5.9E–05	D
Cumene	5.3E–06	E
Cyanide	2.5E–03	D
2,4-Dinitrotoluene	2.8E–07	D
Dimethyl sulfate	4.8E–05	E
Ethyl benzene	9.4E–05	D
Ethyl chloride	4.2E–05	D
Ethylene dichloride	4.0E–05	E
Ethylene dibromide	1.2E–06	E
Formaldehyde	2.4E–04	A
Hexane	6.7E–05	D
Isophorone	5.8E–04	D
Methyl bromide	1.6E–04	D
Methyl chloride	5.3E–04	D
Methyl ethyl ketone	3.9E–04	D
Methyl hydrazine	1.7E–04	E
Methyl methacrylate	2.0E–05	E
Methyl *tert*-butyl ether	3.5E–05	E
Methylene chloride	2.9E–04	D
Phenol	1.6E–05	D
Propionaldehyde	3.8E–04	D
Tetrachloroethylene	4.3E–05	D
Toluene	2.4E–04	A
1,1,1–Trichloroethane	2.0E–05	E
Styrene	2.5E–05	D
Xylenes	3.7E–05	C
Vinyl acetate	7.6E–06	E

[a]Data from EPA, *Compilation of Air Pollutant Emission Factors*, AP-42, Fifth ed., U.S. Environmental Protection Agency, U.S. Government Printing Office, Washington, D.C., January 1995. Tons are short tons. Factors were developed from emissions data from ten sites firing bituminous coal, eight sights firing subbituminous coal, and one site firing lignite. The emission factors are applicable to boilers using both wet limestone scrubbers or spray dryers and an electrostatic precipitator (ESP) or fabric filter (FF). In addition, the factors apply to boilers utilizing only an ESP or FF.

[b]Pollutants sampled for but not detected in any sampling run include: carbon tetrachloride, two sites; 1,3-dichloropropyene, two sites; *N*–nitrosodimethylamine, two sites; ethylidene dichloride, two sites; hexachlorobutadiene, two sites; hexachloroethane, 1 site; propylene dichloride, 2 sites; 1,1,2,2-tetrachloro-ethane, two sites; 1,1,2-trichloroethane, two sites; vinyl chloride, two sites; and hexachlorobenzene, two sites.

[c]Emission factor should be applied to coal feed, as fired.

TABLE A-23
Emission Factors for Polynuclear Aromatic Hydrocarbons (PAH) from Controlled Coal Combustion[a]

Pollutant	*Emission Factor*[b] (lb/ton)	*Emission Factor Rating*
Biphenyl	1.7E-06	D
Acenaphthene	5.1E-07	B
Acenaphthylene	2.5E-07	B
Anthracene	2.1E-07	B
Benzo(*a*)anthracene	8.0E–08	B
Benzo(*a*)pyrene	3.8E–08	D
Benzo(*b*, *j*, *k*)fluoranthene	1.1E–07	B
Benzo(*g*, *h*, *i*)perylene	2.7E–08	D
Chrysene	1.0E–07	C
Fluoranthene	7.1E–07	B
Fluorene	9.1E–07	B
Indeno(1,2,3-*c*, *d*)pyrene	6.1E–08	C
Naphthalene	1.3E–05	C
Phenanthrene	2.7E–06	B
Pyrene	3.3E–07	B
5-Methyl chrysene	2.2E–08	D

[a]Data from EPA, *Compilation of Air Pollutant Emission Factors*, AP-42, Fifth ed., U.S. Environmental Protection Agency, U.S. Government Printing Office, Washington, D.C., January 1995. Tons are short tons. Factors were developed from emissions data from six sites firing bituminous coal, four sights firing subbituminous coal, and one site firing lignite. Factors apply to boilers utilizing both wet limestone scrubbers or spray dryers and an electrostatic precipitator (ESP) or fabric filter (FF). The factors apply to boilers utilizing only an ESP or FF.

[b]Emission factor should be applied to coal feed, as fired. To convert from lb/ton to lb/MM Btu, multiply by 0.0625. To convert from lb/ton to kg/Mg, multiply by 0.5. Emissions are pounds of pollutant per ton of coal combusted.

TABLE A-24
Emission Factors for Hydrogen Chloride (HCl) and Hydrogen Fluoride (HF) from Coal Combustion (Emission Factor Rating: B)[a]

Firing Configuration	*HCl Emission Factor* (lb/ton)	*HF Emission Factor* (lb/ton)
PC-fired	1.2	0.15
PC-fired, tangential	1.2	0.15
Cyclone furnace	1.2	0.15
Traveling-grate (overfeed stoker)	1.2	0.15
Spreader stoker	1.2	0.15
Fluidized-bed combustion, circulating-bed	1.2	0.15

[a]Data from EPA, *Compilation of Air Pollutant Emission Factors*, AP-42, Fifth ed., U.S. Environmental Protection Agency, U.S. Government Printing Office, Washington, D.C., January 1995. Tons are short tons. The emission factors were developed from bituminous coal, subbituminous coal, and lignite emissions data. To convert from lb/ton to kg/Mg, multiply by 0.5. To convert from lb/ton to lb/MM Btu, multiply by 0.0625. The factors apply to both controlled and uncontrolled sources.

TABLE A-25
Emission Factors for Trace Metals from Controlled Coal Combustion[a]

Pollutant	*Emission Factor*[b] (lb/ton)	*Emission Factor Rating*
Antimony	1.8E–05	A
Arsenic	4.1E–04	A
Beryllium	2.1E–05	A
Cadmium	5.1E–05	A
Chromium	2.6E–04	A
Chromium (VI)	7.9E–05	D
Cobalt	1.0E–04	A
Lead	4.2E–04	A
Magnesium	1.1E–02	A
Manganese	4.9E–04	A
Mercury	8.3E–05	A
Nickel	2.8E–04	A
Selenium	1.3E–03	A

[a]Data from EPA, *Compilation of Air Pollutant Emission Factors*, AP-42, Fifth ed., U.S. Environmental Protection Agency, U.S. Government Printing Office, Washington, D.C., January 1995. Tons are short tons. The emission factors were developed from emissions data at 11 facilities firing bituminous coal, 15 facilities firing subbituminous coal, and from 2 facilities firing lignite. The factors apply to boilers utilizing Venturi scrubbers, spray dryer absorbers, or wet limestone scrubbers with an electrostatic precipitator (ESP) or fabric filter (FF). In addition, the factors apply to boilers using only an ESP, a FF, or a Venturi scrubber. Firing configurations include pulverized-coal-fired, dry-bottom boilers; pulverized-coal, dry-bottom, tangentially fired boilers; cyclone boilers; and atmospheric fluidized-bed combustors, circulating-bed.

[b]Emission factor should be applied to coal feed, as fired. To convert from lb/ton to kg/Mg, multiply by 0.5.

TABLE A-26
Emission Factor Equations for Trace Elements from Coal Combustion[a] (Emission Factor Equation Rating: A)[b]

Pollutant	*Emission Equation* ($lb/10^{12}Btu$)[c]
Antimony	$0.92\ [(C/A)PM]^{0.63}$
Arsenic	$3.1\ [(C/A)PM]^{0.85}$
Beryllium	$1.2\ [(C/A)PM]^{1.1}$
Cadmium	$3.3\ [(C/A)PM]^{0.5}$
Chromium	$3.7\ [(C/A)PM]^{0.58}$
Cobalt	$1.7\ [(C/A)PM]^{0.69}$
Lead	$3.4\ [(C/A)PM]^{0.80}$
Manganese	$3.8\ [(C/A)PM]^{0.60}$
Nickel	$4.4\ [(C/A)PM]^{0.48}$

[a]Data from EPA, *Compilation of Air Pollutant Emission Factors*, AP-42, Fifth ed., U.S. Environmental Protection Agency, U.S. Government Printing Office, Washington, D.C., January 1995. The equations were developed from emissions data from bituminous coal combustion, subbituminous coal combustion, and lignite combustion. The equations may be used to generate factors for both controlled and uncontrolled boilers. The emission factor equations are applicable to all typical firing configurations for electric generation (utility), industrial, and commercial/industrial boilers for bituminous coal, subbituminous coal, and lignite.

[b]AP-42 criteria for rating emission factors were used to rate the equations.

[c]The factors produced by the equations should be applied to heat input. To convert from $lb/10^{12}$ Btu to kg/J, multiply by 4.31×10^{-16}. C = concentration of metal in the coal (parts per million by weight [ppmwt]). A = weight fraction of ash in the coal; for example, 10% ash is 0.1 ash fraction. PM = site-specific emission factor for total particulate matter ($lb/10^6$ Btu).

TABLE A-27
Emission Factors for SO_X and NO_X Compounds from Uncontrolled Anthracite Coal Combustors[a]

Source Category	SO_X		NO_X	
	Emission Factor (lb/ton)	*Emission Factor Rating*	*Emission Factor* (lb/ton)	*Emission Factor Rating*
Stoker-fired boilers	39S[b]	B	9.0	C
FBC boilers[c]	2.9	E	1.8	E
Pulverized coal boilers	39S[b]	B	18	B

[a]Data from EPA, *Compilation of Air Pollutant Emission Factors,* AP-42, Fifth ed., U.S. Environmental Protection Agency, U.S. Government Printing Office, Washington, D.C., January 1995. Tons are short tons. Units are pounds of pollutant per ton of coal burned. To convert from lb/ton to kg/Mg, multiply by 0.5.

[b]S = weight percent sulfur; for example, if the sulfur content is 3.4%, then S = 3.4.

[c]Fluidized-bed combustion (FBC) boilers burning culm fuel; all other sources burning anthracite coal.

TABLE A-28
Emission Factors for CO and Carbon Dioxide (CO_2) from Uncontrolled Anthracite Coal Combustors[a]

Source Category	CO		CO_2	
	Emission Factor (lb/ton)	*Emission Factor Rating*	*Emission Factor* (lb/ton)	*Emission Factor Rating*
Stoker-fired boilers	0.6	B	5680	C
FBC boilers[b]	0.6	E	ND	NA

[a]Data from EPA, *Compilation of Air Pollutant Emission Factors,* AP-42, Fifth ed., U.S. Environmental Protection Agency, U.S. Government Printing Office, Washington, D.C., January 1995. Tons are short tons. Units are pounds of pollutant per ton of coal burned. To convert from lb/ton to kg/Mg, multiply by 0.5. ND, no data; NA, not applicable.

[b]Fluidized-bed combustion (FBC) boilers burning culm fuel; all other sources burning anthracite coal.

TABLE A-29
Emission Factors for Speciated Organic Compounds from Anthracite Coal Combustors (Emission Factor Rating: E)[a]

Pollutant	*Stoker-Fired Boilers Emission Factor* (lb/ton)
Acenaphthene	ND
Acenaphthylene	ND
Anthrene	ND
Anthracene	ND
Benzo(*a*)anthracene	ND
Benzo(*a*)pyrene	ND
Benzo(*e*)pyrene	ND
Benzo(*g, h, i*)perylene	ND
Benzo(*k*)fluoranthrene	ND
Biphenyl	2.5E–02
Chrysene	ND
Coronene	ND
Fluoranthrene	ND
Fluorene	ND
Indeno(123-*c, d*)perylene	ND
Naphthalene	1.3E–01
Perylene	ND
Phenanthrene	6.8E–03
Pyrene	ND

[a]Data from EPA, *Compilation of Air Pollutant Emission Factors*, AP-42, Fifth ed., U.S. Environmental Protection Agency, U.S. Government Printing Office, Washington, D.C., January 1995. Tons are short tons. Units are pounds of pollutant per ton of coal burned. To convert from lb/ton to kg/Mg, multiply by 0.5. ND, no data.

TABLE A-30
Emission Factors for Total Organic Carbon (TOC) and Methane (CH_4) from Anthracite Coal Combustors (Emission Factor Rating: E)[a]

Source Category	*TOC Emission Factor* (lb/ton)	*CH_4 Emission Factor* (lb/ton)
Stoker fired boilers	0.30	ND

[a]Data from EPA, *Compilation of Air Pollutant Emission Factors*, AP-42, Fifth ed., U.S. Environmental Protection Agency, U.S. Government Printing Office, Washington, D.C., January 1995. Tons are short tons. Units are pounds of pollutant per ton of coal burned. To convert from lb/ton to kg/Mg, multiply by 0.5. ND, no data.

TABLE A-31
Emission Factors for Speciated Metals from Anthracite Coal Combustion in Stoker-Fired Boilers (Emission Factor Rating: E)[a]

Pollutant	*Emission Factor Range* (lb/ton)	*Average Emission Factor* (lb/ton)
Arsenic	BDL to 2.4E–04	1.9E–04
Antimony	BDL	BDL
Beryllium	3.0E–05 to 5.4E–04	3.1E–04
Cadmium	4.5E–05 to 1.1E–04	7.1E–05
Chromium	5.9E–03 to 4.9E–02	2.8E–02
Manganese	9.8E–04 to 5.3E–03	3.6E–03
Mercury	8.7E–05 to 1.7E–04	1.3E–04
Nickel	7.8E–03 to 3.5E–02	2.6E–02
Selenium	4.7E–04 to 2.1E–03	1.3E–03

[a]Data from EPA, *Compilation of Air Pollutant Emission Factors*, AP-42, Fifth ed., U.S. Environmental Protection Agency, U.S. Government Printing Office, Washington, D.C., January 1995. Tons are short tons. Units are pounds of pollutant per ton of coal burned. To convert from lb/ton to kg/Mg, multiply by 0.5. BDL, below detection limit.

TABLE A-32
Emission Factors for Particulate Matter (PM) and Lead (Pb) from Uncontrolled Anthracite Coal Combustors[a]

Source Category	*Filterable PM*		*Condensable PM*		*Pb*	
	Emission *Factor* (lb/ton)	Emission *Factor Rating*	*Emission Factor* (lb/ton)	*Emission Factor Rating*	*Emission Factor* (lb/ton)	*Emission Factor Rating*
Stoker-fired boilers	0.8A[b]	C	0.08A[b]	C	8.9E–03	E
Hand-fired units	10	B	ND	NA	ND	NA

[a]Data from EPA, *Compilation of Air Pollutant Emission Factors*, AP-42, Fifth ed., U.S. Environmental Protection Agency, U.S. Government Printing Office, Washington, D.C., January 1995. Tons are short tons. Units are pounds of pollutant per ton of coal burned. To convert from lb/ton to kg/Mg, multiply by 0.5. ND, no data; NA, not applicable.
[b]A = ash content of fuel (wt.%); for example, if the ash content is 5%, then A = 5.

TABLE A-33
Cumulative Particle Size Distribution and Size-Specific Emission Factors for Dry-Bottom Boilers Burning Pulverized Anthracite Coal (Emission Factor Rating: D)[a]

Particle size[b] (μm)	*Cumulative Mass % ≤ Stated Size*			*Cumulative Emission Factor As Fired*[c] (lb/ton)		
	Uncontrolled	Controlled[d]		*Uncontrolled*	Controlled[d]	
		Multiple-Cyclone	*Baghouse*		*Multiple-Cyclone*	*Baghouse*
15	32	63	79	3.2A[e]	1.26A	0.016A
10	23	55	67	2.3A	1.10A	0.013A
6	17	46	51	1.7A	0.92A	0.010A
2.5	6	24	32	0.6A	0.48A	0.006A
1.25	2	13	21	0.2A	0.26A	0.004A
1.00	2	10	18	0.2A	0.20A	0.004A
0.625	1	7	—[f]	0.1A	0.14A	—[f]
Total	100	100	100	10A	2A	0.02A

[a]Data from EPA, *Compilation of Air Pollutant Emission Factors*, AP-42, Fifth ed., U.S. Environmental Protection Agency, U.S. Government Printing Office, Washington, D.C., January 1995. Tons are short tons.
[b]Expressed as aerodynamic equivalent diameter.
[c]Units are pounds of pollutant per ton of coal burned. To convert from lb/ton to kg/Mg, multiply by 0.5.
[d]Estimated control efficiency for multiple cyclone is 80%; for baghouse, 99.8%.
[e]A = coal ash weight %, as fired; for example, if ash content is 5%, then A = 5.
[f]Insufficient data.

APPENDIX B

Original List of Hazardous Air Pollutants

CAS Number	Chemical Name
75070	Acetaldehyde
60355	Acetamide
75058	Acetonitrile
98862	Acetophenone
53963	2-Acetylaminofluorene
107028	Acrolein
79061	Acrylamide
79107	Acrylic acid
107131	Acrylonitrile
107051	Allyl chloride
92671	4-Aminobiphenyl
62533	Aniline
90040	*o*-Anisidine
1332214	Asbestos
71432	Benzene (including benzene from gasoline)
92875	Benzidine
98077	Benzotrichloride
100447	Benzyl chloride
92524	Biphenyl
117817	*bis*(2-Ethylhexyl)phthalate (DEHP)
542881	*bis*(Chloromethyl)ether
75252	Bromoform
106990	1,3-Butadiene
156627	Calcium cyanamide
105602	Caprolactam[a]
133062	Captan
63252	Carbaryl
75150	Carbon disulfide
56235	Carbon tetrachloride
463581	Carbonyl sulfide
120809	Catechol

(continued)

(continued)

CAS Number	*Chemical Name*
133904	Chloramben
57749	Chlordane
7782505	Chlorine
79118	Chloroacetic acid
532274	2-Chloroacetophenone
108907	Chlorobenzene
510156	Chlorobenzilate
67663	Chloroform
107302	Chloromethyl methyl ether
126998	Chloropene
1319773	Cresols/cresylic acid (isomers and mixture)
95487	*o*-Cresol
108394	*m*-Cresol
106445	*p*-Cresol
98828	Cumene
94757	2,4-D, salts and esters
3547044	DDE
334883	Diazomethane
132649	Dibenzofurans
96128	1,2-Dibromo-3-chloropropane
84742	Dibutylphthalate
106467	1,4-Dichlorobenzene(*p*)
91941	3,3-Dichlorobenzidene
111444	Dichloroethyl ether (*bis*(2-chloroethyl)ether)
542756	1,3-Dichloropropene
62737	Dichlorvos
111422	Diethanolamine
121697	*N,N*-Diethylaniline (*N,N*-dimethylaniline)
64675	Diethyl sulfate
119904	3,3-Dimethoxybenzidine
60117	Dimethylaminoazobenzene
119937	3,3′-Dimethylbenzidine
79447	Dimethyl carbamoyl chloride
68122	Dimethylformamide
57147	1,1-Dimethyl hydrazine
131113	Dimethyl phthalate
77781	Dimethyl sulfate
534521	4,6-Dinitro-*o*-cresol and salts
51285	2,4-Dinitrophenol
121142	2,4-Dinitrotoluene
123911	1,4-Dioxane (1,4-diethyleneoxide)
122667	1,2-Diphenylhydrazine
106898	Epichlorohydrin (l-chloro-2,3-epoxypropane)
106887	1,2-Epoxybutane
140885	Ethyl acrylate
100414	Ethyl benzene

(continued)

(continued)

CAS Number	*Chemical Name*
51796	Ethyl carbamate (urethane)
75003	Ethyl chloride (chloroethane)
106934	Ethylene dibromide (dibromoethane)
107062	Ethylene dichloride (1,2-dichloroethane)
107211	Ethylene glycol
151564	Ethylene imine (aziridine)
75218	Ethylene oxide
96457	Ethylene thiourea
75343	Ethylidene dichloride (1,1-dichloroethane)
50000	Formaldehyde
76448	Heptachlor
118741	Hexachlorobenzene
87683	Hexachlorobutadiene
77474	Hexachlorocyclopentadiene
67721	Hexachloroethane
822060	Hexamethylene-1,6-diisocyanate
680319	Hexamethylphosphoramide
110543	Hexane
302012	Hydrazine
7647010	Hydrochloric acid
7664393	Hydrogen fluoride (hydrofluoric acid)
7783064	Hydrogen sulfide[b]
123319	Hydroquinone
78591	Isophorone
58899	Lindane (all isomers)
108316	Maleic anhydride
67561	Methanol
72435	Methoxychlor
74839	Methyl bromide (bromomethane)
74873	Methyl chloride (chloromethane)
71556	Methyl chloroform (1,1,1-trichloroethane)
78933	Methyl ethyl ketone (2-butanone)
60344	Methyl hydrazine
74884	Methyl iodide (iodomethane)
108101	Methyl isobutyl ketone (hexone)
624839	Methyl isocyanate
80626	Methyl methacrylate
1634044	Methyl *tert*-butyl ether
101144	4,4-Methylene *bis*(2-chloroaniline)
75092	Methylene chloride (dichloromethane)
101688	Methylene diphenyl diisocyanate (MDI)
101779	4,4-Methylenedianiline
91203	Naphthalene
98953	Nitrobenzene
92933	4-Nitrobiphenyl
100027	4-Nitrophenol

(continued)

(continued)

CAS Number	*Chemical Name*
79469	2-Nitropropane
684935	*N*-Nitroso-*N*-methylurea
62759	*N*-Nitrosodimethylamine
59892	*N*-Nitrosomorpholine
56382	Parathion
82688	Pentachloronitrobenzene (quintobenzene)
87865	Pentachlorophenol
108952	Phenol
106503	*p*-Phenylenediamine
75445	Phosgene
7803512	Phosphine
7723140	Phosphorus
85449	Phthalic anhydride
1336363	Polychlorinated biphenyls (Aroclor®)
1120714	1,3-Propane sultone
57578	β-Propiolactone
123386	Propionaldehyde
114261	Propoxur (Baygon®)
78875	Propylene dichloride (1,2-dichloropropane)
75569	Propylene oxide
75558	1,2-Propylenimine (2-methyl aziridine)
91225	Quinoline
106514	Quinone
100425	Styrene
96093	Styrene oxide
1746016	2,3,7,8-Tetrachlorodibenzo-*p*-dioxin
79345	1,1,2,2-Tetrachloroethane
127184	Tetrachloroethylene (perchloroethylene)
7550450	Titanium tetrachloride
108883	Toluene
95807	2,4-Toluene diamine
584849	2,4-Toluene diisocyanate
95534	*o*-Toluidine
8001352	Toxaphene (chlorinated camphene)
120821	1,2,4-Trichlorobenzene
79005	1,1,2-Trichloroethane
79016	Trichloroethylene
95954	2,4,5-Trichlorophenol
88062	2,4,6-Trichlorophenol
121448	Triethylamine
1582098	Trifluralin
540841	2,2,4-Trimethylpentane
108054	Vinyl acetate
593602	Vinyl bromide
75014	Vinyl chloride

(continued)

(continued)

CAS Number	*Chemical Name*
75354	Vinylidene chloride (1,1-dichloroethylene)
1330207	Xylenes (isomers and mixture)
95476	*o*-Xylenes
108383	*m*-Xylenes
106423	*p*-Xylenes
0	Antimony compounds
0	Arsenic compounds (inorganic, including arsine)
0	Beryllium compounds
0	Cadmium compounds
0	Chromium compounds
0	Cobalt compounds
0	Coke oven emissions
0	Cyanide compounds[c]
0	Glycol ethers[d]
0	Lead compounds
0	Manganese compounds
0	Mercury compounds
0	Fine mineral fibers[e]
0	Nickel compounds
0	Polycylic organic matter[f]
0	Radionuclides (including radon)[g]
0	Selenium compounds

[a]Caprolactam was delisted on June 18, 1996.

[b]Hydrogen sulfide was inadvertently added to the Section 112(b) list of HAPs through a clerical error. A joint resolution was passed by Congress and approved by the President on December 4, 1991, to remove hydrogen sulfide from Section 112(b). Hydrogen sulfide is included in Section 112(r) and is subject to accidental release provisions.

[c]X′CN where X = H′or any other group where a formal dissociation may occur—for example, KCN or $Ca(CN)_2$.

[d]Includes mono- and diethers of ethylene glycol—diethylene glycol and triethylene glycol R–$(OCH_2CH_2)n$–OR′ where $n = 1, 2$, or 3; R = alkyl or aryl groups; R′ = R, H, or groups that, when removed, yield glycol ethers with the structure R–$(OCH_2CH)n$–OH. Polymers are excluded from the glycol category.

[e]Includes mineral fiber emissions from facilities manufacturing or processing glass, rock, or slag fibers (or other mineral-derived fibers) of average diameter 1 μm or less.

[f]Includes organic compounds with more than one benzene ring and which have a boiling point greater than or equal to 100°C.

[g]A type of atom that spontaneously undergoes radioactive decay.

Note: For all of the listings that contain the word "compounds" and for glycol ethers, the following applies: Unless otherwise specified, these listings are defined as including any unique chemical substance that contains the named chemical (*i.e.*, antimony, arsenic, etc.) as part of the infrastructure of that chemical.

APPENDIX C

Initial 263 Units Identified in Phase I (SO_2) of the Acid Rain Program

Plant Name	*Unit Number*	*State*
Albright	3	West Virginia
Allen	1, 2, 3	Tennessee
Armstrong	1, 2	Pennsylvania
Asbury	1	Missouri
Ashtabula	7	Ohio
Avon Lake	11, 12	Ohio
BL England	1, 2	New Jersey
Bailly	7, 8	Indiana
Baldwin	1, 2, 3	Illinois
Big Bend	BB01, BB02, BB03	Florida
Bowen	1BLR, 2BLR, 3BLR, 4BLR	Georgia
Breed	1	Indiana
Brunner Island	1, 2, 3	Pennsylvania
Burlington	1	Iowa
C P Crane	1, 2	Maryland
Cardinal/Tidd	1, 2	Ohio
Cayuga	1, 2	Indiana
Chalk Point	1, 2	Maryland
Cheswick	1	Pennsylvania
Clifty Creek	1, 2, 3, 4, 5, 6	Indiana
Coffeen	1, 2	Illinois
Colbert	1, 2, 3, 4, 5	Alabama
Coleman	C1, C2, C3	Kentucky
Conemaugh	1, 2	Pennsylvania

(continued)

(continued)

Plant Name	*Unit Number*	*State*
Conesville	1, 2, 3, 4	Ohio
Cooper	1, 2	Kentucky
Crist	6, 7	Florida
Cumberland	1, 2	Tennessee
Des Moines	11	Iowa
Dunkirk	3, 4	New York
EC Gaston	1, 2, 3, 4, 5	Alabama
EW Brown	1, 2, 3	Kentucky
Elmer W Stout	50, 60, 70	Indiana
Eastlake	1, 2, 3, 4, 5	Ohio
Edgewater	4	Wisconsin
Edgewater	13	Ohio
Elmer Smith	1, 2	Kentucky
FB Culley	2, 3	Indiana
Fort Martin	1, 2	West Virginia
Frank E Ratts	1SG1, 2SG1	Indiana
Gallatin	1, 2, 3, 4	Tennessee
Gen JM Gavin	1, 2	Ohio
Genoa	1	Wisconsin
George Neal North	1	Indiana
Ghent	1	Kentucky
Gibson	1, 2, 3,4	Indiana
Grand Tower	09	Illinois
Green River	5	Kentucky
Greenidge	6	New York
HL Spurlock	1	Kentucky
HMP&L Station 2	H1, H2	Kentucky
HT Pritchard	6	Indiana
Hammond	1, 2, 3, 4	Georgia
Harrison	1, 2,3	West Virginia
Hatfield's Ferry	3	Pennsylvania
Hennepin	2	Illinois
High Bridge	6	Minnesota
JH Campbell	1, 2	Michigan
Jack McDonough	MB1, MB2	Georgia
Jack Watson	4, 5	Mississippi
James River	5	Missouri
Johnsonville	1, 2, 3 4, 5, 6, 7, 8, 9, 10	Tennessee
Joppa Steam	1, 2, 3, 4, 5, 6	Illinois
Kammer	1, 2, 3	West Virginia
Kincaid	1, 2	Illinois
Kyger Creek	1, 2, 3, 4, 5	Ohio
Labadie	1, 2, 3, 4	Missouri
Martins Creek	1, 2	Pennsylvania
Meredosia	05	Illinois

(continued)

(continued)

Plant Name	*Unit Number*	*State*
Merrimack	1, 2	New Hampshire
Miami Fort	5-1, 5-2, 6, 7	Ohio
Michigan City	12	Indiana
Milliken	1, 2	New York
Milton L Kapp	2	Iowa
Mitchell	1, 2	West Virginia
Montrose	1, 2, 3	Missouri
Morgantown	1, 2	Maryland
Mt. Storm	1, 2, 3	West Virginia
Muskingum River	1, 2, 3, 4, 5	Ohio
Nelson Dewey	1, 2	Wisconsin
New Madrid	1, 2	Missouri
Niles	1, 2	Ohio
North Oak Creek	1, 2, 3,	Wisconsin
Northport	1, 2, 3	New York
Paradise	3	Kentucky
Petersburg	1, 2	Indiana
Picway	9	Ohio
Port Jefferson	3, 4	New York
Portland	1, 2	Pennsylvania
Prairie Creek	4	Iowa
Pulliam	8	Wisconsin
Quindaro	2	Kansas
RE Burger	5, 6, 7, 8	Ohio
R Gallagher	1, 2, 3, 4	Indiana
Riverside	9	Iowa
Shawnee	10	Kentucky
Shawville	1, 2, 3, 4	Pennsylvania
Sibley	3	Missouri
Sioux	1, 2	Missouri
South Oak Creek	5, 6, 7, 8	Wisconsin
Sunbury	3, 4	Pennsylvania
Tanners Creek	U4	Indiana
Thomas Hill	MB1, MB2	Missouri
Vermillion	2	Illinois
WH Sammis	5, 6, 7	Ohio
Wabash River	1, 2, 3, 4, 5, 6	Indiana
Walter C. Beckjord	5, 6	Ohio
Wansley	1, 2	Georgia
Warrick	4	Indiana
Yates	Y1BR, Y2BR, Y3BR, Y4BR, Y5BR, Y6BR, Y7BR	Georgia

APPENDIX D

Commercial Gasification Facilities Worldwide

Country/Application	*Number of Projects*		*Number of Gasifiers*		*Syngas Capacity (MM scf/d)*
	Active	*Planned*	*Active*	*Planned*	
Australia					
Chemicals: hydrogen	1	—	2	—	28.4
Power: electricity	1	—	1	—	0.7
Brazil					
Chemicals: ammonia	1	—	3	—	116.5
China					
Chemicals: ammonia	14	—	36	—	967.9
Chemicals: methanol	2	1	5	1	79.2 (74.1 planned)
Chemicals: oxochemicals	3	—	3	—	20.8
Chemicals: syngas	—	2	—	3	327.0
Gaseous fuels: fuel gas	1	—	8	—	105.9
Gaseous fuels: town gas	1	—	1	—	53.0
Czech Republic					
Chemicals: methanol	1	—	6	—	127.1
Power: electricity	1	1	26	2	165.9 (189.2 planned)
Dominican Republic					
Gaseous fuels: reducing gas	1	—	12	—	50.8
Egypt					
Chemicals: ammonia	1	—	3	—	27.5
Finland					
Chemicals: syngas	1	—	1	—	10.6
Gaseous fuels: syngas	3	—	3	—	20.7
Power: electricity	1	—	1	—	12.4
France					
Chemicals: CO	1	—	1	—	9.8
Chemicals: oxochemicals	3	—	3	—	57.9

(continued)

(continued)

Country/Application	*Number of Projects*		*Number of Gasifiers*		*Syngas Capacity (MM scf/d)*
	Active	*Planned*	*Active*	*Planned*	
Gaseous fuels: syngas	1	—	1	—	0.5
Power: electricity	1	—	3	—	268.3
Germany					
Chemicals: ammonia	2	—	8	—	310.6
Chemicals: CO	2	—	1	—	9.3
Chemicals: hydrogen	1	—	10	—	288.8
Chemicals: methanol	4	—	10	—	265.9
Chemicals: oxochemicals	5	—	5	—	153.9
Gaseous fuels: fuel gas	1	—	1	—	25.8
Gaseous fuels: syngas	1	—	1	—	53.0
Gaseous fuels: town gas	1	—	7	—	219.0
Power: electricity	3	—	3	—	133.4
India					
Chemicals: ammonia	9	—	24	—	619.5
Chemicals: syngas	1	—	2	—	28.2
Power: electricity	1	1	1	1	28.4 (187.8 planned)
Italy					
Chemicals: CO	1	—	2	—	24.7
Chemicals: hydrogen	1	—	2	—	40.6
Power: electricity	4	2	8	5	701.6 (286.1 planned)
Japan					
Chemicals: ammonia	1	—	4	—	75.9
Chemicals: CO	2	—	3	—	24.7
Chemicals: methanol	1	—	2	—	7.1
Chemicals: syngas	1	—	2	—	14.1
Power: electricity	1	2	3	3	162.0 (225.3 planned)
Malaysia					
Fischer–Tropsch processing liquids: mid-distillates	1	—	6	—	266.6
Netherlands					
Chemicals: hydrogen	1	—	3	—	164.6
Gaseous fuels: fuel gas	1	—	1	—	21.7
Power: electricity	1	1	1	1	120.3 (12.4 planned)
Poland					
Power: electricity	—	1	—	1	238.3
Portugal					
Chemicals: ammonia	2	—	5	—	106.6
Gaseous fuels: syngas	1	—	1	—	3.1
Russia					
Chemicals: methanol	1	—	1	—	2.4
Singapore					
Chemicals: hydrogen	1	—	2	—	56.8
Power: electricity	1	—	2	—	93.9

(continued)

(continued)

Country/Application	*Number of Projects*		*Number of Gasifiers*		*Syngas Capacity (MM scf/d)*
	Active	*Planned*	*Active*	*Planned*	
South Africa					
Chemicals: ammonia	1	—	6	—	75.9
Fischer–Tropsch processing liquids	3	—	99	—	2580.4
South Korea					
Chemicals: ammonia	1	—	1	—	17.7
Chemicals: CO	1	—	1	—	21.2
Chemicals: oxochemicals	1	—	1	—	13.6
Spain					
Chemicals: ammonia	1	—	1	—	1.5
Chemicals: CO	1	—	1	—	5.7
Power: electricity	2	—	3	—	578.9
Sweden					
Chemicals: oxochemicals	1	—	1	—	7.1
Gaseous fuels: syngas	1	—	1	—	5.6
Power: electricity	1	—	1	—	2.8
Taiwan					
Chemicals: hydrogen	1	—	2	—	75.6
Ukraine					
Gaseous fuels: syngas	—	1	—	1	114.7
United Kingdom					
Chemicals: acetyls	1	—	1	—	32.2
Chemicals: oxochemicals	1	—	3	—	21.2
Gaseous fuels: syngas	1	—	1	—	7.8
Power: electricity	2	2	2	2	68.2 (199.2 planned)
United States					
Chemicals: acetic anhydride	1	—	1	—	56.5
Chemicals: ammonia	1	—	1	—	75.6
Chemicals: hydrogen	4	—	7	—	271.8
Chemicals: methanol	1	—	2	—	169.4
Chemicals: oxochemicals	7	—	13	—	182.9
Chemicals: syngas	—	1	—	2	74.1
Gaseous fuels: synthetic natural gas	1	—	12	—	490.7
Gaseous fuels: syngas	1	—	2	—	1.6
Power: electricity	7	4	10	4	546.6 (890.2 planned)
Unspecified Asian country					
Chemicals	—	1	—	2	258.2
Unspecified European country					
Chemicals: ammonia	—	1	—	2	258.2
Power: electricity	—	1	—	2	258.2

(continued)

(continued)

Country/Application	*Number of Projects*		*Number of Gasifiers*		*Syngas Capacity (MM scf/d)*
	Active	*Planned*	*Active*	*Planned*	
Yugoslavia, Former					
Chemicals: ammonia	1	—	1	—	4.2
Chemicals: methanol	1	—	1	—	54.4
Zambia					
Chemicals: ammonia	1	—	2	—	31.0

Note: Syngas capacity is reported as million standard cubic feet per day.
Source: Adapted from Gasification Technologies Council, *Gasification: A Growing, World-wide Industry* (Gasification Technologies Council, Arlington, VA), www.gasification.org.story (accessed October 2, 2003).

Index

A

Abraham, Spencer – 161, 455
Acenaphthene – 106
Acenaphthylene – 106
Acid gases – 107
Acid mine drainage (AMD) – 77, 82, 84, 87–88
Acid rain – 98, 175, 184
Acid Rain Program – 142, 145, 147, 184, 286, 322, 505–507
Advanced Coal Conversion Process (ACCP) – 245–246
Advanced combustion technology – 459
Aerosol Research and Inhalation Epidemiology Study (ARIES) – 99, 102, 104
Agglomeration – 91–92
Air-to-cloth ratio (A/C) – 358
Alkalis – 234
Allen, John – 197
Alstom Power/ Combustion Engineering – 200
Air pollution – 92–93, 95, 97–98, 124
Air Pollution Control Act of 1955 – 126
Air Pollution Control Act Amendments of 1960 – 126
Air Pollution Control Act Amendments of 1962 – 126
Air quality – 77, 173, 175
Air Quality Control Act of 1967 – 129
Air Quality Control Regions (AQCR) – 127
Air Quality Criteria (AQC) – 127, 129
Air quality monitoring – 104
Air toxics – 144–145
Allochthonous – 2
Aluminum – 96, 109
American Indian Religious Freedom Act – 89
Ammonia – 100–101, 338–342
Ammonia slip – 342
Ammonium bisulfate – 339
Ammonium sulfate – 339
Anhydrous ammonia – 341
Anaerobic decomposition – 115
Anthracite – 2, 4, 16, 19–25, 41
Anthropogenic activities – 97, 115–116
Antimony – 107
Antiquities Act of 1906 – 89
Archeological and Historical Preservation Act of 1974 – 89
AP-42 – 137–138, 473–498
Archeological Salvage Act – 89
Aristotle – 29
Arsenic – 96–97, 104, 107, 109–110, 145
Ash characteristics – 231
Ash fusibility – 229, 232
Ash fusion temperatures – 8, 215, 255
Ash viscosity – 229
ASTM, American Society of Testing Materials – 8
ASTM Coal classification system – 9
Autochthonous – 2
Avian Influenza (AI) – 464

B

Babcock, George – 199
Babcock & Wilcox, B&W – 199–200

Bakeries – 32
Bald Eagle Protection Act of 1969 – 89
Barium – 96, 107
Beehive ovens – 239–240
Bergius, Friedrich – 269, 273
Berthelot – 269
Benz[a]anthracene – 106
Benzo[b]fluoranthene – 106
Benzo[ghi]perylene – 106
Benzo[a]pyrene – 106
Beryllium – 96, 107, 110
Best available control technology (BACT) – 138, 140–141
Bituminous coal – 2, 4, 9, 16, 19–26, 41, 47–48
Bhopal, India – 145
Blacksmiths – 31
Blackwater – 93
Blakely, William – 198
Blast furnace – 32, 34, 237, 239
Blue corona effect – 349
Boiler design – 229
Boiler types:
- Bubbling-bed – 221–222
- Cell-burner – 148
- Chain grate – 217
- Circulating fluidized-bed – 221–222
- Cyclone – 103, 105, 148, 199, 200, 203, 365
- Firetube – 198
- Fluidized bed – 201, 203, 219, 319, 343, 365, 464
- Industrial – 201, 203–204, 225
- Mass burning – 217
- Overfeed – 213, 217
- Packaged – 203
- Pulverized coal – 103, 105, 148, 199–200, 203, 225, 322
- Spreader – 213, 217
- Stokers – 103, 105, 199, 200, 203, 214, 344, 365
- Subcritical – 201
- Supercritical – 201
- Tangentially-fired – 226
- Traveling – 217–218
- Ultra-supercritical – 201
- Underfeed – 213–214
- Utility – 201, 203–204
- Wall-fired – 148
- Water-cooled, traveling grate – 217
- Watertube – 198

Boron – 107, 110
Bovine Spongiform Encephalopathy (BSE) – 462
Bottom ash – 95–96, 107–108
Boudard reaction – 248
Boulton, Matthew – 32, 197
Breweries – 32
Bronze Age – 30
Brown coal – 21–22, 24
Buffalo Creek, West Virginia – 94
Bunson, R.W. – 33
Bush, George W. President – 72, 74, 141, 159, 393, 424, 435, 448
By-product recovery oven – 240

C

Cadmium – 96, 107, 109–110, 170
Calcium – 96, 109
Calcium hydroxide – 296
Calcium sulfate – 299, 303, 311
Calcium sulfate dihydrate – 296, 298
Calcium sulfite – 95, 299, 303
Calcium sulfite hemihydrate – 296, 298
Carbon dioxide – 81–82, 97, 103, 114–117, 394
Carbon dioxide (CO_2) capture – 337
Carbon dioxide capture/control technologies:
- Advanced conversion processes – 378
- Amine absorption – 377
- CO_2 adsorption – 377
- Fluor Daniel Econamine FG CO_2 Recovery process – 377
- Fuel switching – 376
- Oxygen-fired combustion – 378
- Post-combustion capture – 378
- Pre-combustion de-carbonization – 378

Carbon dioxide sequestration:
- Geologic – 379–380
- Oceans – 379, 381
- Terrestrial – 379, 381

Carbonic acid – 98
Carboniferous Period – 2, 19, 21–25
Carbonization – 4, 33, 237–238, 241, 248

Carbonization, types:
  High-temperature – 237–238, 244
  Low-temperature - 237, 243–244
Carbon Sequestration Leadership Forum – 161
Carbon Sequestration Program – 376–377, 379
Carbon tetrachloride – 115
Cardiovascular disease – 99, 110
Carnegie Steel Company – 239
Carter, J.E. Jr., President – 73, 456
Cap-and-trade programs – 145, 158
Catalysts – 341
Central Electricity Generating Board – 219
40 CFR Part 60, Subpart D – 132
40 CFR Part 60, Subpart Da – 132
40 CFR Part 60, Subpart Db – 134
40 CFR Part 60, Subpart Dc – 134–135
Charcoal – 30–32
Chemical treatment system – 84
Cheney, R., Vice President – 73
Child labor – 78
Chlorine – 107, 115
Chlorofluorocarbons – 115
Chromium – 96, 107, 109, 170
Chronic lung disease – 99
Chronic Wasting Disease (CWD) – 462–463
Chrysene – 106
City gas – 247
Clayton, John – 247
Clean Air Act – 89
Clean Air Act of 1963 – 126–127
Clean Air Act Amendments of 1970 – 124–125, 128, 130
Clean Air Act Amendments of 1977 – 139–141, 144
Clean Air Act Amendments of 1990 – 106–107, 126, 134, 137, 139, 142–143, 149
Clean Air Planning Act – 155
Clean coal – 450
Clean Coal Power Initiative (CCPI) – 422, 424–425, 431, 450
Clean Coal Power Initiative Projects:
  Advanced coal beneficiation processing – 427
  Clean coal power – 430
  Co-production – 431
  Fuel processing – 427
  Multipollutant control - 427
  On-line optimization systems – 427
  TOXECON – 427
Clean Coal Technology Demonstration Program (CCT) – 74, 245, 382, 394–397, 418
Clean Coal Technology Demonstration Program Projects:
  ACFB technology – 407
  Advanced coal conversion process (ACCP) – 416
  Advanced combustion/heat engine technology – 414
  Advanced electric power generation – 406
  Alstom Power LNCFS™ – 404
  B&W coal reburning – 403
  B&W LNCB® – 403
  Blast Furnace Granular-Coal Injection System Demonstration – 418
  Chiyoda Thoroughbred-121 AFGD process – 402
  Clean coal combustion system – 414
  Coal-fired diesel engine – 415
  Coal processing - 415
  Coal Quality Expert™ – 416
  Combined SO_2/NO_x Control Technologies – 404
  Confined-zone dispersion (CZD) FGD – 402
  EERC GR/LNB – 403
  FLS milfo, Inc.'s Gas Suspension Absorption system – 401
  Fluidized-bed combustion – 406, 421
  Foster Wheeler's LNB and OFA/EPRI's GNOCIS) – 403
  Fuel cell – 408
  IGCC – 408, 410–411, 414
  Industrial applications – 418
  LFC® Process – 415
  LIFAC sorbent injection process – 402
  LIMB – 405
  LMPEOH™ – 416
  Micronized coal reburning – 403
  Molten carbonate fuel cell – 408
  Multipollutant control – 405–406
  NGR/FSI – 405

Clean Coal Technology Demonstration Program Projects: (*continued*)
  NO_x emissions control – 401–402
  Passamaquoddy Technology Recovery Scrubber^{TM} – 418
  PFBC technology – 407
  PCFB technology – 406
  Pure Air's AFGD process – 402
  SCR catalysts – 404
  SNOX^{TM} – 404
  SNRB^{TM} – 405
  SO_2 emissions control – 401
  Solid oxide fuel cell – 408
Clean Power Act – 115
Clean Water Act – 89
Clear Skies Act of 2002 – 155, 158, 161, 383, 424, 427
Climatic changes – 115
Coal – 21–22, 35, 54, 57, 63, 66, 68, 451
Coal analyses:
  Calorific determination – 5
  Heating value – 5–6, 8
  Proximate – 5
  Ultimate – 5
Coal analysis bases:
  As received – 5
  Dry – 5
  Dry, ash-free – 5
  Dry, mineral-matter free – 4–6
  Moist, ash-free – 6
  Moist, mineral matter-free - 6
Coal Basins:
  Appalachian – 19, 292
  Bowen – 24
  Central German – 25
  Dneiper – 22
  Dontesk – 21–22
  Ekibastuz – 22
  Gippsland – 24
  Great Karoo – 25
  Illinois – 19
  Jharia – 23
  Kansk-Achinsk – 21–22
  Karaganda – 22
  Kuznetsk – 21–22
  Lower Silesian – 22
  Lublin – 22
  Lusatian – 25
  Pechora – 21
  Powder River – 20
  Raniganj – 23
  Rhineland – 25
  Ruhr – 25
  Saar – 25
  Sidney – 24
  Upper Silesian – 22
Coal beneficiation – 77
Coal cleaning – 91, 287, 295
Coal coking – 8
Coal combustion – 8
Coal combustion by-products (CCB) – 77, 95–96
Coal combustion products (CCP) – 95
Coal consumption – 8, 48–50, 60, 63–69, 71
Coal deposits – 1, 12, 41
Coal grade – 2, 8, 10
Coal exports – 50, 65, 67–68, 72
Coal gas – 32, 34, 247
Coal imports – 50
Coal maturation – 6
Coal mineralogy – 8
Coal mining – 78–80, 84, 89, 94, 115
Coal Power Program – 394, 437, 467
Coal preparation – 8, 90–91, 94
Coal production – 34, 41, 43–44, 46–47, 65–66, 68, 72, 86
Coal properties (gasification):
  ash – 254
  fixed carbon – 254
  moisture – 253
  reactivity – 255
  volatile matter – 254
Coal Provinces:
  Alaskan – 19–20
  Anhui – 23
  Eastern - 17, 19
  Guizhou – 23
  Gulf – 17, 20
  Henan – 23
  Heilongjiang – 23
  Inner Mongolia – 23
  Interior – 17, 19
  Jiangxi – 23
  Northern Great Plains – 19–21
  Pacific Coast – 19–20
  Rocky Mountain – 19–20

Shaanxi - 23
Shandong – 23
Shanxi – 23
Coal quality – 91
Coal rank – 2, 6, 12, 19, 21, 80
Coal refuse – 93–94
Coal regions:
Appalachian – 19, 41, 292
Central (Appalachian Basin) – 19
Eastern (Interior Province) – 19
Fort Union (Northern Great Plains Province) – 20
Interior – 46
Mississippi (Gulf Province) – 20
Southern (Appalachian Basin) – 19
Texas (Gulf Province) – 20
Western – 47
Western (Interior Province) – 19
Coal reserves – 1, 12–14, 17, 20–26
Coal resources – 12–13
Coal seam – 1, 3, 41, 80
Coal slurry – 95
Coal transportation – 95
Coal type – 1
Coalbed methane (CBM) – 81
Coalbeds – 41
Coalfields:
Appalachian – 82
Green River – 20
Hanna – 20
Hanna Fork – 20
Neyveli – 23
Coalification – 2–4, 8
Coal mine methane (CMM) – 81–82
Cobalt – 96, 107, 109
Codification – 8, 10
Coke – 8, 22–25, 32–33, 64, 72, 254, 306
Coke production – 34, 241
Combustion – 33, 91, 97, 195, 208–209, 211
Committee on Coal Waste Impoundments – 94
Conditioning agents – 354
Confined Zone Dispersion (CZD) – 315
Continuous emissions monitors (CEMs) – 146
Concentrating tables – 92
Conservation – 447
Consumer Price Index (CPI) – 451
Copper – 96, 109–110, 170
Copper casting – 30
Corrosion – 98
Cretaceous Period – 21, 26
Criteria air pollutants – 106, 109, 128, 152, 173, 175
Critical viscosity – 233
Crude oil – 35, 39, 54
Crude oil imports – 35
Culm bank recovery – 86
Cunningham correction factor – 350
Cyclones – 92

D

Darby, Abraham – 32, 238
Darcy's equation – 359
DB Riley Inc./ Riley Stoker Inc. – 200
Degasification systems – 81
Dense medium separation – 91
Deutsch-Anderson equation – 351
Diagentic – 2, 8
Dibasic acid – 302
Dibenz[ah]anthracene – 106
Dinitrogen pentoxide – 100
Direct impaction – 356
Direct interception – 357
Directive on Controlling Emissions from Large Combustion Plants – 166
Directive on Integrated Pollution Prevention and Control – 164
Directive on National Emission Ceilings for Certain Atmospheric Pollutants – 168
Directive on the Limitation of Emissions of Certain Pollutants into the Air from Large Combustion Plants -166
Direct liquefaction – 267, 273
Dolomite – 221, 319
Donora, Pennsylvania – 124
Dredging – 86
Drift velocity – 349, 351
Dust – 78–79, 85, 88, 91
Dust cake – 357
Dust loading – 357

E

Earth Summit – 172
ECE, Economic Commission for Europe – 8
ECE International Codification of Higher Rank Coals – 10
Economic growth – 451–452, 468
Economic security – 451–462
Edward I, King – 124
Elanor, Queen – 31, 124
Electricity – 39, 62, 66, 68, 71–73
Electricity cost – 452
Electricity generation – 19–20, 23–25, 33, 61, 64–65, 67, 69, 73, 81, 393, 452, 454, 467
Electric power – 34, 48–49, 54, 58, 71
Electric Power Research Institute (EPRI) – 99, 102, 104, 366
Electrification – 38, 196, 467
Electrostatic force – 349, 356
Electrostatic precipitator (ESP) – 93
Elemental mercury – 370, 374
Emergency Planning and Community Right-to-Know Act 1986 – 137
Emissions factors – 136–137
Emissions factors (mining) for:
  Methane – 81, 87
  Underground mining – 81
Emissions Factors and Inventory Group (EFIG) – 137
Emission inventories – 137
Emissions of:
  Carbon dioxide – 78, 146, 172, 175, 189, 375–376
  Carbon monoxide – 78, 97, 106, 140, 143, 175, 182, 284
  Fugitive dust – 77, 95
  Greenhouse gas – 82
  Greenhouse gas (GHG) – 161, 172
  Hazardous air pollutants (HAPs) – 187
  Hydrocarbons – 140, 284
  Lead – 78, 175, 182
  Mercury – 157–158, 175, 188, 369–370
  Nitric oxide – 322
  Nitrogen dioxide – 322, 324
  Nitrogen oxides – 74, 78, 97–98, 103, 133–134, 140, 145–147, 149, 158, 160, 175–177, 185
  Nitrous oxides – 284, 317, 321
  Organic compounds - 78
  Ozone – 78, 107, 143, 160, 175, 179, 284
  Particulate matter – 78, 97, 99, 103–105, 284, 347
  Photochemical oxidants – 140
  Polycyclic organic matter – 145
  Sulfates – 185
  Sulfur dioxide – 74, 78, 95, 97–99, 103, 108, 133, 135, 145–146, 158, 160, 164, 175, 179, 185, 221, 284, 286
  Sulfur trioxide – 97, 286–287, 311
  Volatile organic compounds (VOCs) – 177, 284
Emission rates – 284
Emissions standards – 66, 69, 130, 132, 136, 139, 164, 169
Emissions technology – 459
Endangered Species Act of 1963 – 89
Energy consumption – 34, 38–40, 51, 60–63, 65, 67, 69–72
Energy efficiency – 459
Energy infrastructure – 447–448
Energy plan/policy – 73
Energy prices – 451
Energy production – 35, 40, 51, 70, 72
Energy security – 446–449, 456, 458, 462, 467–468
Energy Security Act of 1980 – 395
Energy sustainability – 468
English Clean Air Act of 1956 – 124
Enteric fermentation – 115
Environmental effects – 83, 98, 102–103, 106–108, 117
Environmental impact – 83, 86, 95
Environmental protection – 450
Ethanol – 58
Ethylene – 106
Eutrophication – 102
Exinite – 7

F

Federal NO_X budget Trading Program – 152
Feldspar jigs – 92
Ferric iron – 83
Ferrous iron – 83

Filter cake – 359
Filters – 93, 362
Fine particulate matter – 74, 100
Fischer, Franz – 269–271
Fischer-Tropsch synthesis – 33, 268, 458, 468
Fish and Wildlife Coordination Act of 1934 – 89
Fixed-bed combustion – 213
Flame oven – 240
Fluidized-bed combustion – 219–221, 406, 421
Flue gas desulfurization (FGD) – 95–96
Flue gas desulfurization process:
  Calcium-based scrubbers – 296
  Circulating fluidized-bed scrubbers – 287, 307, 319, 322
  Combined SO_2/NO_x systems -287
  Convective pass (economizer) injection – 307, 309, 313
  Dry FGD – 307
  Dry sorbent injection processes – 287, 307, 309, 315–316, 321
  Dry scrubbers – 291
  Duct injection – 307
  Furnace sorbent injection (FSI) – 307, 309, 312
  Hybrid methods – 307, 309, 312
  In-duct injection (DI) – 309
  In-duct spray drying (DSD) – 309, 314–315
  LIMB (limestone injection into a multistage burner) – 309
  Lime dual alkali process – 303–304
  Lime scrubbers – 298
  Lime spray drying – 307
  Limestone scrubbing with forced oxidation (LSFO) – 298
  Limestone spray drying – 307–308
  Limestone with diabasic acid – 302
  Limestone with forced oxidation producing a wallboard gypsum by-product process (LS/WB) - 299
  Limestone with inhibited oxidation process – 300
  Magnesium-based scrubbers – 296
  Magnesium enhanced lime process (MagLime) – 301
  Regenerative magnesia scrubbing – 304–305, 322
  Regenerative processes - 287, 304
  Sodium-based scrubbers – 296
  Spray dryers – 287, 307, 321
  Wellman-Lord process – 304
  Wet lime systems – 291
  Wet scrubbers – 287, 291, 295–296, 320
  Wet sodium-based systems – 303
Fluroene – 106
Fluorine – 97, 104, 107, 110
Fluoranthene – 106
Fluorosis – 110
Fly ash – 95–96, 107, 113
Fly ash reinjection – 103
Fly ash reentrainment – 353
Fly ash resistivity – 351–354
Food supply security – 462
Forest and Rangeland Resources Planning Act of 1974 – 89
Formed coke – 243
Fossil fuel – 13, 35, 115, 393
Fossil fuel combustion wastes (FFCW) – 95
Foster, Earnest – 200
Foster, Pell – 200
Fouling – 232
Foundaries – 237
Fuel diversity – 204, 449, 458
Fuel switching – 292, 376
Froth flotation – 92
Fugitive dust – 86
Furnace design – 229
Furnace, types:
  coke – 238
  Cyclone – 228
  Double U-flame – 227
  Dry-bottom – 225
  Horizontal – 226
  Opposed horizontal – 226
  Opposed inclined - 226
  Single U-flame – 227
  Slag-tap – 225
  Stuckofen – 238
  Tangential – 226
  Wet-bottom – 225, 227
FutureGen – 394, 424, 435

G

Gas conditioning – 365
Gasification – 33, 69, 91, 246–247, 421, 458
Gasification facilities – 509–511
Gasification processes:
BGL – 261
ChevronTexaco – 258, 264
E-Gas – 264, 266
Entrained-flow – 249, 253–254, 264–266
Fixed-bed – 249–250, 254–255, 258
Fluidized-bed – 249, 252–254, 262–263
High-Temperature Winkler – 262
Kellogg-Rust-Westinghouse (KRW) – 262–263
Koppers-Totzek – 265
Lurgi – 258–260
Prenflo – 264, 266
Shell – 258, 264–265
Gas ionization – 348
Gas sneakage – 353
Gelification – 4
Geochemical – 3
Geologic Period – 2, 3
Geothermal – 34, 54, 58
Global alliances – 450
Global climate change – 117–118, 161
Global Climate Change Initiative – 378, 424, 427
Global Coal Initiative (GCI) – 464–465
Global energy markets – 447
Global warming – 117–118
Global warming potential (GWP) – 116
Goddard Institute for Space Studies – 117
Great Plains Synfuels Plant – 258
Greeks – 30, 196
Greenhouse effect – 114
Greenhouse gases – 114–117
Gross domestic product (GDP) – 449, 451
Groundwater – 77–78, 82, 84, 87, 110
Gypsum – 95, 296, 299

H

Hadrian's Wall – 30
Halocarbons – 114–115
Halons – 115
Hard coal – 3, 22–25, 68
Hazardous air pollutants – 128, 144, 499–503
Health effects – 78, 84, 89, 99–100, 102, 106, 110, 118
Heavy oil – 240
Helmont, Jan van – 247
Herbaceous plants – 4
Hero of Alexandria – 196
Heterogeneous reaction – 207
Homogenous reaction – 207–208
Historic Preservation Act of 1966 – 89
Hoof-and-mouth disease – 463
Horizontal slot-type coke oven – 240
Human health – 77, 96, 104
Humic substances – 3
Hydrated calcium sulfite – 302
Hydrated lime – 307, 312, 315, 319
Hydraulic separation – 91–92
Hydroelectric – 34, 50, 54, 58, 68–69, 71
Hydrogen – 74, 447, 462, 468
Hydrogenation – 33
Hydrogen chloride – 307
Hydrogen cyanide – 101
Hydrogen economy – 69, 74
Hydrogen production – 69, 394, 425, 459
Hydrogen storage – 459
Hydrogen utilization – 459
Hydrolysis – 83
Hydrobromofluorocarbons – 115
Hydrochlorofluorocarbons – 114
Hydrofluorocarbons – 115
Hydrorotators – 92

I

I.G. Farben – 269
Illuminating gas – 247
Indirect liquefaction – 267–268, 271
Industrial Revolution – 195
Information Collection Request (ICR) – 157
Interstate Air Quality Rule – 160
Impoundment – 77, 91, 93–95
Indeno[1, 2.3-cd]pyrene – 106
Indirect liquefaction – 33
Industrial boilers – 66
Industrial Revolution – 31–32, 39, 117

Inertial impaction – 356–357
Inertinite – 7
Information collection request (ICR) – 369–370
Integrated gasification combined cycle (IGCC) – 257
Intermittent energization – 354
Iron – 82, 96, 107, 110
Iron industry – 32
Iron making – 238–239
Iron smelting -29, 32
ISO (International Organization for Standardization) codification of brown coals and lignites – 10

J

Joliet, Louis – 30
Jurassic Period – 19, 21–23

K

Kilns – 239–240
Kyoto Protocol – 173–174

L

Labor relations – 78
Large Combustion Plants Directive – 164
Launders – 92
Lead – 96–97, 107, 109, 113, 170
Light oil – 240
Lignin – 3
Lignite – 2–4, 16, 20, 22–26, 47–48, 68
Lime – 96, 296, 307
Lime burners – 31, 124
Lime kilns – 31
Limestone – 85, 96, 221, 296, 299, 319
Liptinite – 7
Liquid effluents – 82, 87, 93
Liquefaction – 33, 267
Liquefaction processes:
  Bergius – 274
  Costeam – 275
  Exxon Donor Solvent (EDS) – 277
  H-Coal – 276
  High-pressure synthesis – 271
  Iso-synthesis – 273
  Oxo-synthesis – 273
  Solvent-Refined Coal (SRC) – 275
  Synthol – 273
Liquids-From-Coal Process (LFC) – 245
Liquefied natural gas – 72
London – 125
Lowest achievable emission rate (LAER) – 138, 141, 149
Low-rank coal – 9
Low-sulfur coal/fuel – 17, 287, 291

M

Macerals – 7
Magnesium – 96, 109
Magnesium bisulfite – 301
Magnesium sulfate – 306
Magnesium sulfite – 301, 306
Manganese – 96, 107, 109–110, 112
Manifest Destiny – 34
Manufactured gas – 33
Maximum achievable control technology (MACT) – 145
Mechanical separators – 93
Mercuric chloride – 373
Mercuric mercury – 112
Mercury – 74, 78, 96–97, 107, 109–110, 112, 145, 156, 170
Mercury control technologies:
  Activated carbon injection – 373
  Coal cleaning - 373–374
  Sorbent injection – 373
  TOXECON – 374
  Wet Flue gas desulfurization (FGD) – 373–374
Mercury Study Report – 156
Metallurgical coal – 21, 24, 50–51, 65
Metallurgical coke – 8, 237
Metallurgical coke production – 19–20, 22
Metalworking – 30
Metamorphism – 2, 6, 8
Methane – 77, 79–81, 85, 87, 114–116
Methyl bromide – 115
Methyl chloroform – 115
Methyl mercury – 109, 112
Middle oil – 240
Migratory Bird Treaty Act of 1918 – 89
Mineral matter – 8, 103

Mine/Mining types:
- Area – 86–88
- Auger – 86
- Continuous - 79
- Contour – 86–88
- Conventional – 79, 87–88
- Drift – 79
- Longwall – 79, 87–88
- Mountain top removal – 87–88
- Pitch – 79
- Room and pillar – 79
- Shaft – 79
- Slope – 79
- Strip – 86, 91
- Surface – 41, 77, 79, 82, 86–90
- Underground – 41, 79, 81–82, 84–88, 91, 94

Mine permit – 78
Miner training – 78
Mine Safety & Health Administration (MSHA) – 85–86, 94
Mining and Minerals Policy Act of 1970 – 89
Mining Enforcement and Safety Administration – 94
Molybdenum – 107, 109, 112
Multiple Use – Sustained Yield Act of 1960 – 89
Multipollutant control – 104, 382–383
Multipollutant control technologies:
- Airborne Pollution Control – 384
- Electro-Catalytic Oxidation (ECO) – 383
- Integrated Dry NO_x/SO_2 Emissions Control – 382
- Low-temperature oxidation process (LoTOx) – 385
- Mobotec systems – 385
- SNOX™ – 382
- SNRB™ – 382

Multipollutant legislation – 155
Murdock, William – 247

N

Nacholite – 317
Naphthalene – 106
National Air Quality Control Act 1967 – 127
National Ambient Air Quality Standards (NAAQS) – 128, 130, 139, 141–142, 155
National Coal Board – 219
National Emission Standards for Hazardous Air Pollutants (NESHAP) – 128
National Energy Technology Laboratory (NETL) – 105
National energy plan – 449–450
National Energy Policy – 72, 425, 448, 467
National Energy Policy Development (NEPD) Group – 72–74
National Environmental Policy Act – 89
National Forests Management Act of 1976 – 90
National Mining Association – 86
National Research Council – 94
National Trails System Act – 90
Natural gas – 33–35, 38–39, 49–50, 54, 61, 63–64, 66–68, 71–73, 115, 448, 454, 462, 467
Natural gas availability – 445
Natural gas crisis – 454–455
Natural gas plants liquids – 35, 54
Natural gas prices – 445, 454
Natural gas reserves – 13
Navigation Acts – 32
Near-zero emissions – 394, 425
Newcomen, Thomas – 197
New Source Performance Standards (NSPS) – 128, 132, 134, 140, 142
New Source Review (NSR) – 152–155
Nickel – 96, 107, 109, 112, 170
Nitrate components – 100
Nitrate radical – 100
Nitric acid – 100, 102
Nitric oxide – 100–101, 114
Nitrogen dioxide – 100, 102
Nitrous acid – 100
Nitrous oxide – 100, 102, 115
Nixon, R., President – 127
Noise Control Act of 1976 – 90
Noise pollution – 78
Nonattainment area – 143–144, 149
Non-coking coal – 23
Northern Great Plains – 72

NO_x control technologies:
Burners out of service (BOOS) – 329
Close-coupled overfire air (CCOFA) – 334
Cofiring – 336, 345
Combustion modifications – 329, 344
Flue gas recirculation (FGR) – 329, 334–335
Fuel staging/ reburn – 329, 334–336, 344–345
Furnace air staging – 329, 333–334, 344
High-dust SCR – 339
Hybrid SNCR/SCR – 342–343, 346
Low-dust SCR – 339–340
Low excess air (LEA) – 329, 344
Low-NO_x burners – 329–330, 332–333, 344
Natural gas reburn (NGR) – 344
Overfire air (OFA) – 333–334, 336, 344
Process optimization – 329, 338, 345
Rich reagent injection (RRI) - 343, 346
Selective catalytic reduction (SCR) – 338–339, 344–345
Selective non-catalytic reduction (SNCR) - 338, 341, 344–345
Tail-end SCR – 339–340
NO_x/NO formation:
Fuel – 100–102, 322, 324
Thermal – 100, 102, 322
Prompt – 100, 322
NO_x reduction – 421
NO_x SIP Call – 152
NO_y – 100–101
NSR Equipment Replacement Rule – 155
Nuclear – 35, 54, 58, 65, 67, 69, 71–72

O

Office of Surface Mining – 90
Occupational Safety and Health Administration (OSHA) – 85, 94
Oil – 61, 66–67, 72, 456
Oil prices – 445, 447
Oil reserves – 13
Opacity – 146
OPEC (Organization of Petroleum Exporting Countries) – 36, 447
Organic compounds – 97, 105
Overburden – 77, 86, 88
Overburden blasting – 86
Oxidized mercury – 370, 373–374
Ozone standards – 130, 149
Ozone Transport Commission (OTC) – 150
Ozone transport region (OTR) – 150

P

1999 Protocol to Abate Acidification, Eutrophication, and Ground-Level Ozone – 166
Papin, Denis – 197
Particulate matter control technologies:
Advanced Hybrid Particulate Collector (AHPC) – 366–367
Baghouses/fabric filters – 347, 356, 368
Cold-side ESP – 354
Compact Hybrid Particulate Collector (COHPAC) – 366
Electrostatic precipitators (ESPs) – 347–348, 367–368
Hot-side ESP – 354
Hybrid systems – 365–366, 368
Plate and wire ESPs – 348
Pulse-jet baghouse/fabric filter – 359, 361–362
Reverse-gas baghouse/fabric filter – 359–360
Shake-deflate baghouse/fabric filter – 359, 361
Wet ESP (WESP) – 355
Particulate mercury – 373
Passive treatment system – 84
Peat – 2, 4
Perfluorocarbons – 114–115
Perkins, Jacob – 199
Permian Period – 23–25
Permitting – 148
Peroxyacetyl nitrate (PAN) – 100
Petrographic analysis – 7, 8, 10
Petroleum – 34–35, 38–39, 49, 54, 71, 115
Phenanthrene – 106
Photochemical smog – 102, 108

Plantwide Applicability Limits (PALs) – 153, 155
Pliney – 29, 195
$PM_{2.5}$ standards – 130, 155
PM_{10} standards – 130
Pneumoconiosis – 85
Polo, Marco – 29
Polycyclic aromatic compounds (PACs) – 106
Polycyclic aromatic hydrocarbons (PAHs) – 106
Polycyclic organic matter (POM) – 105
Potassium – 96, 109, 234
Power generation – 20–22, 24, 33, 63, 67, 69, 103, 105, 107, 110, 449, 452, 458, 462
Power Plant Improvement Initiative (PPII) – 394, 422–423
Power Plant Improvement Projects:
  Advanced hybrid particulate generator (AHPC) – 423
  Aggregate manufacturing plant – 423
  Boiler fouling – 423
  Combustion modification/control system – 423
  Hybrid NO_x system – 423
  Multipollutant control system – 424
Preparation plant – 92
Pre-Industrial Revolution – 29
Pressure drop – 358–359
Prevention of Significant Deterioration – 139–141, 143
Primary energy – 54, 57–58, 61
Primary pollutants – 96–97
Protocol on Further Reductions of Sulfur Emissions – 162
Pulse energization – 354
Pulmonary disease – 102
Pyrene – 106
Pyrite – 82–83, 88, 93
Pyrolysis – 245, 267

R

Radially stratified fuel core burner – 330
Radiative forcing – 116
Radionuclides – 97, 107, 113
Radon – 107
Reasonably available control measures (RACM) – 144
Reasonable available control technology (RACT) – 149
Reclamation – 78, 86–90
Regional Haze Rule – 155
Renewable energy – 58, 69, 71–72
Renewable energy portfolios – 69
Resource Conservation and Recovery Act – 90, 96, 107
Respiratory disease – 78, 85, 89, 99, 102, 104, 110
Ridge, T. – 458
Roman Empire – 30
Romans – 30, 196, 238
Rome – 124
Run-of-mine (ROM) coal – 91

S

Safe Drinking Water Act of 1974 – 90
Safety valve – 197
Salts – 88
Sasol – 258
Saturated steam – 204
Savery, Thomas – 196
Sea-coal – 30
Secondary Pollutants – 97
Section 29 – 48
Selenium – 96–97, 107, 109–110, 112, 170
Semi-anthracite – 25
Seneca – 124
Shelley – 124
Silica – 96
Silicosis – 85
Slag – 95–96, 232
Slagging – 232, 234
Slagging and fouling indices – 233–236
Slag viscosity – 232
Slaked magnesium oxide – 306
Smithies – 32
Smiths – 30
Smokeless fuels – 244
SO_2 Helsinki Protocol – 162
SO_2 reduction – 421
Soda ash – 314
Sodium – 96, 109, 234
Sodium bicarbonate – 303, 316–317

Sodium bisulfite – 303
Sodium carbonate – 303, 306
Sodium hydroxide – 303
Sodium sesquicarbonate – 317
Sodium sulfate -303
Sodium sulfite – 303–304
Sofia NO_x Protocol – 166
Soil and Water Resources Conservation Act of 1977 – 90
Sorbent – 221
Specific collection area (SCA) – 351
State Implementation Plan (SIP) – 139, 149–150
Steam – 97
Steam coal – 22, 24–25, 50, 71
Steam-driven pump – 197
Steam engine – 32, 196–197
Steam production – 19–20, 22
Steel industry – 67–68
Steel production – 64
Stevens, John – 199
Stevens, John Cox – 199
Stone coals – 113
Strategic Petroleum Reserve – 395
Study of Hazardous Air Pollutant Emissions from Electric Utility Steam Generating Units – 187
Solar – 34, 54, 58
Subbituminous coal – 2, 4, 16, 20–22, 26, 48
Subsidence – 78–80, 84
Sugar refineries – 32
Sulfate components – 97, 100
Sulfur – 17, 286
Sulfur dioxide emissions allowances – 147
Sulfur emissions regulations – 17
Sulfur hexafluoride – 114–115
Sulfuric acid – 98, 305–306
Superheated steam – 204
Surface Mining Control and Reclamation Act (SMRCA) – 89, 94, 97
Surface water – 78, 87
Sustainable energy policy – 466
Synfuel plant – 48
Syngas – 256, 258
Synthetic coal – 48
Synthetic Fuels Corporation (SFC) – 395
Synthetic liquid fuels – 48
Synthetic natural gas (SNG) – 257–258

T

Tanneries – 32
Tectonic – 2
Tertiary Period – 19, 21–26
Thallium – 112
Theophrastus -29, 195
Thermal coal – 24
Thermal dryers – 92
Thermal upgrading – 245
Thermoplastic – 8
Thiosulfate – 300
Thorium – 107, 113
Town gas – 247
Toxic elements – 96, 107
Trace elements – 8, 97, 108, 110, 175
Trevithick, Richard – 197
Trona – 317
Tropsch, Hans – 269

U

1998 United Nations Protocol on Heavy Metals – 170
Unburned carbon – 333
UNECE Gothenburg Protocol – 162
United Nations Economic Commission for Europe's (UNECE) Convention on Long Range Transboundary Air Pollution (LRTAP) – 162, 165
United Nations Environment Report – 189
United Nations Framework Convention on Climate Change – 172–173
United States Department of Energy – 95, 219
United Sates Department of Interior, Office of Coal Research – 219
United States Environmental Protection Agency (EPA) – 95, 100, 116, 128
Uranium – 107, 113
Urea – 341
Utility Hazardous Air Pollutant Report to Congress – 157, 188

V

Vanadium – 107, 109, 113
Vanadium pentoxide – 107, 113
Ventilation systems – 81
Vibration – 78
Visibility – 98, 103
Vision 21 – 394, 432, 434–435, 450
Vitrification – 4
Vitrinite – 7
Volatile matter – 8, 208
Volatile organic compounds
(VOCs) – 102–103

W

Water gas – 249
Water pollution – 92
Water vapor – 114–115
Watt, James – 32, 197
Wet scrubbers – 93
Wilcox, Stephen – 199
Wild and Scenic Rivers Act – 90
Wilderness Act of 1964 – 90
Wind – 34, 54, 58
Wood – 4, 30, 32, 34, 54, 58, 62
World Bank – 66, 88, 168, 456
World Summit on Sustainable
Development – 456, 466

Z

Zeldovich mechanism -100, 322
Zero emissions – 74
Zinc – 96, 109–110, 170